Dynamics and Evolution of Galactic Nuclei

Princeton Series in Astrophysics

Edited by David N. Spergel

Dynamics and Evolution of Galactic Nuclei

David Merritt

PRINCETON UNIVERSITY PRESS

PRINCETON AND OXFORD

Published by Princeton University Press, 41 William Street, Princeton, New Jersey 08540

In the United Kingdom: Princeton University Press, 6 Oxford Street, Woodstock, Oxfordshire OX20 1TW

press.princeton.edu

Jacket Photograph: Centaurus A (NGC 5128) is the closest active galaxy to the Earth. This galaxy appears to be the result of a collision between two otherwise normal galaxies; near the galaxy's center, cosmic debris is being consumed by a central, supermassive black hole.
Courtesy of Dr Tim Carruthers, Cairns, Australia

ISBN 978-0-691-12101-7 (cloth)

ISBN 978-0-691-15860-0 (pbk)

Library of Congress Control Number: 2013936624

British Library Cataloging-in-Publication Data is available

This book has been composed in Times

Printed on acid-free paper ∞

Typeset by S R Nova Pvt Ltd, Bangalore, India

Printed in the United States of America

10 9 8 7 6 5 4 3 2 1

To the memory of David Axon

Contents

Preface

Since the 1960s, when the connection between quasar energetics and supermassive black holes was first established, galactic nuclei have remained objects of intense interest to astrophysicists. Nowadays, the community of scientists with an interest in galactic nuclei is enormous: not just the astronomers who study so-called active galaxies, but also the relativists who hope to detect gravitational waves, cosmologists concerned with the role of feedback in structure formation, and particle physicists searching for radiation produced by the annihilation of dark matter particles clustered around supermassive black holes. This book is intended as a basic resource for all of these researchers, and for graduate students who are planning to work in a field that is related, directly or indirectly, to galactic nuclei.

Supermassive black holes are sometimes accompanied by gas, but as near as we can tell, they are always associated with stars: the stars of the galactic nuclei in which they sit. Furthermore, this association appears to be more than a casual one since there are strong correlations between black-hole mass and the large-scale properties—mass, velocity dispersion, central concentration—of the host stellar systems. Correlations on smaller scales exist as well; for instance, with the so-called "mass deficits" observed at the centers of bright galaxies. While the origin of these correlations is still debated, they suggest a deep connection between supermassive black holes and the stellar components of galaxies.

Partly for this reason, the emphasis in this book is on dynamical interactions involving stars: either interactions of (single or binary) stars with (single or binary) supermassive black holes, or interactions of stars with each other in the vicinity of supermassive black holes. Gas dynamics is covered in a much less comprehensive way, and only in contexts where its role appears essential: in the late evolution of binary supermassive black holes (chapter 8), or in theories that attempt to explain the tight empirical correlations by invoking radiative feedback (chapter 2). The reader who needs to know more about nuclear gas dynamics, or emission mechanisms related to black holes, is directed toward the "Suggestions for Further Reading" at the end of this book.

On the other hand, the treatment here of stellar dynamics is as complete and as self-contained as I could make it in a book of prescribed length. Chapter 3, on collisionless models of nuclei, may be slightly more terse than the other chapters, but this is due to the availability of *Galactic Dynamics*, the comprehensive (if largely black-hole-free) text by J. Binney and S. Tremaine. Likewise, some parts of chapters 5 and 7 overlap with L. Spitzer's superbly succinct, and sadly out-of-print, *Dynamical Evolution of Globular Clusters*. But chapters 6 (Loss-cone dynamics), 7 (Collisional evolution of nuclei), and 8 (Binary and multiple supermassive black holes) deal with topics that appear not to have been treated in any detail in textbooks before now.

From a dynamical point of view, galactic nuclei occupy an interesting middle ground. They are denser than the other parts of the galaxies in which they sit; but not so dense that they are likely to be "collisionally relaxed," in the way that most globular clusters appear to be. That is, gravitational encounters between the stars in most galactic nuclei do not appear to be frequent enough to have established a statistically "most likely" distribution around the supermassive black hole—even in nuclei as dense as that of the Milky Way. One consequence is that we probably cannot trust intuition derived from the relaxed models to predict the distribution of stars on subparsec scales around supermassive black holes. This is a relatively new insight, and one that is still resisted in some circles, perhaps because it complicates the calculation of event rates. On the other hand, the much larger variety of steady states associated with "collisionless" nuclei implies more freedom for the theorist to construct models—a positive development, at least in the author's eyes. The collisionless nature of galactic nuclei is a recurring theme in this book. If I sometimes seem to press the point a little too strongly, it is with the good intention of motivating others to think carefully about this important question.

It would be natural in a book like this to devote a separate chapter to the Galactic center—the nucleus of our own galaxy, the Milky Way. Instead, the decision was made to spread the discussion of the Galactic center among several chapters, using the data and theoretical models to illustrate concepts as they arise. So, for instance, the use of stellar kinematics to infer gravitational potentials is illustrated in chapter 3 via proper motion studies of stars in the inner parsec of the Milky Way; the "clockwise disk" at the Galactic center is introduced in chapter 5 in the context of spin-orbit torques; the interaction of binary stars with supermassive black holes is presented in chapter 6 together with a discussion of the Milky Way data that seem to verify such interactions. The reader who is interested in specific topics related to the Galactic center is directed to the index.

The author has always felt that there are two, equally important sorts of textbook: those that are intended as reference works, to be dipped into as needed, and those that are meant to be read from cover to cover. This book belongs to the second category. What it lacks in comprehensiveness, it hopefully makes up for in quality of exposition. There are no appendices, or problems for the reader to work out; all the material that is deemed important is included in the main body of the text, and derivations are presented from first principles whenever feasible. Readers who wish to get a quick feeling for the topics covered are invited to begin by reading chapters 1 and 2 and the introductory sections of chapters 3–8.

Finally, the acknowledgments. Much of what I know about galactic nuclei and black holes is owed to conversations over the years with colleagues in Rochester who are knowledgable about such things: D. Axon, S. Baum, J. Faber, D. Figer, P. Kharb, R. Mittal, B. Mundim, H. Nakano, C. O'Dea, and A. Robinson. My understanding of post-Newtonian dynamics has benefited enormously from my collaborations with C. Will. Some sections of this book are based in part on review articles that I wrote with various collaborators: chapter 1 is a revised and extended version of an article on supermassive black holes written with L. Ferrarese and published in *Physics World* (vol. 15N6, pp. 41–46, June 2002); section 3.1.1 is based on a review of torus construction written with M. Valluri (*Astronomical Society of the Pacific Conference Series*, vol. 182, pp. 178–190); section 8.4.5, on interactions of binary black holes with gas, is adapted from material in a review article with M. Milosavljevic in *Living Reviews in Relativity* (vol. 8, no. 8, 2005). M. Milosavljevic, E. Vasiliev, and C. Will kindly gave their permissions to reproduce unpublished calculations in sections 6.1.2, 4.4.2.2, and 4.6, respectively. T. Alexander, H. Cohn, M. Colpi, A. Graham, M. Kesden, A. King, A. Marconi, H. Perets, and E. Vasiliev were kind enough to read substantial parts of the manuscript and to make detailed suggestions for improvements. A draft version of the manuscript was used as the basis for a course on galactic nuclei taught at the Rochester Institute of Technology in the winter of 2011–2012. I thank the students in that course, M. Freeman, D. Lena, P. Peiris, I. Ruchlin, C. Trombley, and S. Vaddi, for checking many of the derivations and identifying typos. I thank S. Vaddi also for her assistance in making many of the figures. Parts of this book were written during a sabbatical semester that was taken at various places, including Leiden University, and the Weizmann Institute; I thank, respectively, Simon Portegies Zwart and Tal Alexander for hosting me during these visits. Last but far from least, I thank my wife for putting up with me during the hectic year in which this book was written.

Rochester, NY   D. M.
December, 2011

Dynamics and Evolution of Galactic Nuclei

Chapter One

Introduction and Historical Overview

Of all the legacies of Einstein's general theory of relativity, none is more fascinating than black holes. While we now take their existence almost for granted, black holes were viewed for much of the 20th century as mathematical curiosities with no counterparts in nature. Einstein himself had reservations about the existence of black holes. In 1939 he published a paper with the daunting title "On a stationary system with spherical symmetry consisting of many gravitating masses" [135]. In it, Einstein sought to prove that black holes—objects so dense that their gravity prevents even light from escaping—were impossible. Einstein's resistance to the idea is understandable. Like most physicists of his day, he found it hard to believe that nature could permit the formation of objects with such extreme properties. Ironically, in making his case, Einstein used his own general theory of relativity. That same theory was used, just a few months later, to argue the opposite case: a paper by J. Robert Oppenheimer and Hartland S. Snyder, entitled "On continued gravitational contraction" [407], showed how black holes might form.

The modern view—that black holes are the almost inevitable end result of the evolution of massive stars—arose from the work of Oppenheimer, Subrahmanyan Chandrasekhar, Lev Landau, and others, in the first half of the 20th century. However, it was not until the discovery in 1963 of extremely luminous distant objects called quasars that the existence of black holes was taken seriously. What is more, black holes appeared to exist on a scale far larger than anyone had anticipated.

Quasi-stellar objects, or quasars,[1] belong to a class of galaxies known as active galactic nuclei, or AGNs. What makes these galaxies "active" is the emission of large amounts of energy from their nuclei. Moreover, the luminosities of AGNs fluctuate on very short timescales—within days or sometimes even minutes. The time variation sets an upper limit on the size of the emitting region. For this reason we know that the emitting regions of AGNs are only light-minutes or light-days across; far smaller than the galaxies in which they sit. At the time, astronomers were faced with a daunting task: to explain how a luminosity hundreds of times that of an entire galaxy could be emitted from a volume billions of times smaller. Of all proposed explanations, only one survived close scrutiny: the release of gravitational energy by matter falling toward a black hole. Even using an energy source as efficient as gravity, the black holes in AGNs would need to be enormous—millions or even billions of times more massive than the Sun—in order to produce the luminosities of quasars. To distinguish these black holes from the stellar-mass black holes

---

[1]Purists reserve the term "quasar" for the subset of quasi-stellar objects that are radio loud.

left behind by supernova explosions, the term "supermassive black hole" (SBH) was coined.

For nearly three decades after quasars were discovered, SBHs continued to be viewed as exotic phenomena and their existence was accepted only out of necessity. However, by the late 1980s, a crisis was brewing. Surveys with optical telescopes had shown that the number of quasars per unit volume is not constant with time. By studying the redshift of the light emitted by the quasar on its journey to the Earth, astronomers found that the number density of quasars peaked when the universe was only about 2.5 billion years old and has been declining steadily ever since.

The reason for this evolution is still not completely understood. But whatever its explanation, the evolution presents astronomers with a challenge. Many of the quasars with large redshifts simply disappear at lower redshifts. Indeed, of the quasars that populated the skies almost 10 billion years ago, only one in 500 can be identified today—but we know of no way to destroy the SBHs that powered the quasar activity. The unavoidable conclusion is that the local universe is filled with "dead" quasars, SBHs that have exhausted the fuel supply that made the quasars shine so brightly 10 billion years ago.

Where are these dead quasars? A reasonable place to look is at the centers of AGNs. But while the AGNs almost certainly contain SBHs, there are far too few of them—only a few percent of all galaxies are considered to be active—to account for the SBHs that once powered the quasars. By the early 1990s, astronomers were faced with the prospect that an SBH might have to be located at the center of almost every galaxy, which would make them as fundamental a component of galactic structure as stars.

This idea—though natural enough—did not come easily, since most galaxies show no evidence for the emission associated with a central SBH. But the gravitational field of an SBH is strong enough to imprint a characteristic signature on the motion of surrounding matter at distances that are millions of times greater than the event horizon. Stars, gas, and dust moving around a black hole—or any compact object—have orbital velocities that follow the same laws discovered by Johannes Kepler in the 17th century for the solar system. Moreover, the mass of the compact object is easily computed once this Keplerian rotation has been mapped. These arguments have been applied to measure the mass of the SBH in the core of our own galaxy, the Milky Way. But almost a decade earlier, the same was done for a distant galaxy, called NGC 4258.

NGC 4258 is a spiral galaxy, like the Milky Way, but containing an active nucleus. Sufficient radiation is produced in the nucleus to excite water molecules in the gas clouds that orbit around it, resulting in strong, stimulated emission at radio wavelengths. These so-called water masers can be studied with very high spatial and velocity resolution using radio interferometric techniques. In 1994, it was reported that the maser clouds trace a very thin disk, which made their dynamics easy to interpret. It was found that the motion of the clouds followed Kepler's law to 1 part in 100, reaching a velocity of $1100 \, \mathrm{km \, s^{-1}}$ at a distance of about one parsec from the center (figure 1.1). Only by assuming that the nucleus of NGC 4258 hosts a central body with a mass 40 million times greater than the Sun could these observations be explained.

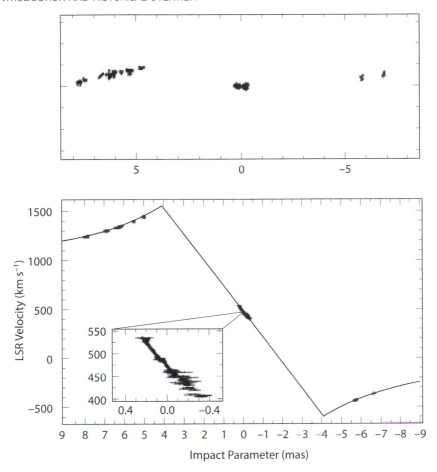

Figure 1.1 The top panel shows the distribution on the plane of the sky of the water masers
in NGC 4258. The units are milliarcseconds (mas), one mas corresponding to
0.035 pc at the distance of the galaxy. The bottom panel shows the rotation curve
traced by the maser clouds [393].

Perhaps even more remarkable is the case of the SBH at the center of the Milky
Way. The Galactic center has long been known to host a radio source, called Sagittarius A* (Sgr A*), that is at rest, indicating that it must be very massive. But because of the high visible-wavelength extinction toward the Galactic center, almost
70 years elapsed between the discovery of Sgr A* and the demonstration, using
ground-based telescopes, that Sgr A* is in fact an SBH. Beginning around 1992,
two groups, at the University of California at Los Angeles and the Max Planck
Institute in Garching, have monitored the positions and velocities of over a thousand stars within a parsec of Sgr A*. The stellar motions have been reconstructed
by combining the projected motion on the plane of the sky (the "proper motion")
with the velocity along the line of sight; the latter was measured from the Doppler
shifts of absorption lines in the stellar spectra. These data revealed the unmistakable

fingerprint of an SBH: stars closer to Sgr A* move faster than stars farther away in the exact ratio predicted by Kepler's law. Stars only a few light-days away from the source move at fantastic speeds, in excess of $1000 \, \mathrm{km \, s^{-1}}$. Such velocities can only be maintained if Sgr A* is roughly four million times more massive than our Sun. The current, best estimate of its mass is about 4.3×10^6 solar masses.

The Galactic center is 100 times closer than the next large galaxy, Andromeda, and 2000 times closer than the nearby association of galaxies, the Virgo Cluster. In no other galaxy do we have the opportunity to study the dynamics of individual stars orbiting a central SBH in such exquisite detail. To make matters worse, water masers like the one that populates the nucleus of NGC 4258 are very rare, and even more rarely are they organized in simple dynamical structures that can be easily interpreted.

In each of these two cases, the data probe regions in which the stellar or gas motions are completely dominated by the gravitational force from the SBH, just as the motions of planets in the solar system are dominated by the force from the Sun. If we were to look further from the center of these galaxies, we would find that the motion of the stars and gas clouds is influenced more by all the other nearby stars than by the central black hole.

In this regard, it is useful to define the "sphere of influence" of an SBH as the region of space within which the gravitational force from the SBH dominates that of the surrounding stars. The Galactic center stars, and the water masers in NGC 4258, lie well inside the respective spheres of influence. Measuring the mass of an SBH from data that do not resolve the sphere of influence is a bit like judging the weight of a turkey that may, or may not, be lurking in a distant bush. The rustling of the leaves may indicate a turkey; or it might be a bevy of quails; or maybe it's just the wind. All one can say for certain is that the bush isn't hiding an ostrich, or an elephant.[2]

Largely with the help of the Hubble Space Telescope, we have now resolved the spheres of influence of the SBHs at the centers of a handful of nearby galaxies. The most massive SBH detected to date, with a mass of about four billion solar masses, belongs to a giant elliptical galaxy M87 at the center of the Virgo Cluster.

Confident of the existence of SBHs, we can begin to ask more fundamental questions about them: How are SBHs related to their host galaxies? How did they form? What role do they play in galaxy evolution?

A partial answer to the first question emerged in 2000. A strong correlation turns out to exist between SBHs and the properties of their host galaxies. The mass of an SBH can be predicted with remarkable accuracy by measuring a single number—the velocity dispersion, σ, of the stars in the galaxy (figure 1.2). What is so surprising about this relation, aside from its precision, is that the stars whose velocities are measured are too far from the SBH to be influenced by its gravitational field. In other words, SBHs appear to "know" about the motion of stars that lie well outside of their sphere of mutual influence.

The origin of this relation is still being debated by theorists. But whatever its ultimate meaning, the relation is an extremely valuable tool, because it links

---

[2]In fact, in order to place useful constraints on the mass of an SBH, such data must be resolved on scales about one tenth of the influence radius, as discussed in chapters 2 and 3.

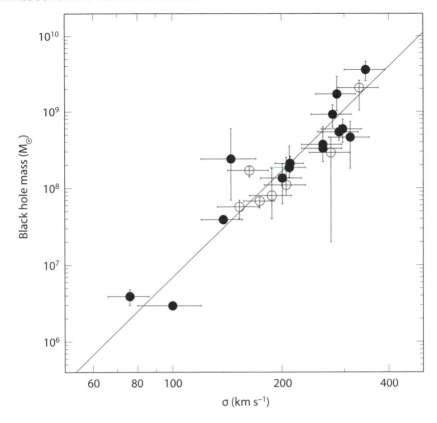

Figure 1.2 The $M_\bullet$–σ relation. The masses of SBHs are tightly correlated with the velocity dispersion of the stars in their host galaxies. The scatter of points about the best-fit line is consistent with that expected on the basis of measurement errors alone (shown by the error bars), implying that the underlying correlation is essentially perfect. However, as discussed in chapters 2 and 3, the number of galaxies with reliably measured SBH masses is still quite small.

something that is difficult to measure (the mass of an SBH) to something that is easy to measure (the stellar velocity dispersion far from the SBH). This makes it possible to determine the masses of SBHs in large samples of galaxies, much larger than the sample for which the techniques described previously can be applied. The result was a new field of endeavor: black-hole demographics. One finds that about 0.1% of a galaxy's stellar mass[3] is associated with its SBH, and the average density of SBHs in the local universe agrees remarkably well with the density inferred from observations of quasars. For the first time, the SBHs that powered the distant quasars were fully accounted for.

Current models of galaxy formation suggest that most large galaxies have experienced at least one "major merger" during their lifetime: a close collision between

---

[3]Dark matter, if it exists, far outweighs luminous matter in galaxies, but most of it is thought to lie far beyond the nucleus.

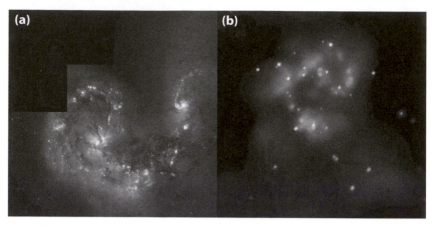

Figure 1.3 (a) An optical image of the "antennae" galaxies NGC 4038/4039, believed to be
in the early stages of a merger. (Image courtesy of NASA/STScI/ B. Whitmore.)
(b) The bright pointlike sources in this X-ray image of the same galaxies
may be intermediate-mass black holes. (Image courtesy of NASA/SAO/CXC/G.
Fabbiano et al.)

two galaxies that results in a coalescence (figure 1.3a). Computer simulations suggest that when two spiral galaxies merge, the result is an elliptical galaxy: most of
the gas is converted into stars, and the ordered rotation of stars in a disk is converted
into the more random motions observed in elliptical galaxies. In a galactic merger,
the SBHs at the centers of the two galaxies would sink rapidly to the center of the
merged system through a process called dynamical friction. Once at the center, they
would form a bound pair—a binary SBH—separated by about a parsec. It has long
been known that some active galaxies emit radio jets that twist symmetrically on
either side of the nucleus, suggesting that the SBH producing the jets is wobbling
like a precessing top. This is exactly what would happen in a binary SBH: the spinning SBH that produces the jets would precess as it orbits around the other SBH,
just as the Earth's axis wobbles due to the gravitational pull of the Sun and the
Moon.

Other active galaxies show periodic shifts in the amplitude or Doppler shift of
their emission. The best-studied case, a quasar called OJ 287, has experienced several major outbursts every 12 years since monitoring began in 1895. These flares
could be produced by a smaller SBH (10^8 solar masses) passing through the accretion disk of a larger one (10^9 solar masses) once every 12 years. In the last few
years, a number of other candidate binary SBHs have been found, some with apparent separations as small as 10 parsecs. While the number of such cases is still
very small, astronomers believe that most or all of the SBHs currently observed at
the centers of nearby galaxies must have been preceded by massive binaries.

As the two black holes in a binary system orbit each other, they emit energy
in the form of gravitational waves, ripples in space-time that propagate outward
at the speed of light. Any accelerating mass produces this kind of radiation, but
the only systems that can produce gravitational waves of appreciable amplitude
are pairs of relativistically compact objects—black holes or neutron stars—in orbit

about each other. Gravitational waves carry away energy, and so a system emitting gravitational radiation must lose energy—in the case of a binary black hole, this means that the two black holes must spiral in toward each other. The infall would be slow at first, but would accelerate until the final plunge when the two black holes coalesced into a single object. The coalescence of a binary SBH would be one of the most energetic events in the universe. However, virtually all of the energy would be released in the form of gravitational waves, which are extremely difficult to detect; there would be little if any of the electromagnetic radiation (light, heat, etc.) that make supernova explosions or quasars so spectacular.

No direct detection of gravitational radiation has ever been achieved, but the prospect of detecting gravitational waves from coalescing black holes is extremely exciting to physicists: it would constitute robust proof of the existence of black holes and it would permit the first real test of Einstein's relativity equations in the so-called strong-field limit. Furthermore, by comparing the gravitational waves of coalescing black holes with detailed numerical simulations, the masses, spins, orientations, and even distances of the two black holes could in principle be derived.

The prospect of observing the coalescence of a binary SBH is one of the primary motivations behind building a gravitational-wave detector in space. Existing, Earth-based gravitational-wave detectors are not able to detect the long-wavelength gravitational waves that would be generated by binary SBHs. A detailed design exists for a space-based detector, called the Laser Interferometer Space Antenna or LISA, which would consist of three spacecraft separated by five million kilometers flying in an equilateral triangle in the Earth's orbit. A passing gravitational wave would stretch and squeeze the space between the spacecraft, causing very slight shifts in their separations. Although such shifts are tiny—some 10^{-12} m across—they could be detected by laser interferometers.

The scientists who propose instruments like LISA must address one important question: How frequently will the instrument detect a signal from coalescing black holes? Space-based interferometers will have the sensitivity to detect mergers of SBHs out to incredible distances, essentially to the edge of the observable universe. One way to estimate the event rate is to calculate how frequently galaxies merge within this enormous volume. On this basis, a telescope like LISA should detect at least one event every few years. However, the situation is more complicated than this, since it is only the final stages of black-hole coalescence that produce an observable signal. In order to reach such small distances—less than 0.01 pc— the black holes must first spiral together from their initial separation of several parsecs. Gravitational radiation itself is too inefficient to achieve this; some other mechanism must first extract energy from the binary or else the decay will stall at a separation too great to generate a measurable signal for gravitational-wave tele-scopes. The prospect that some binaries might fail to close this gap has been called the "final-parsec problem." (Of course, it is a "problem" only from the standpoint of the physicists who hope to detect gravitational waves.)

As the merging galaxies come together, the two SBHs fall rapidly to the cen-ter of the merged system, dragged by the dynamical friction force acting on the galaxies as a whole. Once the binary is in place, a new mechanism comes into play called the gravitational slingshot. Any star that passes near to the massive binary

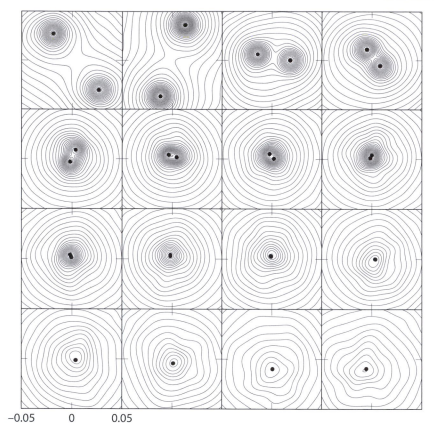

−0.05   0   0.05

Figure 1.4 One of the first computer simulations of a merger between two galaxies contain-
          ing supermassive black holes [386]. Time progresses from upper left to lower
          right; the contours indicate the projected density of stars, and the SBHs are indi-
          cated by the two filled circles.

is accelerated to high velocities and ejected, taking energy away from the binary
and causing its orbit to decay slightly. As a result, the separation between the SBHs
gradually shrinks, although perhaps not enough to place the massive binary into the
gravitational-wave regime.

There is good observational support for this model. Ejection of stars from a
galactic nucleus by a massive binary would drastically lower the density of stars
there, on a spatial scale roughly equal to the gravitational influence radius of the
larger SBH (figure 1.4). Just such a feature—a low-density core, or "mass deficit"—
is always found in bright elliptical galaxies, which statistically should have experienced the most mergers.

These cores are formed at an early stage in the evolution of the binary, and their
presence does not necessarily imply that the two SBHs coalesced. If coalescence
were delayed, long enough for a *third* SBH to be brought in by a subsequent galaxy
merger, the three massive objects could undergo a gravitational slingshot. The

resulting violent interaction could eject one or more of the black holes from the nucleus, and possibly from the entire galaxy. In this way, rogue SBHs might be created that drift forever between the galaxies.

Another prediction concerns the spins of black holes. If two SBHs coalesce, their orbital motion during the final plunge is converted into rotation of the resulting object. This means that SBHs at the centers of galaxies should be rotating rapidly. Furthermore, the directions of their spin axes should be essentially random, since the mergers responsible for imparting the spin take place from random directions. This prediction is consistent with observations of the orientations of radio jets in active galaxies, which are thought to point in the same direction as the spin axis of the black hole: the jet orientations are random with respect to the orientations of their host galaxies.

The presence of an SBH in the nucleus of every or nearly every galaxy has become a standard paradigm (even though the number of secure dynamical detections of SBHs is still quite small—depending on whom is asked, somewhere between 5 and 50). However, all of the SBHs detected so far have masses above about one million solar masses, while black holes created in supernova explosions are believed to be much smaller, probably no more than about 20 solar masses. Nothing definite is known about the existence of "intermediate-mass black holes," with masses between 10^2 and 10^6 solar masses. Empirical scaling relations like the one plotted in figure 1.2 suggest that a natural place to look for such objects might be at the centers of dwarf galaxies. However, the galaxies would need to be quite near in order for the SBH influence spheres to be resolved. There are a few such galaxies in the Local Group, the collection of three large, and many smaller, galaxies that contains the Milky Way. Hubble Space Telescope data have been obtained for two of these galaxies. Only an upper limit on the mass of a putative black hole could be found—either because no massive black hole was present, or because the sphere of influence was too small to be resolved. Frustratingly, the upper mass limits are close to what the scaling relations predict.

Resolving the issue of whether these intermediate objects exist goes beyond mere bookkeeping: it might be crucial for understanding how SBHs form. X-ray images of galaxies that are the sites of recent mergers, like the "antennae" galaxies, reveal pointlike X-ray sources that are too bright to have been produced by accretion onto a ten-solar-mass black hole (figure 1.3b). It is possible that these X-ray sources are produced by intermediate-mass black holes that formed, perhaps via runaway stellar mergers, in dense star clusters. With time, such star clusters might spiral toward the center of their host galaxy—thanks to the same dynamical friction process already described. The intermediate-mass black holes deposited at the galaxy's center might subsequently merge to form supermassive ones; or at least contribute to their mass; or, they might continue to orbit for long periods of time around the SBH.

Before the presence of SBHs at the center of every galaxy became accepted, theoretical studies of galactic nuclei usually invoked the same physical mechanisms—stellar encounters and collisions in a dense stellar system—as a source of nuclear evolution. In these models, the density of a galactic nucleus gradually increases as stars exchange kinetic energy in near collisions—or "gravitational encounters," as they are called. The increase in density leads to a higher rate of *physical* collisions between stars; when collision velocities exceed $1000 \, \mathrm{km \, s^{-1}}$, roughly the escape velocity from a star, collisions liberate gas that falls to the center of the system and condenses into new stars which undergo further collisions. Later it was argued that the evolution of a sufficiently dense nucleus might lead to the formation of a massive black hole at the center, either by runaway stellar mergers or by creation of a massive gas cloud which subsequently collapses.

A fundamental timescale in these models is the relaxation time: the time over which gravitational (not physical) encounters between stars cause them to exchange orbital kinetic energy. In order for the evolutionary models to be viable, the relaxation time must be less than about 100 million years, implying very high stellar densities—much higher than could be confirmed via direct observation at the time, or indeed now. The necessity of attaining high densities in order for these evolutionary models to work was clearly recognized. For instance, W. Saslaw [475] wrote,

> It is an extrapolation from the observations of galaxies we have discussed to the idea that even more dense stellar systems exist.... Yet this follows naturally enough from the observations of quasars and the realization that the central density of massive compact stellar systems increases with age.

L. Spitzer [501] speculated,

> The rate of dynamical evolution will depend on how compact is the stellar system resulting from initial gas inflow. If this rate of evolution is slow, activity will not begin for a long time. In fact, in some systems there might be a wait of 10^{12} years before the fireworks begin.

We now know that relaxation times near the centers of galaxies are much longer than 100 million years: partly as a result of the presence of SBHs, which increase the mean velocities of stars and reduce the rate of gravitational encounters (which have a strongly velocity-dependent cross section), and, in larger galaxies, because of the low-density cores. The long relaxation times imply that nuclear structure will still reflect to a large extent the details of the nuclear formation process; and indeed the persistence of the cores is probably an example of this.

The fact that most galactic nuclei are essentially "collisionless"—that is, their relaxation times are longer than the age of the universe—has important implications for their allowed dynamical states. In a collisional nucleus—that is, one that is many relaxation times old—the distribution of stars will have had time to evolve to a more-or-less unique form, a so-called Bahcall–Wolf cusp, in which the stellar density rises steeply with radius into the SBH. We have recently learned that the Milky Way probably does not contain such a cusp, in spite of the fact that it has one of the shortest measured relaxation times. In a collisionless nucleus, on the other hand, the range of allowed stellar distributions is much larger: anything from rapidly rotating disks like the solar system, to spheroidal configurations in which the stellar motions are essentially random, to lopsided configurations. It is the possibility of such a wide variety of dynamical states that makes the interpretation of observational data so frustrating, and the theoretical study of nuclei so fascinating.

Chapter Two

Observations of Galactic Nuclei and Supermassive Black Holes

2.1 STRUCTURE OF GALAXIES AND GALACTIC NUCLEI

Supermassive black holes (SBHs) appear always to be associated with **stellar spheroids**: the approximately spherical, or ellipsoidal, groupings of stars that constitute the luminous parts of elliptical galaxies, or the central components of some disk (spiral) galaxies. There are important systematic differences between the composition and internal kinematics of elliptical galaxies and the bulges of disk galaxies; but for the purposes of this book, it will almost always be adequate to lump the two sorts of system together. The terms "stellar spheroid" and "bulge" will be used interchangeably to describe these components of galaxies.

We are most concerned in this book with the distribution of mass near the centers of galaxies, but in practice, far more is known about the distribution of light. It is common to assume as a working hypothesis that "mass follows light," that is, that the density of starlight is proportional to the mass density. (An exception is naturally made in the case of SBHs. Other important exceptions are noted elsewhere in this book.) To the extent that galaxies are transparent—a good approximation for the stellar spheroids in which SBHs reside[1]—the observed intensity I of starlight at a given position in the image of a galaxy is an integral along the line of sight of the luminosity density j, and the problem of determining the galaxy's mass density ρ can be broken into two pieces: deprojecting the observed distribution of intensity to find the intrinsic luminosity density; and assigning a value to ρ/j, the **mass-to-light ratio**, often written simply as M/L.

Determination of mass-to-light ratios requires information in addition to photometric data—for instance, kinematical data—from which the gravitational acceleration produced by the mass can be measured. This problem is discussed at some length in chapter 3, and more briefly later in this chapter, when we review techniques for weighing SBHs.

The distribution of intensity across the image of a stellar spheroid is often found to be well approximated by

$$I(X, Y) = I(\xi), \quad \xi^2 = \frac{X^2}{A^2} + \frac{Y^2}{B^2}; \tag{2.1}$$

---

[1] The center of the Milky Way is an exception, since we observe it along a line of sight through the dusty Galactic disk.

in other words, the isophotes are elliptical—hence the name "elliptical galaxy." In equation (2.1), X and Y are spatial coordinates on the plane of the sky, and the coordinate axes have been aligned with the principal axes of the elliptical figure. A projected density of the form (2.1) is consistent with an intrinsic (three-dimensional) luminosity distribution of the form

$$j(x, y, z) = j\left(\frac{x^2}{a^2} + \frac{y^2}{b^2} + \frac{z^2}{c^2}\right), \qquad (2.2)$$

that is, a density stratified on triaxial ellipsoids. However, the two axis ratios that define the intrinsic figure are not uniquely derivable from the (single) observed axis ratio, unless one is willing to assume that the orientation of the ellipsoid is known. The deprojection problem becomes even more strongly underdetermined if the space density is allowed to have a more general functional form than in equation (2.2); for instance, if the ellipsoids have radially varying axis ratios or orientations [468].

For the remainder of this chapter, we will generally ignore the complications arising from the unknown three-dimensional shapes of stellar spheroids, and simply replace ξ in equation (2.1) by the projected radius R, the distance from the projected center of the galaxy. The function $I(R)$ is called the **intensity profile** or **surface brightness profile** of the galaxy. At least from a purely mathematical point of view, knowledge of $I(R)$ at all projected radii in the image of a spherical galaxy is equivalent to knowledge of $j(r)$. In practice, $I(R)$ might be constructed by averaging the two-dimensional surface brightness over the azimuthal angle at every R.

It is possible to deal nonparametrically with observed intensity profiles, but in practice, astronomers usually prefer to fit simple, parametrized functions to $I(R)$. Probably the most important of these is the **Sérsic profile** [485, 486]:

$$\ln I(R) = \ln I_e - b(n)\left[(R/R_e)^{1/n} - 1\right]. \qquad (2.3)$$

The constant b is normally chosen such that R_e is the projected radius containing one half of the total light—the **effective radius**. Aside from R_e, there are then two remaining parameters: $I_e \equiv I(R_e)$ and the **Sérsic index** n. Equation (2.3) may seem an unlikely expression, but it can be cast into a slightly more appealing form by differentiation:

$$\frac{d\ln I}{d\ln R} = -\frac{b}{n}\left(\frac{R}{R_e}\right)^{1/n}. \qquad (2.4)$$

In other words, the slope of I versus R on a log–log plot varies continuously as a power of R. This might be seen as the simplest generalization of a power law, which has a constant slope on a log–log plot. The popularity of Sérsic's law is due to two facts: first, it fits the intensity profiles of many individual galaxies over a very wide radial range, often the entire range for which there are data, two or three decades in radius; and second, it describes galaxies with a wide variety of types [68]. Aside from the "scaling" parameters R_e and I_e, Sérsic's law has just one "shape" parameter, n; it turns out that the best-fit value of n varies systematically with spheroid size or luminosity, in the sense that larger galaxies typically have larger n. Setting

$n = 4$ gives the **de Vaucouleurs profile** [106, 107], which is a good representation of bright elliptical galaxies. Setting $n = 1$ yields the exponential law, which approximates the intensity profiles of many dwarf galaxies and the disk components of spiral galaxies.

Unfortunately, the intrinsic (three-dimensional) density that projects to Sérsic's law does not have a simple mathematical form [342]. It can be shown that at small radii the deprojected density varies as

$$j(r) \propto r^{(1-n)/n}, \quad r \ll R_e \tag{2.5}$$

[86], a power law. In spite of the good behavior of Sérsic's law in projected space, the deprojected density diverges at the origin when $n > 1$; as steeply as r^{-1} in the limit of large n.

Various expressions have been proposed as approximations to the full, deprojected $j(r)$. Perhaps the most widely used is the **Prugniel–Simien model** [439]:

$$j(r) = j_0 \left(\frac{r}{R_e} \right)^{-p} e^{-b(r/R_e)^{1/n}}. \tag{2.6}$$

The quantities R_e, b, and n that appear in equation (2.6) are understood to be the same quantities that appear in equation (2.3). The additional parameter p is given by

$$p = 1 - \frac{0.6097}{n} + \frac{0.05563}{n^2}. \tag{2.7}$$

The Prugniel–Simien model is said to be a good approximation to the deprojected Sérsic profile over the radial range $10^{-2} \lesssim r/R_e \lesssim 10^3$, for Sérsic indices in the range $0.6 \lesssim n \lesssim 10$. Note that its asymptotic, small-radius behavior differs slightly from that of the exact, deprojected Sérsic profile.

While Sérsic's profile is a very good fit overall to most galaxies, systematic deviations do appear; typically at the largest or smallest radii. The former do not concern us here. Deviations at small radii are of two general kinds. Stellar spheroids fainter than $\sim 10^{10.3} \, L_\odot$ are often observed to have higher central surface brightnesses than predicted by Sérsic's law. These central enhancements are now often called **nuclear star clusters**, or **NSCs**, a name that reflects their very compact form, and also the fact that these components often appear to contain a young, intrinsically luminous stellar population. Radii of NSCs are a few parsecs at most, giving them an unresolved, pointlike appearance in galaxies much beyond the Local Group. To the extent that they can be spatially resolved, NSCs are sometimes seen to be flattened (disklike) in morphology, possibly reflecting the recent formation of stars in a gaseous disk (figure 2.1).

As a class, NSCs are the only distinct components of galaxies that might be called "nuclei." However, they are not present in all galaxies. Spheroids brighter than $\sim 10^{10.3} \, L_\odot$ generally exhibit central *deficits* in the intensity, compared with an inward extrapolation of the best-fitting Sérsic profile (figure 2.2). The deviations typically begin fairly sharply at a **break radius** or **core radius**, R_b or R_c, that is tens or hundreds of parsecs in extent—often large enough to be very well resolved, even by ground-based observations. Galaxies with this feature are called **core galaxies**.

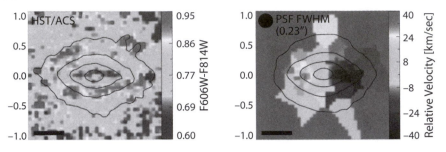

Figure 2.1 The NSC in NGC 4244, a nearly edge-on spiral galaxy [489]. Both panels display
a region that is two seconds of arc, or roughly 40 pc, on a side; the black bar at
the lower left in each panel has a length of about 10 pc. The left image is a "color
map," i.e., the ratio of intensities measured in two different filters; darker shades
indicate spectral energy distributions that are weighted toward shorter wavelengths, presumably indicating the presence of younger stars. The right panel
is a map showing the mean line-of-sight velocity of the stars. Rotation of the
nucleus is clearly visible, with a maximum amplitude of $\sim$30 km s^{-1}. In both
panels, the solid contours follow the intensity of starlight as measured at a single (near-infrared) wavelength. Gray areas in the panel on the right are regions
where the data quality is poor.

A relatively nearby example is the giant elliptical galaxy M87 in the Virgo Galaxy
Cluster; the core radius is a few hundred parsecs.[2] Essentially all bright elliptical
galaxies are core galaxies.

The surface-brightness profiles of core galaxies are often well fit by the **core-
Sérsic profile** [210]:

$$I(R) = \begin{cases} I_b \left(\dfrac{R_b}{R} \right)^{\Gamma}, & R \le R_b, \\[2ex] I_b e^{b(R_b/R_e)^{1/n}} e^{-b(R/R_e)^{1/n}}, & R > R_b. \end{cases} \tag{2.8}$$

In addition to the break radius R_b, equation (2.8) also contains Γ, the logarithmic
slope of the inner intensity profile; for core galaxies, one typically finds $0 \le \Gamma \lesssim 1$.
A more general version of equation (2.8) exists, with extra parameters, that allows
for a smoother transition of $I(R)$ from $R < R_b$ to $R > R_b$ [523]. As in the case
of the Sérsic model, there exist simple approximations to a deprojected core-Sérsic
model; for instance, the **core-Prugniel–Simien model** [513],

$$j(r) = \begin{cases} j_b \left(\dfrac{r}{R_b} \right)^{-\gamma}, & r \le R_b, \\[2ex] j_b \left(\dfrac{r}{R_b} \right)^{-p} \left(\dfrac{R_b}{R_e} \right)^{p} e^{b(R_b/R_e)^{1/n}} e^{-b(r/R_e)^{1/n}}, & r > R_b. \end{cases} \tag{2.9}$$

[2]M87 is an active galaxy, and its intensity profile at visual wavelengths also exhibits an unresolved
central *enhancement*, presumably due to light from gas in the accretion disk orbiting very near the SBH.

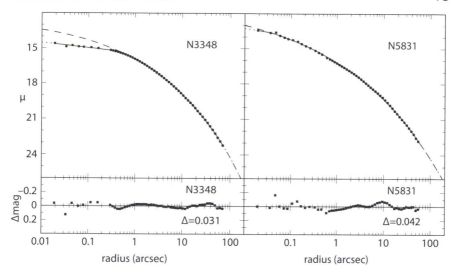

Figure 2.2 Observed intensity profiles of two elliptical galaxies at visual wavelengths, showing fits of the data to standard model profiles [210]. The galaxy on the right is
well fit by Sérsic's law, equation (2.3), shown as the dashed curve. The galaxy on
the left has a well-resolved core; the solid curve shows a fit of the data to a core-
Sérsic model, equation (2.8), while the dashed curve is the best-fitting Sérsic
profile. The lower panels show the deviations between data and models. Intensities are expressed in "magnitudes," which are proportional to the logarithm of
the inverse of the surface brightness.

As in the Prugniel–Simien model, the parameters R_e, R_b, b, and n are identified
with the same parameters in the core-Sérsic profile, and $\gamma = \Gamma + 1$.

As discussed in chapter 8, a leading hypothesis for the origin of cores is the
"scouring" effect of binary SBHs—binaries that were created, presumably, during the same galactic merger event that formed the spheroid. The idea that the
cores of bright elliptical galaxies are due to an extrinsic modification of an underlying, "universal" density profile gains support from another argument. If one
plots central intensity versus total galaxy luminosity, a break appears in the relation
at $L \approx 10^{10}\, L_\odot$, between "dwarf" and "giant" elliptical galaxies [296]. "Dwarf"
ellipticals define a continuous sequence spanning some four decades in luminosity, along which the central surface brightness increases with increasing L. This
relation reverses for "giant" galaxies, which exhibit *lower* central intensities as L
increases. But if one replaces the actual central intensity by the inward extrapolation of the best-fitting Sérsic law, one finds that bright ellipticals smoothly continue
the sequence defined by fainter ellipticals: central density increases with increasing
luminosity, with no sign of a break in the relation [259, 211].

Various generalizations of Sérsic's profile have also been proposed for fitting the
excess nuclear light observed in fainter spheroids. But because NSCs are typically
poorly resolved, many different functional forms are found to do an equally good
job. Local Group galaxies provide the only exceptions: at least two of these contain

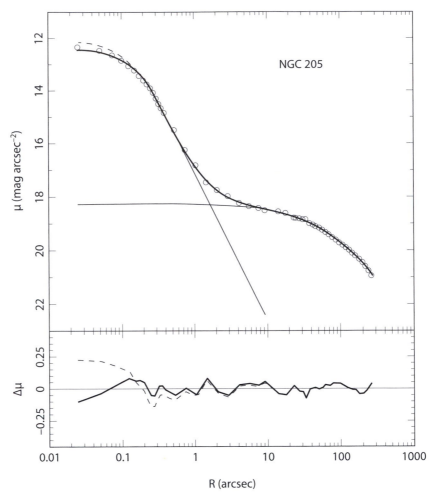

Figure 2.3 Surface brightness data of the Local Group dwarf galaxy NGC 205 showing the resolved NSC [356]. The observations, shown as the open circles, were made in the *I*-band, which centers around $\lambda \approx 8000\,\text{Å}$ [527, 280]. The data have been fit by a two-component model; the dashed and solid curves show the model before and after convolution with the instrumental point-spread function. The lower panel shows the fit residuals. This galaxy is roughly 800 kpc away, and one arcsecond corresponds to about 3 pc. Surface brightness units (magnitudes per square arcsecond) are explained in the caption to figure 2.2.

NSCs that are near enough to be resolved. One is the dwarf elliptical galaxy NGC 205, a satellite companion to the "Andromeda nebula," the giant spiral galaxy M31. Figure 2.3 shows the intensity profile measured at near-infrared wavelengths. The nuclear cluster appears as a very distinct component, with a half-light radius of about 2 pc and an inner core. The intensity profile of this component was fit by a model with intrinsic luminosity density

$$j(r) = j_0 \left(1 + \frac{r^2}{r_c^2} \right)^{-\gamma}, \tag{2.10}$$

sometimes called a "modified Hubble model," with $\gamma = 3.2$ and $r_c = 0.3$ pc. The prominence of this galaxy's NSC is due in part to the presence of a population of bright, blue, and presumably young stars [241]. A young stellar component turns out to be common in NSCs; population synthesis suggests that many have experienced an extended star formation history, with luminosity-weighted mean stellar ages that range from ~ 10 Myr in the nuclei of late-type spiral galaxies to ~ 10 Gyr in elliptical galaxies [553]. However, it appears that the mass is typically dominated by an old stellar population even when most of the light comes from young stars. The assumption of a constant "mass-to-light ratio" breaks down in galaxies with young NSCs since young stars have a much smaller M/L than old stars.

The Milky Way also contains an NSC. At a distance of only about 8 kpc, it is by far the best-resolved example, although the strong and spatially varying extinction toward the Galactic center presents serious obstacles to inferring its structure, and indeed it was only in the last few years that its resemblance to nuclear clusters in other galaxies became clear [479]. Fits of the intensity data at near-infrared wavelengths reveal that this component of our galaxy can be represented approximately in terms of a power law in the space density, $j(r) \sim r^{-1.8}$, within ~ 10 pc of Sgr A* [310]. The total mass is perhaps 30 million solar masses, or roughly ten times the mass of the SBH. As in NGC 205, the Milky Way NSC contains a population of apparently young stars, which dominate the light inside of a parsec or so. When these young stars are excluded, the number counts of the old, and presumably dominant, population exhibit an inner *core* of radius ~ 0.5 pc [66], similar in size to the core in NGC 205.

In the early days of the search for SBHs it was commonly assumed that the presence of an SBH would reveal itself via an enhancement in the central stellar density, roughly at the radius where the gravitational force from the SBH begins to rise above that from the galaxy as a whole [416]. For instance, in the "adiabatic growth" model (section 3.3), the density near the SBH is predicted to rise as $\rho \sim r^{-3/2}$; another example is the "Bahcall–Wolf cusp"[3] which forms in response to exchange of orbital energy between stars (section 5.5.2) and has $\rho \sim r^{-7/4}$. Photometric data were sometimes interpreted to imply the presence of such density cusps in nearby galaxies. For instance, M87, the giant elliptical galaxy in the Virgo Cluster, contains a well-resolved core, but fits of the intensity profile to the standard ("isothermal") models that were in use at the time (the 1970s) seemed to require an extra, cusplike component to fit the small-radius data. This was taken as indirect evidence for the presence of an SBH [575].

To the author's knowledge, there is no galaxy in which the presence of a density cusp can convincingly be connected to the dynamical influence of an SBH. In the case of galaxies like M87, it was realized [295] that "nonisothermal" cores are

---

[3]"Cusp" refers, apparently, to the fact that a plot of density versus radius that extends from one side of the galaxy to the other, on linear axes, has a cusplike appearance.

the norm, and that their intensity profiles could be easily fit by simple parametric models like the core-Sérsic model, without the need for an additional component representing a cusp. The Milky Way is a more interesting case, since an SBH-induced density cusp could easily be resolved if it were present. Indeed, the steep rise in luminosity density observed in the NSC continues well inside the influence radius, roughly as a power law, $j \sim r^{-\gamma}, \gamma \approx$ 1–1.5 [183]. But as noted above, most of the light in this region is due to the puzzling young stars, which cannot have been present long enough for any of the proposed cusp-formation mechanisms to apply. The number counts of the older stars are flatter, and in fact their density is observed to be nearly constant inside $\sim$0.5 pc, which is just where a density cusp would be expected to show itself [357].

It is rather ironic that the morphological feature most commonly associated nowadays with SBHs is not a density cusp, but just the opposite—a low-density core.

2.2 TECHNIQUES FOR WEIGHING BLACK HOLES

As discussed in chapter 1, there is compelling, albeit often circumstantial, evidence for SBHs at the centers of many galaxies. But even when the existence of an SBH is all but incontrovertible, one would still like to know its mass; and for many astronomers, "detection" of an SBH is considered tantamount to a measurement of its mass.

With few exceptions, convincing mass determinations in astronomy are dynamical: one uses the observed motions of stars or gas to infer a force, then applies Newton's laws of gravitation to calculate the mass that is responsible for the force. In the case of an SBH at the center of a galaxy, stars will always be present near the SBH, and their velocities can be measured via individual Doppler shifts or proper motions (in the nearest galaxies), or Doppler broadening of stellar absorption lines in the integrated spectra (in galaxies beyond the Local Group). Gas may or may not be present, at least in amounts, or in a form, that is easily observed. But luckily, some galaxies happen to contain ionized gas that orbits, more-or-less coherently, in a thin disk near the SBH, allowing the interior mass to be computed from a measurement of the emission-line Doppler shifts.

Force is related to mass via distance, and techniques that determine mass from Newton's laws require a measurement also of the distance between the SBH and the objects whose velocities are being measured. At the very least, this requires an estimate of the distance to the galaxy; the inferred $M_\bullet$ will scale linearly with that distance. Less trivially, the kinematical data must be spatially resolved, on angular scales small enough that the gravitational force is dominated by the SBH. We will use the adjective **primary** to describe mass estimation methods that are based on such data. **Secondary** mass estimation methods are based on empirical correlations that were calibrated using primary mass estimates; and one can define tertiary and even quaternary methods, in terms of the number of empirical correlations that separate the data from the inferred mass.

An SBH of mass $M_\bullet$ embedded in a galactic nucleus strongly affects the motion of gas or stars within a certain distance, called the **gravitational influence radius**, or simply the "influence radius." Since the gravitational force in a spherical galaxy is determined by the enclosed mass, a natural definition of the influence radius is the radius of a sphere that encloses a mass in stars similar to $M_\bullet$. This idea motivates the definition of r_m:

$$M_\star(r < r_m) = 2M_\bullet. \tag{2.11}$$

At $r = r_m$ in a spherical galaxy, $1/3$ of the gravitational force comes from the SBH and $2/3$ from the stars. The numerical factor in this relation is somewhat arbitrary; the reason for choosing a value of two for this factor is discussed below.

Equation (2.11) can be difficult to apply to real galaxies, given that the stellar mass density is rarely well determined inside the influence radius. A second definition of the influence radius is based on a quantity that is, at least in principle, easier to measure. Let v_{rms} be the rms stellar velocity near the center of a galaxy. If the velocity distribution is isotropic, the velocity dispersion along any direction is $\sigma = v_{rms}/\sqrt{3}$. The latter quantity is closely related to the line-of-sight velocity dispersion that would be measured from the integrated spectrum of stars near the galaxy's projected center. The second influence radius, r_h, is defined such that the velocity of a circular orbit around the SBH at r_h, $v_c = (GM_\bullet/r_h)^{1/2}$, is equal to σ:

$$r_h \equiv \frac{GM_\bullet}{\sigma^2} \approx 10.8 \left(\frac{M_\bullet}{10^8\, M_\odot} \right) \left(\frac{\sigma}{200\,\mathrm{km\,s^{-1}}} \right)^{-2} \mathrm{pc}. \tag{2.12}$$

Of course, sufficiently close to the SBH, velocities must increase as $r^{-1/2}$, and σ will become a steep function of radius. However, only a handful of galaxies[4] are observed at sufficient angular resolution that the $r^{-1/2}$ behavior of σ near the SBH is apparent. In practice, observers often define σ as the rms, line-of-sight velocity of stars within an aperture centered (hopefully) on the SBH—an **aperture dispersion**. That practice is based on the expectation, which is probably justified in all but a handful of galaxies, that the presence of the SBH has very little effect on the measured velocities.

Which of these two definitions of "influence radius" is most relevant depends on the physical question being addressed. Since the rms velocity near the center of a galaxy contains contributions from stars that move far from the center, the second definition compares the local gravitational effects of the SBH with those from the galaxy as a whole. The first definition is based on a local comparison of forces, which may be more appropriate when discussing the motion of gas, or of test particles moving in nearly circular orbits around the SBH, since these are unaffected by the distribution of matter farther out.

In a nucleus with stellar density $\rho \propto r^{-2}$, and no SBH, the velocity dispersion can be shown, using the techniques presented in chapter 3, to be independent of radius:

$$\sigma^2 = 2\pi G r^2 \rho(r), \tag{2.13}$$

[4]The Milky Way, M31, and possibly M32.

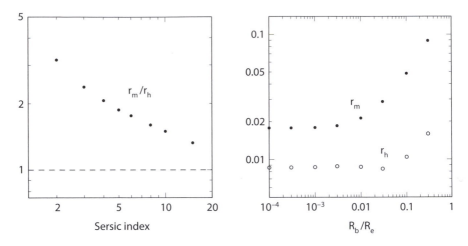

Figure 2.4 Comparisons between the two definitions given in the text, $r_{\rm m}$ and $r_{\rm h}$, for the gravitational influence radius of an SBH. The left panel shows the ratio of the two radii in model galaxies that follow the Prugniel–Simien density law, equation (2.6), assuming that $M_\bullet = 0.002 M_{\rm gal}$. In the right panel, the core-Prugniel–Simien model, equation (2.9) was used, setting the Sérsic index n to 4. In this panel, radii are expressed in units of the effective radius R_e.

and the enclosed mass is

$$M_\star(< r) = 4\pi \int^r dr\, r^2 \rho(r) = \frac{2\sigma^2}{G} r. \tag{2.14}$$

In this model, called the **singular isothermal sphere**,[5] the radius at which $M_\star = 2M_\bullet$ is $GM_\bullet/\sigma^2$; in other words, $r_{\rm m} = r_{\rm h}$. (Of course, we are ignoring the fact that addition of the SBH would change σ at $r \lesssim r_{\rm m}$.) This is the motivation for the numerical factor in equation (2.11). The singular isothermal sphere is a reasonable description of the mass distribution in the inner few parsecs of the Milky Way, and indeed for our galaxy, both $r_{\rm h}$ and $r_{\rm m}$ are approximately equal to 2.5 pc.

In nuclei that are not well described by the singular isothermal sphere model, $r_{\rm m}$ and $r_{\rm h}$ can be substantially different. This is illustrated in figure 2.4, based on spherical models that follow the Prugniel–Simien, or core-Prugniel–Simien, density profiles defined above. In making this figure, $r_{\rm h}$ was defined as the root of

$$\sigma^2(r) - \frac{GM_\bullet}{r} = 0, \tag{2.15}$$

and $\sigma(r)$ was computed from $\rho(r)$ and $M_\bullet$ using the isotropic Jeans equation (3.58); that equation yields the unique dependence of σ on r in a spherical, isotropic, steady-state galaxy with the specified mass distribution and SBH mass. As the central density profile becomes flatter, the figure shows that $r_{\rm m}$ increases compared with $r_{\rm h}$. When the core radius exceeds $\sim r_{\rm h}$, as appears to be the case in some luminous elliptical galaxies, $r_{\rm m} \approx R_b$ and $r_{\rm m} \gg r_{\rm h}$.

[5]A nonsingular version of this model is presented in section 7.5.1.

The observed relation between $M_\bullet$ and r_m for "core" galaxies (i.e., galaxies with $M_\bullet \gtrsim 10^8 \, M_\odot$) is [374]

$$r_m \approx 35 \left(\frac{M_\bullet}{10^8 \, M_\odot} \right)^{0.56} \text{pc}. \tag{2.16}$$

The form of this relation at smaller $M_\bullet$ is probably similar, but it is less well determined due to the difficulty of resolving the region $r \lesssim r_m$ in smaller galaxies.

2.2.1 Primary mass determination methods: Stellar and gas kinematics

Primary SBH mass estimates are most often based on velocities that have been affected to some degree by the gravitational force from the distributed mass (stars, gas) in the nucleus, as well as the SBH. When stars are used as the dynamical tracers, the relation between the gravitational potential and the kinematical quantities is described by an equation like

$$n \frac{\partial \Phi}{\partial r} = -\frac{\partial (n\sigma^2)}{\partial r} + \frac{n}{r} \bar{v}_\phi^2. \tag{2.17}$$

This is essentially equation (3.112b), the Jeans equation for an axisymmetric galaxy; Φ is the gravitational potential, n is the number density of stars, $\bar{v}_\phi$ is the mean, or "streaming," velocity of stars about the galaxy center and σ is the one-dimensional velocity dispersion as defined above. The exact equation has been simplified here by (1) expressing it in the galaxy's equatorial plane; (2) replacing ϖ, the cylindrical radius, by r, the distance from the center; and (3) assuming isotropy of the stellar velocities with respect to their mean motions.

Without any additional approximations, equation (2.17) can be rewritten in the form

$$G \left[M_\star(r) + M_\bullet \right] = r \left(\sigma^2 + \bar{v}_\phi^2 \right) + r\sigma^2 \left[-\left(\frac{\partial \log n}{\partial \log r} + 1 \right) - \frac{\partial \log \sigma^2}{\partial \log r} \right]. \tag{2.18}$$

Typically, the two terms inside the brackets on the right-hand side of this relation are small: because $n \sim r^{-1}$ (as in a Sérsic galaxy of high index); and because σ is a slowly varying function of radius (unless one is well inside r_h). The enclosed mass is then determined essentially by $(\sigma^2 + \bar{v}_\phi^2)$. If the galaxy is observed from a direction not too far off from its equatorial plane, the latter quantity is not too different from the rms, line-of-sight velocity of stars, as measured via the broadening of absorption lines in an integrated stellar spectrum.

Figure 2.5 plots just this quantity,[6] for 12 galaxies in which the presence of an SBH has been claimed based on stellar kinematical data. The galaxies have been ordered by the angular size of the influence radius r_h; the latter was computed from the published $M_\bullet$ value and from the measured σ. The computed value of r_h is indicated by the dotted, vertical line in each frame.

---

[6]There is an additional justification for combining $\bar{v}$ and σ, having to do with instrumental limitations; see section 2.3.3.

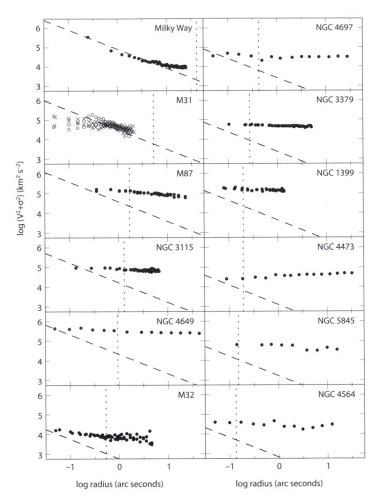

Figure 2.5 Stellar kinematical data for galaxies with putative SBHs. The horizontal axis is the angular distance from the center of the galaxy in seconds of arc. The vertical axis is the mean square line-of-sight stellar velocity. In the case of M31, which has a lopsided nucleus, data are plotted with respect to two possible centers: the point of peak velocity dispersion (circles) and the point of zero mean velocity (crosses). Galaxies are ordered in terms of the angular size of the SBH influence radius $r_h = GM_\bullet/\sigma^2$, shown as the vertical dotted line; r_h was calculated using the published value of $M_\bullet$. In the case of the Milky Way and M87, the SBH mass was derived from other data than shown here: individual stellar orbits in the case of the Milky Way [193], and rotation of a gas disk in the case of M87 [332]. Well inside r_h, the mean square velocity should rise inversely with projected radius; this is plotted as the dashed line. Only the Milky Way and the Local Group galaxy M31 exhibit prima facie evidence for an SBH in the form of a Keplerian rise in the stellar velocities near the center. (Data sources: Milky Way [481]; M31 [12]; M87 [181, 538]; NGC 3115 [141]; NGC 4649 [180]; M32 [262, 547]; NGC 4697 [180]; NGC 3379 [490]; NGC 1399 [246]; NGC 4473, 5845, 4564 [180].)

Close enough to the SBH, $M_\bullet \gg M_\star$, and we expect to see a kinematical signature associated with the SBH. Suppose for sake of simplicity that $\bar{v}_\phi^2 + \sigma^2 = K\sigma^2$ with K some constant; since most of the kinetic energy near the centers of galaxies is in the form of random motions, $K \gtrsim 1$. Returning to the more general expression (2.17), and assuming that $n(r) \propto r^{-\gamma}$, the solution is

$$\sigma^2 + \bar{v}_\phi^2 \equiv v_{\text{rms}}^2 = \frac{K}{K+\gamma} \frac{GM_\bullet}{r} \approx \frac{1}{1+\gamma} \frac{GM_\bullet}{r}. \qquad (2.19)$$

The $r^{-1/2}$ dependence of velocity on radius is the signature of an SBH; this dependence is sometimes called "Keplerian" since, of course, it is the same dependence exhibited by the orbital velocities of planets in the solar system. The constant term in equation (2.19) is modified slightly if one replaces σ by the true, line-of-sight velocity dispersion and r by the projected distance from the center; the proper relation is plotted as the dashed lines in figure 2.5.

Beyond the Local Group, figure 2.5 reveals that no galaxy exhibits a convincing Keplerian rise in stellar velocities near the center. Indeed, for a few of the galaxies in the figure, the rms velocities are seen to *drop* in the inner few resolution elements. Data like these do not provide much reassurance that an SBH is present, and the value of $M_\bullet$ inferred from such data will depend critically on how much of the gravitational force near the center is attributed to the stars [181].

In the Local Group, two, and perhaps three, galaxies exhibit a convincing rise in v_{rms} near the center: the Milky Way, M31, and (possibly) M32. In the case of the Milky Way, the best estimates of $M_\bullet$ come not from the data plotted in figure 2.5, but from the detailed astrometric (positional) data of a handful of stars, as discussed in chapter 4. Nevertheless, the rms stellar velocities that are plotted in figure 2.5 (from a sample of $\sim$6000 stars with measured proper motions in the inner parsec) show a very convincing Keplerian rise inside a projected radius of $\sim 10''\approx 0.4$ pc, or roughly $0.2r_{\text{h}}$. Indeed, one could obtain quite an accurate estimate of $M_\bullet$ in the Milky Way by simply laying a ruler on figure 2.5—although the mass of Sgr A* has been inferred from these data in a slightly more careful way [481], and found to be consistent with the more accurate value based on stellar astrometry [192].

M31, the Andromeda nebula, also shows a reasonably Keplerian rise in stellar velocities near the center, although the nucleus of this galaxy is highly asymmetric, making the interpretation of data like those in figure 2.5 problematic (section 2.3).

Finally, M32, the dwarf companion to M31, shows what might be called the beginnings of a velocity rise near the center.

As noted above, in the Milky Way, the Keplerian rise in the stellar velocities begins to become apparent inside a projected radius of $\sim 0.2r_{\text{h}}$. If one hopes to fit the observed rise in rms velocities to a relation like (2.19), spatially resolved data would need to extend somewhat farther in; let us say, to at least $0.1r_{\text{h}}$. If we assume that roughly the same is true for other galaxies, we can write an approximate criterion for detectability of SBHs: the measured velocities must be resolved on an angular scale corresponding to a linear distance at least as small as $0.1r_{\text{h}}$, or

$$\theta_{\text{det}} \lesssim 0''.02 \left(\frac{M_\bullet}{10^8 \, M_\odot}\right) \left(\frac{\sigma}{200 \, \text{km s}^{-1}}\right)^{-2} \left(\frac{D}{10 \, \text{Mpc}}\right)^{-1}, \qquad (2.20)$$

where D is the distance to the galaxy. By comparison, the resolution of the Space Telescope Imaging Spectrograph (STIS) on the Hubble Space Telescope (HST), which was the source for much of the data in figure 2.5, is $\sim 0''.1$. If we apply equation (2.20) to galaxies in the Virgo Cluster (of which five appear in figure 2.5) at a distance of $\sim 16\,\mathrm{Mpc}$, we find the following condition for detectability with HST:

$$\frac{M_\bullet}{10^8\,M_\odot} \gtrsim 8\left(\frac{\sigma}{200\,\mathrm{km\,s^{-1}}}\right)^2 \quad \text{(Virgo)}. \qquad (2.21)$$

If one accepts the claimed, tight correlation between $M_\bullet$ and σ discussed later in this chapter (the "$M_\bullet$–σ relation"), one finds that the condition (2.21) requires $M_\bullet \gtrsim 2 \times 10^9\,M_\odot$ for galaxies in the Virgo Cluster. Only the giant galaxy M87 at the cluster center satisfies this condition. However, figure 2.5 shows only a very gradual rise in the rms stellar velocities toward the center of M87. This is probably a consequence of the low central density (large core) in this galaxy, which means that the measured, line-of-sight velocities are more weakly weighted by stars that are intrinsically close to the SBH. Most of the other galaxies in the figure have cores similar to M87's, and it is likely that the resolution required to detect SBHs in these galaxies via stellar motions is likewise higher than in galaxies like the Milky Way—in other words, that a resolution of $\sim 0.1 r_\mathrm{h}$ is barely sufficient.

M87 is notable not just for having the most massive SBH with a primary mass determination, it is also the nearest galaxy for which gas-dynamical data were used to measure $M_\bullet$ [332]. Figure 2.6 shows the data. The prima facie case for an SBH in this galaxy, based on the gaseous rotation curve, is clearly much stronger than the stellar-dynamical case (figure 2.5).

As this example suggests, estimates of $M_\bullet$ based on the motion of gas tend to be inherently superior to stellar-dynamical estimates, for a number of reasons:

1. The rotational velocity of the gas measures the interior mass directly:

$$v_c^2(r) = \frac{G\,(M_\star + M_\bullet)}{r}. \qquad (2.22)$$

   In the case of stellar motions, velocities measured near the SBH are "contaminated" by stars that orbit to much greater distances.
2. With gas, there is less diminution of the signal due to an averaging along the line of sight.
3. Stellar motions are inherently complex; for instance, the rms velocity can be different in different directions. Gas moving in a regular disk is characterized by just one velocity at every radius.

There are disadvantages to gas-dynamical mass estimates as well. Foremost is that relatively few galaxies contain ionized gas orbiting in a more-or-less regular disk near the center. Correcting the measured rotation velocities for the inclination of the disk is also a problem, although stellar-dynamical mass estimates can also be strongly inclination dependent, and the inclination of a (circular) gas disk is more easily constrained than the shape and orientation of a galaxy's three-dimensional figure [536]. Gas motions can also deviate from the ballistic trajectories assumed in writing equation (2.22) due to pressure gradients.

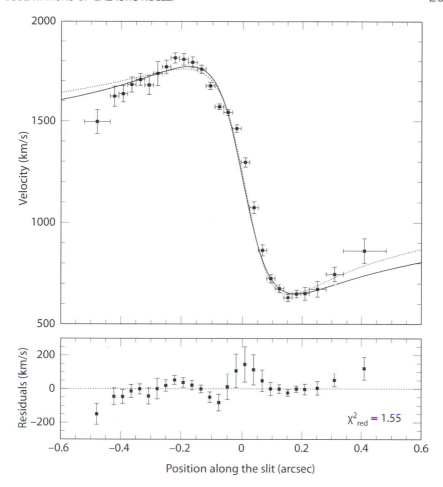

Figure 2.6 The rotation curve of the ionized gas disk near the center of the giant elliptical galaxy M87 [332]. The data are from the Faint-Object Camera on the Hubble Space Telescope. The solid and dotted curves are from models that assume two different orientations for the gas disk; all such models are found to imply the presence of an SBH of a few $10^9 \, M_\odot$. The best-fit value is claimed to be $3.2 \pm 0.9 \times 10^9 \, M_\odot$. The stellar-dynamical data for this galaxy (figure 2.5) show only a weak indication of a central rise and the value of $M_\bullet$ derived from those data is accordingly less certain.

Assuming that a gas disk is present, the criterion for detectability of the SBH is the availability of velocity data that extends inside $\sim r_{\rm m}$; this guarantees that $M_\bullet > M_\star$ in equation (2.22). This is a less stringent condition than in the case of stellar velocity data, and indeed there are several galaxies beyond the Local Group in which the gas velocities are clearly seen to rise near the center.

There is a small, but important, class of active galaxies in which rotation can be measured at much higher spatial resolutions. These are galaxies containing **water masers**, believed to originate from dense molecular clouds that are excited into stimulated emission by X-rays from the accretion disk around the SBH [215]. The 22 GHz maser emission can be studied using radio interferometric techniques at angular resolutions of milliarcseconds, roughly 100 times better than can be achieved with optical telescopes. Observations of the active galaxy NGC 4258 in the mid 1990s using the newly commissioned Very Long Baseline Array revealed the presence of water masers with large velocities relative to the galaxy; the maser sources turned out to be arrayed in a thin, regular disk extending only a fraction of a parsec from the central source. These data provide one of the finest Keplerian rotation curves observed for any SBH, yielding a mass of $M_\bullet = 3.9 \times 10^7 \, M_\odot$ and an influence radius of $\sim 0''.15$ (figure 1.1) [392]. Furthermore, because the innermost clouds sit at a distance of only ~ 0.13 pc from the central source, the lower limit on the mean *density* of the enclosed mass is $\sim 4 \times 10^9 \, M_\odot \, \text{pc}^{-3}$, several orders of magnitude larger than in any other nucleus, with the exception of the Milky Way. Unfortunately, galaxies with water masers are rare, and none has been found to exhibit as regular a rotation curve as NGC 4258 [303].

The technique of **spectroastrometry**, first developed in the context of binary star observations [16], also has the potential to overcome the $\sim 0''.1$ limit of optical telescopes in observations of gas rotation curves. Consider an unresolved source consisting of two stars at different velocities. If the stars have different spectra, the position measured for the centroid of the light will differ as a function of wavelength. It turns out that relative shifts in the centroid position can be measured on scales much smaller than the resolution limit of the telescope, and modeling of simulated data from gas disks around SBHs suggests that information about the gravitational potential can be obtained on angular scales of $\sim 0''.01$ [195]. This potentially important technique has just begun to be applied to the determination of $M_\bullet$ in galactic nuclei [196].

2.2.2 Primary mass determination methods: Reverberation mapping

The techniques that allow us to measure SBH masses in quiescent galaxies are difficult to apply to the host galaxies of bright AGN. Even in active galaxies near enough that the SBH's sphere of influence has some chance of being resolved, the presence of a bright, pointlike nucleus due to the gaseous accretion disk tends to dilute the very features that are necessary for dynamical studies. An alternative method, called **reverberation mapping**, can be applied to a subset of AGN.

Spectra of AGN at optical and ultraviolet wavelengths exhibit **broad emission lines**. The strongest lines are from the hydrogen Balmer series (Hα, $\lambda = 6563$ Å; Hβ, $\lambda = 4861$ Å; etc.) The line widths are assumed to reflect Doppler broadening; the inferred velocity widths lie in the range $500 \, \text{km s}^{-1} \lesssim \Delta V \lesssim 10^4 \, \text{km s}^{-1}$, where ΔV is the full width at half maximum of the velocity broadening function. The fact that the emission-line fluxes vary strongly in response to changes in the continuum (i.e., the light from the accretion disk near the SBH) implies that ionizing photons from the central source are responsible for the emission lines. Furthermore, the emission-line response is found to be delayed with respect to changes in the continuum; assuming that the delay is due to light travel times, the implied size of the **broad emission-line region** (BLR) works out to be of order 0.01–0.1 pc.

In principle, this "reverberation" response of the BLR to variations in the continuum could be used to map out the three-dimensional structure of the BLR [53]. In
practice, the amount and quality of data required to carry out a model-independent
data deconvolution is prohibitive, and applications of this technique are usually directed toward the more modest goal of recovering an estimate of the size of the
BLR [424]. Because the emission-line gas lies well inside the SBH sphere of influence [427], a determination of R_{BLR}, together with the known velocity width,
yields a direct estimate of the SBH mass:

$$GM_\bullet = f \, R_{BLR}(\Delta V)^2, \tag{2.23}$$

where f is a constant of order unity. Unfortunately, the proper value to take for
f depends on unknowns like the geometry of the BLR (sphere, disk), the radial
emissivity of the gas, etc. Until about 2004, f was estimated ab initio based on
simple models for the structure of the BLR. These early studies indicated that for
galaxies of comparable magnitude, masses derived from reverberation mapping and
those obtained using stellar dynamics differed by as much as a factor of 50, with
the AGN masses systematically lower [554]. The blame for the discrepancy fell
initially on the reverberation mapping results; however, these were vindicated when
it was realized that the stellar dynamical masses used in the comparison [335] were
seriously flawed. Further discussion of this issue and its resolution can be found
later in this chapter.

In spite of systematic uncertainties due to the unknown structure of the BLR,
reverberation mapping mass estimates have an important advantage compared with
those based on stellar kinematics, since the ΔV that appears in equation (2.23) is
due almost entirely to the gravitational force from the SBH. A devil's advocate
could reconcile almost all existing stellar dynamical data from galaxies beyond the
Local Group with zero SBH masses by, for instance, allowing modest changes in
the stellar mass-to-light ratio inside the unresolved region.

2.2.3 Mass determination based on empirical correlations

Primary methods for determining $M_\bullet$ are limited to galaxies near enough that r_h or
r_m is well resolved. With the exception of a few special cases, like NGC 4258 with
its maser disk, there are almost no galaxies beyond the Local Group that satisfy
this criterion, and astronomers who wish to enlarge the sample of SBH masses are
forced to fall back on less secure techniques.

If one had complete faith in the phenomenological relations described in the next
section—for instance, the $M_\bullet$–σ or $M_\bullet$–L relations—the simplest way to estimate
$M_\bullet$ would be to insert a measured σ or L into those relations. That is probably not
an unreasonable way to proceed; but it rules out any possibility of detecting *changes*
of those relations over cosmological time. A less drastic alternative is possible in
AGN, by combining a measured ΔV with an estimate of R_{BLR}, the latter based on
empirical correlations derived from reverberation mapping studies.

Radii of the BLR inferred from reverberation mapping correlate with the observed luminosity of the nuclear source [272]. A recent determination of this

$R_{BLR}-L$ relation is [41]

$$R_{BLR} = (34.4 \pm 4.5) \left[\frac{\lambda L_\lambda (5100 \, \text{Å})}{10^{44} \, \text{erg s}^{-1}} \right]^{0.519 \pm 0.064} \text{light-days}, \qquad (2.24)$$

where R_{BLR} is the radius of the BLR as measured from the Hβ emission line. As a proxy for the continuum luminosity, the flux density (per unit of wavelength) is measured at a rest-frame wavelength of $\lambda = 5100 \, \text{Å}$ and multiplied by λ; this region of the spectrum is relatively free of contamination by strong emission lines. When restricted to the highest quality reverberation data, it is found that the $R_{BLR}-L$ relation has a scatter of about 0.1 dex [425].[7] The relation has tremendous appeal because it yields an estimate of the BLR size from a quick, simple measurement of the continuum luminosity, bypassing the need for long monitoring programs. Once the BLR size is known, the SBH mass follows from equation (2.23):

$$M_\bullet \approx f \left(1.96 \times 10^5 \right) \left(\frac{R_{BLR}}{\text{light-days}} \right) \left(\frac{\Delta V}{10^3 \, \text{km s}^{-1}} \right)^2 M_\odot, \qquad (2.25)$$

where ΔV is the velocity dispersion of the variable part of the broad Hβ emission line. Since the variable part of the emission line cannot be isolated in a single spectrum, it generally suffices to use [93]

$$M_\bullet \approx 2.3 \times 10^5 \left(\frac{R_{BLR}}{\text{light-days}} \right) \left(\frac{\Delta V}{10^3 \, \text{km s}^{-1}} \right)^2 M_\odot, \qquad (2.26)$$

where ΔV is the full width at half maximum flux of the Hβ emission line. The value of this relation is that it is easily applicable to large samples of AGN, for which direct reverberation mapping measurements would be infeasible.

As discussed in more detail below, since about 2004, the form factor f in equation (2.23) has generally been calibrated by requiring consistency with the $M_\bullet-\sigma$ relation of quiescent galaxies. For this reason, it is partly a matter of taste whether reverberation mapping masses as they are currently computed should be labeled "primary" or "secondary" estimates; if the latter, then methods for determining $M_\bullet$ based on relations like equation (2.24) would become "tertiary" methods etc.

In some low-luminosity AGN, the only broad emission line observed is Hα. It turns out that the velocity width ΔV measured from the Hα line correlates with that measured from Hβ; furthermore, there is an empirical relation between the luminosity in the Hα line and L_{5100}. Based on these correlations, versions of equations (2.24) and (2.25) can be written that use only measurements of the Hα broad emission line [213]. This tertiary (quaternary?) technique has been used to estimate SBH masses in large samples of low-luminosity AGN [214]. Similar relations based on the C IV $\lambda 1549$ [548] and Mg II $\lambda 2798$ [346] emission lines have been used to estimate masses of SBHs in large numbers of distant quasars.

---

[7]In astronomy, "dex" is a contraction of "decimal exponent." A scatter of 0.1 dex denotes a range of ± 0.1 in the base-10 logarithm of the measured quantity.

2.3 SUPERMASSIVE BLACK HOLES IN THE LOCAL GROUP

The Local Group is the collection of three giant and numerous dwarf galaxies that includes the Milky Way. The **Andromeda Galaxy** (also called M31, or NGC 224[8]) is a spiral galaxy with roughly the same size and mass as the Milky Way; its distance is estimated at about 780 kpc [345]. The **Triangulum Galaxy** (M33, NGC 598) is also spiral but with perhaps one tenth the mass of the Milky Way or M31, and lies at a distance estimated to be between 800 and 900 kpc [344]. Beyond the Local Group, giant galaxies are sparse within a sphere of radius 10 Mpc: the Centaurus Galaxy (NGC 5128) at 3–5 Mpc; M83 at 4.5 Mpc; M95 and M96 at about 10 Mpc; and a few others. The next big grouping of galaxies, the **Virgo Galaxy Cluster**, has a center (coincident with the giant elliptical galaxy M87) that is about 16.4 Mpc from the Local Group.

There are three galaxies in the Local Group for which the (stellar) kinematical data show evidence of a central massive object: the Milky Way, M31, and M32; the latter is a dwarf elliptical companion of M31. Not surprisingly, the dynamical case is strongest—most astronomers would say "incontrovertible"—in the case of the Milky Way. Because the Galactic center is so much closer than the nucleus of any other galaxy, the data that are available—for example, proper motions of individual stars—are qualitatively different than in other galaxies. For this reason, the determination of $M_\bullet$ in the Milky Way is discussed separately, in chapters 3 and 4.

The dynamical case for an SBH is somewhat weaker in M31, due primarily to a significant asymmetry in the nucleus which makes standard mass estimation techniques difficult to apply. In the case of M32, the *detection* is reasonably secure, but estimates of the SBH *mass* cover a wide range, apparently because the influence radius is not sufficiently well resolved. The Local Group also contains two galaxies in which the upper limits on the mass of a putative SBH are interestingly low: M33 and NGC 205. Each of these galaxies is discussed separately below, from largest to smallest.

2.3.1 M31 / NGC 224 (the Andromeda Galaxy)

Photographs taken in 1974 using the balloon-borne, 36-inch telescope Stratoscope II revealed that the nucleus of M31 was asymmetric, having a low-intensity extension on one side of the bright peak [322]. The authors noted that "the observed asymmetry is an intrinsic property of the nucleus and will probably require a dynamic explanation." Observations with the Hubble Space Telescope confirmed the asymmetry, resolving the nucleus into two components [309]. The two brightness peaks, denoted P1 and P2, have an angular separation of $0''.5$, or roughly 2 pc at the distance of M31. P2, the fainter peak, is located near what appears to be the dynamical center of the bulge while P1 is offset. The combined luminosities of P1 and P2 are about $3 \times 10^6 \, L_\odot$ and the combined mass is perhaps $2 \times 10^7 \, M_\odot$.

---

[8]"M31" means "the 31st object in the catalog of Charles Messier" (1730–1817). "NGC" stands for the "New General Catalogue of Nebulae and Clusters of Stars," compiled by J.L.E. Dreyer (1852–1926).

P1 and P2 are similar spectroscopically although P2 is bluer; in fact it is brighter than P1 in the ultraviolet [398]. This difference is due to a population of A stars with ages of about 200 Myr embedded in P2 [39]. This population, called P3, appears to consist of a disk of stars with a mass of $\sim$4000 $M_\odot$ and a radial extent of about one parsec that surrounds the central SBH, roughly in the same plane as the P1/P2 disk. The absorption lines of the A stars are kinematically broadened to almost 1000 km s^{-1} within the inner 0$\!''$.02. Assuming that this broadening is due to unresolved circular motion of the stars in a circular disk, the implied, deprojected circular velocity is about 1700 km s^{-1} at 0.19 pc, implying an SBH mass of $1.4 \pm 0.9 \times 10^8 \, M_\odot$ [39].

The asymmetry has been argued to be a result of stars in P1 and P2 orbiting in an eccentric disk, with P1 near apoapsis and P2 near periapsis [520]. However, the persistence of the asymmetry in the face of phase mixing is difficult to demonstrate [256] and the feature may be transient, perhaps even a result of a recent infall event [140].

2.3.2 M33 / NGC 598 (the Triangulum Galaxy)

M33, situated about 850 kpc from the Milky Way, is the third brightest galaxy in the Local Group. Like its two more massive neighbors, M33 is a spiral galaxy; it is classified as "late type" (ScII-III), meaning that it has almost no bulge. It does contain an NSC, with structural properties (mass, size) similar to those of the most massive globular clusters [297]. One might hypothesize that the nucleus *is* a globular cluster that managed to find its way into the center of the galaxy. However, the small mass-to-light ratio of the nucleus, coupled with the fact that the stars are metal rich, implies a much younger stellar population than in globular clusters [535], and the nucleus presumably formed from gas that accumulated, relatively recently, at the bottom of the galaxy's potential well.

The rms line-of-sight stellar velocities show no evidence of a central rise; in fact they drop toward the center, to a value of about 20 km s^{-1}. Dynamical modeling of the kinematical data place only upper limits on the mass of a putative SBH [178, 365]. But because the nuclear cluster is poorly resolved, the inferred upper limit on $M_\bullet$ depends strongly on what assumptions are made about the distribution of (stellar) mass and light inside the inner resolution element. Values of $M_\bullet$ as large as $1 \times 10^4 \, M_\odot$ are consistent with the data if the stellar M/L is left unconstrained, or if the character of the stellar orbits is allowed to change suddenly at a radius unresolvable by the telescope. Such models are physically permissible but seem unlikely. More reasonable models require $M_\bullet \lesssim 3000 \, M_\odot$.

2.3.3 M32

This dwarf galaxy is the brightest satellite of M31. It is classified as a "compact elliptical galaxy" (cE): a rare galaxy type exhibiting high central surface brightness compared with normal elliptical galaxies of the same total luminosity. M32 also has a high central velocity dispersion, $\sigma \approx 60$ km s^{-1}, that places it well off of the mean relation between L and σ. So extreme are M32's properties that a number of authors have explored speculative models for its origin; for instance, that it is the tidally stripped remnant of a once much larger galaxy [285]. It has also been suggested that M32 is actually a normal elliptical galaxy that lies three times farther away than M31, that is, well beyond the Local Group [570]; however, the most recent analyses [166] yield distance estimates that are consistent with that of M31.

M32 was one of the first galaxies for which an attempt was made to obtain spatially resolved stellar kinematics in the search for an SBH [518, 519], using data from the 5 meter Hale Telescope on Mount Palomar. Those data revealed that the stellar rotational velocity near the center of M32 is $\sim 50\,\mathrm{km\,s^{-1}}$ (line of sight), comparable with σ. Subsequent observations were made at higher angular resolutions with other ground-based telescopes and with the Hubble Space Telescope (figure 2.7). The velocity dispersion profile exhibits an impressive central "spike"; however, the rotational velocities exhibit an equally impressive *drop* near the center, with the result that the rms line-of-sight velocity is nearly constant going into the center (figure 2.5). This is an instrumental effect [518]: a finite-width slit positioned near the center of the galaxy includes light from stars orbiting in both directions, and some fraction of the mean motion along the line of sight is converted into an apparent dispersion.[9]

While there is no prima facie signature of an SBH in M32, just maintaining a constant v_{rms} near the center implies an increase in the mass-to-light ratio in the inner parsec, at least if the stellar velocity distribution is assumed to be isotropic about $\overline{v}_\phi$ [519]. If this M/L increase is attributed to a central dark mass, then $M_\bullet \gtrsim 3 \times 10^6\,M_\odot$. However, modeling that accounts for the nonspherical shape of M32, and for the possibility of velocity anisotropy, can reproduce the Hubble Space Telescope data equally well with a range of assumed SBH masses that extends almost to zero: $1.5 \times 10^6 \leq M_\bullet \leq 5 \times 10^6\,M_\odot$ [527]. Taking $M_\bullet$ from the upper end of this range, the implied radius of influence has an angular size of $\sim 1''.4$, and the detectability criterion (2.20) would demand an instrumental resolution of $\sim 0''.14$—roughly correct for the Hubble Space Telescope, and consistent with the fact that the constraints placed by these data on $M_\bullet$ are weak.

The estimation of $M_\bullet$ in M32 is discussed in more detail in chapter 3.

2.3.4 NGC 205

Like M32, NGC 205 is a dwarf elliptical companion to M31. But it has roughly half the luminosity of M32, and its surface brightness is also lower, consistent with the mean relation between luminosity and surface brightness defined by "normal" elliptical galaxies. Its central velocity dispersion is about $20\,\mathrm{km\,s^{-1}}$, compared with $60\,\mathrm{km\,s^{-1}}$ in M32. NGC 205 contains an NSC (figure 2.3); the prominence of the nucleus is due in part to its young stellar population, which is estimated to have formed in the last $\sim 0.5\,\mathrm{Gyr}$ [312]. The nuclear mass is estimated at $\sim 9 \times 10^4\,M_\odot$ [261] which places a firm upper limit on the value of $M_\bullet$. Dynamical modeling of stellar kinematical data obtained from the Hubble Space Telescope observations [527] yields an upper limit on $M_\bullet$ of about $4 \times 10^4\,M_\odot$.

---

[9]It is common practice, in papers on stellar-dynamical SBH detections, to plot $\sigma(r)$ and $\overline{v}(r)$ separately. This practice can give a reader the false impression that a sharp rise in velocities has been observed.

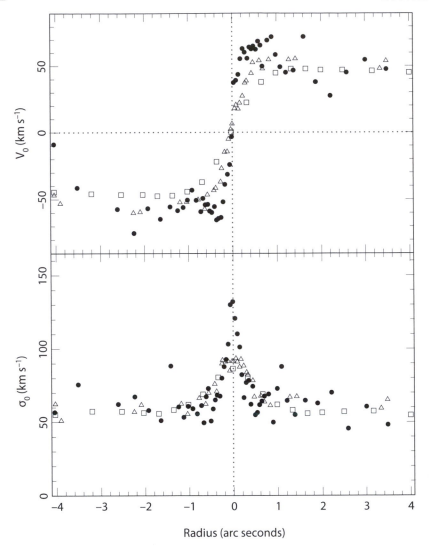

Figure 2.7 Measurements of the line-of-sight mean velocity, V_0, and velocity dispersion, σ_0, in the Local Group dwarf elliptical galaxy M32 [262]. Filled circles are data from the Space Telescope Imaging Spectrograph [262]. Squares and triangles are ground-based measurements from the William Herschel Telescope [542] and the Canada–France–Hawaii Telescope [40], respectively. As discussed in the text, the impressive peak in the velocity dispersion profile is an instrumental artifact and is not due to the SBH.

2.3.5 Summary of Local Group observations

The ability to estimate SBH masses in galactic nuclei is critically dependent on spatial resolution, and from this standpoint, Local Group galaxies have an advantage over almost all others. It is significant therefore that the constraints on $M_\bullet$

are fairly weak in all these galaxies, with the sole exception of the Milky Way. Even in the Milky Way, early estimates of $M_\bullet$ using stellar velocities (as opposed to fitting orbital solutions to astrometric data) yielded masses for Sgr A* that were consistently lower by factors of two or more than what we now believe to be the correct mass, in spite of the fact that these data extended well inside the influence radius [76, 182, 190]; the reason turned out to be the presence of an unexpectedly low space density of stars in the inner fraction of a parsec, which had not been accounted for in the dynamical models [481]. One could argue that Local Group galaxies are all "difficult" cases: M31 with its lopsided nucleus; NGC 205 with its spatially varying mass-to-light ratio; the Milky Way with its inner core; etc. But it would be more reasonable to assume that many galaxies, if observed with the same degree of resolution, would exhibit the same sorts of troubling detail.

2.4 PHENOMENOLOGY

By the late 1990s, the number of primary mass estimates for SBHs had reached ten or so, and it was natural to begin considering phenomenological relations between $M_\bullet$ and other galaxy properties. In astronomy, the discovery of tight correlations between measured properties of some class of object has often marked a turning point in the study of those objects, even (or especially) in cases where the physical basis for the relation was not evident prior to the discovery. Famous examples include the main sequence for stars, the Tully–Fisher relation for spiral galaxies, and the redshift–distance relation. The discovery of such relations can motivate the search for theoretical explanations. But even when the underlying physics remains uncertain, tight correlations can be extremely useful tools for the observational astronomer, by relating a distance-independent quantity (e.g., color) to an intrinsic property (e.g., luminosity) that would otherwise be difficult to measure. For instance, "main-sequence fitting" exploits the observed narrowness of the main sequence to assign intrinsic luminosities to stars based on their spectral types, thus yielding distances to star clusters that are too far for geometrical distance determinations to work.

In the case of SBHs, the natural correlations to search for relate $M_\bullet$ to the properties of the stellar spheroid. It turns out—perhaps not surprisingly—that $M_\bullet$ correlates well with many such properties, although with various degrees of scatter. In the following sections, the most important of the phenomenological relations involving $M_\bullet$ are described.

2.4.1 Relations with bulge luminosity and mass

One expects larger galaxies to contain larger SBHs. On the other hand, there are luminous galaxies which appear to contain no SBH—or at least, for which the upper limits on $M_\bullet$ are small. The best example is the Local Group galaxy M33. The luminosity of M33's disk is about $3 \times 10^9 \, L_\odot$ at visual wavelengths [535], and as discussed above, the upper limit on $M_\bullet$ is about $3000 \, M_\odot$—very roughly, one part in 10^6 of the galaxy's mass, a ratio that is about a thousand times smaller than in the Milky Way.

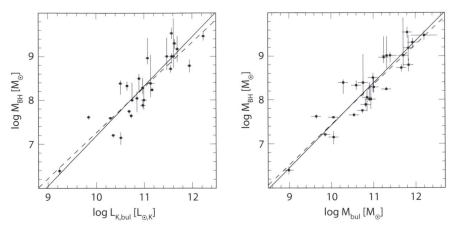

Figure 2.8 Relations between SBH mass and the total luminosity and mass of the host spheroid [338]. On the left, $M_\bullet$ is plotted against the K-band luminosity of the bulge,
and on the right, against an estimate of the bulge mass. The solid and dashed
lines are least-squares fits using two different algorithms.

The apparent lack of an SBH in M33 would be natural if $M_\bullet$ were correlated
strongly with L_{bulge}, the luminosity of the stellar spheroid, rather than L_{gal}, since
M33 contains essentially no bulge. Figure 2.8 verifies that such a correlation ex-
ists. The luminosities plotted in this figure were based on K-band (near-infrared)
images; infrared intensities are less affected by extinction than measurements in
visual bands, and they are also less influenced by light from the rare, but bright,
blue stars that are present in regions of recent star formation. In the case of disk
galaxies, the luminosity associated with the bulge was identified by carrying out a
"bulge-disk decomposition", that is, by fitting two-component, parametrized models to the two-dimensional surface brightness data, one meant to represent the bulge
and the other the disk [338].

The relation between $M_\bullet$ and $L_{K,\text{bulge}}$, the K-band bulge luminosity, is well fit
by a power law:

$$\log_{10}(M_\bullet/M_\odot) = a + b[\log_{10}(L_{K,\text{bulge}}/L_{K,\odot}) - 10.9] \qquad (2.27)$$

with $a = 8.21 \pm 0.07$, $b = 1.13 \pm 0.12$ [338]. This is one version of the **black-hole
mass–bulge luminosity relation** (or $M_\bullet$–L relation). Other versions are based on
luminosities measured in different wavelength bands, for example, at visual wavelengths [298]; the slope is found to be similar, but the scatter tends to increase in
the bluer passbands.

The right panel of figure 2.8 plots the same SBH masses against a crude estimate
of the bulge mass:

$$M_{\text{bulge}} = 3G^{-1}R_e\sigma_e^2. \qquad (2.28)$$

Here, R_e is the effective radius defined above, σ_e is the velocity dispersion measured at R_e, and the factor 3 was chosen to maximize the correspondence of M_{bulge}

with galaxy masses derived from the more careful dynamical modeling that has been carried out for a few galaxies. Again a good correlation is observed:

$$\log_{10}(M_\bullet/M_\odot) = a + b[\log_{10}(M_{\mathrm{bulge}}/M_\odot) - 10.9] \qquad (2.29)$$

with $a = 8.28 \pm 0.06$, $b = 0.96 \pm 0.07$, and a vertical scatter of 0.25 dex [338]. The estimate of b is consistent with unity; setting $b = 1$ implies a strict proportionality between $M_\bullet$ and the mass of the bulge:

$$M_\bullet \approx 2.4 \times 10^{-3} M_{\mathrm{bulge}}. \qquad (2.30)$$

This is one version of the **black-hole mass–bulge mass relation** (or $M_\bullet$–M_{bulge} relation).

2.4.2 Mass–velocity dispersion relation

Galaxy luminosities have long been known to correlate well with some measure of their internal motions. The "Faber–Jackson law" [152] is an empirical relation between L_{gal}, the luminosity of an elliptical galaxy, and the central stellar velocity dispersion σ:

$$L_{\mathrm{gal}} \propto \sigma^\alpha \qquad (2.31)$$

with $\alpha \approx 4$. The "Tully–Fisher relation" [524] states that the luminosities of disk galaxies correlate with ΔV, the rotation curve amplitude, roughly as

$$L_{\mathrm{disk}} \propto (\Delta V)^\beta, \qquad (2.32)$$

with $\beta \approx 4$ also. The existence of these well-known correlations, coupled with the fact that σ is an easily measured quantity (and one that would necessarily be known in any galaxy having a dynamically determined $M_\bullet$), might tempt anyone with a few free minutes to try plotting $M_\bullet$ versus σ. But the $M_\bullet$–σ relation[10] [160, 176] was not published until 2000, some five years after the first attempts were made at constructing the $M_\bullet$–L relation [298].

The reasons for this delay are interesting and instructive [364, 159]. If one plots $M_\bullet$ versus σ, including all galaxies for which SBH mass estimates had been published prior to 2000, the relation is not very striking: the scatter is large, comparable with the scatter in the visual $M_\bullet$–L relation, and much greater than the scatter in the infrared $M_\bullet$–L relation. Progress was made only after it was realized that the scatter in the $M_\bullet$–σ relation depends greatly on sample selection [160]. If the sample is restricted to galaxies showing clear evidence of a central velocity rise, the scatter in the relation drops to a value that is consistent with zero *intrinsic* scatter (figure 2.9b). The mean relation defined by the secure subsample is [159]

$$\frac{M_\bullet}{10^8 \, M_\odot} = (1.66 \pm 0.24) \left(\frac{\sigma}{200 \, \mathrm{km \, s^{-1}}} \right)^{4.86 \pm 0.43}. \qquad (2.33)$$

By contrast, the scatter in the visual $M_\bullet$–L relation is almost unchanged when the restricted sample is used (figure 2.9a). Interestingly, the secure sample contains

---

[10]Originally called the "Faber–Jackson law for black holes" [351].

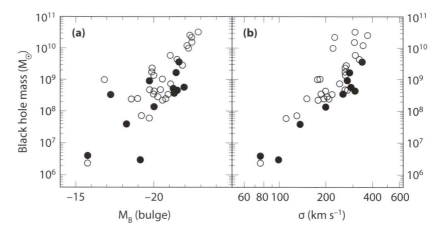

Figure 2.9 Discovery of the $M_\bullet$–σ relation. (a) $M_\bullet$ versus bulge visual luminosity, expressed as M_B, or absolute B (blue) magnitude (see caption to figure 2.2 for an explanation of magnitudes). (b) $M_\bullet$ versus bulge central velocity dispersion. SBH masses derived from data that exhibit a central velocity increase are indicated with filled circles; masses derived from lower quality—mostly stellar-dynamical—data are the open circles. (Adapted from [160].)

galaxies with a wide range of types, while the less-secure measurements were almost all derived from stellar kinematics in elliptical galaxies.

Based on these findings, the $M_\bullet$–σ relation was proposed as the more fundamental of the two relations, a point of view that is still widely held.

The scatter in the $M_\bullet$–σ relation is so small that it is reasonable to use the relation to predict SBH masses, even in galaxies for which determinations of $M_\bullet$ based on detailed modeling had previously been published. When this was done [363] it was found that essentially all of the SBH mass estimates derived from ground-based, stellar-dynamical data were too large,[11] by factors ranging from ~3 to ~100, and that the errors correlated with the ratio of instrumental resolution to r_h. Discarding those masses led to a severe revision of the $M_\bullet$–M_{bulge} relation: the mean ratio of SBH mass to bulge mass dropped from ~6×10^{-3} (the "Magorrian relation" [335]) to its currently accepted value of ~1×10^{-3} [362].

This downward revision resolved at a stroke two long-standing controversies. The mass density of SBHs at large redshifts can be estimated by requiring the optical luminosity function of quasars to be reproduced by accretion onto SBHs (the **Soltan argument** [499]). Assuming a standard accretion efficiency of ~10%, the mean mass density in SBHs works out to be $\rho_\bullet \approx 2 \times 10^5\ M_\odot\,\mathrm{Mpc}^{-3}$ [85]. A similar argument based on the X-ray background gives consistent results, $\rho_\bullet \approx 3 \times 10^5\ M_\odot\,\mathrm{Mpc}^{-3}$ [153]. By comparison, the local SBH mass density implied by all the primary $M_\bullet$ mass estimates published prior to 2000 was five to ten times higher [454]. A similar discrepancy existed with regard to SBH masses derived

---

[11]The reasons for the systematic error are explored in chapter 3.

from reverberation mapping in AGN. While no galaxies having both sorts of mass measurement existed—the bright, nonstellar light that makes reverberation mapping possible tends to swamp the stellar spectra—masses derived from reverberation mapping were lower by factors as large as 50 compared with the claimed values of $M_\bullet$ in quiescent galaxies having comparable bulge luminosities [554]. Both of these discrepancies disappeared when the stellar dynamical mass estimates were shown to be substantially in error [362].

The newfound consistency of SBH masses in active and quiescent galaxies motivated observational programs designed to measure accurate stellar velocity dispersions in the host galaxies of AGN for which reverberation mapping masses existed [161, 177]. A natural next step was to use the $M_\bullet$–σ relation to "calibrate" the reverberation mapping measurements, that is, to fix the geometrical factor f in equation (2.23) so as to bring the $M_\bullet$–σ relation for AGN into the best possible agreement with the $M_\bullet$–σ relation for quiescent galaxies. The result [405] was $\langle f \rangle \approx 5 \pm 2$. Since about 2004, most AGN researchers have adopted this approach to the determination of f.

As of this writing, published tabulations of primary SBH mass estimates include as many as 50 values, most derived from the modeling of stellar kinematical data [221]. Based on the history outlined above, the reader might be excused for wondering whether all of the published $M_\bullet$ values are based on bona fide dynamical detections. As figure 2.5 illustrates, there are no galaxies beyond the Local Group for which the rms stellar velocities show a central upturn on spatial scales that can plausibly be associated with an SBH. Gas dynamical observations fare better, but not a great deal; careful investigators emphasize the large systematic uncertainties [337], and independent analyses of the same data can yield estimates of $M_\bullet$ that differ by amounts that are many times greater than the claimed uncertainties [333]. Perhaps all that can be said for certain is that SBH masses cannot be much *greater* than the published values: if they were, the kinematical signatures would be unambiguous. This argument has led to suggestions that the $M_\bullet$–σ and $M_\bullet$–M_{bulge} relations may only trace the upper envelope of the SBH mass distribution [32]. The author is not aware of any compelling counterargument to this hypothesis.

Uncertainties about which of the published SBH masses are based on bona fide dynamical detections are reflected in uncertainties about the slope and scatter of the $M_\bullet$–σ relation. As one broadens the sample of primary SBH mass estimates to include masses that are less and less secure (based on the ratio of instrumental resolution to r_h, the latter computed using the published $M_\bullet$ value—an admittedly circular procedure), the inferred slope of the $M_\bullet$–σ relation falls and the scatter increases (figure 2.10). Studies that accept most or all of the published masses as bona fide measurements typically find $4.0 \lesssim \alpha \lesssim 4.3$ in the best-fit relation $M_\bullet \propto \sigma^\alpha$, and an intrinsic vertical scatter of ~ 0.5 dex; while more conservative studies that include only galaxies with clear kinematic signatures find slopes $4.5 \lesssim \alpha \lesssim 5$ and an intrinsic scatter consistent with zero.

If the more conservative estimates of slope and scatter are adopted, the upper limits on $M_\bullet$ in the Local Group galaxies M33 and NGC 205 are both consistent, within the uncertainties, with the $M_\bullet$–σ relation, as shown in figure 2.11.

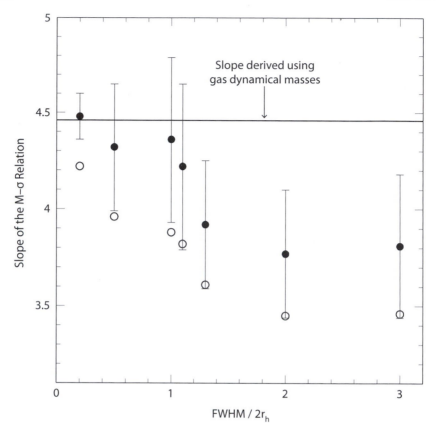

Figure 2.10 The slope of the $M_\bullet$–σ relation as derived from stellar dynamical data as a function of data quality [364]. Here FWHM is the full width at half maximum of the instrumental point spread function. Solid circles are slopes derived from a regression algorithm that accounts for errors in both variables; open circles are from a standard least-squares routine. The solid line is the slope derived from gas dynamical data.

2.4.3 Relation with galaxy concentration

As noted above, larger galaxies tend to be fit by Sérsic models with larger values of the index n in equation (2.3). The mean relation for elliptical galaxies is [208]

$$n \approx 3.6 \left(\frac{L_B}{10^{10} \, L_{\odot, B}} \right)^{0.27}, \tag{2.34}$$

where L_B indicates total luminosity in the B ("blue") passband. Since SBH masses also correlate with L, one expects a correlation between $M_\bullet$ and n. It turns out [209] that such a correlation can be made tighter if n is replaced by an alternate measure of "central concentration": the ratio of the light inside a radius equal to $R_e/3$ to the light inside R_e. Calling this parameter $C_{1/3}$, the correlation is

$$\log_{10}\left(\frac{M_\bullet}{M_\odot}\right) = (6.81 \pm 0.95)C_{1/3} + (5.03 \pm 0.41). \qquad (2.35)$$

Most of the SBH masses that were used in deriving this relation were themselves derived from the $M_\bullet$–σ relation, and so the **black-hole mass–galaxy concentration relation** might be more properly interpreted as a relation between concentration and σ. Nevertheless, the correlation is impressively tight, with a scatter of about 0.3 dex in $M_\bullet$.

2.4.4 Relations involving nuclear star clusters

There is no evidence of an SBH in either of the Local Group galaxies M33, a fairly luminous disk galaxy [178, 352]; or NGC 205, a dwarf elliptical companion to the Andromeda Galaxy [527]. But both galaxies do contain a compact stellar nucleus. The mass of the NGC 205 nucleus (shown in figure 2.3) has been estimated at about $10^5 \, M_\odot$ [261]. That is not too different from the mass predicted by the $M_\bullet$–σ relation for an SBH in this galaxy (figure 2.11).

In fact, as one moves to fainter galaxies, NSCs become increasingly common, and they appear to be almost ubiquitous in stellar spheroids with central velocity dispersions below about $100 \, \text{km s}^{-1}$ [71, 96]. That is roughly the value of σ at the Galactic center; and the Milky Way's SBH is the smallest with a secure dynamical detection. These facts suggest that NSCs might be "complementary" to SBHs, in the sense that SBHs are "replaced" by compact nuclei in stellar spheroids below a certain mass.

Of course, the Milky Way itself is an exception to this hypothesis since it contains both an SBH and an NSC. Another clear counterexample, discussed in more detail in section 2.5, is the active galaxy NGC 4395. Other galaxies with spheroid luminosities comparable to, or less than, that of the Milky Way might also contain undetected SBHs. Nevertheless, it is natural to wonder whether SBHs and NSCs might constitute two members of a single category of objects that reside—sometimes together, sometimes alone—at the centers of galaxies. The name **central massive object**, or **CMO**, is sometimes used to describe both compact stellar nuclei and SBHs [157, 557].

Figure 2.12 provides some support for the idea that NSCs and SBHs belong to a single class. Both categories of CMO contain, on average, about 0.2% of the mass of the stellar spheroid. Nuclear star clusters also obey a relation similar to the $M_\bullet$–σ relation for SBHs, although with substantially more scatter, and with a vertical offset.

It is entirely possible that NSCs were present, at one time or another, in all galaxies, but that they were destroyed in the more massive galaxies by the same mechanism that created cores. And even if SBHs were present in all galaxies, they would have gone undetected in low-mass spheroids due to instrumental limitations. For these reasons, the significance of the apparent "replacement" of SBHs by NSCs in galaxies below a certain mass is currently unclear.

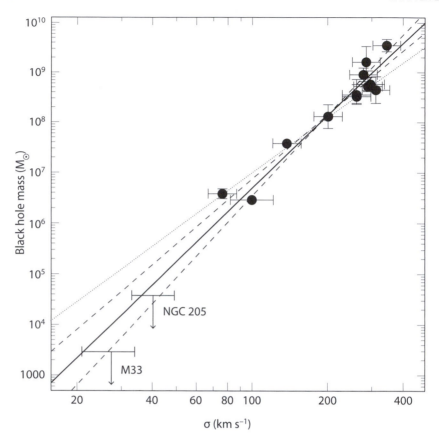

Figure 2.11 The $M_\bullet$–σ relation, extended to low σ values in order to plot the upper limits on $M_\bullet$ in the Local Group galaxies M33 [178, 365] and NGC 205 [527]. The thick solid line is equation (2.33), with $1-\sigma$ confidence intervals shown by the dashed lines. The two upper limits are consistent with this relation, but marginally inconsistent with the shallower relation (thin dotted line) that one obtains by using all published SBH mass estimates regardless of their quality [176].

2.4.5 Significance of the phenomenological relations

Like other tight, empirical correlations in astronomy, the $M_\bullet$–σ and $M_\bullet$–L relations must be telling us something fundamental about origins, and in particular, about the connection between SBHs and the stellar spheroids in which they reside. But the precise nature of that connection remains uncertain.

One could take the view that all of the phenomenological relations discussed above are manifestations of a fixed ratio of SBH mass to bulge mass. The masses of stellar spheroids scale with their luminosities as $M_{\text{bulge}} \sim L^{5/4}$ [151] and the Faber–Jackson law states that $L \sim \sigma^4$, hence $M_{\text{bulge}} \propto \sigma^5$. Setting $M_\bullet \propto M_{\text{bulge}}$

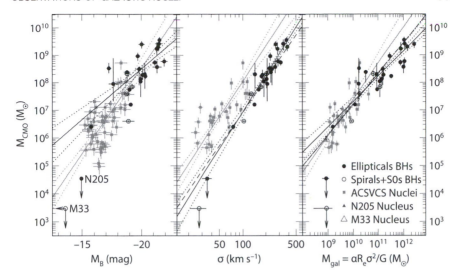

Figure 2.12 Correlations involving central massive objects (CMOs) [157]. The left panel
plots the mass of the CMO against the absolute blue magnitude of the host
spheroid. (See the caption of figure 2.2 for a definition of magnitudes.) Gray
squares are NSCs; SBHs in elliptical and spiral galaxies are shown as the filled
and open circles, respectively. The middle and right panels show CMO mass
as a function of galaxy velocity dispersion and total galaxy mass, respectively.
The solid gray and black lines show the best-fit relations fit to the nuclei and to
the SBHs, respectively, with $1 - \sigma$ confidence levels on the slope shown by the
dotted lines. In the middle panel, the dashed line is the $M_\bullet$–σ relation; in the
right panel, the dashed line is the fit obtained for the combined (nuclei+SBH)
sample.

then implies

$$M_\bullet \propto \sigma^5, \quad M_\bullet \propto L^{5/4}, \tag{2.36}$$

consistent with the observed $M_\bullet$–σ and $M_\bullet$–L relations.

There is nothing wrong with this argument; but the tightness of the $M_\bullet$–σ relation
suggests to most researchers that something else must be going on. After all, even
if a perfect correlation were set up between SBH mass and spheroid mass in the
early universe, it is hard to see how it could survive galaxy mergers, which convert
disks to bulges and may also channel gas into the nucleus, producing (presumably)
uncorrelated changes in $M_\bullet$ and M_{bulge}. In order to maintain a close connection
between $M_\bullet$ and σ or $M_\bullet$ and M_{bulge}, some sort of "negative feedback" would seem
to be required, allowing the SBH to regulate its own growth.

The Soltan argument summarized in section 2.4.2 suggests that SBHs acquired
most of their mass via accretion of gas. The radiation field from an accreting quasar
must drive a wind from the vicinity of the SBH, which would collide with the ambient gas in the galaxy and produce shocks. Perhaps an outflow driven by the accretion could be the source of the feedback. In this picture, the SBH would eventually

reach a mass such that further accretion is prevented because the outflow sweeps away the ambient gas.

It is easy to show that this argument is not falsifiable on *energetic* grounds. The energy released in growing an SBH of mass $M_\bullet$ through the accretion of gas is $\eta M_\bullet c^2$, where η is the **accretion efficiency**, defined as

$$L = \eta \dot{M} c^2, \tag{2.37}$$

with L the accretion-driven luminosity and $\dot{M}$ the mass accretion rate. The accretion efficiency is calculated by comparing the energy of an accreting mass element at infinity with its energy at the last stable orbit around the SBH; while the result is dependent on the SBH spin (chapter 4), $\eta \approx 0.1$ is commonly assumed. The ratio of the energy released to the gravitational binding energy of the bulge is then

$$\frac{\eta M_\bullet c^2}{G M_{\text{bulge}}^2 / R_{\text{bulge}}} \approx \eta \frac{M_\bullet}{M_{\text{bulge}}} \frac{c^2}{\sigma^2}$$

$$\approx 225 \left(\frac{\eta}{0.1}\right) \left(\frac{M_\bullet / M_{\text{bulge}}}{10^{-3}}\right) \left(\frac{\sigma}{200\,\text{km s}^{-1}}\right)^{-2} \gg 1. \tag{2.38}$$

In other words, there is more than enough energy released in the formation of an SBH to unbind the entire mass of a galactic bulge.

There is no obvious way to couple that energy to the stars, but there is a well-known way to couple it to the gas: Thomson scattering, the scattering of electromagnetic radiation by charged particles—in this case, the electrons in the ionized gas. In a transparent plasma, both the radiation intensity from the central source, and its gravitational force, drop off as $1/r^2$. The **Eddington luminosity** is defined such that the outward radiation force on a single electron is equal to the gravitational force on an electron–proton pair:

$$L_{\text{E}} = \frac{4\pi G M_\bullet m_p c}{\sigma_e} \approx 3.2 \times 10^{12} \left(\frac{M_\bullet}{10^8\,M_\odot}\right) L_\odot, \tag{2.39}$$

where σ_e is the Thomson scattering cross section and m_p the mass of the proton. When $L > L_{\text{E}}$, the net force on the ions is outward and accretion halts. Equating equations (2.37) and (2.39) gives the accretion rate at which $L = L_{\text{E}}$, the **Eddington accretion rate**:

$$\dot{M}_{\text{E}} = \frac{4\pi G M_\bullet m_p}{\eta \sigma_e c} \approx 2.6 \left(\frac{\eta}{0.1}\right)^{-1} \left(\frac{M_\bullet}{10^8\,M_\odot}\right) M_\odot\,\text{yr}^{-1}. \tag{2.40}$$

A number of lines of evidence suggest that SBHs accrete at roughly the Eddington rate during their most luminous, quasar phase; for instance, accretion rates that were much lower would not allow them to grow to their observed masses in the available time.

We can ask what mass of SBH, radiating at the Eddington limit, produces enough energy to unbind the bulge in one crossing time—the minimum time for infall to occur [225, 497]? Approximating the crossing time as R_{bulge}/σ, this condition is

$$L_{\text{E}} \times \frac{R_{\text{bulge}}}{\sigma} \approx \frac{G M_{\text{bulge}}^2}{R_{\text{bulge}}}. \tag{2.41}$$

Writing $G M_{\text{bulge}} \approx \sigma^2 R_{\text{bulge}}$ via the virial theorem, this becomes

$$M_\bullet \approx \frac{\sigma_e \sigma^5}{4\pi G^2 m_p c} \approx 3 \times 10^5 \left(\frac{\sigma}{200\,\text{km s}^{-1}}\right)^5 M_\odot. \qquad (2.42)$$

This has roughly the same functional form as the $M_\bullet$–σ relation, but the constant of proportionality is too small by about three orders of magnitude.

The argument so far has implicitly assumed that all of the energy produced by the SBH is available to drive the gas. This would be the case in an "energy-driven flow"; one condition for such a flow is that the gas does not cool. At the other extreme, a "momentum-driven flow" is one in which the cooling time is so short that essentially all the energy in the flow is in the form of bulk motion. In a momentum-driven flow, most of the energy released by the SBH is lost to radiation, and only a small fraction (a few percent) is left to affect the bulge gas mechanically.[12] It can in fact be argued [281] that flows driven by accreting SBHs are more likely to be momentum driven than energy driven. This is because there are inevitable sources of cooling: for instance, the radiation field from the central source is efficient at cooling the shocked gas out to kiloparsec distances ("inverse Compton cooling" [87]).

In a momentum-driven flow, if the optical depth is of order unity, the momentum of the outflow is comparable to the photon momentum:

$$\dot{M}v \approx \frac{L_E}{c}. \qquad (2.43)$$

Equating $\dot{M}$ in this expression with $\dot{M}_E$ implies a velocity for the gas of

$$v \approx \eta c \approx 0.1c. \qquad (2.44)$$

Winds with these properties are in fact observed [284, 517].

Consider a shell of gas, of radius $R(t)$, that has been swept up by the flow. The mass of that shell is $f_g M(R)$, where $M(R)$ is the total mass (stars plus gas) within radius R and f_g is the gas fraction. The equation of motion of the shell is [281]

$$\frac{d}{dt}\left[f_g M(R)\dot{R}\right] + \frac{G f_g M(R)\left[M_\bullet + M(R)\right]}{R^2} = \frac{L_E}{c}. \qquad (2.45)$$

Adopting as a simple model for the stellar bulge the singular isothermal sphere defined in section 2.2, we can write $M(R) = 2\sigma^2 R/G$, and equation (2.45) becomes

$$\frac{d}{dt}(R\dot{R}) + \frac{G M_\bullet}{R} = -2\sigma^2\left(1 - \frac{M_\bullet}{M_\sigma}\right), \qquad (2.46)$$

where

$$M_\sigma \equiv \frac{f_g \sigma_e}{\pi G^2 m_p}\sigma^4 \approx 2 \times 10^8 \left(\frac{f_g}{0.1}\right)\left(\frac{\sigma}{200\,\text{km s}^{-1}}\right)^4 M_\odot. \qquad (2.47)$$

It is easy to show that equation (2.46) has no solution at large R if $M_\bullet < M_\sigma$; in other words, if $M_\bullet$ is too small, the force on the shell is unable to lift it beyond a certain radius. On the other hand, if $M_\bullet > M_\sigma$, $\dot{R}^2 \to \sigma^2$ for large R and the shell

---

[12]The shock that must be cooled to produce a momentum-driven flow is the shock decelerating the wind, not the one accelerating the ambient gas.

can be expelled completely. Finally, setting $f_g \approx 0.16$—the accepted value for the "cosmic baryon fraction," that is, the ratio of ordinary matter to total (ordinary plus dark) matter—yields a relation that is quite close to the observed $M_\bullet$–σ relation [281, 282].

To summarize, if outflows communicate with the ambient gas by exchange of momentum, the $M_\bullet$–σ relation emerges naturally as the SBH mass for which radiation at the Eddington limit can drive a significant outward flow. At smaller $M_\bullet$, the SBH can only drive flows which recollapse, failing to interrupt the gas supply that is (presumably) responsible for its growth.

Note that the critical mass in equation (2.47) is larger by a factor $\sim c/\sigma \approx 10^3$ than the critical mass derived under the assumption of an energy-driven flow, equation (2.42). This is because the former condition assumed that all the mechanical energy was available to drive the gas, rather than assuming that the gas cools efficiently.

The argument just presented really implies only an upper limit to $M_\bullet$. For instance, in active galaxies—which must still be in the process of accreting—the argument implies that $M_\bullet < M_\sigma$, and the same is likely to be true in other galaxies as well [283]. But as discussed earlier in this chapter, and in more detail in chapter 3, the observations are not able to rule out the possibility that many galaxies contain "underweight" SBHs.

Arguments that explain the $M_\bullet$–σ relation in terms of gas-mediated feedback are found convincing by many researchers. But SBHs can also grow by consumption of stars: either directly through capture of stars that pass within the event horizon, or indirectly by accreting the gas from tidally disrupted stars. In the most optimistic scenario, capture would occur at the so-called "full-loss-cone" rate, which assumes that orbits are (somehow) repopulated at a rate equal to or higher than their rate of depletion due to capture by the SBH. The rate at which stars carry mass past a sphere of radius r as they move along their orbits in a singular-isothermal-sphere nucleus is

$$\sim 4\pi r^2 \sigma \rho \approx \frac{2\sigma^3}{G}. \tag{2.48}$$

A fraction $\sim r_{lc}/r$ are moving on orbits that will take them within a distance r_{lc} of the SBH. Setting r (rather arbitrarily) to the influence radius r_h, and taking for r_{lc} some multiple of the SBH gravitational radius $r_h \equiv GM_\bullet/c^2$—for capture by massive SBHs, $r_g \lesssim r_{lc} \lesssim 10 r_g$ is appropriate (section 4.6)—the accretion rate becomes

$$\dot{M} \approx \frac{2\sigma^3}{G} \frac{r_{lc}}{r_h} \approx 10 \frac{\sigma^5}{Gc^2}. \tag{2.49}$$

After $\sim 10\,\text{Gyr}$, the accumulated mass would be

$$M_\bullet \approx 1 \times 10^8 \left(\frac{\sigma}{200\,\text{km s}^{-1}} \right)^5 M_\odot, \tag{2.50}$$

again consistent with the observed relation [581]. Arguments like this one (which is presented in more detail in chapter 6) are perhaps not as compelling as arguments based on feedback, but they do suggest that capture of stars may compete with gas accretion as a mechanism for growing SBHs.

2.5 EVIDENCE FOR INTERMEDIATE-MASS BLACK HOLES

The Milky Way SBH is the smallest having a secure dynamical detection. It is possible that its mass—approximately four million solar masses—represents the lower limit of the SBH mass distribution. But that would make our galaxy and its SBH special; and given the difficulties associated with dynamical detections, particularly at the lower values of $M_\bullet$, it seems more reasonable to suppose that black holes with masses smaller than the Milky Way's exist: at the centers of low-luminosity galaxies, or perhaps in smaller systems like globular star clusters, or even in intergalactic space.

Of course, black holes with masses of about 5–20 $M_\odot$—so-called **stellar-mass black holes**—have long been accepted as a likely end state of the evolution of upper-main-sequence stars [173], and there is strong observational evidence for the existence of stellar-mass black holes in binary star systems in our galaxy [83]. But there is a big gap between ten solar masses and a million solar masses. In the absence of compelling theoretical models for how massive black holes come into existence, it is unclear how much of that gap Nature manages to fill.

Black holes in the mass range $10^2 \, M_\odot \lesssim M_\bullet \lesssim 10^6 \, M_\odot$ are called **intermediate-mass black holes**, or IBHs. It is fair to say that IBHs are only hypothetical objects: as of this writing, there are no unambiguous detections. But the indirect evidence from various directions is tantalizing, as reviewed briefly here.

There exist bright, pointlike, off-center X-ray sources in several nearby galaxies: the sources are less luminous than AGN, but brighter than can easily be explained by accretion onto stellar-mass black holes. These **ultra-luminous X-ray sources**, or **ULXs**, have X-ray luminosities in the range $2 \times 10^{39} \, \mathrm{erg \, s^{-1}} \lesssim L_X \lesssim 10^{41} \, \mathrm{erg \, s^{-1}}$ [149]. The numbers just quoted assume that the emission is isotropic. By comparison, the Eddington luminosity of stellar-mass black holes is given by equation (2.39) as

$$L_E \approx 1.2 \times 10^{38} \left(\frac{M_\bullet}{10 \, M_\odot} \right) \mathrm{erg \, s^{-1}}. \tag{2.51}$$

If ULXs are the result of accretion onto compact objects, their masses would need to be at least as large as 15–1000 $M_\odot$, or higher if the accretion is sub-Eddington. On the other hand, if the X-ray emission from these objects is directionally collimated, then the observed fluxes would imply smaller *total* luminosities and hence smaller masses. One possibility is that ULXs belong to the class of "X-ray binaries," binary systems in which one component is a normal star and the other is a compact object. Such models need to invoke beaming, or super-Eddington accretion, to reconcile the observed fluxes with accretor masses of 10 $M_\odot$ or less. The fact that ULXs are often associated with regions of ongoing star formation is consistent with an alternate model in which runaway stellar mergers in dense star clusters create very massive stars that leave behind remnants in the IBH mass range [435]. Such models are still rather speculative since they are based on the poorly understood evolution of massive stars.

As discussed above, the upper limits on the masses of dark central objects in the Local Group galaxies M33 and NGC 205 are $\sim 3000 \, M_\odot$ and $\sim 2 \times 10^4 \, M_\odot$, respectively. These upper limits are only slightly below what one would have predicted for $M_\bullet$ in these galaxies using the $M_\bullet$–σ relation (figure 2.11). There is no good reason to suppose that that relation extends to spheroidal systems of such low mass, much less of even lower mass. But if we apply the relation (2.33) to stellar systems as small as globular clusters, the predicted central masses would be

$$M_\bullet \approx 2.3 \times 10^3 \left(\frac{\sigma}{20\,\mathrm{km\,s^{-1}}} \right)^{4.86} M_\odot, \tag{2.52}$$

and the influence radii

$$r_\mathrm{h} = \frac{GM_\bullet}{\sigma^2} \approx 24 \left(\frac{\sigma}{20\,\mathrm{km\,s^{-1}}} \right)^{2.86} \mathrm{mpc} \tag{2.53}$$

(mpc = milliparsecs), corresponding to an angular size

$$\theta_h \approx 0''\!.5 \left(\frac{\sigma}{20\,\mathrm{km\,s^{-1}}} \right)^{2.86} \left(\frac{D}{10\,\mathrm{kpc}} \right)^{-1} \tag{2.54}$$

suggesting that IBHs might be detectable in globular clusters belonging to the Milky Way, which have typical distances of 10 kpc. Just this claim has been made in a handful of objects, based on fits of dynamical models to stellar velocity data. The putative masses, followed by the implied influence radii, are ω Centauri [401] ($4.0 \times 10^4 \, M_\odot$, 15''.0); M15 [188] ($3.9 \times 10^3 \, M_\odot$, 2''.9); and the Andromeda Galaxy globular cluster G1 [179] ($1.7 \times 10^4 \, M_\odot$, 0''.031). None of the claimed detections has stood up to scrutiny. In the case of ω Centauri, the putative influence radius is well resolved; however, a reanalysis [5] found that the center of the cluster had been misidentified, and a modeling study based on a more extensive sample of proper motions [540] inferred only an upper limit on a central dark mass, $M_\bullet \lesssim 1.2 \times 10^4 M_\odot$. Reanalysis of the data for M15 and G1 also showed that they could be fit equally well without massive central objects [34, 35].

Reasonably convincing evidence for the existence of IBHs comes from reverberation mapping studies of a few, low-luminosity AGN. NGC 4395 is an active spiral galaxy at a distance of about 4 Mpc; it does not contain a bulge, and the upper limit on the velocity dispersion of its nuclear cluster is $\sim$30 km s^{-1}, which would suggest an $M_\bullet$–σ mass of $\sim$$1.6 \times 10^4 \, M_\odot$. In fact, reverberation mapping yields $M_\bullet = (3.6 \pm 1.1) \times 10^5 \, M_\odot$ [426]; a firm upper limit of $\sim$$6 \times 10^6 \, M_\odot$ is set by the nuclear cluster's size combined with the limit on σ [164]. As discussed above, tertiary and quaternary mass-estimation techniques exist that are based on empirical correlations derived from reverberation mapping data. These techniques have been applied to spectral data of large samples of low-luminosity AGN lacking the variability data required for reverberation mapping; the results have been interpreted to imply a spectrum of black-hole masses that extends down to a few times $10^5 \, M_\odot$ [214].

Simply establishing the presence of compact massive accretors in galaxies like these, which often have small or nonexistent bulges, would constitute prima facie evidence for IBHs. This becomes progressively more difficult as one considers AGN with lower and lower luminosities. Such galaxies often lack the broad emission lines that are uniquely associated with a massive black hole; the (narrow) emission lines in these galaxies might be due exclusively to gas heated by young stars, or to other purely stellar processes. A standard technique for distinguishing between these possibilities [18, 545] consists of plotting the galaxy on a "line-ratio diagram" that compares the fluxes in two emission lines at similar wavelengths (to avoid reddening effects); certain regions in these diagrams are said to be associated with AGN, that is, with the presence of a source of hard ionizing photons produced by gas near an SBH; and others with "star-forming regions," that is, sources of ionization due to massive young stars. The division is partly empirical and partly based on theoretical modeling, and galaxies often fall in an intermediate region where the classification is uncertain.

In summary, while there is no good reason to suppose that the Milky Way's SBH defines the low-$M_\bullet$ limit of the mass distribution, evidence for lower-mass black holes is not yet compelling.

As noted above, one model for the formation of IBHs invokes the runaway merger of massive stars in dense clusters. The Milky Way contains a number of very dense star clusters in the inner few parsecs: the "Arches" cluster, the "Quintuplet," and a few others [163]. It has been suggested that such clusters, or similar ones that existed in the past, might be natural places for IBHs to form [222]. The cluster would eventually spiral into the nucleus, carrying the IBH toward the SBH until two black holes formed a binary system [226]. Models like these raise an interesting question: Could there be an IBH at the center of our own galaxy? Figure 2.13 summarizes the constraints that can currently be placed on the mass and location of a second black hole at the Galactic center. The most stringent constraints derive from limits on the astrometric "wobble" of the radio source Sgr A* due to its motion about the binary center of mass [453]; these are shown as the region labeled "VLBA" in figure 2.13. Another constraint comes from the requirement that the center of mass of the binary coincides with the peak of the stellar distribution, within observational uncertainties; this is the region labeled "CoM". As discussed in chapter 4, the orbit of a binary black hole will gradually shrink due to emission of gravitational waves; demanding a lifetime of at least 10 Myr excludes the lowest region in figure 2.13. The star S2 has an orbit about the SBH that is well fit by a Keplerian ellipse; the resulting constraints on the amount of additional mass within its orbit [193] yield the box labeled "S2". None of these constraints is rock solid; and even if they were, there is a large region of $M_{\rm IBH}$–semimajor axis space that is not yet excluded.

2.6 EVIDENCE FOR BINARY AND MULTIPLE SUPERMASSIVE BLACK HOLES

The constraints summarized in figure 2.13 on a second, massive black hole near the center of the Milky Way are mostly dynamical in character. In external galaxies, dynamical detection of a single SBH is hard enough, even if its location is known (or assumed) to be precisely at the center of its host galaxy. Detecting a second (smaller) black hole dynamically would be extremely difficult, and essentially

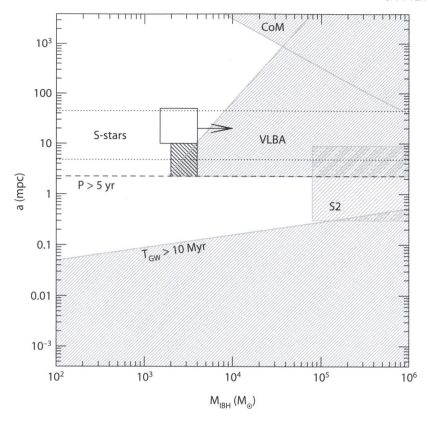

Figure 2.13 Constraints on the existence of a hypothetical binary system at the center of
the Milky Way, consisting of the known SBH, and a second black hole of mass
M_{IBH}; a is the semimajor axis of the binary. Shaded areas represent regions
of parameter space that can be excluded based on observational or theoretical
arguments, as discussed in the text. The shaded box near the center is excluded
due to the perturbations that would be exerted by the second black hole on
the orbits of the S-stars [219]; the dotted lines mark the distances at which the
S-stars are currently observed. The dashed line represents the 5 yr binary orbital
period corresponding to discoverable systems. (Adapted from [219].)

impossible if it were located well inside the influence sphere of the larger SBH. But
if the goal is simply to establish the presence of a second SBH, active galaxies are
an obvious place to look: some of them might contain a source of activity that is
spatially or kinematically offset with respect to the primary AGN.

Figure 2.14 shows what was probably the first clear example of two SBHs in one
"system": in this case, a pair of interacting galaxies near the center of the galaxy
cluster Abell 400. The associated radio source, 3C 75,[13] consists of a pair of twin
radio lobes originating from the centers of the two galaxies, which have a projected

[13]"3C" refers to the "Third Cambridge Catalogue of Radio Sources," a catalog published in 1959 by the
Radio Astronomy Group at the University of Cambridge.

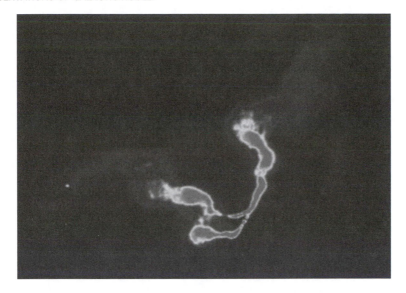

Figure 2.14 An image at radio wavelengths of the radio source 3C 75 near the center of the
galaxy cluster Abell 400: a "dual SBH." There are two, twin-lobe radio sources
associated with each of two elliptical galaxies; the galaxies themselves are not
visible at these wavelengths. The jets bend and appear to be interacting. The
projected separation of the radio cores—presumably indicating the location of
the two SBHs—is about 7 kpc. (Image credit: NRAO/AUI and F. N. Owen, C. P.
O'Dea, M. Inoue, and J. Eilek.)

separation of about 7 kpc [408]. Kiloparsec-scale lobes are common components
of so-called **radio-loud quasars** and **radio galaxies**; the emission consists of synchrotron radiation from the plasma that makes up the lobes, which is thought to
be energized by **jets**, beams of particles originating from near the accretion disk
around the SBH. The lobes typically come in more-or-less symmetric pairs that
are aligned through the center of the galaxy; in 3C 75, the lobes are bent or de-
formed, presumably due to pressure from the intracluster gas as the two galaxies
move about their common center of mass. Systems like 3C 75—containing two
SBHs separated by a distance much greater than their influence radii—are called
dual supermassive black holes or **dual AGN**.

The two galaxies that are the source of the radio emission in 3C 75 will probably merge, although it may take a very long time. Galaxies in the late stages of
a merger are plausible sites for dual SBHs, and many of these exhibit double nuclei in the optical or infrared range. However, few show unambiguous evidence of
AGN activity in both nuclei. There are a handful of clear exceptions; the best case
is probably NGC 6240, shown in figure 2.15. Both nuclei exhibit the flat X-ray
spectra characteristic of AGNs [292] and the projected separation is 1.4 kpc. Other
likely candidates are Arp 299 [19] and NGC 3393 [150]. A possible example of a
merging system containing *three* SBHs is shown in figure 2.16.

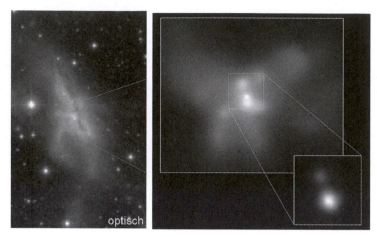

Figure 2.15 The galaxy NGC 6240, which appears to contain a dual SBH. The left panel shows an optical image of the galaxy [274], which appears to be in the late stages of a merger. The right panel is an X-ray image of the galaxy's center [292]; the two "nuclei" are compact, hard-spectrum X-ray sources believed to be associated with SBHs. Projected separation of the nuclei is about 1.4 kpc. (Image courtesy of NASA/CXC/MPE/S. Komossa et al.)

Spatially resolving dual AGN at optical or X-ray wavelengths becomes difficult if their separation is much smaller than in NGC 6240. Radio-interferometric techniques can do much better, and in fact there is one (elliptical) galaxy that is observed to contain two, compact radio sources with spectra indicative of AGN, and a projected separation of only 7.3 pc [460]. The likely combined mass of the two SBHs is of the order of $10^8 \, M_\odot$, which would imply a separation smaller than the influence radius—a true **binary supermassive black hole**.

Even in cases where the binary separation is too small to be resolved spatially, one can still hope to measure a *kinematical* offset between two emission-line systems, one or both of which is associated with an SBH. If it should happen that each SBH has its own accretion disk and associated broad emission-line region, motion about the binary center of mass would cause the lines to shift periodically, making the combined spectrum analogous to that of a single- or double-lined, spectroscopic binary star. In fact, many so-called "double-peaked emitters" are known; however, an interpretation in terms of a binary system has fallen out of favor since the candidate systems tend not to show the predicted radial velocity variations that would be present in a binary [143]. Even if only one broad emission-line region is present, binary motion might still be detected by a velocity offset between these lines, and the lines from the **narrow emission-line region** associated with excited gas that is far from the galaxy center [375]. A number of such cases are known; a striking example is the quasar SDSS J092712.65+294344.0, which exhibits a velocity offset between broad and narrow emission lines of 2650 km s^{-1} [294]. However, the researchers who discovered this object preferred a different interpretation: a *single*

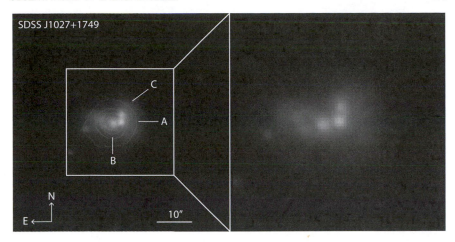

Figure 2.16 The galaxy SDSS J1027+1749 at a redshift $z = 0.066$ contains three active
nuclei, possibly indicating the presence of three SBHs [324]. The left panel is
a radio image of the galaxy, showing distortions that may be due to a recent
merger event; the contours indicate flux densities in the radio, at 1.4 GHz. The
system contains three nuclei, labeled as A, B, and C; based on their optical line
ratios, all three appear to be AGN. The projected separation between nucleus B
(C) and nucleus A is 3.0 (2.4) kpc and the line-of-sight velocity separations are
450 (110) km s^{-1}. The right panel is an optical image of the central region.

SBH that is recoiling from the center of its host galaxy. A number of similar cases
have since been found; however, the available data are generally consistent with
interpretations other than the binary SBH hypothesis.

Many active galaxies exhibit periodic variability with periods of days or years,
consistent with the orbital periods of binary SBHs. The most famous example is
probably OJ 287, a **blazar**, that is, an AGN in which the jet is believed to be
oriented nearly parallel to the line of sight. Optical variability of OJ 287 has been
recorded since 1890 [440] and it has a well-defined period of 11.86 yr (or ~9 yr in
the galaxy's rest frame; the difference is a result of the galaxy's considerable distance and corresponding cosmological redshift). This period can be associated with
the orbital period of a second black hole; in such models, the variability is ascribed
to variations in the accretion rate as the smaller SBH passes through the accretion
disk surrounding the larger SBH [498]. Many other examples of variability in AGN
at optical, radio and even TeV energies are documented, some of which have periods as short as days; evidence for periodic variability has even been claimed for
the Milky Way SBH, at radio wavelengths; the ostensible period is 106 days [582].
However, none of these examples exhibits as clear a periodicity as OJ 287.

Radio lobes in active galaxies provide a fossil record of the orientation history
of the jets powering the lobes. Many examples of sinusoidally or helically distorted jets are known, and these observations are often interpreted via a binary
SBH model. The wiggles may be due to physical displacements of the SBH emitting the jet [464] or to precession of the larger SBH induced by orbital motion of

the smaller SBH [462]. In the radio galaxy 3C 66B, the position of the radio core shows well-defined elliptical motions with a period of just 1.05 yr [508].

A number of other radio galaxies exhibit sharp changes in the orientation of their radio lobes, producing a "winged" or X-shaped morphology [311]. While originally interpreted via a precession model [138], an alternative explanation is that the SBH producing the jet has undergone a "spin flip," a sudden reorientation of the SBH's spin axis, due perhaps to capture of a second SBH following a merger [361].

A number of other possibilities exist for detecting binary SBHs, including the use of space-based interferometers to measure the astrometric reflex motion of AGN photocenters due to orbital motion of the jet-producing SBHs [558]; measurement of periodic shifts in pulsar arrival times due to passage of gravitational waves from binary SBHs [119]; and—as discussed in the next section—the ultimate confirmation, direct detection of gravitational waves.

2.7 GRAVITATIONAL WAVES

Electromagnetic radiation originates in the acceleration of charged particles; for instance, the electrons and ions of a hot plasma. According to the general theory of relativity, radiation is also produced by the acceleration of masses. Since the "gravitational charge" (i.e., mass) only comes with one sign, **gravitational waves** (GWs) are attributable at lowest order to the time-changing quadrupole moment of the mass, and the waves themselves are likewise quadrupolar in nature: squeezing along one axis while stretching along the other, like tides. (Monopole waves would violate mass-energy conservation; dipole waves would violate momentum conservation.)

To order of magnitude, the quadrupole moment, Q, of a source is given by (source mass)$\times$(source size)^2. Dimensional analysis then tells us that the amplitude, h, of a GW is[14]

$$h \approx \frac{G}{c^4} \frac{\ddot{Q}}{D} \approx \frac{G M_Q}{c^2 D} \frac{v^2}{c^2}. \tag{2.55}$$

Here, v is the internal velocity of the source, M_Q is the portion of the source's mass that participates in quadrupolar motions, and D is the distance to the source. The constant that couples the radiation amplitude, h, to the source, Q, is very small:

$$\frac{G}{c^4} = \frac{6.673 \times 10^{-8} \, \text{cm}^3 \, \text{s}^{-2} \, \text{g}^{-1}}{(2.998 \times 10 \, \text{cm}^4 \, \text{s}^{-4})} \approx 8.26 \times 10^{-50} \, \text{s}^2 \, \text{g}^{-1} \, \text{cm}^{-1}. \tag{2.56}$$

On the other hand, since the observable is the wave amplitude, rather than the energy flux as in the case of electromagnetic radiation, the signal falls off only linearly with distance. Furthermore, the weak interaction strength means that GWs propagate from source to observer with almost no absorption.

---

[14]Of course, this is hand waving of a high order. Consensus that GWs are a consequence of Einstein's theory arose only after decades of theoretical work [98] and the discovery of a relativistic binary pulsar [249].

GWs act by changing the distance between widely separated objects:

$$\Delta L(t) \equiv L_1(t) - L_2(t) = h(t)L. \tag{2.57}$$

GW detectors measure this tidal field by measuring the change in distance between two or more masses, which are either freely floating (in space) or which are isolated from local forces (on the Earth). Typically, the masses are laid out in an "L" shape to take advantage of the alternate stretching and squeezing along orthogonal axes. To get an idea of the size of a detector required to observe a GW, consider a binary SBH of mass M_{12} at distance D, and with separation a between the components of the binary. Optimistically assigning all of the binary's mass to M_Q, and setting $v^2 \approx GM_{12}/a$, equation (2.58) implies

$$h \approx 2 \times 10^{-16} \left(\frac{M_{12}}{10^8 \, M_\odot} \right)^2 \left(\frac{a}{\text{mpc}} \right)^{-1} \left(\frac{D}{100 \, \text{Mpc}} \right)^{-1}. \tag{2.58}$$

For every kilometer of baseline L, we must measure a shift $\Delta L \approx 10^{-10}$ cm! Remarkably, such measurement is possible using laser interferometry, even though the displacements being measured may be smaller than the amplitude of thermal fluctuations in the detectors.

GWs differ in important ways from electromagnetic waves. Electromagnetic radiation typically arises from the incoherent superposition of waves produced by many emitters; GWs are *coherent* superpositions arising from the bulk dynamics of a dense source of mass-energy, and their frequency is determined by the frequency of the bulk motion; for instance, the orbital frequency of two SBHs. This means that GWs directly probe the dynamical state of a system, rather than, say, its thermodynamical state. The wavelength of the GWs produced by a binary SBH with orbital period P is given roughly by

$$\lambda \approx cP \approx c\frac{a^{3/2}}{\sqrt{GM_\bullet}} \approx \frac{c}{v}a > a, \tag{2.59}$$

that is, greater than the physical separation between the bodies. Unlike electromagnetic waves, GWs cannot be used to image a source; a closer analogy would be with sound waves.

GW sources associated with an SBH will have frequencies no higher than the frequency of the most tightly bound, stable orbit around the SBH. This frequency is measured in mHz (millihertz, 10^{-3} s^{-1}), far too low a frequency to be detected on the Earth given the presence of local sources of noise (e.g., geothermal motions). Figure 2.17 shows one proposed design for a space-based interferometer, called the Laser Interferometer Space Antenna, or LISA. LISA would use free-floating masses at separations of $\sim 10^7$ km to detect GWs in the mHz part of the spectrum.[15]

At least two types of GW source involving massive black holes would be detectable with a telescope like LISA. The first is the coalescence of binary SBHs, the

---

[15] In 2011 it was announced that the US National Aeronautics and Space Administration would be unable to continue its participation in the LISA project. As this book goes to print, the European Space Agency is still committed to launching LISA Pathfinder, a technology test mission, in 2014.

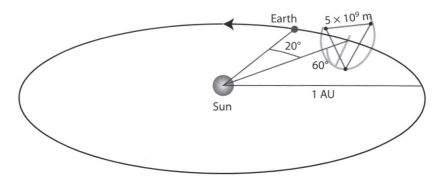

Figure 2.17 One design proposal for a space-based gravitational-wave observatory, called LISA (Laser Interferometer Space Antenna) [103]. Each of the three spacecraft contains two optical assemblies, each containing a 1 W laser and a 30 cm telescope. The three spacecraft which delineate the ends of LISA's arms are placed into orbits such that they form a triangular "constellation" orbiting the Sun, inclined 60° with respect to the plane of the ecliptic and following the Earth with a 20° lag. Since it essentially shares Earth's orbit, the constellation orbits the Sun once per year, "rolling" as it does so. This orbital motion plays an important role in pinpointing the position of gravitational-wave sources by modulating the measured waveform—the modulation encodes source location and makes position determination possible.

same objects that were discussed in the previous section. Depending on the mass of the binary, GWs from the coalescence would be detectable to large redshifts, $z \approx 5$–15, possibly probing an early epoch in the formation of the universe's structure. For a telescope like LISA, the optimal mass of the binary is near 10^5–$10^6 \, M_\odot$: the GWs from smaller systems are weak, whereas instrumental noise at the lower frequencies is likely to swamp the signal. The other major class of sources consists of relatively small bodies—black holes with masses $\sim 10 \, M_\odot$, neutron stars, or white dwarfs—that are captured by SBHs. These are called **extreme-mass-ratio inspirals**, or **EMRI**s. Such events would be measurable to a distance of a few Gigaparsecs if the inspiraling body is a $10 \, M_\odot$ black hole. For both types of event, the observable volume is so large that the rate of measurable events is likely to be interestingly high even if the intrinsic event rate (events per galaxy per year) is low.

Detection of GWs will be important as a check on theories of gravity in the strong-field regime, and this is a major motivation for building detectors. But GW signals also contain information of astrophysical interest about the masses and orbits of the interacting bodies. In the case of binary SBHs, a signal could be detected roughly one year before the coalescence. Information about the mass of the binary is encoded in the rate at which the GW frequency changes (increases) over time. Combining equation (4.240), the rate of change of a binary's separation due to GW emission, with Kepler's third law gives an expression for the evolution of the orbital frequency:[16]

$$f(t) = f_0 \left(1 - \frac{t}{t_{\text{ch}}}\right)^{-3/8}, \tag{2.60a}$$

$$t_{\text{ch}} = \frac{5}{256} \frac{c^5}{G^{5/3}} \frac{M_{12}^{1/3}}{M_1 M_2} f_0^{-8/3} \tag{2.60b}$$

$$= \frac{5}{256} \frac{c^5}{G^{5/3}} M_{\text{ch}}^{-5/3} f_0^{-8/3}. \tag{2.60c}$$

The time t_{ch} is called the **chirp time** (since a chirp is a signal whose frequency changes with time) and $M_{\text{ch}} = (M_1 M_2)^{3/5}/M_{12}^{1/5}$ is the **chirp mass**. Since both f_0 and t_{ch} can be determined through analysis of the GW signal, the chirp mass can be measured. Combining this with a measurement of h yields the distance to the source via equation (2.58). The other parameters defining the binary—its mass ratio, orbital eccentricity, spins of the component SBHs, etc.—can in principle be determined by comparing the observed waveform to precomputed template waveforms derived from detailed numerical simulations. (Similar techniques are used to detect pulsars in radio or X-ray data.) If a source is sufficiently long lived that an antenna like LISA can complete a large fraction of an orbit, it is possible also to infer the binary's position on the sky with some degree of accuracy, possibly allowing identification of the GW source with an electromagnetic counterpart.

In the case of EMRIs, the extreme mass ratio implies that the smaller body takes much longer to spiral in, remaining in the passband of a telescope like LISA for $\sim 10^5$ orbits. In effect, the small body moves gradually through a sequence of orbits around the SBH, spending a long time close to any orbit in the sequence. Studying this evolution via the GWs emitted has been called **bothrodesy**,[17] by analogy to geodesy, the mapping of Earth's gravity with satellite orbits. One goal of bothrodesy is to confirm (or falsify) the predictions of general relativity by "mapping out" the space-time around the SBH. But assuming that the theory is correct, bothrodesy allows one to measure the masses of both objects, and the spin of the SBH, to exquisite (by astronomical standards) precision, roughly one part in 10^4 [20].

Because of the smaller masses involved, a typical EMRI will generate GWs with amplitudes an order of magnitude below the instrumental noise of a detector like LISA, and as much as several orders of magnitude below the GW background from sources like compact normal binary stars in the Milky Way. Extracting the EMRI signal from beneath this torrent of noise is a challenging computational problem, and it is currently unclear how effectively it can be solved.

---

[16]Defined here as the angular frequency, i.e., 2π divided by the Kepler period.
[17]From the Greek word "bothros" ($\beta o \theta \rho o \sigma$), meaning "garbage pit," or so the author is told.

Chapter Three

Collisionless Equilibria

It is often useful to approximate galactic nuclei as steady-state systems in which the gravitational potential is a smooth and continuous function of the position. This approximation is valid if two conditions are met. First, the nucleus must be much older than the orbital period of a star, so that processes like phase mixing (defined below) have had sufficient time to distribute stars uniformly around their orbits. The period of a circular orbit of radius r around the center of a spherical galaxy containing a supermassive black hole (SBH) is

$$P = \frac{2\pi r}{v_c}$$

$$\approx 2.96 \times 10^5 \left(\frac{M_\bullet}{10^8 \, M_\odot}\right)^{-1/2} \left(\frac{r}{10 \, \mathrm{pc}}\right)^{3/2} (1-f)^{1/2} \, \mathrm{yr}, \qquad (3.1)$$

where $f(r)$ is the fraction of the total mass within r that is due to stars. At a radius $r = r_\mathrm{m}$, the gravitational influence radius defined in chapter 2, the enclosed stellar mass is twice the mass of the SBH and $(1-f)^{1/2} = 0.58$. Any nucleus that is more than a few hundred million years old is likely to satisfy this condition at all $r \lesssim r_\mathrm{m}$.

Second, if the effects of close encounters between stars are to be ignored, the nuclear **relaxation time**, defined as the time for encounters between stars to change orbital energies and angular momenta, must be longer than other timescales of interest. (The timescale for physical collisions to occur between stars is generally much longer than the relaxation time.) A standard definition for the relaxation time is [502]

$$T_r = \frac{0.34\sigma^3}{G^2 m_\star \rho \ln \Lambda} \qquad (3.2)$$

$$\approx 0.95 \times 10^{10} \left(\frac{\sigma}{200 \, \mathrm{km \, s^{-1}}}\right)^3 \left(\frac{\rho}{10^6 \, M_\odot \, \mathrm{pc}^{-3}}\right)^{-1} \left(\frac{m_\star}{M_\odot}\right)^{-1} \left(\frac{\ln \Lambda}{15}\right)^{-1} \mathrm{yr}.$$

Here ρ is the stellar density, σ is the one-dimensional velocity dispersion of the stars, $m_\star$ is the mass of a single star, and $\ln \Lambda$, the "Coulomb logarithm," is a "fudge factor" that corrects for the divergent total perturbing force that would be expected in an infinite homogeneous medium (chapter 5). Within the SBH's sphere of influence,

$$\ln \Lambda \approx \ln (M_\bullet/m_\star) \approx \ln(N_h), \qquad (3.3)$$

with $N_h \equiv M_\bullet/m_\star$ the number of stars whose mass equals $M_\bullet$ (section 5.2.3.3). For $m_\star = M_\odot$ and $M_\bullet = (0.1, 1, 10) \times 10^8 \, M_\odot$, $\ln \Lambda \approx (15, 18, 20)$. Even in

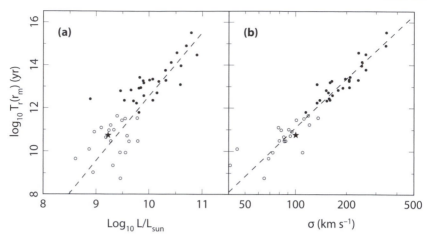

Figure 3.1 Estimates of the relaxation time at the SBH influence radius $r_{\rm m}$ in a complete
sample of early-type galaxies in the Virgo Cluster (circles) and the Milky Way
(star). Panel (a) plots T_r versus the total blue luminosity of the galaxy bulge in
units of the solar luminosity [368]. Panel (b) plots T_r versus the bulge central
velocity dispersion σ [369]. A stellar mass of 1 $M_\odot$ was assumed when computing T_r. Filled circles indicate galaxies in which $r_{\rm m}$ is resolved, i.e., $r_{\rm m} \geq 0''.1$.
Open circles indicate galaxies in which $r_{\rm m}$ is unresolved. In the latter galaxies,
the stellar density at $r_{\rm m}$ is inferred from an extrapolation of the density at larger
radii and is highly uncertain. SBH masses were computed from the $M_\bullet$–σ relation, except in the case of the Milky Way. Dashed lines are least-squares fits to
the data, equations (3.4) and (3.5). Many of the galaxies in this figure for which
$r_{\rm m}$ is unresolved contain nuclear star clusters (NSCs); figure 7.3 plots estimates
of T_r in these galaxies assuming no SBH.

nuclei that are older than one relaxation time, the stellar distribution at any given
moment must satisfy the equations describing a collisionless steady state. Many of
the models discussed in this chapter will therefore apply, for some interval of time,
to nuclei that are gradually evolving due to encounters.

Figure 3.1 shows estimates of T_r at $r = r_{\rm m}$ in a sample of early-type (elliptical
and lenticular) galaxies. The stellar mass in equation (3.2) was set to $M_\odot$, the
mass of the Sun, which is a reasonable guess for the typical mass in old stellar
populations (table 7.1). There are well-defined trends of $T_r(r_{\rm m})$ with spheroid
luminosity L:

$$T_r(r_{\rm m}) \approx 3.8 \times 10^9 \left(\frac{L}{10^9\, L_\odot} \right)^{3.0} {\rm yr}, \tag{3.4}$$

with bulge central velocity dispersion σ:

$$T_r(r_{\rm m}) \approx 1.2 \times 10^{11} \left(\frac{\sigma}{100\,{\rm km\,s^{-1}}} \right)^{7.47} {\rm yr}, \tag{3.5}$$

and with SBH mass:

$$T_r(r_\mathrm{m}) \approx 8.0 \times 10^9 \left(\frac{M_\bullet}{10^6 \, M_\odot} \right)^{1.54} \, \mathrm{yr}. \tag{3.6}$$

The mean relaxation time at r_m drops below 10^{10} yr for spheroids fainter than $\sim 10^9 \, L_\odot$, a little fainter than the bulge of the Milky Way. Relaxation times in the nuclei of brighter galaxies—or equivalently, galaxies with SBH masses greater than $\sim 10^7 \, M_\odot$—are probably always longer than $10 \, \mathrm{Gyr}$.

Figure 3.1 suggests a distinction between **collisionless nuclei**, which have $T_r(r_\mathrm{m}) \gtrsim 10^{10}$ yr, and **collisional nuclei** for which $T_r(r_\mathrm{m}) \lesssim 10^{10}$ yr. Among galaxies with well-determined SBH masses, only the Milky Way comes close to having a nucleus that is in the "collisional" regime.

The morphology and dynamical state of a collisionless nucleus is constrained only by the requirement that the stellar phase-space density satisfy Jeans's theorem (section 3.1.3), that is, that f be constant along orbits. This weak condition is consistent with a wide variety of possible equilibrium configurations, including nonaxisymmetric nuclei, and nuclei in which the majority of orbits are chaotic. In a nucleus with $T_r(r_\mathrm{m}) \lesssim 10^{10}$ yr, on the other hand, the stellar distribution will have had time to evolve to a more strongly constrained, collisionally relaxed steady state, as discussed in chapters 5 and 7.

The collisionless dynamics discussed in this chapter are relevant to the problem of SBH mass estimation, particularly in galaxies where $M_\bullet$ is inferred from the observed motions of stars. The mass estimation problem is significantly hampered by our current inability to resolve structure and kinematics on scales much smaller than r_h or r_m in galaxies beyond the Local Group. As a result, estimates of $M_\bullet$ can be highly uncertain, or even degenerate.

Throughout this chapter, Newtonian mechanics are assumed. Relativistic effects on the motion are discussed in chapter 4.

3.1 ORBITS, INTEGRALS, AND STEADY STATES

3.1.1 Basic concepts

Consider a nucleus that contains distributed matter in the form of stars, stellar remnants, dark matter, etc., and possibly also a massive black hole. Assume furthermore that the granularity in the distributed component is sufficiently small that the gravitational potential $\Phi(x, t)$ can be approximated as a smoothly varying function of position and time.

Focus attention on one component of the distributed mass; call these objects "stars." We define the **distribution function** of the stars, $f(x, v, t)$, such that the number of stars at time t within the phase-space volume element $dx \, dv$, centered at (x, v), is $f(x, v, t) dx \, dv$. Evidently, f is the density of stars in phase space, and it will often be referred to as such. If the stellar trajectories are smooth and continuous, f obeys a continuity equation:

$$\frac{Df}{Dt} = \frac{\partial f}{\partial t} + \sum_i v_i \frac{\partial f}{\partial x_i} + \sum_i a_i \frac{\partial f}{\partial v_i} = 0, \tag{3.7}$$

where a_i is the acceleration in the ith coordinate direction. Equation (3.7) is called the **collisionless Boltzmann equation**, and it states simply that the phase-space density is conserved following the flow. In vector notation,

$$\frac{\partial f}{\partial t} + \boldsymbol{v} \cdot \nabla f - \nabla \Phi \cdot \frac{\partial f}{\partial \boldsymbol{v}} = 0, \tag{3.8}$$

where the accelerations have been written in terms of the potential as

$$\boldsymbol{a} = -\nabla \Phi. \tag{3.9}$$

In a steady state ($\partial f/\partial t = 0$), the phase-space density is the same, at all times, at any phase-space point. Since equation (3.7) says that the value of f is "carried along" with the flow, a steady state demands that f have the same value, at any *given* time, at every point along a trajectory. The condition $\partial f/\partial t = 0$ is therefore equivalent to the statement that f is constant along trajectories. A steady state also implies $\partial \Phi/\partial t = 0$; hence these trajectories are just the orbits defined by the time-independent potential $\Phi(\boldsymbol{x})$.

The requirement that f be constant along orbits does not imply anything special about the character of the motion in the potential $\Phi(\boldsymbol{x})$. At one extreme, the motion could be essentially random, with every trajectory moving chaotically over the phase-space volume enclosed by an energy surface $\Phi(\boldsymbol{x}) \leq E_0$. A steady state would then demand a constant $f = f_0(E_0)$ within every such region. At the other extreme, orbits could behave very regularly, remaining confined to small parts of the energy surface for all times. In this case, a steady-state f could have a different value in each distinct piece of the energy surface.

The generic case lies somewhere between these two extremes. An important distinction can be made between motion that is **regular** or **integrable** and motion that is **chaotic** or **stochastic**. Regular motion is defined as motion that respects at least N_{dof} **isolating integrals of the motion**, where N_{dof} is the number of degrees of freedom (d.o.f.) of the motion—in the present context, $N_{\mathrm{dof}} = 3$, the number of spatial dimensions. An integral of motion is any function $I(\boldsymbol{x}, \boldsymbol{v})$ of the phase-space coordinates that is constant along an orbit; isolating integrals are those that—in some transformed coordinate system $(\boldsymbol{p}, \boldsymbol{q})$—make the Hamiltonian independent of one of the "velocity" coordinates p_i, so that $dq_i/dt = \partial H/\partial p_i = f(q_i)$ can be solved by quadratures. Each additional isolating integral reduces by one the dimensionality of the phase-space volume traversed by an orbit.

Isolating integrals are often associated with symmetries in the potential. The following three classes of potential exhibit a high degree of symmetry, and the motion in them always respects at least one isolating integral in addition to the energy:

1. The Kepler potential, $\Phi(r) = -GM_\bullet/r$. This is the (Newtonian) potential of a point mass $M_\bullet$ fixed at the origin. As is well known (and discussed in detail in chapter 4), all trajectories respect five isolating integrals, which can be

identified with the classical, or Keplerian, orbital elements. In configuration space, bound orbits ($E < 0$) take the form of ellipses with one focus at the origin.

2. Spherical potentials, $\Phi(\boldsymbol{x}) = \Phi(r)$. Four isolating integrals exist: the energy and the three components of the angular momentum. Expressing these per unit mass,

$$E = \frac{1}{2}v^2 + \Phi(r), \quad \boldsymbol{L} = \boldsymbol{x} \times \boldsymbol{v}. \tag{3.10}$$

An orbit defined by a given (E, $\boldsymbol{L}$) is restricted to a planar annulus in configuration space.

3. Axisymmetric potentials, $\Phi(\boldsymbol{x}) = \Phi(\varpi, z)$ where $\varpi^2 = x^2 + y^2$ and the symmetry axis is parallel to z. Two isolating integrals exist for every orbit: the energy and the component L_z of the angular momentum parallel to the z-axis. Orbits in axisymmetric potentials often exhibit a third integral; in configuration space, such orbits typically resemble tori that surround the z-axis. Low-L_z orbits may lack a third integral, particularly if the central force is steeply rising. Such orbits are still approximately toroidal in shape due to the conservation of L_z.

3.1.2 Action-angle variables

Regular orbits—orbits that respect three or more isolating integrals—are **quasiperiodic** or **conditionally periodic**: the dependence of the coordinates and velocities on time can be expressed as

$$\boldsymbol{x}(t) = \sum_{k=1}^{\infty} \boldsymbol{X}_k \exp\left[i \left(l_k \nu_1 + m_k \nu_2 + n_k \nu_3\right) t\right], \tag{3.11a}$$

$$\boldsymbol{v}(t) = \sum_{k=1}^{\infty} \boldsymbol{V}_k \exp\left[i \left(l_k \nu_1 + m_k \nu_2 + n_k \nu_3\right) t\right], \tag{3.11b}$$

with $\{l_k, m_k, n_k\}$ integers. The Fourier transform of $\boldsymbol{x}(t)$ or $\boldsymbol{v}(t)$ for a regular orbit will therefore consist of a set of spikes at discrete frequencies $\nu_k = l_k \nu_1 + m_k \nu_2 + n_k \nu_3$ that are linear combinations of the **fundamental frequencies** $\{\nu_1, \nu_2, \nu_3\}$ for that orbit. Each regular orbit has its own set of fundamental frequencies, and in general, the three ν_i for a given orbit are distinct.

Using Hamilton–Jacobi theory, it can be shown [59] that quasiperiodic motion is always derivable from a Hamiltonian that is "cyclic in"—that is, independent of— the coordinate variables, w_i, if the latter are defined in a particular way. For this special choice of the w_i, Hamilton's equations imply motion that is very simple: the coordinate variables increase linearly with time, and the conjugate momentum variables are conserved. The special set of canonically conjugate variables for which the motion has this simple representation is called the **action-angle variables**. The action (i.e., momentum) variables are commonly written as J_i, and the angle

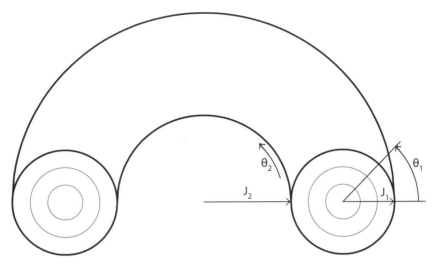

Figure 3.2 Invariant torus defining the motion of a regular orbit in a two-dimensional potential. The torus is determined by the values of the actions J_1 and J_2; the position
of the trajectory on the torus is defined by the angles θ_1 and θ_2, which increase
linearly with time, $\theta_i = v_i t + \theta_i^0$.

(i.e., coordinate) variables as θ_i. In terms of $\{J_i, \theta_i\}$, the equations of motion are

$$J_i = \text{constant}, \tag{3.12a}$$

$$\theta_i = v_i t + \theta_i^0, \quad v_i = \frac{\partial H}{\partial J_i}, \ i = 1, \ldots, N. \tag{3.12b}$$

The frequencies v_i are just the fundamental frequencies that appear in equation (3.11), and the amplitudes $\{X_k, V_k\}$ in that equation are functions only of J.

The motion described by equations (3.12) can be said to lie on a "torus"
(figure 3.2). In dynamics, one speaks most often of the **invariant torus** associated with a regular orbit—invariant, since a trajectory that lies on a torus at any
time will remain on it forever. The first d.o.f. is represented by the angle θ_1 along
a circle of radius J_1. The second d.o.f. is included by adding the angle θ_2 at right
angles to θ_1, forming a "2-torus" whose second dimension is J_2; and similarly for
the third d.o.f. The period of oscillation about each dimension of the torus is given
by $2\pi/v_i$.

As noted above, the dimensionality of a regular orbit is reduced by one for every
additional integral that it respects: a regular orbit moves on a torus of dimension
$(6 - N_{\text{int}})$ where $N_{\text{int}} \geq 3$ is the number of isolating integrals. More trivially, a
reduction of dimensionality also occurs for any regular orbit whose fundamental
frequencies happen to satisfy a **commensurability condition**:

$$m_1 v_1 + m_2 v_2 + m_3 v_3 = 0 \tag{3.13}$$

with the m_i integers (not all of which are zero). These **resonant orbits** (figure 3.3)
are dense in the phase space, in the same sense that the rational numbers are dense

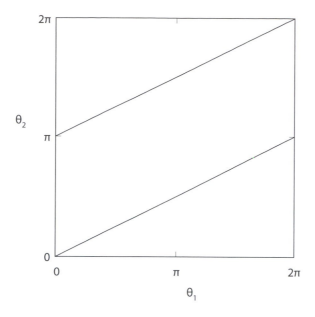

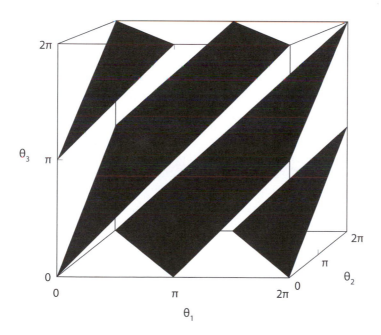

Figure 3.3 Illustrating resonant tori, i.e., orbits that satisfy a commensurability condition like equation (3.13) [377]. The top image shows a two-dimensional torus, plotted as a square with identified edges. The trajectory satisfies a 2:1 resonance between

Figure 3.3 Continued. the fundamental frequencies, $\nu_1 - 2\nu_2 = 0$. The trajectory repeats
after one rotation in θ_1 and two rotations in θ_2. The corresponding orbit is closed
in configuration space and confined to a one-dimensional curve. The bottom image illustrates a three-dimensional torus, represented as a cube with identified
sides. The shaded region is covered densely by a resonant trajectory for which
$2\nu_1 + \nu_2 - 2\nu_3 = 0$. This trajectory is not closed, but it is restricted by the
resonance condition to a two-dimensional subset of the torus. The orbit in configuration space is thin, i.e., confined to a membrane.

in the space of real numbers. However, a commensurability of very high order leads
to motion that is, for practical purposes, incommensurable.

If it should happen that a regular trajectory satisfies *two* independent commensurability conditions, the motion is reduced in dimensionality yet again, to a one-
dimensional curve—a closed orbit.[1] One expects the set of orbits satisfying two
commensurability conditions to be smaller than the set satisfying just one; in other
words, that thin orbits should be more common than closed orbits. Exceptions to
this rule occur in potentials for which there is global commensurability: for instance, in which all orbits are closed. The Kepler potential is an example; another
is the potential of a uniform-density sphere, in which all orbits are ellipses centered
on the origin.

Action-angle variables are useful primarily as a starting point for perturbation
theory. In the case of perturbed Keplerian motion, a useful set of action-angle variables are the "Delaunay variables" defined in chapter 4. The Delaunay variables
can be expressed fairly simply in terms of the Cartesian coordinates and velocities,
or in terms of the Keplerian elements of the "osculating ellipse," the Keplerian orbit
that is tangent to the instantaneous trajectory.

But in practice, the relation between the action-angle variables and the Cartesian
variables is rarely analytic. The process of determining the maps $\{x, v\} \leftrightarrow \{J, \theta\}$
is called **torus construction**. There are a number of contexts in which it is useful
to have expressions, analytic or otherwise, for the $\{J, \theta\}$. One example is when
calculating the response of orbits to slow changes in the potential, which leave the
actions unchanged [59]. Another is the behavior of weakly chaotic orbits, which
may be approximated as regular orbits that slowly diffuse from one torus to another. A third example is galaxy modeling, where regular orbits are most efficiently
represented and stored via the coordinates that define their tori.

Even in potentials that are not globally integrable, regular orbits may still exist;
indeed these are the orbits for which torus construction machinery is most useful.
One expects that for a regular orbit in a nonintegrable potential, a canonical transformation $\{x, v\} \to \{J, \theta\}$ can be found such that

$$\dot{J}_i = 0, \quad \dot{\theta}_i = \Omega_i, \quad i = 1, \ldots, N_{\text{dof}}. \tag{3.14}$$

However, there is no guarantee that the full Hamiltonian will be expressible as a
continuous function of the J_i. In general, the maps $\{x, v\} \leftrightarrow \{J, \theta\}$ will be different

---

[1] In the galaxy dynamics literature, many authors do not distinguish between resonant and closed orbits;
in other words, they implicitly assume that resonant orbits are closed.

for each orbit and will not exist for those trajectories that do not respect N isolating integrals.

Two general approaches to the torus construction problem have been developed. So-called trajectory-following algorithms [418, 58] are based on Fourier decomposition of the numerically integrated trajectories, while iterative approaches [82, 447] begin from some initial guess for $x(\theta)$, then iterate via Hamilton's equations with the requirement that the θ_i increase linearly with time. We discuss the two approaches briefly here; they are to a large degree complementary.

3.1.2.1 Trajectory-following approaches

The Fourier decomposition of a quasiperiodic orbit, equation (3.11), yields a discrete frequency spectrum. The precise form of this spectrum depends on the coordinates in which the orbit is integrated, but certain of its properties are invariant, including the N fundamental frequencies ν_i from which every line is made up,[2] $\nu_k = l_k \nu_1 + m_k \nu_2 + n_k \nu_3$. Typically the strongest line in a spectrum lies at one of the fundamental frequencies; once the ν_i have been identified, the integer vectors (l_k, m_k, n_k) corresponding to every line ν_k are uniquely defined, to within computational uncertainties [47]. Approximations to the actions can then be computed using Percival's formulas [418]; for example, the action associated with θ_1 in a 3 d.o.f. system is

$$J_1 = \sum_k l_k \left(l_k \nu_1 + m_k \nu_2 + n_k \nu_3 \right) |\mathbf{X}_k|^2. \tag{3.15}$$

Finally, the maps $\boldsymbol{\theta} \rightarrow \boldsymbol{x}$ are obtained by making the substitution $\nu_i t \rightarrow \theta_i$ in the spectrum; for example,

$$\begin{aligned}
x(t) &= \sum_k X_k(J) \exp\left[i \left(l_k \nu_1 + m_k \nu_2 + n_k \nu_3 \right) t \right] \\
&= \sum_k X_k(J) \exp\left[i \left(l_k \theta_1 + m_k \theta_2 + n_k \theta_3 \right) \right] \\
&= x(\theta_1, \theta_2, \theta_3). \tag{3.16}
\end{aligned}$$

Trajectory-following algorithms are easily automated; for instance, integer programming can be used to recover the vectors (l_k, m_k, n_k) [528].

Since Fourier techniques focus on the frequency domain, they are particularly well suited to identifying resonances and studying the effect of resonances on the structure of phase space. Resonant tori are places where perturbation expansions of integrable systems break down, due to the "problem of small denominators" [320]. In "perturbed" (i.e., nonintegrable) potentials, one expects stable resonant tori to generate finite regions of regular motion and unstable resonant tori to give rise to chaotic regions. Trajectory-following algorithms allow one to construct a **frequency map** of the phase space: a plot of the ratios of the fundamental frequencies $(\nu_1/\nu_3, \nu_2/\nu_3)$ for a large a set of orbits selected from a uniform grid in initial

---

[2]This statement ignores the fact that any linear combination of fundamental frequencies can also be defined as a fundamental frequency.

condition space [308]. Resonances appear on the frequency map as lines, either densely filled lines in the case of stable resonances, or gaps in the case of unstable resonances. Examples of frequency maps constructed from galactic potentials are presented in sections 3.5 and 6.2.2.

In constructing a frequency map, one is applying the torus construction machinery to orbits not all of which are restricted to tori. But many chaotic orbits have properties similar to those of regular orbits, at least for some restricted time. The frequency spectrum of a weakly chaotic orbit will typically be close to that of a regular orbit, with most of the lines well approximated as linear combinations of three "fundamental frequencies" Ω_i. However, these frequencies will change with time as the orbit diffuses from one "torus" to another. The diffusion rate can be measured via quantities like $|\Omega_1 - \Omega_1'|$, the change in a "fundamental frequency" over two consecutive integration intervals [414, 528]. While such "tori" clearly do not describe the motion of chaotic orbits over arbitrarily long times, they are useful for understanding the onset of chaos and its relationship to resonances.

3.1.2.2 Iterative approaches

Iterative approaches to torus construction consist of finding successively better approximations to the map $\theta \to x$ given some initial guess $x(\theta)$; canonical perturbation theory is a special case, and in fact iterative schemes often reduce to perturbative methods in appropriate limits. Iterative algorithms were first developed [82] in the context of semiclassical quantization for computing energy levels of bound molecular systems, and they are still best suited to assigning energies to actions, $H(J)$. Most of the other quantities of interest to galactic dynamicists—for example, the fundamental frequencies ν_i—are not recovered with high accuracy by these algorithms. Iterative schemes also tend to be numerically unstable unless the initial guess is close to the true solution. On the other hand, iterative algorithms can be more efficient than trajectory-following methods for orbits that are near resonance.

The equations of motion of a 2 d.o.f. regular orbit,

$$\ddot{x} = -\frac{\partial \Phi}{\partial x}, \quad \ddot{y} = -\frac{\partial \Phi}{\partial y}, \tag{3.17}$$

can be written in the form

$$\left(\nu_1 \frac{\partial}{\partial \theta_1} + \nu_2 \frac{\partial}{\partial \theta_2} \right)^2 x = -\frac{\partial \Phi}{\partial x},$$

$$\left(\nu_1 \frac{\partial}{\partial \theta_1} + \nu_2 \frac{\partial}{\partial \theta_2} \right)^2 y = -\frac{\partial \Phi}{\partial y}. \tag{3.18}$$

If one specifies ν_1 and ν_2 and treats $\partial \Phi / \partial x$ and $\partial \Phi / \partial y$ as functions of the θ_i, equations (3.18) can be viewed as nonlinear differential equations for $x(\theta_1, \theta_2)$ and $y(\theta_1, \theta_2)$ [447]. Expressing the coordinates as Fourier series in the angle variables,

$$x(\theta) = \sum_n X_n e^{in \cdot \theta}, \tag{3.19}$$

and substituting (3.19) into (3.18) gives

$$\sum_n (n \cdot v)^2 X_n e^{in \cdot \theta} = \nabla \Phi. \qquad (3.20)$$

Solutions can be found numerically on a grid of points around the torus by truncating the Fourier series after a finite number of terms and solving for the X_n by iterating from an initial guess.

Another iterative approach is based on generating functions [82]. One begins by dividing the Hamiltonian H into separable and nonseparable parts H_0 and H_1, then one seeks a generating function S that maps the known tori of H_0 into tori of H. For a generating function of the F_2 type [201], one has

$$J(\theta, J') = \frac{\partial S}{\partial \theta}, \quad \theta'(\theta, J') = \frac{\partial S}{\partial J'}, \qquad (3.21)$$

where (J, θ) and (J', θ') are the action-angle variables of H_0 and H, respectively. The generator S is determined, for a specified J', by substituting the first of equations (3.21) into the Hamiltonian and requiring the result to be independent of θ. One then arrives at $H(J')$. A sufficiently general form for S is

$$S(\theta, J') = \theta \cdot J' - i \sum_{n \neq 0} S_n(J') e^{in \cdot \theta}. \qquad (3.22)$$

Iterative schemes for finding the S_n can be set up that recover the results of first-order perturbation theory after a single iteration [82]. Having computed the energy on a grid of J' values, one can interpolate to obtain the full Hamiltonian $H(J')$. If the system is not in fact completely integrable, this H can be rigorously interpreted as a smooth approximation to the true H [556] and can be taken as the starting point for secular perturbation theory [265].

The generating function approach is not naturally suited to deriving the other quantities of interest to galactic dynamicists. For instance, equation (3.21) gives $\theta'(\theta)$ as a derivative of S, but since S must be computed separately for every J' its derivative is likely to be ill conditioned.

I. C. Percival [419] described a variational principle for constructing tori. His technique has apparently not yet been implemented in the context of galactic dynamics.

3.1.3 Jeans's theorem

If all orbits in a galaxy respected three isolating integrals, the condition given above for a steady state—that f be constant along trajectories—would become

$$f = f(J_1, J_2, J_3), \qquad (3.23)$$

or equivalently

$$f = f(E, I_2, I_3), \qquad (3.24)$$

where $\{I_2, I_3\}$ are the isolating integrals in addition to the energy. In other words,

> The phase-space density of a stationary stellar system with a globally
> integrable potential can be expressed in terms of the isolating integrals
> in that potential.

This is commonly referred to as **Jeans's theorem**.

If not all orbits are regular, steady states are still possible (think of a steady-state gas in which all the trajectories are chaotic), but Jeans's theorem must be cast into a more general form:

> The phase-space density of a stationary stellar system is constant within
> every well-connected region.

A well-connected region is one that cannot be decomposed into two finite regions such that all trajectories lie, for all time, in either one or the other. Invariant tori are such regions, but so are the more complex parts of phase space associated with stochastic trajectories.

Jeans [257] first stated condition (3.24) in the context of motion in axisymmetric potentials, assuming that the motion respected only the two classical integrals E and L_z. Writing $f = f(E, L_z)$ implies that the phase-space density is constant on hypersurfaces of constant E and L_z. But not all orbits in axisymmetric potentials are characterized by a third isolating integral, and parts of those hypersurfaces are likely to be associated with chaotic trajectories. Assigning a constant density to such regions—as Jeans (and many others since him) implicitly did by writing $f = f(E, L_z)$—thus depends on the more general form of the theorem for its justification. One could construct axisymmetric models in which the surfaces of constant E and L_z are not sampled uniformly; for instance, by excluding all chaotic orbits, or by assigning different densities to different chaotic regions on the same (E, L_z) hypersurface. Models like these do not appear to have been constructed.[3]

3.1.4 Mixing

Jeans's theorem is often assumed to be satisfied in any stellar system that is many crossing times old. However, Jeans's theorem begs the question of *how* f came to have a constant value throughout the regions of phase space associated with individual orbits. A set of phase points on a given torus does not evolve toward a uniform distribution; it simply translates forever, unchanged, around the torus. The strongest sort of "mixing" that can occur in a fully integrable potential is **phase mixing**: points that lie on *different*, but nearby, tori gradually move apart, due to the (generally) different orbital frequencies associated with different tori. After many revolutions, the density of points in such a filament, averaged over a small but finite phase-space volume (i.e., a volume that intersects different tori)—the **coarse-grained phase-space density**—will be independent of position on the torus, even though the fine-grained f never reaches a steady state. Stochastic trajectories are more obliging in this regard: their extreme sensitivity to initial conditions implies

---

[3] At least, not by theorists. There is no reason why real galaxies with such properties should not exist.

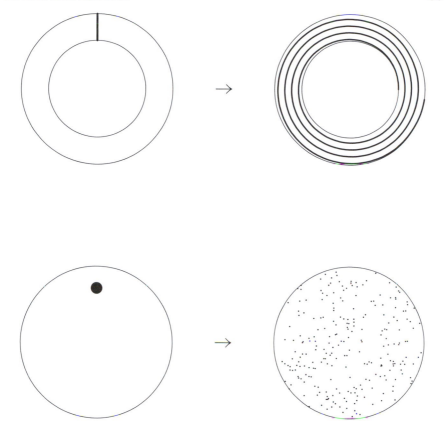

Figure 3.4 Schematic representations of phase mixing (top) and chaotic mixing (bottom) [376].

a stronger sort of mixing, **chaotic mixing**, in which a small patch of phase space evolves so as to uniformly cover, at a later time, a much larger region (figure 3.4).

Phase mixing is an intrinsically slow process—in fact it has no well-defined timescale. The rate at which a group of trajectories shears depends on the range of orbital frequencies in the group. If the maximum and minimum frequencies are Ω_1 and Ω_2, respectively, we expect phase mixing to take place in a time of order $|\Omega_1 - \Omega_2|^{-1}$. This time is never less than a dynamical time and can be much longer. For instance, near the half-mass radius r_e of a de Vaucouleurs-law galaxy, the phase mixing time $2\pi/(\Omega_1 - \Omega_2)$ for two stars on circular orbits with separation Δr is roughly $0.85/(\Delta r/r_e)$ times the orbital period. Inhomogeneities on a scale of $0.1r_e$—roughly 100 pc in a real galaxy—damp out over $\sim$10 dynamical times, and shorter length scales reach equilibrium even more slowly. Chaotic mixing, on the other hand, has a well-defined timescale, roughly the **Lyapunov time**, the mean time associated with the exponential divergence of initially nearby trajectories. The chaotic mixing time is often short, of order the crossing time or even less [270, 376].

These two sorts of mixing are idealized models for what probably happens in real stellar systems. Galaxy formation is a chaotic process, characterized by density inhomogeneities with a wide spectrum of masses, and by a gravitational potential that changes substantially on timescales of single orbital periods. These properties of the potential would have promoted mixing by providing enhanced gravitational perturbations, at least until such a time as the mixing process itself managed to smooth out the lumps. The oxymoron "violent relaxation" [330] is popularly used to describe this poorly understood phenomenon.

In any case, when applying Jeans's theorem, it is commonplace to make no distinction between fine-grained and coarse-grained f's; the effects of incomplete mixing are simply ignored and Jeans's theorem is assumed, without real justification, to be a statement about the fine-grained f. Unless otherwise noted, that practice will be followed here.[4]

3.1.5 Self-consistency

In a steady state, the density of stars whose distribution function is f is

$$n(x) = \int \int \int f(x, v) \, dv. \tag{3.25}$$

In the special case that all orbits respect three isolating integrals, this becomes

$$n(x) = \int \int \int f(E, I_2, I_3) \frac{\partial(v_1, v_2, v_3)}{\partial(E, I_2, I_3)} dE \, dI_2 \, dI_3. \tag{3.26}$$

In the more general case of a nonintegrable potential, a similar relation holds, except that f is allowed to have the more complicated form corresponding to a constant value in every well-connected phase-space region. In all of these cases, the region of integration extends over the entire velocity-space volume traversed by orbits that pass through point x.

The configuration-space density $n(x)$ in equations (3.25) and (3.26) need not bear any relation to the density of the matter that generates the gravitational potential. Indeed, in a nucleus containing an SBH, only a small fraction of the gravitational force may come from the stars. However, one sometimes seeks a steady-state model that describes a population of stars, under the assumption that the gravitational potential in which the stars move is generated (at least in part) by the stars themselves. This is the **self-consistency problem**. For instance, in a nucleus containing a central SBH and stars with number density $n(x)$ and mass $m_\star$, one can write

$$\Phi(x) = -\frac{G M_\bullet}{r} + \Phi_\star(x), \tag{3.27a}$$

$$\nabla^2 \Phi_\star = 4\pi G m_\star n(x) = 4\pi G m_\star \int f \, dv, \tag{3.27b}$$

where f is the phase-space density of the stars; the second equation, **Poisson's equation**, relates $\Phi_\star$ to the stellar mass density.

---

[4]Differences between the fine-grained and coarse-grained f are important in loss-cone theory; see section 6.1.2.

A useful way of thinking about the self-consistency problem is to imagine that the gravitational potential in equation (3.27) is known. Each orbit then generates a "partial density" $n_i(x)$, which can be computed by integrating a trajectory for a long time and computing the time-averaged density at any point in configuration space.[5] For self-consistency, the total density must be representable as a superposition of those partial densities; that is, $n(x) = \sum_i C_i n_i(x)$, $C_i \geq 0$. This may or may not be possible [482]. For instance, if most of the orbits are stochastic, their time-averaged shapes will be similar to the shapes of equipotential surfaces; since the latter are generally rounder than the equidensity surfaces, no self-consistent solution may exist. On the other hand, if most orbits respect three or more integrals, many different orbital "shapes" will be available, increasing the chance that self-consistent solutions can be found.

3.1.6 Modeling and the Jeans equations

In the context of galactic nuclei, a common goal of dynamical modeling is to determine the gravitational potential, and hence the mass, given observations of a sample of stars that are moving in that potential. We call this the **potential estimation problem**; it is a subset of the **dynamical inverse problem**, the problem of jointly determining f and Φ for a stellar system given kinematical and positional data of its member stars.

In exceptionally favorable cases, one can hope to measure the accelerations of individual stars. This method has been applied, with great success, to determination of the mass of the Galactic center SBH (chapter 4). However, in most galaxies, the best one can hope to do is to measure some set of **moments of the velocity distribution** determined by large samples of stars. The velocity moments can be defined generally as

$$n \, \overline{v_x^\alpha v_y^\beta v_z^\gamma} = \int f(x, v) \, v_x^\alpha v_y^\beta v_z^\gamma \, dv. \tag{3.28}$$

For instance, if z is defined as the distance along the line of sight (l.o.s.), the lowest l.o.s. velocity moments ($\alpha = \beta = 0$) are

$$n = \int f \, dv, \quad n\overline{v}_z = \int f v_z \, dv, \quad n\sigma_z^2 = \int f \, (v_z - \overline{v}_z)^2 \, dv, \tag{3.29}$$

which respectively define the number density, mean l.o.s. velocity $\overline{v}_z$, and l.o.s. velocity dispersion σ_z. The number density is accessible via number counts or measurement of the distribution of starlight, while the velocity moments are accessible, in principle, via Doppler-broadened stellar spectra. Since f will in general be a nonlinear function of the isolating integrals, equations like (3.28) will generally be nonlinear in their dependence on Φ. One consequence is that the existence and uniqueness of Φ, given kinematical data of a certain type, can be difficult to prove.

When modeling a galaxy in terms of velocity moments, it is often convenient to work with moments of the collisionless Boltzmann equation, rather than with

---

[5]The ergodic theorem, which applies very generally to regular and stochastic motion, guarantees the equivalence of time-averaged and steady-state densities [320].

integral expressions like equation (3.28). The **Jeans equations** are derived by taking progressively higher moments over velocity of equation (3.8).[6] The simplest Jeans equation is obtained by integrating equation (3.8) over velocities; expressed in Cartesian coordinates, the result is

$$\frac{\partial n}{\partial t} + \sum_{i=1}^{3} \frac{\partial (n \bar{v}_i)}{\partial x_i} = 0. \tag{3.30}$$

Equation (3.30) is just a conservation equation for stars; it is trivially satisfied in a steady-state system in which $\bar{v} = 0$. Multiplying equation (3.8) by v_j and again integrating over all velocities, one finds the Jeans equation of next higher order:

$$\frac{\partial (n \bar{v}_j)}{\partial t} + \sum_{i=1}^{3} \frac{\partial (n \overline{v_i v_j})}{\partial x_i} + n \frac{\partial \Phi}{\partial x_j} = 0, \tag{3.31}$$

where

$$n \overline{v_i v_j} = \int v_i v_j f \, d\boldsymbol{v}. \tag{3.32}$$

Equations like these relate Φ to moments of f in a straightforward way; the problem is that one can rarely measure *all* of the moments that appear in the Jeans equations, and as a result, Φ will be underdetermined, unless ad hoc assumptions are made about the unobserved moments or about the functional form of f. In practice, the degeneracy in Φ is often quite severe, as discussed in more detail below.

3.2 SPHERICAL NUCLEI

The integrals of motion in a spherically symmetric potential are the energy E and the three components of the angular momentum $\boldsymbol{L}$. The most general form of f that describes a time-independent distribution of stars in a spherical potential is

$$f = f(E, \boldsymbol{L}) = f \left[v^2/2 + \Phi(r), \boldsymbol{x} \times \boldsymbol{v} \right]. \tag{3.33}$$

Because this f depends in different ways on the different components of the velocity, it implies a distribution of velocities that is **anisotropic**, in general, at any r. In addition, since f can depend on the orientation of $\boldsymbol{L}$, equation (3.33) can describe a population with a nonspherical mass distribution; for instance, a disk.

When solving the self-consistency problem in spherical symmetry, one wants to choose a form for f that implies $n = n(r)$. For instance, $f(E, \boldsymbol{L})$ can be assumed to have the restricted form

$$f = f(E, L^2) = f \left[v^2/2 + \Phi(r), r^2 v_t^2 \right] = f(r, v_r, v_t), \tag{3.34}$$

with $\{v_r, v_t\}$ the components of the velocity parallel and orthogonal to a radius vector. This form for f implies no streaming motions (i.e., $\bar{v} = 0$) and the first Jeans equation (3.30), is trivially satisfied. The second Jeans equation (3.31) becomes

$$\frac{d (n \sigma_r^2)}{dr} + \frac{2n}{r} \left(\sigma_r^2 - \sigma_t^2 \right) + n \frac{d \Phi}{dr} = 0, \tag{3.35}$$

---

[6]The Jeans equations can also be derived directly from expressions like (3.28) [117].

where

$$\sigma_r^2 = \overline{v_r^2}, \quad \sigma_t^2 = \overline{v_\theta^2} = \overline{v_\varphi^2}. \tag{3.36}$$

One sometimes defines the anisotropy in terms of a parameter β where

$$\beta(r) \equiv 1 - \frac{\sigma_t^2}{\sigma_r^2}; \tag{3.37}$$

evidently $-\infty \leq \beta \leq 1$. In terms of β, equation (3.35) becomes

$$\frac{d\left(n\sigma_r^2\right)}{dr} + \frac{2n\beta\sigma_r^2}{r} + n\frac{d\Phi}{dr} = 0 \tag{3.38}$$

and the potential can be expressed in terms of the mass density $\rho(r) = m_\star n(r)$ by solving Poisson's equation (3.27b):

$$\Phi(r) = -\frac{4\pi G}{r} \int_0^r \rho(r')r'^2 dr' - 4\pi G \int_r^\infty \rho(r')r' dr'. \tag{3.39}$$

A spherical model described by $f(E, L^2)$ can be modified into a model that "rotates" by selectively changing the sign of v for some of the stars, leaving the density unchanged ("orbit flipping"). Discussion of rotating models is postponed until the next section on axisymmetric models; unless otherwise noted, the spherical models discussed in this section are assumed to have $\overline{v} = 0$ everywhere.

It is useful to define $N(E, L) dE dL$ as the number of stars with energy and angular momentum in the intervals dE and dL centered at E and L. The configuration-space volume accessible to stars of a particular E and L has the form of a spherical shell, with inner and outer radii equal to r_- and r_+, the turning points of the motion; these are the roots of $v_r = 0$ or

$$0 = 2[\Phi(r) - E] - \frac{L^2}{r^2}. \tag{3.40}$$

At each radius within this shell, a given E and L corresponds to both positive and negative values of v_r; we define v_r to be positive and double the number of stars in what follows. Integrating over the shell, the relation between N and f is

$$N(E, L) dE dL = 2 \int_{r_-}^{r_+} f(E, L) \times 4\pi r^2 dr \times v^2 dv \times 2\pi \sin\theta\, d\theta, \tag{3.41}$$

where θ is the angle between v and r:

$$v_r = v\cos\theta, \quad v_t = v\sin\theta, \tag{3.42}$$

and $0 \leq \theta \leq \pi/2$. We can write $dE = v\, dv$, and since $L = rv_t = rv\sin\theta$,

$$\sin\theta\, d\theta = \frac{v_t\, dL}{rvv_r} = \frac{L\, dL}{r^2 vv_r}. \tag{3.43}$$

Equation (3.41) then becomes

$$N(E, L) = 16\pi^2 Lf(E, L) \int_{r_-}^{r_+} \frac{dr}{v_r} \tag{3.44a}$$

$$= 8\pi^2 Lf(E, L)P(E, L), \tag{3.44b}$$

where $P(E, L)$ is the radial period.

3.2.1 Isotropic models

After a time of order the relaxation time, the velocity distribution in a nonrotating galactic nucleus is expected to be nearly isotropic. In the spherical geometry, an isotropic velocity distribution requires

$$f = f(E) = f\left[v^2/2 + \Phi(r)\right] = f(r, v). \tag{3.45}$$

Based on figure 3.1, one expects isotropy to be necessary only in the nuclei of galaxies with spheroids that are fainter than the Milky Way, that is, galaxies with nuclear relaxation times shorter than $\sim 10\,\text{Gyr}$. However, it is common to assume $f = f(E)$ as a convenient starting point, even in cases where no a priori case for isotropy can be made.

The relation between n and f for an isotropic population in a spherical potential is

$$n(r) = 4\pi \int f(r, v)v^2 dv = 4\pi \int_{\Phi(r)}^{0} f(E)\sqrt{2\left[E - \Phi(r)\right]}\, dE. \tag{3.46}$$

Writing $n(r) = n(\Phi(r)) = n(\Phi)$, this equation can be converted into an Abel integral equation, with solution

$$f(E) = \frac{\sqrt{2}}{4\pi^2} \frac{d}{dE} \int_{E}^{0} \frac{d\Phi}{\sqrt{\Phi - E}} \frac{dn}{d\Phi}. \tag{3.47}$$

Equation (3.47) is **Eddington's formula** [128]. While equation (3.47) shows that f is uniquely determined by n and Φ in the isotropic case, it does not guarantee that f will be finite and nonnegative. Indeed, the latter condition requires that

$$\int_{E}^{0} \frac{d\Phi}{\sqrt{\Phi - E}} \frac{dn}{d\Phi}$$

be an increasing function of energy at all E.

An important application of Eddington's formula is to the case of a power-law distribution of stars around an SBH. Suppose that $n(r) = \rho(r)/m_\star$, with $m_\star$ the mass of a single star, and that

$$n(r) = \frac{3 - \gamma}{2\pi} \frac{M_\bullet}{m_\star} \frac{1}{r_{\rm m}^3} \left(\frac{r}{r_{\rm m}}\right)^{-\gamma}. \tag{3.48}$$

Recall from chapter 2 that $r_{\rm m}$ is the radius containing a mass in stars equal to twice $M_\bullet$. Restricting attention to small radii, $r \ll r_{\rm m}$, where the contribution to the gravitational potential from the stars is negligible, the potential is simply $\Phi \approx -GM_\bullet/r$, and $n \propto (-\Phi)^\gamma$. Eddington's formula then gives

$$f(E) = \frac{3 - \gamma}{8} \sqrt{\frac{2}{\pi^5}} \frac{\Gamma(\gamma + 1)}{\Gamma(\gamma - \frac{1}{2})} \frac{M_\bullet}{m_\star} \frac{\varphi_0^{3/2}}{(GM_\bullet)^3} \left(\frac{|E|}{\varphi_0}\right)^{\gamma - 3/2}, \tag{3.49}$$

where $\varphi_0 \equiv GM_\bullet/r_{\rm m}$.

This solution is unphysical for $\gamma \leq 1/2$. Stated differently, an isotropic distribution of stars in a point-mass potential requires a stellar density that increases faster than $r^{-1/2}$ toward the center. The reason is that the low-angular-momentum orbits at each energy imply a density that rises at least as fast as $r^{-1/2}$ as $r \to 0$. The stars in some galaxies (including, perhaps, the Milky Way) are observed to have flatter density profiles than $n \sim r^{-1/2}$ within the SBH influence radius. Achieving a steady-state stellar distribution in this case requires a depopulation of the eccentric orbits compared with the isotropic case, hence an anisotropic velocity distribution.

In the same way that we defined $N(E, L)$, we can define $N(E)$, the distribution of energies, such that $N(E) dE$ is the total number of stars with an energy in the interval dE centered at E. Clearly,

$$N(E)\,dE = \int 4\pi\, r^2 dr \int f(E)\, 4\pi\, v^2 dv. \tag{3.50}$$

Since $dE = v\,dv$, this becomes

$$N(E) = 4\pi f(E) \int 4\pi\, r^2 v\, dr$$
$$= 4\pi^2 p(E) f(E), \tag{3.51}$$

where[7]

$$p(E) = 4 \int_0^{\Phi^{-1}(E)} v(r, E)\, r^2 dr$$
$$= 4\sqrt{2} \int_0^{\Phi^{-1}(E)} \sqrt{E - \Phi(r)}\, r^2 dr \tag{3.52}$$

and $\Phi^{-1}(E)$ is the radius at which $\Phi(r) = E$. Alternatively, we can derive $N(E)$ from $N(E, L)$, equation (3.44), after replacing $f(E, L)$ by $f(E)$:

$$N(E) = \int_0^{L_c(E)} N(E, L)\, dL \tag{3.53a}$$

$$= 8\pi^2 f(E) \int_0^{L_c(E)} P(E, L) L\, dL, \tag{3.53b}$$

which yields a different (but equivalent) definition of $p(E)$. The period $P(E, L)$ is a function of only E near the SBH, and it turns out that $P(E, L)$ is typically a weak function of L at all energies.[8] Setting $P \approx P(E)$, equation (3.53b) can be written approximately as

$$N(E) \approx 8\pi^2 P(E) f(E) \int_0^{L_c(E)} L\, dL \tag{3.54a}$$

$$\approx 4\pi^2 L_c^2(E) P(E) f(E), \tag{3.54b}$$

showing that

$$N(E, L)dE\, dL \approx N(E)dE\, d(L^2/L_c^2) \tag{3.55}$$

---

[7] Some authors, e.g., Spitzer [502], omit the factor of 4 from the definition of $p(E)$.
[8] The "isochrone" model [234] has $P = P(E)$, and its $\rho(r)$ does not differ too strongly from observed density profiles of giant galaxies.

and

$$p(E) \approx L_c^2(E)P(E). \tag{3.56}$$

.

Continuing with the isotropic case, and defining $\sigma \equiv \sigma_r = \sigma_t$, the Jeans equation (3.35) becomes

$$\frac{d\left(n\sigma^2\right)}{dr} + n\frac{d\Phi}{dr} = 0. \tag{3.57}$$

Interpreted as a differential equation for $n\sigma^2$, the solution is

$$n\sigma^2 = \int_r^\infty dr' \frac{GM(r')n(r')}{r'^2}, \tag{3.58}$$

where $M(r) = (r^2/G)d\Phi/dr$ is the total mass within r. Typically one does not observe n or σ directly, but rather their projections along the line of sight. The projected density, or **surface density**, is

$$\Sigma(R) = \int_{-\infty}^\infty n(z)dz = 2\int_R^\infty \frac{n(r)r\, dr}{\sqrt{r^2 - R^2}}, \tag{3.59}$$

where z defines the distance along the l.o.s. and R is the distance from the projected center. (We have assumed in writing equation (3.59) that the galaxy is free from differential absorption.) The l.o.s. velocity dispersion σ_p is given by a similar expression:

$$\left(\Sigma\sigma_p^2\right)(R) = \int_{-\infty}^\infty n(z)\sigma^2(z)dz = 2\int_R^\infty \frac{r\, dr}{\sqrt{r^2 - R^2}}n\sigma^2. \tag{3.60}$$

Combining (3.58) and (3.60),

$$\begin{aligned}
\left(\Sigma\sigma_p^2\right)(R) &= 2G\int_R^\infty \frac{dr\, r}{\sqrt{r^2 - R^2}}\int_r^\infty \frac{dr'n(r')M(r')}{r'^2}\\
&= 2G\int_R^\infty \frac{dr}{r^2}\sqrt{r^2 - R^2}\, n(r)\, M(r)
\end{aligned} \tag{3.61}$$

and substituting for $\Sigma(R)$ from equation (3.59),

$$\sigma_p^2(R) = G\frac{\int_R^\infty dr\, r^{-2}\left(r^2 - R^2\right)^{1/2} n(r)M(r)}{\int_R^\infty dr\, r\left(r^2 - R^2\right)^{-1/2} n(r)}. \tag{3.62}$$

This expression gives the line-of-sight, projected velocity dispersion of an isotropic, spherical, nonrotating stellar system in terms of the mass distribution $M(r)$ and the number density $n(r)$ of the observed population.

We can apply this formula to the case considered at the start of this subsection: a power-law distribution of stars, $n \propto r^{-\gamma}$, in the point-mass potential of an SBH of mass $M_\bullet$. The results are

$$\sigma^2(r) = \frac{1}{1+\gamma}\frac{GM_\bullet}{r}, \tag{3.63a}$$

$$\sigma_p^2(R) = F(\gamma)\frac{GM_\bullet}{R}, \quad F(\gamma) \equiv \frac{1}{2}\frac{\left[\Gamma(\gamma/2)\right]^2}{\Gamma(\frac{\gamma+3}{2})\Gamma(\frac{\gamma-1}{2})} \quad (\gamma > 1). \tag{3.63b}$$

Not surprisingly, the observed velocity dispersion increases as $R^{-1/2}$ toward the SBH. The function F is weakly dependent on γ for $\gamma \gtrsim 1$; for $1.5 \le \gamma \le 3.5$, $0.18 \le F(\gamma) \le 0.21$.

The top panel of figure 3.5 shows the velocity dispersion profile predicted by equation (3.63b) for stars at the center of the Milky Way, after setting $F = 0.2$ and taking several values for $M_\bullet$. The predicted σ_p falls below the observed velocity dispersions outside of $\sim$0.5 pc, suggesting that the stellar motions are being affected at these radii by the distributed mass [481].

The lower panel of figure 3.5 shows the prediction of equation (3.62) when a distributed component is added:

$$M(r) = M_\bullet + 4\pi \int_0^r dr' \, r'^2 \rho(r'), \tag{3.64}$$

with $\rho = \rho_0 (r/r_0)^{-1}$. Setting $M_\bullet = 4.0 \times 10^6 \, M_\odot$ and varying ρ_0, the best fit from this series of models is obtained for $M_\star(< 1 \, \text{pc}) \approx 1.1 \times 10^6 \, M_\odot$.

Comparisons of models to data like these are useful, but give little sense of the uniqueness of the fitted parameters. In the isotropic case, equations (3.59) and (3.60) are both of Abel form and have *unique* inversions:

$$n(r) = -\frac{1}{\pi} \int_r^\infty \frac{d\Sigma}{dR} \frac{dR}{\sqrt{R^2 - r^2}}, \tag{3.65a}$$

$$\left(n\sigma^2\right)(r) = -\frac{1}{\pi} \int_r^\infty \frac{\left(d\Sigma\sigma_p^2\right)}{dR} \frac{dR}{\sqrt{R^2 - r^2}}, \tag{3.65b}$$

independent of any assumptions about $\Phi(r)$. Having determined n and σ, the potential then follows uniquely from the Jeans equation (3.57) in the form

$$\Phi(r) = \Phi_0 + \int_r^\infty \frac{1}{n} \frac{d\left(n\sigma^2\right)}{dr} dr, \tag{3.66}$$

and the distribution function $f(E)$ follows uniquely from n and Φ via equation (3.47). It would generally be ill advised to try deriving Φ and f from equations like (3.47) and (3.66), since they require a numerical differentiation of data, and data tend to be noisy and incomplete. However, these equations demonstrate that—under the spherical isotropic assumption—one can in principle infer the gravitational potential of a nucleus, and the mass of the SBH, in a model-independent way from the lowest observable velocity moments.

3.2.2 Anisotropic models and the problem of degeneracy

Unfortunately, the uniqueness of the solutions in the spherical isotropic case is an exception. Allowing the velocity distribution to have the more general form $f = f(E, L^2)$ means that the Jeans equation will contain two velocity dispersions, σ_r and σ_t, too many to be determined uniquely from a single observed velocity dispersion profile $\sigma_p(R)$. And since the distribution function now depends on two variables (E, L^2), it cannot be derived uniquely from functions of one variable as in Eddington's formula (3.47).

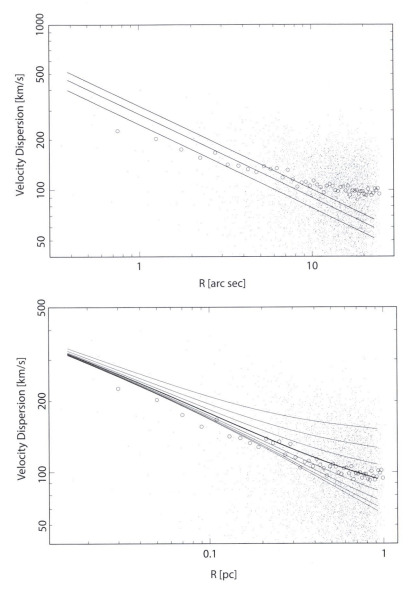

Figure 3.5 A spherical, isotropic model for the center of the Milky Way [481]. Points are measured velocities of late-type stars from a sample of proper motions. Open circles are binned estimates of the one-dimensional velocity dispersion based on the plotted data points. The upper panel shows predictions of equation (3.63b) for three values of the SBH mass, $M_\bullet = (3, 4, 5) \times 10^6 \, M_\odot$. In the bottom panel, equation (3.62) has been used to compute the projected velocity dispersion profile, assuming an enclosed mass of the form given by equation (3.64) with $M_\bullet = 4 \times 10^6 \, M_\odot$; the lines correspond to different values of the distributed mass within one parsec of the SBH, from $2 \times 10^5 \, M_\odot$ to $5 \times 10^6 \, M_\odot$. The thick line has $M_\star(< 1\,\mathrm{pc}) = 1.25 \times 10^6 \, M_\odot$.

In terms of (v_r, v_t), the components of the velocity parallel and orthogonal to a radius vector, the velocity space volume element is

$$d\mathbf{v} = 2\pi v_t \, dv_t \, dv_r = 2\pi v_t \frac{\partial (v_r, v_t)}{\partial (E, L)} dE \, dL = \frac{2\pi}{r^2 v_r} dE \, dL^2. \tag{3.67}$$

The configuration-space density is given in terms of f by

$$n(r) = \frac{2\pi}{r^2} \int_E^0 dE \int_0^{2r^2 E - \Phi(r)} dL^2 \frac{f(E, L^2)}{\sqrt{2[E - \Phi(r)] - L^2/r^2}}, \tag{3.68}$$

and the (one-dimensional) velocity dispersions in the directions parallel and perpendicular to the radius vector are

$$n(r)\sigma_r^2(r) = \frac{2\pi}{r^2} \int_E^0 dE \int_0^{2r^2 E - \Phi(r)} dL^2 f(E, L^2) \sqrt{2[E - \Phi(r)] - L^2/r^2}, \tag{3.69a}$$

$$n(r)\sigma_t^2(r) = \frac{\pi}{r^2} \int_E^0 dE \int_0^{2r^2 E - \Phi(r)} dL^2 \frac{L^2}{r^2} \frac{f(E, L^2)}{\sqrt{2[E - \Phi(r)] - L^2/r^2}}. \tag{3.69b}$$

In the spherical geometry, $n(r)$ will still be uniquely recoverable from a measured $\Sigma(R)$. But the measured velocity dispersion profile, $\sigma_p(R)$, will contain contributions from both $\sigma_r(r)$ and $\sigma_t(r)$. Unless more information is available than $\Sigma(R)$ and $\sigma_p(R)$, many solutions for f and Φ (and many values for the mass of the SBH) will be equally consistent with the data. This situation is sometimes referred to as **mass-anisotropy degeneracy**, and it accounts for much of the systematic uncertainty associated with estimates of $M_\bullet$ in galactic nuclei based on stellar kinematical data.

The degeneracy is most easily illustrated using the Jeans equations [46, 117]. Suppose one observes the projected density and velocity dispersion profiles $\Sigma(R)$ and $\sigma_p(R)$ of the stars in a spherical, nonrotating galaxy. The spatial density $n(r)$ is obtained from $\Sigma(R)$ via equation (3.65a). The relation between the l.o.s. velocity dispersion $\sigma_p(R)$, and the two velocity dispersions $\{\sigma_r, \sigma_t\}$ is

$$\Sigma \sigma_p^2 = \int_{-\infty}^{\infty} \overline{n(v_r \cos \varphi - v_\theta \sin \varphi)^2} dz \tag{3.70a}$$

$$= 2 \int_R^{\infty} (\sigma_r^2 \cos^2 \varphi + \sigma_t^2 \sin^2 \varphi) \frac{n r \, dr}{\sqrt{r^2 - R^2}} \tag{3.70b}$$

$$= 2 \int_R^{\infty} \left(1 - \frac{R^2}{r^2} \beta\right) \frac{n \sigma_r^2 r \, dr}{\sqrt{r^2 - R^2}}, \tag{3.70c}$$

where $\sin \varphi = R/r$. This equation has a known left-hand side, but the right-hand side contains two unknown functions (σ_r, β). We can formally remove the indeterminacy by assuming for the moment that the potential $\Phi(r)$ is known and using the

Jeans equation (3.38) to write β in terms of Φ. The result is

$$\Sigma\sigma_p^2 - R^2 \int_R^\infty \frac{d\Phi}{dr} \frac{n\,dr}{\sqrt{r^2 - R^2}} =$$

$$2 \int_R^\infty \frac{n\sigma_r^2 r\,dr}{\sqrt{r^2 - R^2}} + R^2 \int_R^\infty \frac{d\left(n\sigma_r^2\right)}{dr} \frac{dr}{\sqrt{r^2 - R^2}}. \qquad (3.71)$$

We denote the left-hand side by

$$g(y) = \left(\Sigma\sigma_p^2\right)(y) - y \int_{\sqrt{y}}^\infty \frac{d\Phi}{dr} \frac{n\,dr}{\sqrt{r^2 - y}}, \qquad y \equiv R^2, \qquad (3.72)$$

which is (by assumption) a known function. Writing $x \equiv r^2$, equation (3.71) then becomes

$$g(y) = \int_y^\infty n\sigma_r^2 \frac{dx}{\sqrt{x - y}} + y \int_y^\infty \frac{d\left(n\sigma_r^2\right)}{dx} \frac{dx}{\sqrt{x - y}} \qquad (3.73a)$$

$$= \frac{d}{dy} \left(y \int_y^\infty n\sigma_r^2 \frac{dx}{\sqrt{x - y}} \right). \qquad (3.73b)$$

This equation is easily integrated:

$$\int_{y_0}^y g(y')dy' = y \int_y^\infty n\sigma_r^2 \frac{dx}{\sqrt{x - y}}, \qquad (3.74)$$

where y_0 is an integration constant; regularity of $n\sigma_r^2$ at $r = 0$ dictates $y_0 = 0$. Defining $G(y) \equiv \int_0^y g(y')dy'$—also a known function—equation (3.74) can be inverted to give

$$\left(n\sigma_r^2\right)(x) = -\frac{1}{\pi} \int_x^\infty \frac{d}{dy} \left[\frac{G(y)}{y} \right] \frac{dy}{\sqrt{y - x}}. \qquad (3.75)$$

Returning to the original variables $\{r, R\}$, this becomes

$$\left(n\sigma_r^2\right)(r) = \frac{G(\infty)}{2r^3} - \frac{2}{\pi r^3} \int_r^\infty \left[\frac{r}{\sqrt{R^2 - r^2}} + \cos^{-1}\left(\frac{r}{R}\right) \right] g(R) R\,dR$$

$$= \frac{G(\infty)}{2r^3}$$

$$- \frac{2}{\pi r^3} \int_r^\infty \left[\frac{r}{\sqrt{R^2 - r^2}} + \cos^{-1}\left(\frac{r}{R}\right) \right] \left(\Sigma\sigma_p^2\right)(R) R\,dR$$

$$+ \frac{2}{3r^3} \int_r^\infty \left(r'^3 + \frac{r^3}{2} \right) n \frac{d\Phi}{dr'} dr'. \qquad (3.76)$$

Once $n\sigma_r^2$ has been computed from equation (3.76), the tangential velocity dispersion follows from the Jeans equation (3.38):

$$\sigma_t^2 = \sigma_r^2 + \frac{r}{2} \left[\frac{d\Phi}{dr} + \frac{1}{n} \frac{d\left(n\sigma_r^2\right)}{dr} \right]. \qquad (3.77)$$

As $r \to \infty$ (or equivalently, at the edge of a finite system), these equations imply

$$n\sigma_r^2 \to \frac{G(\infty)}{2r^3}, \qquad n\sigma_t^2 \to -\frac{G(\infty)}{4r^3}. \qquad (3.78)$$

Unless $G(\infty) = 0$, then either σ_r^2 or σ_t^2 must become negative at large radii, which is unphysical. Thus we require

$$G(\infty) = \int_0^\infty g(y)dy = 0 \tag{3.79}$$

or, using equation (3.72),

$$3 \int_0^\infty \Sigma(R)\, \sigma_p^2(R)\, R\, dR = 2 \int_0^\infty n(r) \frac{d\Phi}{dr} r^3 dr. \tag{3.80}$$

This is an expression of the **virial theorem** [201]: the left-hand side is the mean square stellar velocity, and the right-hand side is $\langle x \cdot \nabla\Phi \rangle$, or twice the virial of Clausius. Equation (3.80) sets the normalization of the potential.

According to equations (3.76) and (3.77), *any* assumed functional form for $\Phi(r)$ that respects the virial condition (3.80) yields a mathematically well-defined solution for $\sigma_r(r)$ and $\sigma_t(r)$, that is, a solution that satisfies the Jeans equation (3.38) and that generates the observed number density and velocity dispersion profiles in projection. The degeneracy of the potential estimation problem in the spherical geometry is a consequence of this result.

In fact the situation is not quite as bad as this. An additional constraint on Φ exists, due to the requirement that the derived σ_r^2 and σ_t^2 be everywhere non-negative. Only some choices for $\Phi(r)$ will satisfy the nonnegativity constraint. This means that, if one characterizes the potential via some set of parameters, one expects to find a finite (rather than unbounded) range of parameter values that are equally consistent with the data.

A common parametrization is

$$\Phi(r) = -\frac{GM_\bullet}{r} + \left(\frac{M}{L}\right) \Phi_L(r), \tag{3.81}$$

where M/L is the **mass-to-light ratio** of the stars, assumed here to be independent of radius, and Φ_L is the "potential" corresponding to the stellar luminosity density $j(r)$:

$$\Phi_L(r) = -4\pi G \left(\frac{1}{r} \int_0^r j(r')r'^2 dr' + \int_r^\infty j(r')r' dr' \right). \tag{3.82}$$

Denoting the projected luminosity density as $\Sigma_L(R)$, the virial theorem (3.80) requires

$$\frac{M}{L} = \frac{3 \int_0^\infty dR\, R\Sigma_L(R)\sigma_p(R)^2 - 2GM_\bullet \int_0^\infty dr\, r\, j(r)}{8\pi G \int_0^\infty dr\, r\, j(r) \int_0^r dr'r'^2\, j(r')}. \tag{3.83}$$

Given an assumed $M_\bullet$, equation (3.83) allows M/L to be computed from the data. The set of allowed $M_\bullet$ values is then determined, at least in principle, by computing the functions $\{\sigma_r, \sigma_t\}$ from equations (3.76) and (3.77) and rejecting values of $(M_\bullet, M/L)$ that imply $\sigma_r^2 < 0$ and/or $\sigma_t^2 < 0$.

Figure 3.6 shows the results of carrying out this program using data from the giant elliptical galaxy M87, the dominant galaxy in the Virgo Galaxy Cluster [370]. M87 is an illustrative case, since the value of $M_\bullet$ is reasonably well constrained

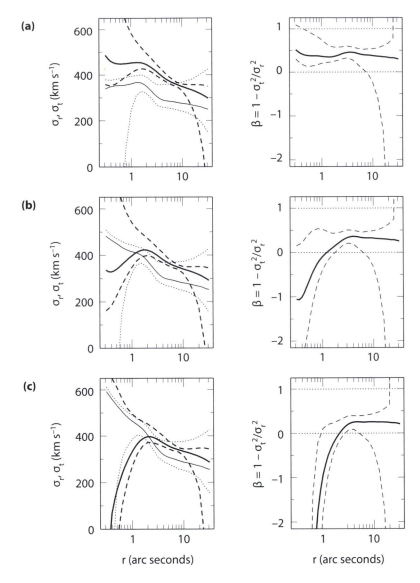

Figure 3.6 Mass-anisotropy degeneracy in stellar dynamical models of the giant elliptical galaxy M87 [370]. The variations of σ_r and σ_t with radius were derived from the observed surface density and line-of-sight velocity dispersion profiles assuming the expression (3.81) for the gravitational potential. Three different values were chosen for the mass of the SBH: (a) $M_\bullet = 1.0 \times 10^9 \, M_\odot$, $r_\mathrm{h} \approx 0''.3$; (b) $M_\bullet = 2.4 \times 10^9 \, M_\odot$, $r_\mathrm{h} \approx 0''.8$; (c) $M_\bullet = 3.8 \times 10^9 \, M_\odot$, $r_\mathrm{h} \approx 1''.3$. On the left, thick lines are σ_r and thin lines are σ_t; dashed lines show 90% confidence intervals. Each of the three models produces the same projected profiles $\Sigma(R)$, $\sigma_p(R)$. Models with $M_\bullet \gtrsim 1 \times 10^9 \, M_\odot$ are characterized by tangential anisotropy, $\sigma_t > \sigma_r$, at $r \lesssim r_\mathrm{h}$.

based on the kinematics of a parsec-scale gas disk: $M_\bullet = 3.4 \pm 1.0 \times 10^9\, M_\odot$ [332]. Furthermore, at a distance of only $\sim 16\,$Mpc, M87 is close enough that the SBH sphere of influence is well resolved: adopting $\sigma = 330\,$km s^{-1} for the stellar velocity dispersion gives $r_{\rm h} \equiv G M_\bullet / \sigma^2 \approx 120\,$pc, corresponding to an angular size of $\sim 1''\!.5$, substantially greater than the $\sim 0''\!.5$ resolution of the data on which the modeling is based [538]. Nevertheless, figure 3.6 shows that values for $M_\bullet$ in the range $1.0 \times 10^9 \leq M_\bullet/M_\odot \leq 3.9 \times 10^9$ are equally consistent with the stellar kinematical data, and indeed setting $M_\bullet = 0$ also works [46].

As this example suggests, some of the degeneracy in stellar dynamical estimates of $M_\bullet$ arises from the fact that a central rise in velocities can be due either to the gravitational force from a central point mass, or to the force from a distributed mass, as long as orbits that pass near the center also feel the force from the distributed mass. Unless the spatial resolution of the velocity data substantially exceeds $r_{\rm h}$, the Keplerian rise in velocities that is characteristic of a central point mass, equation (3.63b), will not be observed, and it will not be clear whether the stellar motions measured within the central resolution elements are attributable to the presence of an SBH, to the stellar mass, or to a combination of the two.

The degree of degeneracy to be expected in estimates of $M_\bullet$ will depend on the size $r_{\rm res}$ of the resolved region compared with the radius $r_{\rm h}$ of the SBH sphere of influence. Writing $\theta_{\rm res}$ for the smallest angular separation that can be resolved, we can define a "figure of merit" $\mathcal{R}$ such that

$$\mathcal{R} \equiv \frac{r_{\rm h}}{r_{\rm res}} = \frac{G M_\bullet}{\sigma^2}\, \frac{1}{\theta_{\rm res}\, D}$$

$$\approx 2.2 \left(\frac{M_\bullet}{10^8\, M_\odot} \right) \left(\frac{\sigma}{200\,{\rm km\,s^{-1}}} \right)^{-2} \left(\frac{D}{10\,{\rm Mpc}} \right)^{-1} \left(\frac{\theta_{\rm res}}{0''\!.1} \right)^{-1} \qquad (3.84)$$

where, D is the distance to the galaxy. One expects large values of $\mathcal{R}$ to be associated with a small degree of degeneracy in $M_\bullet$. In the case of the M87 stellar data plotted in figure 3.6, $\theta_{\rm res} \approx 0''\!.5$ and $\mathcal{R} \approx 3$; since the SBH mass is essentially unconstrained by these data, one concludes that even values of $\mathcal{R}$ greater than one may not be sufficient to remove the degeneracy. Recall that in chapter 2, a resolution of $\sim 0.1 r_{\rm h}$ was suggested as the critical value for an SBH to be detected; the corresponding figure of merit is $\mathcal{R} = 10$.

In practice, other factors will affect the degree to which $M_\bullet$ can be recovered from kinematical data, including the steepness of the stellar density profile, the signal-to-noise ratio of the stellar spectra, etc. We return to this question in subsequent sections of this chapter, where the results of axisymmetric modeling including rotation are discussed. But before doing so, we discuss two sorts of additional information that can help to reduce the degeneracy in stellar dynamical estimates of the nuclear mass:

Proper motions. If the two velocity dispersion profiles $\sigma_r(r)$, $\sigma_t(r)$ were known for a spherical galaxy, the Jeans equation (3.35) would yield a unique solution for $\Phi(r)$. This can be achieved via measurement of **proper motion** velocities of a set of stars [316]. Proper motion is defined as the change in apparent position of a star over time due to its motion relative to the Earth. If the distance to the star is known, the two components $\{\Delta_x, \Delta_y\}$ of its proper motion can be converted into the components of its velocity parallel to the plane of the sky. Knowledge of the Earth's motion relative to the frame of the galaxy then allows two components of the star's velocity with respect to the galaxy to be determined.

Let v_R and v_T be the components of this velocity in directions parallel to, and perpendicular to, the radius vector $\mathbf{r}_p$ on the plane of the sky; $\mathbf{r}_p = 0$ is the apparent center of the galaxy (and, presumably, the location of the SBH). Ignoring rotation, we can write

$$\overline{v_R^2} = \sigma_r^2 \sin^2 \varphi + \sigma_t^2 \cos^2 \varphi, \quad \overline{v_T^2} = \sigma_t^2, \tag{3.85}$$

with $\sin \varphi = R/r$ as above. The velocity dispersion perpendicular to $\mathbf{r}_p$, weighted along the l.o.s. by the stellar density, is

$$\Sigma \sigma_T^2 = \int_{-\infty}^{\infty} n\sigma_t^2 \, dz = 2 \int_R^{\infty} \frac{n\sigma_t^2 r \, dr}{\sqrt{r^2 - R^2}}. \tag{3.86}$$

This can immediately be inverted:

$$\left(n\sigma_t^2\right)(r) = -\frac{1}{\pi} \int_r^{\infty} \frac{d\left(\Sigma\sigma_T^2\right)}{dR} \frac{dR}{\sqrt{R^2 - r^2}}. \tag{3.87}$$

The velocity dispersion parallel to $\mathbf{r}_p$ is given by

$$\Sigma \sigma_R^2 = \int_{-\infty}^{\infty} n \left[\sigma_r^2 \sin^2 \varphi + \sigma_t^2 \cos^2 \varphi\right] dz \tag{3.88a}$$

$$= 2 \int_R^{\infty} \left[\frac{R^2}{r^2}\sigma_r^2 + \left(1 - \frac{R^2}{r^2}\right)\sigma_t^2\right] \frac{nr \, dr}{\sqrt{r^2 - R^2}}. \tag{3.88b}$$

Rearranging,

$$2 \int_R^{\infty} \frac{n\sigma_r^2}{r\sqrt{r^2 - R^2}} dr = R^{-2}\Sigma \left[\sigma_R^2 - \sigma_T^2\right] + 2 \int_R^{\infty} \frac{n\sigma_t^2}{r\sqrt{r^2 - R^2}} dr$$

$$= R^{-2}\Sigma\sigma_R^2 - \frac{1}{R} \int_R^{\infty} \Sigma(R')\sigma_T^2(R') \frac{dR'}{R'^2}$$

$$\equiv A(R), \tag{3.89}$$

which is a known function. The inversion is

$$\left(n\sigma_r^2\right)(r) = -\frac{r^2}{\pi} \int_r^{\infty} \frac{dA}{dR} \frac{dR}{\sqrt{R^2 - r^2}}. \tag{3.90}$$

Equations (3.87)–(3.90) demonstrate the sufficiency of proper-motion velocity dispersions for determining the two functions $\{\sigma_r(r), \sigma_t(r)\}$ in a spherical galaxy. The gravitational potential then follows from equation (3.35).

Only the nucleus of the Milky Way is close enough that stellar proper motions within the SBH influence sphere can be accurately measured. Figure 3.7 shows the results of a study based on proper motion velocities of roughly 6000 stars within the central parsec of Sgr A* [481]. Attempting to directly "deproject" the data, via equations like (3.87)–(3.90), would be ill advised. Instead, smooth, nonparametric representations of $\sigma_r(r)$ and $\sigma_t(r)$ were constructed and varied until the deviations

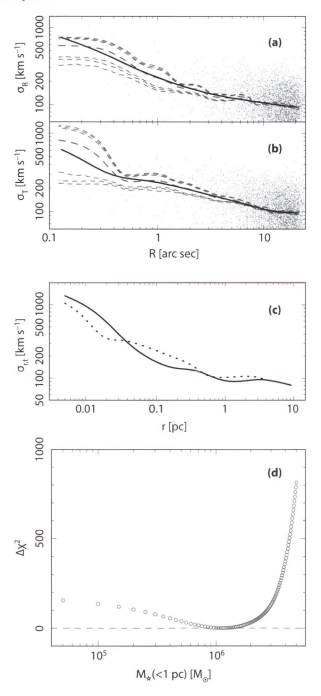

Figure 3.7 Constraining the distribution of mass at the center of the Milky Way using proper
          motions [481]. The points in panels (a) and (b) are the observed proper motion
          velocities, measured (a) parallel, and (b) perpendicular, to the radius vector. The

Figure 3.7 Continued. dashed lines show smoothed velocity dispersion profiles derived from these data and associated (90%, 95%, 98%) confidence intervals (dashed lines). The solid lines are the best-fit models, computed as described in the text. Panel (c) shows the intrinsic velocity dispersions corresponding to the solid curves in (a) and (b): radial (solid) and tangential (dashed) lines. Panel (d) shows the results of fitting a model of the stellar potential to the proper motion data, assuming an SBH mass of $4.0 \times 10^6 \, M_\odot$.

of the projected velocity dispersions from the measured values were minimized, under various assumptions about the form of the gravitational potential, and with the requirement that $\{\sigma_r, \sigma_t, \Phi\}$ satisfy the Jeans equation (3.35). Figure 3.7 shows that the velocity distribution is mildly anisotropic, in the sense $\sigma_t > \sigma_r$, at $0.05 \, \text{pc} \lesssim r \lesssim 0.5 \, \text{pc}$. The best-fit mass for the SBH from these data is $M_\bullet = 3.6^{+0.2}_{-0.4} \times 10^6 \, M_\odot$, consistent with the current best estimate from the orbit of the star S2 [192]. The distributed mass density implied by these data is consistent with $\rho \propto n$ (i.e., a constant mass-to-light ratio), with a total distributed mass inside one parsec of $1.1 \pm 0.5 \times 10^6 \, M_\odot$. A dependence of density on radius as steep as $\rho \sim r^{-2}$ in the central parsec can be securely excluded.

Line-of-sight velocity distributions. If the stellar velocities in a galaxy followed the same Maxwellian distribution that characterizes molecules in a steady-state gas, measurement of the velocity dispersion ("temperature") at any point would be equivalent to measurement of the full $f(v)$, and the isotropic Jeans equation (3.57) would contain a complete description of the dynamics. But in a collisionless nucleus, the velocity distribution can in principle be very different from Maxwellian.

These arguments suggest that there might be useful information about the dynamical state of galactic nuclei in the full **line-of-sight velocity distribution**, or LOSVD. The LOSVD (sometimes also called the **line profile**) is accessible to observation even in distant galaxies, via measurement of Doppler-broadened stellar spectra, although in practice its recovery requires high quality data.

We define the local LOSVD $\mathcal{L}(v_z, r)$ such that $\mathcal{L}(v_z, r)dv_z$ is the number of stars at radius r with l.o.s. velocities in the range v_z to $v_z + dv_z$. Adopting the convention that $\mathcal{L}$ is normalized to unit area at each r, we have, for a spherical galaxy,

$$\mathcal{L}(v_z, r) = \frac{1}{n(r)} \int \int_{f \neq 0} f(E, L^2) dv_x \, dv_y, \tag{3.91}$$

where v_x, v_y are velocity components in the plane of the sky. In most cases, we observe the LOSVD integrated along the line of sight through the galaxy, weighted by the local density:

$$\mathcal{L}_p(v_z, R) = \int_R^{r_{\max}(v_z)} \frac{n(r)r \, dr}{\sqrt{r^2 - R^2}} \mathcal{L}(v_z, r) \bigg/ \int_R^{r_{\max}(v_z)} \frac{n(r)r \, dr}{\sqrt{r^2 - R^2}}, \tag{3.92}$$

where $\Phi(r_{\max}) = -v_z^2/2$ if all stars are bound.

Consider the case of an isotropic distribution of stars around an SBH. We can simplify the expression for the local LOSVD, (3.91), by changing the integration

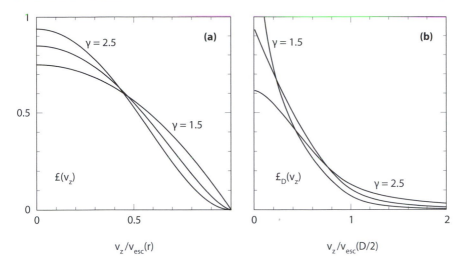

Figure 3.8 Line-of-sight velocity distributions (LOSVDs) for stars following a power-law density profile, $n \propto r^{-\gamma}$, $\gamma = (1.5, 2, 2.5)$, and with an isotropic velocity distribution, in the gravitational potential of an SBH. (a) Local LOSVDs; (b) LOSVDs integrated along the line of sight and averaged over an aperture of diameter D centered on the SBH.

variable to $E = (v_s^2 + v_z^2)/2 + \Phi(r)$ with $v_s^2 = v_x^2 + v_y^2$; then $dv_x dv_y = 2\pi v_s dv_s = 2\pi \, dE$ and

$$\mathcal{L}(v_z, r) = \frac{2\pi}{n(r)} \int_{\Phi(r) + v_z^2/2}^{0} f(E) dE, \quad v_z \leq \sqrt{-2\Phi(r)}. \tag{3.93}$$

Setting $\Phi = -GM_\bullet/r$, and adopting equation (3.49) for f, we find the local LOSVD corresponding to a power-law density profile ($n \propto r^{-\gamma}$) of stars around an SBH:

$$\mathcal{L}(v_z, r) = \frac{1}{\sqrt{2\pi}} \frac{1}{\gamma - 1/2} \frac{\Gamma(\gamma + 1)}{\Gamma(\gamma - 1/2)} \left(\frac{GM_\bullet}{r} \right)^{-1/2} \left(1 - \frac{r v_z^2}{2GM_\bullet} \right)^{\gamma - 1/2},$$
$$\tag{3.94}$$

$$v_z \leq \sqrt{\frac{2GM_\bullet}{r}}.$$

This function is plotted in figure 3.8a for several values of γ. The non-Gaussian nature is apparent.

Suppose we observe the nucleus of a galaxy containing an SBH through an aperture of diameter D centered on the SBH. The (normalized) LOSVD of all the stars in the aperture is

$$\mathcal{L}_D(v_z) = \int_0^{D/2} dR\, R\, \Sigma(R)\, \mathcal{L}_p(v_z, R) \bigg/ \int_0^{D/2} dR\, R\, \Sigma(R)$$

$$\tag{3.95}$$

$$= \int_0^{r_{\max}(v_z)} dr\, r^2 n(r)\, \mathcal{L}(v_z, r)\, H(D/2r) \bigg/ \int_0^\infty dr\, r^2 n(r)\, H(D/2r),$$

where $H(x) = 1 - \sqrt{\max(0, 1 - x^2)}$; the second expression follows after substitution of equation (3.92). Because the aperture contains stars moving very near to the SBH, the aperture-averaged LOSVD extends to arbitrarily high velocities. Adopting equation (3.94) for $\mathcal{L}$, it is easy to show that as $v_z \to \infty$, $\mathcal{L}_D \to v_z^{-7+2\gamma}$, independent of D [537].

Figure 3.8b plots $\mathcal{L}_D(v_z)$ as a function of v_z normalized to the escape velocity at radius $r = D/2$. Because of the non-Gaussian shape of these LOSVDs, with narrow central peaks and broad wings, measurement of the central velocity dispersion of a galaxy containing an SBH can be substantially in error (too small) if the spectral modeling algorithm assumes a Gaussian form for $\mathcal{L}_p(v_z)$ [537]. On the other hand, detection of the extended wings would constitute strong evidence for a deep potential well.

So far, we have said little about how LOSVDs are recovered from observations. The optical spectrum at any point in the image of a galaxy is made up of the sum, along the l.o.s., of the spectra of its component stars, each Doppler shifted according to its l.o.s. velocity. If the galaxy spectrum is $I(\lambda)$ and the spectrum of a single star—after convolution with the instrumental response—is $T(\lambda)$ (the "template" spectrum), then

$$I(\ln \lambda, R) = \int_{-\infty}^\infty \mathcal{L}_p(v_z, R) T(\ln \lambda - v_z/c)\, dv_z = \mathcal{L} \circ T, \tag{3.96}$$

where $\circ$ denotes convolution. Given I and T, this is an integral equation whose solution is $\mathcal{L}_p$; in practice, most of the information about $\mathcal{L}_p$ is contained in broadened stellar absorption features, for example, the Calcium triplet at a rest wavelength $\lambda_0 \approx 8600$ Å. Many schemes have been developed for solving equation (3.96) [38, 459, 350]; all of these schemes must deal with the fact that deconvolution is an "ill-conditioned" operation, in the sense of amplifying noise and errors in the data. The method of solution must therefore incorporate some sort of regularization. Systematic uncertainties include the choice of stellar template spectrum, and determining the level of the spectral continuum; the latter is especially important when attempting to recover the extended wings of the LOSVD.

Because of the difficulty of recovering the full LOSVD, it is common to characterize $\mathcal{L}_p$ in terms of the best-fitting Gaussian function, plus a small number of terms that describe the most important deviations from Gaussian form. A natural choice [185, 543] is a **Gauss–Hermite (GH) series**:

$$\mathcal{L}_p(v_z) \approx \frac{\mathcal{L}_0}{\sqrt{2\pi \sigma_0^2}} \sum_{j=0}^N h_j H_j(w) e^{-w^2}, \qquad w \equiv (v_z - v_0)/\sigma_0, \tag{3.97}$$

where the H_j are Hermite polynomials and h_j are the coefficients of the GH expansion. The lowest-order Hermite polynomials are

$$H_0(y) = 1, \quad H_1(y) = 2y, \quad H_2(y) = 4y^2 - 2,$$
$$H_3(y) = 8y^3 - 12y, \quad H_4(y) = 16y^4 - 48y^2 + 12.$$

The parameters $\{\mathcal{L}_0, v_0, \sigma_0\}$ are essentially arbitrary, but it makes sense to choose them to be close to the amplitude, mean velocity, and velocity dispersion of the "best-fitting" Gaussian approximation to $\mathcal{L}_p$. One way is to choose these parameters such that $(h_0, h_1, h_2) = (1, 0, 0)$. Then the first nontrivial expansion coefficients are h_3 and h_4, which are determined, respectively, by the asymmetric and symmetric deviations of $\mathcal{L}_p$ from a Gaussian. The actual mean velocity and velocity dispersion are

$$\bar{v}_z = v_0 + \sqrt{3}\sigma_0 h_3 + \cdots, \quad \sigma_p = \sigma_0(1 + \sqrt{6}h_4 + \cdots). \tag{3.98}$$

Beyond the SBH influence sphere, but near the center of a galaxy, the shape of the LOSVD reflects the degree of velocity anisotropy: radially anisotropic distributions $(\sigma_r > \sigma_t)$ tend to have flat-topped LOSVDs, while tangentially anisotropic distributions $(\sigma_t > \sigma_r)$ have centrally peaked LOSVDs [116, 185]. One hope in measuring LOSVDs is therefore that they contain the additional information needed to break the mass/anisotropy degeneracy.

Unfortunately, the exact constraints that $\mathcal{L}_p(v_z, R)$ place on f and Φ are not known. Only a more limited statement can be made. Rewriting equation (3.92) schematically as

$$\mathcal{L}_p(v_z, R) = \int dz \int\int dv_x dv_y f\left(E, L^2\right) \tag{3.99}$$

we see that both the (unknown) $f(E, L^2)$ and the (observed) $\mathcal{L}_p(v_z, R)$ are functions of two variables. It is tempting to conclude from a naive application of Fredholm theory (integrals as continuous limits of systems of linear equations) that equation (3.99) has a unique solution $f(E, L^2)$ given $\Phi(r)$, and in fact this turns out to be true [117].

The constraints on Φ are less clear. It is obvious that $\mathcal{L}_p$ contains at least *some* information about Φ that is not contained in the low-order velocity moments. For instance, we know that

$$\mathcal{L}_p(v_z, R) = 0 \quad \text{for} \quad |v_z| \geq \sqrt{-2\Phi(R)} \tag{3.100}$$

since stars with higher velocities would escape. In addition, only certain forms for $\Phi(r)$ imply, via equation (3.99), nonnegative f's; this is a more general example of the constraint on Φ, equation (3.80), that follows after imposing nonnegativity on σ_r^2 and σ_t^2. Numerical experiments [117, 373] have demonstrated that adding progressively more information about $\mathcal{L}_p$ leads to increasingly stringent restrictions on the allowed form of $\Phi(r)$. But this is still very much an open problem, and one that deserves a more complete understanding.

3.3 THE ADIABATIC GROWTH MODEL

In the remaining sections of this chapter, the discussion of collisionless nuclei will be extended to include axisymmetric and triaxial models. But before doing so, we take a detour and approach the structure of collisionless nuclei from a different direction.

Given the inherent difficulties associated with inferring the structure and dynamical state of a nucleus from observations, it is tempting to try to derive the distribution of stars around an SBH from first principles [417]. For instance, if the stellar velocity distribution is assumed to be Maxwellian, $f(v) \propto e^{-v^2/2\sigma^2}$, with constant σ, then Jeans's theorem implies $f(E) \propto e^{-E} \propto e^{-(v^2/2+\Phi(r))/\sigma^2}$ and for the stellar density near the SBH, equation (3.46) gives

$$n(r) \propto \int_0^{\sqrt{2GM_\bullet/r}} e^{-(v^2/2+\Phi(r))/\sigma^2} v^2 dv \propto e^{GM_\bullet/\sigma^2 r}. \qquad (3.101)$$

This expression implies a divergent number of stars within every radius. It has other unphysical features as well: for instance, the fact that the velocity dispersion σ is constant implies that typical kinetic energies near the SBH are much smaller than binding energies, hence most stars must be near their apocenters.

A slightly more sophisticated approach consists of starting from a nucleus with no SBH, then increasing the value of $M_\bullet$ from zero and asking what happens to f and n in the process. As the SBH grows (e.g., by accumulation of gas), its gradually increasing gravity will pull in nearby stars, causing the stellar density to grow. The change in the stellar density can be computed straightforwardly if it is assumed that the timescale for growth of the SBH is long compared with stellar orbital periods [416, 571]. This is reasonable, since even Eddington-limited accretion requires $\sim 10^8$ yr to double the SBH mass, and orbital periods throughout the region dominated by the black hole are $\lesssim 10^6$ yr. (It may be less reasonable to assume that spherical symmetry is maintained during this process, since most models for growth of SBHs invoke substantial departures from spherical symmetry.) Under these assumptions, the adiabatic invariants J_i associated with the stellar orbits are conserved and the phase-space density f remains fixed when expressed in terms of the J_i [201]. In this **adiabatic growth** model, computing the final f becomes a straightforward matter of expressing the final orbital integrals in terms of their initial values under the constraint that the adiabatic invariants remain fixed.

In spherical potentials, the adiabatic invariants are the angular momentum L and the radial action $I = 2 \int_{r_-}^{r_+} dr \sqrt{2[E - \Phi(r)] - L^2/r^2}$, where $r_\pm$ are the pericenter and apocenter radii [201]. It can be shown [329] that orbital shapes remain nearly unchanged when L and I are conserved, implying that an initially isotropic velocity distribution $f(E)$ remains nearly isotropic after the black hole grows (though not exactly isotropic—see below). The final f corresponding to an initially isotropic f is then simply

$$f_f(E_f, L) = f_i(E_i, L)$$
$$\approx f_f(E_f), \qquad (3.102)$$

where E_f is related to E_i through the condition $I_f(E_f, L) = I_i(E_i, L)$.

3.4 AXISYMMETRIC NUCLEI

In an axisymmetric nucleus, the gravitational potential and the density can be expressed in terms of ϖ and z, where (ϖ, z, φ) are cylindrical coordinates and the z-axis is the axis of symmetry. Defining the **effective potential**,

$$\Phi_{\text{eff}} \equiv \Phi(\varpi, z) + \frac{L_z^2}{2\varpi^2}, \tag{3.109}$$

where $L_z = \varpi^2 \dot{\varphi} = \text{constant}$, the equations of motion are

$$\ddot{\varpi} = -\frac{\partial \Phi}{\partial \varpi} - \frac{L_z^2}{\varpi^3} = -\frac{\partial \Phi_{\text{eff}}}{\partial \varpi}, \qquad \ddot{z} = -\frac{\partial \Phi}{\partial z} = -\frac{\partial \Phi_{\text{eff}}}{\partial z}, \tag{3.110}$$

and $\dot{\varphi} = L_z/\varpi^2$. These equations describe the two-dimensional motion of a star in the (ϖ, z), or **meridional**, plane which rotates nonuniformly about the symmetry axis. Motion in axisymmetric potentials is therefore a problem with two dynamical degrees of freedom.

Every trajectory in the meridional plane is constrained by energy conservation to lie within the zero-velocity curve, the set of points satisfying $E = \Phi_{\text{eff}}(\varpi, z)$. While the equations of motion (3.110) cannot be solved in closed form for arbitrary $\Phi(\varpi, z)$, numerical integrations demonstrate that most orbits do not densely fill the zero-velocity curve but instead remain confined to narrower, typically wedge-shaped regions [404]; in three dimensions, the orbits are tubes around the short axis.[9] The restriction of the motion to a subset of the region defined by conservation of E and L_z is indicative of the existence of an additional conserved quantity, or third integral I_3, for the majority of orbits. Varying I_3 at fixed E and L_z is roughly equivalent to varying the height above and below the equatorial plane of the orbit's intersection with the zero velocity curve (figure 3.10). In an oblate potential, extreme values of I_3 correspond either to orbits in the equatorial plane, or to "thin tubes," orbits which have zero radial action and which reduce to precessing circles in the limit of a nearly spherical potential. In prolate potentials, two families of thin tube orbits may exist: "outer" thin tubes, similar to the thin tubes in oblate potentials, and "inner" thin tubes, orbits similar to helixes that wind around the long axis [305].

The area enclosed by the zero-velocity curve tends to zero as L_z approaches $L_c(E)$, the angular momentum of a circular orbit in the equatorial plane. In this limit, the orbits may be viewed as perturbations of the planar circular orbit, and an additional isolating integral can generally be found [546]. As L_z is reduced at fixed E, the amplitudes of allowed motions in ϖ and z increases and resonances between the two degrees of freedom begin to appear. Complete integrability is unlikely in the presence of resonances, and in fact one can often find small regions of stochasticity at sufficiently low L_z in axisymmetric potentials, particularly if the force rises steeply toward the center. However, the fraction of phase space associated with chaotic motion typically remains small unless L_z is close to zero [148].

---

[9]Because of their boxlike shapes in the meridional plane, such orbits were originally called "boxes" even though their three-dimensional shapes are more similar to doughnuts.

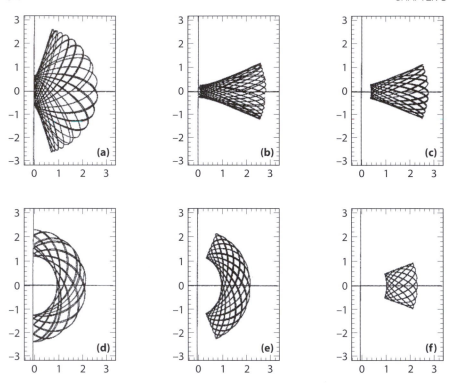

Figure 3.10 Orbits in the meridional plane of an oblate-spheroidal galaxy model [113]. All orbits have the same energy; they differ in terms of the values of the additional integrals L_z and I_3. Orbits b, c, f form a sequence of near-equatorial orbits with increasing L_z, and orbits d, e, f form a sequence of decreasing I_3.

Ignoring stochasticity, and assuming the global existence of a third integral I_3, the distribution function of a steady-state axisymmetric galaxy has the form

$$f = f(E, L_z, I_3) = f\left[v^2/2 + \Phi(r), \varpi v_\varphi, I_3\right],\qquad(3.111)$$

where the dependence of I_3 on the phase-space coordinates is left unspecified. The Jeans equations of second order in velocity are

$$n\frac{\partial\Phi}{\partial z} = -\frac{\partial(n\sigma_z^2)}{\partial z} - \frac{\partial(n\overline{v_\varpi v_z}\varpi)}{\varpi\,\partial\varpi},\qquad(3.112\text{a})$$

$$n\frac{\partial\Phi}{\partial\varpi} = -\frac{\partial(n\sigma_\varpi^2)}{\partial\varpi} - \frac{\partial(n\overline{v_\varpi v_z})}{\partial z} - \frac{n}{\varpi}\left(\sigma_\varpi^2 - \overline{v}_\varphi^2 - \sigma_\varphi^2\right),\qquad(3.112\text{b})$$

where

$$\sigma_\varpi^2 = \overline{v_\varpi^2},\quad \sigma_z^2 = \overline{v_z^2},\quad \sigma_\varphi^2 = \overline{\left(v_\varphi - \overline{v}_\varphi\right)^2}.\qquad(3.113)$$

Rotation is permitted about the symmetry axis, $\overline{v}_\varphi = \overline{v}_\varphi(\varpi, z)$. Nonzero $\overline{v_\varpi v_z}$ corresponds to a tilt of the velocity ellipsoid in the meridional plane.

Well inside the SBH sphere of influence, orbits in axisymmetric nuclei can be described as perturbed Keplerian ellipses. We postpone a detailed discussion of this regime until chapter 4, except to note that the major orbit family in the oblate geometry at $r < r_h$ remains the short-axis tubes. For these orbits, one finds that the angular momentum is relatively constant with time. A second family, the "saucer" orbits, are characterized by large angular momentum variations. Of course, both families must conserve L_z exactly.

The near constancy of L for many orbits in the axisymmetric geometry suggests that I_3 be represented by L, and approximately self-consistent models have been constructed in this way [327]. If the flattening of such a model is not too severe, one expects the distribution of orbital "integrals," $N(E, L)$, to be related to f in much the same way as in equation (3.44). The dependence of N on L_z then follows after noting that in a nearly spherical system, the fraction of orbits at given $\{E, L\}$ with $L_z \equiv L \cos \psi$ in the range L_z to $L_z + dL_z$ is proportional to $d \cos \psi \propto dL_z$, or

$$N(E, L, L_z)dE\,dL\,dL_z \approx 4\pi^2 f(E, L, L_z)P(E, L, L_z)dE\,dL\,dL_z. \quad (3.114)$$

An alternative approach to the unknown form of the third integral is to simply postulate that the phase-space density is constant on hypersurfaces of constant E and L_z, the two classical integrals of motion:

$$f = f(E, L_z) = f\left[v^2/2 + \Phi(r), \varpi v_\varphi\right]. \quad (3.115)$$

Because v_ϖ and v_z appear in the same way as arguments of f in equation (3.115), these models are "isotropic" in the sense $\sigma_\varpi = \sigma_z$.

Just as $f(E)$ for a spherical galaxy is determined uniquely by $n(r)$ and $\Phi(r)$ via Eddington's formula (3.47), so $f(E, L_z)$ is determined uniquely by $n(\varpi, z)$ and $\Phi(\varpi, z)$ in the axisymmetric geometry (up to the choice of which sign to attach to L_z for each orbit). In this sense, two-integral axisymmetric models play a role similar to that of isotropic models in the spherical geometry (although this could equally be said of other, two-integral parametrizations of f). The degeneracy that appears in the spherical inverse problem when f is allowed to depend on a second integral, $f = f(E, L^2)$, appears also in the axisymmetric geometry when f is allowed to depend on a third integral.

Because of these similarities, and also because of the important role that $f(E, L_z)$ axisymmetric models played in the early days of SBH detection, we first discuss two-integral models before turning to the more general three-integral case.

3.4.1 Two-integral models

Ignoring any dependence of f on a third integral, the contribution to the configuration-space density from stars on orbits with classical integrals in the range E to $E + dE$ and L_z to $L_z + dL_z$ is

$$\delta n = f(E, L_z)d\mathbf{v} = 2\pi f(E, L_z)v_m\,dv_m\,dv_\varphi = \frac{2\pi}{\varpi} f(E, L_z)dE\,dL_z, \quad (3.116)$$

where $v_m = \sqrt{v_\varpi^2 + v_z^2}$, the velocity in the meridional plane; δn is defined to be nonzero only at points (ϖ, z) reached by an orbit with the specified E and L_z. The total density contributed by all such phase-space pieces is

$$n(\varpi, z) = \frac{4\pi}{\varpi} \int_{\Phi}^{0} dE \int_{0}^{\varpi\sqrt{2(E-\Phi)}} f_+(E, L_z) dL_z, \qquad (3.117)$$

where f_+ is the part of f even in L_z, $f_+(E, L_z) = (1/2)[f(E, L_z) + f(E, -L_z)]$; the odd part of f affects only the degree of streaming around the symmetry axis. Equation (3.117) is a linear relation between known functions of two variables, $n(\varpi, z)$ and $\Phi(\varpi, z)$, and an unknown function of two variables, $f_+(E, L_z)$; hence one might expect the solution for f_+ to be unique and in fact it is [115, 250, 328], although in practice the inversion can be difficult.

Just as in the spherical case, the "isotropy" of two-integral axisymmetric models allows one to infer a great deal about their internal kinematics without even deriving $f(E, L_z)$. Writing $\sigma = \sigma_\varpi = \sigma_z$, the Jeans equations (3.112) become

$$n\frac{\partial \Phi}{\partial z} = -\frac{\partial(n\sigma^2)}{\partial z}, \qquad (3.118a)$$

$$n\frac{\partial \Phi}{\partial \varpi} = -\frac{\partial(n\sigma^2)}{\partial \varpi} - \frac{n}{\varpi}\left(\sigma^2 - \overline{v_\varphi^2}\right). \qquad (3.118b)$$

Interpreted as differential equations for the velocity moments, these equations have solutions

$$n\sigma^2 = \int_{z}^{\infty} n\frac{\partial \Phi}{\partial z} dz, \qquad (3.119a)$$

$$n\overline{v_\varphi^2} = n\sigma^2 + \varpi \int_{z}^{\infty}\left(\frac{\partial n}{\partial \varpi}\frac{\partial \Phi}{\partial z} - \frac{\partial n}{\partial z}\frac{\partial \Phi}{\partial \varpi}\right) dz, \qquad (3.119b)$$

similar to equation (3.58) in the spherical isotropic case. The uniqueness of the solutions is a consequence of the uniqueness of the even part of f; the only remaining freedom relates to the odd part of f, that is, the division of $\overline{v_\varphi^2}$ into mean motions and dispersion about the mean, $\overline{v_\varphi^2} = \overline{v_\varphi}^2 + \sigma_\varphi^2$. A model with streaming motions adjusted such that $\sigma_\varphi = \sigma_\varpi = \sigma_z$ everywhere is called an "isotropic oblate rotator" and the model's flattening may be interpreted as being due completely to its rotation.

Before continuing, it is important to point out one crucial difference between the spherical and axisymmetric inverse problems. In the spherical geometry, the density profile $n(r)$ is uniquely determined by the projected density $\Sigma(R)$ via equation (3.65a). In the axisymmetric case, $n(\varpi, z)$ is uniquely constrained by the observed surface density $\Sigma(X, Y)$ only if the galaxy is seen edge-on, or if some other restrictive condition applies, for example, if the isodensity contours are assumed to be coaxial ellipsoids with known axis ratios [468]. In general, the range of space densities consistent with a given surface density increases as the inclination varies from edge-on to face-on [187]. Uncertainties in deprojected n's inevitably contribute to uncertainties in computed values of the kinematical quantities [461], thus increasing the degree of degeneracy associated with estimates of the gravitational potential. A fairly common practice is to assume that an observed galaxy is seen edge-on; in point of fact, of course, the observed ellipticity of a galaxy is a lower limit on its true flattening.

In the spherical isotropic case, $f(E)$ and $\Phi(r)$ follow uniquely from $\Sigma(R)$ and $\sigma_p(R)$, independent of assumptions about the relative distribution of mass and light (equations 3.47, 3.65, 3.66). A similar result holds in the two-integral axisymmetric case, assuming that the galaxy is viewed edge-on [348]: complete knowledge of the surface density $\Sigma(x, y)$ of a set of stars, together with their l.o.s. mean velocity and velocity dispersion, $\overline{v}_z(x, y)$ and $\sigma_p^2(x, y)$, is equivalent to knowledge of $f(E, L_z)$ (both odd and even parts) and $\Phi(\varpi, z)$. Just as in the spherical case, this result highlights the difficulty of ruling out "isotropic" (i.e., two-integral) f's for axisymmetric galaxies based on observed moments of the velocity distribution, because the potential can be adjusted in such a way as to reproduce the data without forcing f to depend on a third integral.

A more common approach in modeling axisymmetric galaxies is to parametrize the potential as

$$\Phi(\varpi, z) = -\frac{GM_\bullet}{r} + \left(\frac{M}{L}\right)\Phi_L(\varpi, z), \qquad (3.120)$$

similar to equation (3.81) in the spherical geometry; Φ_L is related to the luminosity density $j(\varpi, z)$ via Poisson's equation; that is, the mass-to-light ratio is assumed to be independent of position. Given any choice for the two parameters $(M/L, M_\bullet)$, the potential and the mass density are known, and the even part of $f(E, L_z)$ is uniquely determined, as discussed above. The velocity moments can then be computed and compared with the data [49]. Note the important point that the kinematical data are not used at all in the *calculation* of f_+, except insofar as they determine the normalization of the potential. Models constructed in this way have been found to reproduce the observations quite well in a few galaxies, notably M32 [112, 441], and the same approach was widely used prior to about 1998 for estimating SBH masses [335, 542].

However, there is a potentially serious pitfall associated with inferring SBH masses from two-integral models [539]. If one imagines making such a model flatter, the velocity dispersion in the meridional plane drops compared with $\overline{v_\varphi^2}$; this is a consequence of their "isotropy," $\sigma_\varpi = \sigma_z$, which links the two meridional plane velocity dispersions to the vertical thickness of the galaxy (equation 3.119a). As a result, the stellar orbits become increasingly circular as the flattening is increased; in the limit of infinite flattening (i.e., a disk), two-integral models contain only circular orbits. One consequence is that the rms velocities near the center are forced to be lower than they would be in a genuinely isotropic galaxy (figure 3.11). Unless the velocity distribution in the observed galaxy is also biased toward circular motions near the center, the value of $M_\bullet$ will need to be artificially increased in order to make up the deficit in the central velocities. A similar effect can be seen in the spherical models of M87 shown in figure 3.6.

In retrospect, it appears likely that most or all of the putative SBH detections based on two-integral models were spurious [539]. This conclusion is based in part on the fact that currently accepted scaling relations like the $M_\bullet$–σ relation imply influence radii r_h that would not have been resolvable given the data that were available at the time [364]. The principal result of these early studies—that SBHs

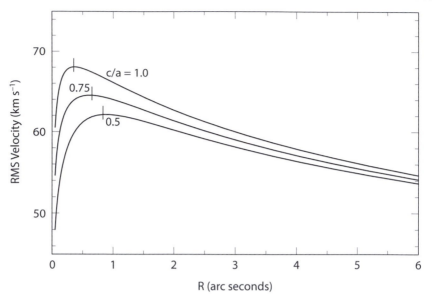

Figure 3.11 Velocity dispersion profiles of two-integral, $f(E, L_z)$ models similar to those that were used as templates for estimating SBH masses prior to about 1998. Model flattening is indicated as c/a; the density increases as $r^{-1.5}$ near the center and there is no central SBH. Ticks mark the point of maximum velocity; this radius moves outward as the flattening is increased. The velocity distribution in two-integral models is forced to become more nearly circular as the flattening is increased.

contain on average ∼0.6% of the mass of their host bulges [454] (the "Magorrian relation")—is now believed to be wrong by a factor of about five.

The shortcomings of the two-integral models would have been immediately apparent if the data being modeled were derived from observations that clearly resolved the SBH sphere of influence. In none of these galaxies (with the possible exception of the Local Group galaxy M31) did the stellar data show anything approximating a Keplerian rise near the center. The observational situation since that time has improved, due primarily to installation of the Space Telescope Imaging Spectrograph (STIS) on the Hubble Space Telescope in 1997 [568]. However, even with the ∼0″.1 resolution of STIS, there are still no galaxies outside of the Local Group that exhibit prima facie evidence for the presence of an SBH in the stellar velocities (figure 2.5).

3.4.2 Three-integral models

In the general axisymmetric case, f is a function of three variables, $f = f(E, L_z, I_3)$ (assuming that all orbits are characterized by three isolating integrals). As discussed above, the third integral can be understood as determining the thickness of the orbit in a direction perpendicular to the equatorial plane of the model (figure 3.10).

A useful way to characterize such orbits is in terms of **shape invariants**, functions of the orbital integrals that relate more directly to their time-averaged shapes [113]. For instance, a radial shape invariant is

$$S_r = (\varpi_+ - \varpi_-)/\varpi_{\max}, \tag{3.121}$$

where (ϖ_+, ϖ_-) are the apo- and pericenter distances of the orbit when it crosses the equatorial plane and $\varpi_{\max}(E)$ is defined by $\Phi(\varpi_{\max}, z = 0) = E$. A second, meridional shape invariant is

$$S_m = 1 - \sin \theta_0, \tag{3.122}$$

where θ_0 is the minimum angle, measured from the symmetry axis, reached by the orbit at apocenter; thus $S_m = 0$ for orbits in the equatorial plane and and $S_m = 1$ for orbits with $L_z = 0$. The radial shape invariant S_r is an approximate measure of the radial extent of an orbit, while the meridional shape invariant S_m measures the extent of the orbit above and below the equatorial plane.

Just as there are many anisotropic distribution functions $f(E, L^2)$ that reproduce a given $n(r)$ in the spherical geometry, so are there many three-integral f's that reproduce a given $n(\varpi, z)$ in an axisymmetric potential. The different solutions correspond to differing degrees of velocity anisotropy, that is, to different σ_ϖ/σ_z and $\sigma_\varpi/\sigma_\varphi$. For instance, distribution functions of the form $f = f(E, S_m)$ assign equal phase-space densities to orbits of all radial extents S_r, leading to roughly equal dispersions in the ϖ and φ directions. As noted above, classical two-integral models, $f = f(E, L_z)$, accentuate the nearly circular orbits to an extent that is probably unphysical.

A common way [455, 165] to construct three-integral axisymmetric models is via numerical integration of orbits in a specified potential, for instance, the potential of equation (3.120). The time spent by each orbit in each of a set of finite cells is recorded. Defining the known mass in the ith cell as M_i, and the mass placed by the lth orbit in the ith cell as $M_{l,i}^{\text{orb}}$, the self-consistency condition (3.27b) becomes

$$M_i = \sum_l^{N_{\text{orb}}} w_l M_{l,i}^{\text{orb}}, \quad i = 1, \ldots, N_{\text{cell}}, \tag{3.123}$$

where w_l is the weight assigned to the lth orbit. The condition $f > 0$ is imposed by requiring the w_l to be nonnegative. The degeneracy in f means that many choices for the w_l will solve equation (3.123). Note however, that this will only be true computationally if $N_{\text{orb}} > N_{\text{cell}}$; otherwise, an algorithm that attempts to invert equations (3.123) will have fewer parameters to vary than the number of equations to be solved. In this case, *no* choice for the w_l is likely to satisfy equations (3.123) exactly. However, an algorithm that minimizes a quantity like

$$\chi^2 = \sum_i \left[M_i - \sum_l^{N_{\text{orb}}} w_l M_{l,i}^{\text{orb}} \right]^2 \tag{3.124}$$

will find a choice for the w_l that minimizes χ^2—a spurious "best-fit" solution.

When using three-integral models to estimate $M_\bullet$ in galactic nuclei, one adds to equation (3.123) or (3.124) an additional set of conditions corresponding to whatever kinematical data are available: l.o.s. mean velocity and velocity dispersion profiles, LOSVDs, etc. Degeneracy in f is still expected, although the degree of degeneracy will depend on the number and quality of the additional constraints. The argument that was made above in the anisotropic spherical case then applies to the three-integral axisymmetric case: changes in the assumed form of $\Phi(\varpi, z)$ can generally be compensated for by changes in f so as to leave the fit to any finite set of data constraints precisely unchanged. However, the degeneracy will only be apparent if the number of orbits used in the solution is large enough—or, stated differently, if the solution algorithm is flexible enough in its representation of f.

The degeneracy is illustrated in figure 3.12 for the galaxy M32, one of the best observed and best resolved of the SBH candidate galaxies (figure 2.5) [530]. When the number of orbits used to represent f is ~3 times the total number of data constraints, a minimum appears in the value of $\chi^2(M_\bullet, M/L)$ at $M_\bullet \approx 3.5 \times 10^6 \, M_\odot$. This is the best-fit value for $M_\bullet$ found in the first three-integral modeling study of M32, which adopted a similar number of orbits [541]. When N_{orb} is increased, this minimum is seen to be spurious: it is gradually replaced by a plateau of nearly constant χ^2. Indeed the range of degeneracy in $M_\bullet$ corresponding to these data is $1.5 \times 10^6 \, M_\odot \lesssim M_\bullet \lesssim 5.0 \times 10^6 \, M_\odot$ [530]. It is discouraging that even the availability of the lowest Gauss–Hermite moments does not remove the degeneracy.

While M32 is currently the only galaxy for which such comprehensive modeling has been carried out, it is likely that $M_\bullet$ as derived from stellar kinematics in other galaxies is comparably degenerate, since most of these galaxies are observed at lower effective resolution than M32 (figure 2.5).

Some studies remove the degeneracy by imposing additional constraints on f; for instance, "maximum entropy" [457]. The additional constraints have the effect of singling out a single solution as "most probable," even (or especially) in cases where the data themselves are unable to do this (figure 3.13).

The Milky Way provides a cautionary example. Estimates of the mass of Sgr A* based on velocities of samples of stars at radii $\lesssim r_h$ [76, 182, 190] gave systematically smaller values for $M_\bullet$ than the (presumably more accurate) mass inferred from the inner S-star orbits [191, 193]. The discrepancy was about a factor of two. The discrepancy was resolved when it was discovered [66] that the distribution of stars at $r \lesssim 0.5 \, pc \lesssim 0.2 r_h$ differs from the inward extrapolation of the density observed at $1 \, pc \lesssim r \lesssim 10 \, pc$: there is a central "hole" [481], hence a lack of stars very near the SBH. No external galaxy has stellar data of a quality remotely comparable with these data for the Galactic center, and any changes in the stellar density, population or kinematics that occur inside r_h would be difficult to detect in these galaxies. The factor of two error that was made in the mass of the Milky Way SBH is probably a lower limit on the systematic uncertainties in estimates of $M_\bullet$ in external galaxies.

3.5 TRIAXIAL NUCLEI

An important shift in our understanding of early-type galaxies took place in 1975, when it was discovered that most elliptical galaxies rotate significantly more slowly

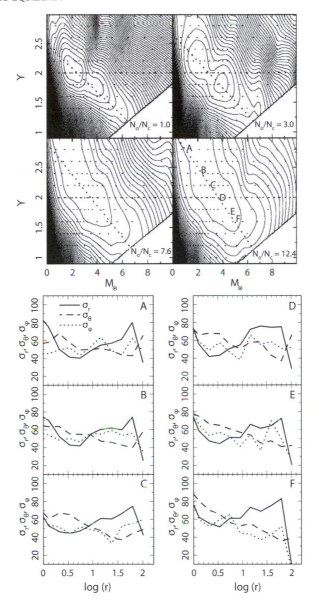

Figure 3.12 Degeneracy in three-integral models of M32 [530]. Top: Contours of constant χ^2 in fits to stellar kinematical data from ground- and space-based observations [541]. M32 is assumed to be edge-on. The four panels show the results using four different sizes of orbit library. Model parameters are the SBH mass $M_\bullet$ in $10^6\ M_\odot$ and the V-band mass-to-light ratio Υ in solar units. Labeled positions are models whose fit to the data is illustrated in detail in the right panel. As the number of orbits is increased, the χ^2 minima merge and broaden into a plateau of constant χ^2. Bottom: Intrinsic three-dimensional kinematics along the major axis for each of the models A–F. SBH masses ranging from $1.4 \times 10^6\ M_\odot$ (model A) to $4.8 \times 10^6\ M_\odot$ (model F) are shown.

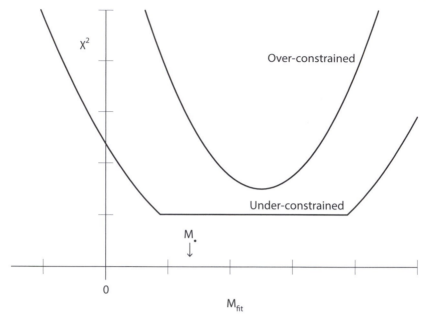

Figure 3.13 Schematic illustration of the way that restrictions on the allowed form of f can lead to spurious "best-fit" solutions when modeling kinematical data. The horizontal axis is the value M_{fit} adopted for the mass of the SBH in equation (3.120); the vertical axis measures the deviations of the model from the data, equation (3.124). The lower curve shows the correct dependence of χ^2 on M_{fit}: a plateau of constant χ^2 that reflects the degeneracy in the solutions. The upper curve, which exhibits a spurious minimum in χ^2, shows the result of modeling when f is restricted in some arbitrary way. This curve could correspond to modeling based on a two-integral f; or a three-integral f represented by too few discrete orbits or phase-space cells; or a three-integral f with the imposition of a "maximum entropy" constraint.

than expected for a fluid body with the same flattening [43]. Elliptical galaxies were revealed to be "hot" stellar systems, in which most of the support against gravitational collapse comes from essentially random motions rather than from ordered rotation. Two questions immediately arose from these observations: First, what produces the observed flattenings? Second, given that rotation plays only a minor role, are elliptical galaxies axisymmetric [45]? Numerical experiments [482] soon showed that many orbits in triaxial potentials are regular, and that self-consistent triaxial equilibria could be constructed via superposition of time-averaged orbits.

Since then, the observational evidence in favor of nonaxisymmetry on large ($\sim$ kpc) scales in early-type galaxies has gradually accumulated [69, 169, 507]. On smaller scales, imaging of the centers of galaxies also reveals a wealth of features in the stellar distribution that are not consistent with axisymmetry, including bars, bars-within-bars, and nuclear spirals [145, 489, 493]. Even if some of these features are transient, they may persist for a significant fraction of a galaxy's lifetime.

Triaxiality—or more precisely, absence of axisymmetry—is potentially of enormous consequence for the dynamics and evolution of galactic nuclei containing SBHs. Motion in nonaxisymmetric potentials does not conserve any component of the angular momentum. As a consequence, at least some orbits—the **centrophilic** orbits—are able to come arbitrarily close to the center after a finite time, even in the absence of gravitational encounters or other collisional perturbations that would otherwise be required to drive stars into the center.[10] Particularly in the largest galaxies, which have very long central relaxation times, centrophilic orbits probably dominate the feeding of stars to the SBH [371].

In a triaxial galaxy containing an SBH, the character of the motion depends strongly on the distance from the center. As a starting point, consider orbits that remain at all times close to the SBH, with radial extent $r_{max} \ll r_m$. (These orbits are discussed in much greater detail in the following chapter.) Such orbits are nearly Keplerian; the force from the distributed mass constitutes a small perturbation which causes the Keplerian elements (semimajor axis, eccentricity, etc.) to gradually change. Typically, the most rapid such change is a precession of the orbit in the plane perpendicular to its instantaneous angular momentum vector, with frequency

$$v_p \approx -v_r \frac{M_\star}{M_\bullet} \sqrt{1 - e^2} \qquad (3.125)$$

(equation 4.88). Here, $v_r = 2\pi/P(a)$ is the frequency of the radial motion, a is the semimajor axis of the (nearly) Keplerian ellipse, and $M_\star(a)$ is the distributed mass within radius $r = a$. (The minus sign indicates that the precession is retrograde, i.e., opposite in sense to the circulation of the orbit about the SBH.) This precession is a consequence of the fact that the gravitational force is not exactly inverse-square due to the distributed mass, so that the degeneracy between radial and angular motions that characterizes the Kepler problem is broken. Since we are assuming $a \ll r_m$, it follows that $M_\star \ll M_\bullet$ and hence $v_p \ll v_r$: many radial periods are required before the accumulated precession is appreciable. But after a sufficiently long time, of order $|2\pi/v_p|$, the orbit will fill an annulus about the SBH.

The rate of this in-plane precession is determined essentially by the radial force law—that is, by the spherical component of the mass distribution. For this reason it is sometimes called simply the **mass precession**. But even if the nucleus is only slightly nonspherical, there will also be a component of the force directed perpendicularly to the radius vector—a *torque*. The effect of the torque will be to change the angular momentum of the orbit: both the amplitude of L (i.e., the eccentricity) and its direction (i.e., the orbital plane).

Two basic types of orbit can result:

1. If the orbital eccentricity is modest, $e \approx 0$, the e-dependent factor in equation (3.125) will be close to unity, and mass precession will continue to define the motion. The orbit will fill an annulus, but the plane of the annulus will

---

[10]The same is true for orbits in axisymmetric potentials if the conserved component of L is zero, i.e., if the orbit lies in the plane containing the long and short axes of the figure.

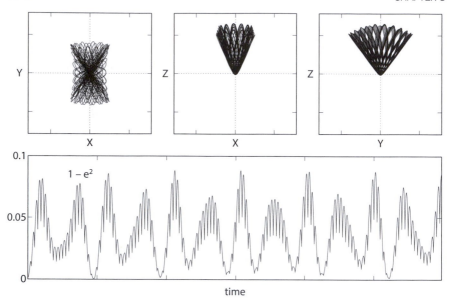

Figure 3.14 Top: A pyramid orbit, seen in three projections. The z-axis is the short axis of the triaxial figure and the SBH is at the origin. Bottom: $1 - e^2$ versus time, where e is the eccentricity. The eccentricity tends to unity when the orbit reaches the corners of the pyramid's base.

gradually change due to the torques, resulting in a orbit that looks qualitatively similar to the tube orbits discussed above. In a triaxial nucleus, it turns out that such tube orbits can circulate about either the long or the short axis of the triaxial figure [472].

2. If the orbital eccentricity is large, $e \approx 1$, equation (3.125) says that the frequency of the mass precession will be very small. As a result, the torques from the nonspherical part of the potential can "build up," causing appreciable changes in the angular momentum—both its amplitude and orientation—in the time it takes the orbit to precess once around. In fact, the eccentricity can reach unity, implying that the orbit goes right into the SBH. A more careful analysis [378] reveals that the orbit reverses its sense of circulation just before this happens; and since the direction of the mass precession is opposite to the sense of circulation, the precession also reverses direction—instead of precessing all the way around, the orbit **librates**. It turns out (as discussed in more detail in the next chapter) that this libration can occur in two orthogonal directions about the short axis of the triaxial figure; in the generic case, both types of libration occur simultaneously (though with different frequencies), and the orbit fills a pyramid-shaped region, with the apex of the pyramid near the SBH and the base perpendicular to the short axis. An example of such a **pyramid orbit** is shown in figure 3.14.

The motion described so far is essentially regular. If we now imagine increasing the size of the orbit, so that $a \gtrsim r_{\mathrm{m}}$, the distributed mass within $r \approx a$ becomes comparable to or greater than the mass of the SBH. Equation (3.125) (which is very approximate in this case) tells us that $|v_p| \approx |v_r|$: the timescales associated with the mass precession and the radial motion are now comparable. The similarity of these two frequencies means that new families of orbits will appear that respect the commensurability conditions (3.13); that is, $m_1 v_r + m_2 v_p = 0$, with m_1, m_2 small integers. For instance, closed, quasi-circular orbits appear in which the radial and angular frequencies are equal.[11] Such orbits avoid the center. In the case of eccentric orbits, chaos begins to rear its interesting head: the SBH now acts like a scattering center, rendering many of the "centrophilic" orbits stochastic [186]. This **zone of chaos** extends outward from a few times r_m to a radius where the enclosed stellar mass is roughly 10^2 times the mass of the SBH [414, 528]. Integrable tube orbits, which avoid the destabilizing center, continue to exist at these radii.

At radii $r \gg r_m$, orbits in triaxial potentials containing SBHs are broadly similar in their properties to those of orbits in triaxial potentials lacking central SBHs. In particular, the tube orbits take almost no notice of the central singularity. This is not quite true of the centrophilic orbits. In the absence of a central SBH, the centrophilic orbits are typically regular, and fill regions that resemble rectangular parallelepipeds with flaring ends. These **box orbits** are found to be a necessary, and often dominant, component of self-consistent triaxial models [482, 483]; for this reason they have been called the "backbone" of triaxial galaxies. The other family of orbits that tend to be highly populated in the self-consistent models are the tube orbits that circulate about the long axis.

In triaxial galaxies with SBHs, box orbits that pass too near the center are generally rendered stochastic [186]. Stochasticity is avoided only if the orbit lies sufficiently close to a resonant orbit; generically, such resonance orbits satisfy a commensurability condition (3.13) involving all three fundamental frequencies and are *thin*, that is, confined to a membrane in configuration space (as opposed to orbits that satisfy two such conditions and are closed). To a good approximation, box orbits in triaxial galaxies are either nonresonant and stochastic, or resonant and thin [377].

A key question is whether self-consistent triaxial equilibria exist for galaxies containing SBHs. While a definitive answer to this question is not available, numerical experiments suggest that triaxiality is supportable within $r \approx r_m$ [434], while on larger scales, the SBH may induce a gradual evolution in shape, toward axisymmetry or sphericity [372]. These uncertainties, coupled with the generally greater degree of degeneracy associated with the triaxial geometry, have kept most galaxy modelers from venturing beyond the axisymmetric paradigm.

The remainder of this chapter focuses on the character of centrophilic orbits in the triaxial geometry, at distances $\gtrsim r_m$ from the center, and on the triaxial self-consistency problem. The reader should note that even many basic questions about triaxial dynamics remain unanswered, particularly in the context of galaxies with central SBHs.

---

[11] In nuclei lacking SBHs, such resonant orbits are the generators of tube orbits, which only exist beyond a certain distance from the center [108, 172].

3.5.1 Regular and chaotic motion in nonrotating triaxial models

A handful of studies [376, 377, 433] have mapped out the major orbit families near the centers of triaxial galaxies containing SBHs, typically via brute-force numerical integration of the equations of motion. Roughly speaking, the set of centrophilic orbits can be identified with the set of orbits that have a **stationary point**, that is, a point where $\Phi(x(t)) = E$. (Tube orbits have no stationary point.) For instance, in the case of pyramid orbits, the stationary points are the (four) points that define the corners of the pyramid's base (figure 3.14). A natural way to choose initial conditions for centrophilic orbits (of a given energy, say) is to locate a set of points on an equipotential surface, $x = \Phi^{-1}(E)$, and to set $v = 0$. Corresponding to each such initial $(x, v = 0)$ will be a single, time-averaged orbit, the important properties of which—its distance of closest approach to the center, its degree of stochasticity, etc.—can be plotted on the equipotential surface as a kind of map.

Figure 3.15 shows several examples. Plotted there are properties of centrophilic orbits near the center of a triaxial galaxy with density

$$\rho(m) = \rho_0 m^{-\gamma} (1 + m)^{\gamma - 4}, \quad m^2 = \frac{x^2}{a^2} + \frac{y^2}{b^2} + \frac{z^2}{c^2}, \tag{3.126}$$

and $\gamma = 0.5$; the central SBH has a mass 0.003 times the total stellar mass, a little larger than the mean value in observed galaxies. The degree of triaxiality of models like these, having fixed axis ratios, is sometimes defined via the index T where

$$T \equiv \frac{1 - (b/a)^2}{1 - (c/a)^2}; \tag{3.127}$$

thus oblate spheroids have $T = 0$ and prolate spheroids have $T = 1$. The models used in figure 3.15 have $c/a = 0.5$, $b/a = 0.791$, which implies $T = 0.5$; in other words, they are "maximally triaxial."

For each orbit in figure 3.15, the fundamental frequencies of the motion were computed as in section 3.1.2.1. Stochastic orbits are not quasiperiodic, and an attempt to recover "fundamental frequencies" will fail; the degree to which the "fundamental frequencies" change as the integration interval is changed is a sensitive measure of stochasticity [414].

Just beyond $r_{\rm m}$, at a radius where the enclosed stellar mass is ~ 3 times $M_\bullet$ (figure 3.15d), almost all orbits with stationary points are still regular; the exceptions are orbits that intersect the equipotential surface near the x–y plane. The unstable orbit that generates this stochastic region is the long-axis orbit. Moving outward to a radius where the enclosed mass is $\sim 4 M_\bullet$ (figure 3.15c), a number of new stochastic regions appear, associated with unstable resonant pyramids of various orders, for example, $(m_1, m_2, m_2) = (2, 4, -5)$, $(6, -2, -3)$, etc. A very rapid transition to almost complete stochasticity then occurs (figure 3.15b,a). The last remaining regular pyramid orbits are associated with a 2:1 resonance. The 2:1 orbits have the shape of parabolas that are elongated parallel to the short axis of the triaxial figure; as one moves outward, to higher energies, the opening angle of the parabola increases, as the stationary point moves along the equipotential surface from the short (z-) to the long (x-) axis.

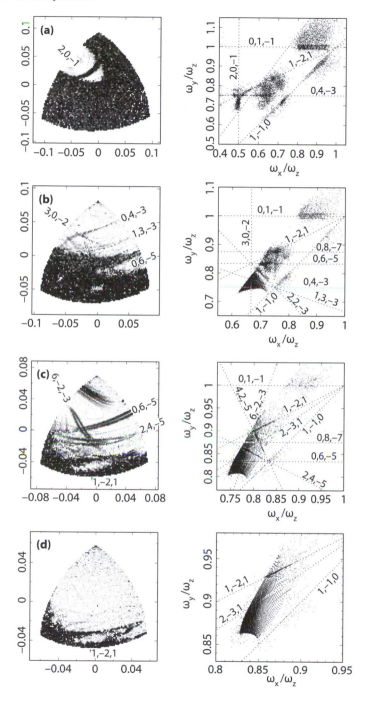

Figure 3.15 Transition to chaos in triaxial galaxies containing SBHs [377]. Panels on the
left show one octant of an equipotential surface on which orbits were started
with zero velocity; the top, left and right vertices correspond to the z- (short),

Figure 3.15 Continued. x- (long), and y- (intermediate) axes. Energy increases upward: the stellar mass enclosed by the equipotential surface is (d) $3.0M_\bullet$, (c) $3.77M_\bullet$, (b) $4.5M_\bullet$ and (a) $5.3M_\bullet$. The gray scale is proportional to the degree of stochasticity; initial conditions corresponding to regular orbits are white. The most important resonant zones are marked with their order (m_1, m_2, m_3). Panels on the right show ratios of the fundamental frequencies; stochastic orbits, which are not quasiperiodic, scatter about the line corresponding to the unstable resonance that generated them.

What is most striking about the transition to stochasticity is its suddenness. Orbits that start from equipotentials that enclose a stellar mass of $\sim 3M_\bullet$ or less are almost all regular. Moving outward just slightly, to a radius where the enclosed stellar mass is $\sim 4M_\bullet$, almost all orbits with stationary points have become stochastic. This "zone of chaos" extends outward to radii where the enclosed mass is $10M_\bullet$–$30M_\bullet$; farther out, regular orbits with stationary points, like the box orbits, begin to reappear.

The extent of the chaotic zone depends somewhat on the parameters of the triaxial potential [433]. For a given triaxiality T, chaos sets in at lower energies (i.e., closer to the SBH) in more highly elongated models, while for a given elongation c/a, the transition to chaos is almost independent of T. The transition is interrupted by the appearance of the 2:1 orbits, particularly in more elongated models. The 2:1 orbits in the most highly flattened models ($c/a \lesssim 0.6$) manage to persist, stably, throughout the chaotic zone, becoming the 2:1 "banana" orbits at high energies [391].

Interestingly, the dynamical roles of the long and short axes of the triaxial figure at low energies (near the SBH) are approximately reversed compared to their role at high energies, or compared to triaxial potentials without SBHs [433]. The pyramid orbits are generated from Keplerian ellipses oriented along the short (z-) axis, while in triaxial potentials without central SBHs, it is the long (x-) axis orbit that generates the box orbits. Similarly, stochastic orbits near the SBH derive mostly from starting points near the x–y plane, while in nonsingular potentials the instability strip lies near the y–z plane [204]. These facts would seem to have important implications for triaxial self-consistency in galaxies containing central SBHs, which however, remain to be worked out in detail.

3.5.2 Motion in rotating triaxial galaxies

In an axisymmetric galaxy, rotation (at least, rotation about the symmetry axis) has no effect on the gravitational force. In the nonaxisymmetric case, rotation of the galaxy's *figure* about any axis implies a significant modification of the equations of motion. As observed from an inertial frame, figure rotation adds sinusoidal terms to the gravitational force, with frequencies that are multiples of the rotational frequency. If the galaxy rotates as a solid body, with frequency Ω_f say, it is natural to transform to a noninertial frame in which the rotation disappears. The equations of motion will no longer contain time-dependent terms but new terms will appear, corresponding to the Coriolis and centrifugal forces:

$$\ddot{x} = -\nabla \Phi - 2(\mathbf{\Omega}_f \times \dot{x}) - \mathbf{\Omega}_f \times (\mathbf{\Omega}_f \times x) \qquad (3.128\text{a})$$

$$= -\nabla \Phi - 2(\mathbf{\Omega}_f \times \dot{x}) + |\mathbf{\Omega}_f|^2 x. \qquad (3.128\text{b})$$

In the rotating frame, the Jacobi integral

$$E_J = \frac{1}{2}|\dot{x}|^2 + \Phi(x) - \frac{1}{2}|\mathbf{\Omega}_f \times x|^2 \qquad (3.129)$$

plays the role of the conserved energy.

How important is figure rotation in real galaxies? Measuring Ω_f is hard; it is much easier to measure the mean velocities of stars integrated along the line of sight. Some of the observed mean motion, $\overline{V}(x, y)$, may be due to figure rotation, while some will be due to streaming with respect to the figure. In the absence of secure knowledge about the three-dimensional shape of a galaxy, a natural measure of the overall strength of the "rotation" is

$$\lambda \equiv \frac{\langle R\overline{V}\rangle}{\left\langle R\sqrt{\overline{V}^2 + \sigma^2}\right\rangle}, \qquad (3.130)$$

where σ is the line-of-sight velocity dispersion, R is the projected distance from the center, and the angle brackets in equation (3.130) denote averages over the image of the galaxy. Since $R V$ has the dimensions of angular momentum per unit mass, λ can be interpreted as a dimensionless angular momentum. In the case that most of the mean motion is due to rotation of the figure, it follows that $\overline{V}$ is of order $\langle R\rangle\Omega_p$; furthermore, by the virial theorem, $\langle\overline{V}^2 + \sigma^2\rangle \sim \langle r\nabla\Phi\rangle \sim \langle r^2\Omega^2\rangle$ where Ω is a typical orbital frequency. Hence $\lambda \sim \Omega/\Omega_f$, and (again assuming that figure rotation is responsible for $\overline{V}$) values of λ approaching unity imply rotation with a period that is comparable to orbital periods in the rotating frame. Of course, a galaxy whose axis of rotation lies along the line of sight will show no evidence of rotation in the Doppler-shifted velocities, and in this sense λ is a lower limit on the degree of rotation.

Figure 3.16 shows estimates of λ for a sample of early type (elliptical and S0) galaxies, plotted versus the apparent axis ratio ϵ. Early-type galaxies appear to fall into two, fairly distinct classes: "slow rotators" with $\lambda < 0.1$, and "rapid rotators" with $\lambda > 0.1$. The slow rotators are believed to be generically triaxial in shape, since they frequently display isophotal and kinematic twists, which are a natural consequence of a nonaxisymmetric system being observed from a random direction. These galaxies are typically luminous, with absolute magnitudes brighter than $M_B \approx -20$. The more numerous, rapid rotators are typically fainter and their shapes appear to be consistent with axisymmetry [522]; furthermore, the distribution of λ with respect to ellipticity for the rapid rotators, as shown in figure 3.16, is statistically consistent with what is expected for the "isotropic oblate rotators" defined above, that is, axisymmetric (oblate) galaxies in which the flattening is due to streaming motions about the symmetry axis. While the origin of this dichotomy is still debated, numerical simulations suggest that "major mergers," that is, mergers

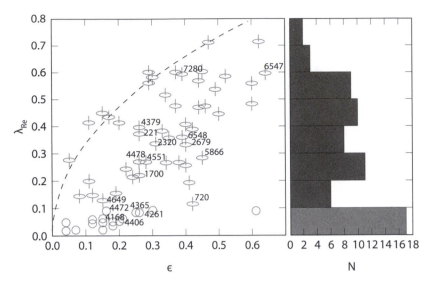

Figure 3.16 Measured values of the parameter λ, equation (3.130), that describes the relative importance of rotational and random motions in galaxies [139]. Slow rotators ($\lambda < 0.1$) are circles and fast rotators ($\lambda > 0.1$) are ellipses. The dashed curve corresponds to "isotropic oblate rotators" seen edge-on. Right panel shows a histogram of the observed λ values.

in which the progenitor galaxies have roughly equal mass, result in slow rotators, particularly if the gas fraction is small [242, 260].

Based on arguments like these, it is reasonable to assume that triaxial galaxies are generically slowly rotating, with $\lambda \lesssim 0.1$. A typical, luminous E galaxy with $\lambda = 0.1$ would have a figure rotation period of $\sim 10^9$ yr and a rotational frequency of $\Omega_f \approx 10 \text{ km s}^{-1} \text{ kpc}^{-1}$. By comparison, the most rapidly rotating (and highly flattened) bars at the centers of some disk galaxies are believed to have frequencies of figure rotation as large as $\sim 100 \text{ km s}^{-1} \text{ kpc}^{-1}$.

Yet another proxy for the amplitude of figure rotation—useful when discussing the behavior of orbits in theoretical models—is the **corotation radius** R_Ω, the radius at which the period of figure rotation, $2\pi / \Omega_f$, is equal to the period $2\pi r / V_c$ of a circular orbit. Strictly circular orbits do not exist in triaxial potentials, but quasi-circular, 1:1 closed orbits typically do; V_c can be defined as the mean velocity of such an orbit, or as the velocity of a circular orbit in a symmetrized model with the same mean radial density law as in the triaxial model. A typical value of R_Ω in a bright E galaxy with $\lambda = 0.1$ would be 10–20 kpc, well beyond the half-light radius at $\sim$1–3 kpc.

In a rotating triaxial potential, orbits that are characterized by a stationary point on the effective potential surface (the surface of constant Jacobi integral) look similar to box orbits in a frame that corotates with the figure. However, there is one important difference. In the absence of figure rotation, a box or pyramid orbit that reaches a stationary point simply reverses velocity and retraces its path. If the figure rotates, the path traced out on the direct (with respect to the figure rotation) segment of the orbit is not retraced during the retrograde segment since the direction of the Coriolis force reverses. The result is **envelope doubling**: the orbit in configuration space appears to be composed of two, partially overlapping orbits with slightly different shapes.

In the case of a nonresonant box orbit, the envelope doubling does not have a significant effect on the character (regular versus stochastic) of the motion. However, a resonant centrophilic orbit, which remains regular by virtue of avoiding the destabilizing center, is converted by figure rotation into a thicker orbit which may intersect the central SBH (figure 3.17, right), rendering it stochastic. This argument does not necessarily imply that the overall degree of chaos increases with the rate of figure rotation: new, thin orbits can crop up that take the place of the thin orbits that were destroyed by the rotation. Nevertheless, one finds via numerical integrations that the fraction of regular centrophilic orbits drops rapidly as the rate of figure rotation is increased (figure 3.17, left). The same turns out to be true for the so-called inner long-axis tube orbits, orbits with the same general shape as box orbits but with enough angular momentum about the long axis to avoid the very center. Scaled to real galaxies, these results suggest that—for pattern speeds in the range 2×10^8 yr $\lesssim T_\Omega \lesssim 5 \times 10^9$ yr—the two orbit families that are most important for maintaining triaxiality in nonrotating models are rendered mostly chaotic [114].

3.5.3 Triaxial self-consistency

In the axisymmetric geometry, self-consistent equilibria exist for essentially all physically interesting choices of the density profile and axis ratio; the only important qualification is the lack of two-integral, $f(E, L_z)$ models for prolate spheroids that are too elongated [33]. But allowing f to depend on a third integral, $f = f(E, L_z, I_3)$, yields self-consistent equilibria for prolate spheroids with arbitrary elongations [251].

The situation is very different for triaxial galaxies. Triaxial self-consistency was considered inherently implausible until the demonstration [482, 483] that most of the orbits in triaxial potentials with large smooth cores do respect three integrals and that self-consistent equilibria could be constructed by superposition of such orbits. The box orbits, which are unique to the triaxial geometry, were found to be especially important. Subsequent support for the existence of self-consistent triaxial equilibria came from N-body simulations of collapse in which the final configurations were often found to be nonaxisymmetric [534, 565]. In addition, the degree of stochasticity in triaxial potentials with central density cusps—but lacking SBHs—was found to be modest [349, 366, 484], implying that most orbits would behave like regular orbits.

As noted above, the box orbits that are so important for triaxial self-consistency tend to disappear in triaxial potentials with central SBHs. Within the influence radius, they are replaced by the pyramid orbits; at larger radii, they are either stochastic, or they are associated with thin resonant orbits that avoid the center. Stochastic orbits are generally considered to be an obstacle for self-consistency, for

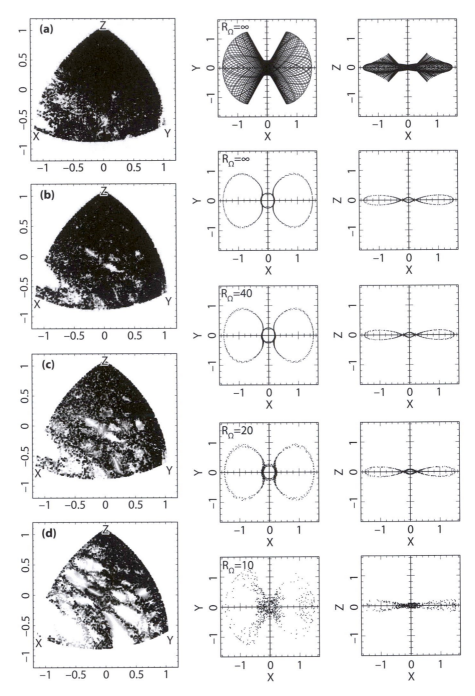

Figure 3.17 Left: Stability maps of orbits launched from an effective (Jacobi) equipotential surface of rotating triaxial models having density law (3.126) with $c/a = 0.5$, $T = 0.58$, $\gamma = 1$, and with a central SBH that contains 0.001 times the galaxy

Figure 3.17 Continued. mass [114]. As in figure 3.15, the gray scale is proportional to the degree of stochasticity. Each equipotential encloses roughly 1/2 the total mass of the model; what varies is the degree of figure rotation, which is zero in panel (d), while $R_\Omega/R_J \approx (25, 12, 6)$ in panels (c), (b), and (a), respectively, where R_J is the approximate radius of the equipotential surface. Right: An orbit associated with the $(3, -1, -1)$ resonance. The top row shows two Cartesian projections of the orbit in the nonrotating model. The next four rows show cross sections of the orbit with the $x-y$ plane (left) and with the $x-z$ plane (right) in rotating models with the same four values of R_Ω/R_J as at left.

two reasons:

1. Stochastic orbits have time-averaged shapes that are rounder than the isodensity contours of the mass model that generated the potential. For instance, an orbit that respects only one integral, the energy, will densely fill the volume enclosed by an equipotential surface.

2. Unless it lies close to a resonance, a regular orbit covers its invariant torus densely in a fairly short time, leading to a constant, time-averaged phase-space density. Stochastic orbits do not behave so predictably; for instance, they can "mimic" a regular orbit for many orbital periods, before suddenly appearing to move to a different "torus." This leads to the expectation that a galaxy containing stochastic orbits would necessarily evolve in shape, perhaps toward axisymmetry.

These concerns are probably less relevant to triaxial *nuclei*. Far from the center of a galaxy, the equipotential surfaces are nearly spherical, and stochastic orbits that fill such surfaces are not very useful for reconstructing an elongated triaxial figure. However, near the center of a galaxy, on scales $r \approx r_{\rm m}$, equipotentials are only slightly rounder than equidensities, particularly if the stellar density rises rapidly toward the center. Because of this, even time-averaged stochastic orbits might be useful building blocks. Furthermore, near the center of a galaxy, evolution timescales for stochastic orbits tend to be short: in part because orbital periods are intrinsically short, but also because the stochastic orbits near the SBH are found to behave in a highly chaotic way, rapidly sampling the full phase-space volume available to them.

A useful way to think about the consequences of stochastic motion is in terms of chaotic mixing as defined in section 3.1.4. If the motion is chaotic, an initially nonuniform distribution of points in phase space evolves to a uniform one, at least in a coarse-grained sense [376]. Because of the exponential instability of stochastic motion, chaotic mixing is essentially irreversible, in the sense that an infinitely fine tuning is required to undo its effects. Chaotic mixing can also be very efficient. In regions of phase space where most of the trajectories are stochastic—for instance, the "zone of chaos" described above—there are few "barriers," in the form of invariant tori, to inhibit the stochastic motion, and chaotic mixing can lead to essentially uniform phase-space distributions in a modest number of crossing times [271, 529].

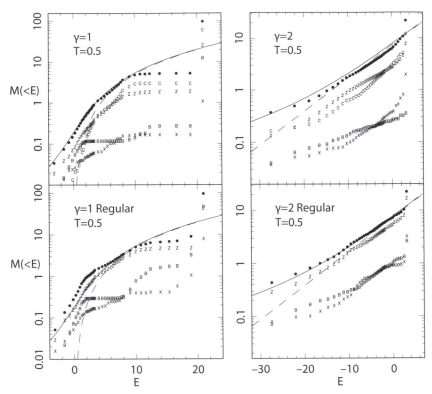

Figure 3.18 Cumulative energy distributions of the various orbit families in self-consistent models of triaxial nuclei with SBHs [434]. The symbols B, X, Z, C denote the mass contributed by box (i.e., pyramid), x-tube, z-tube, and chaotic (i.e., stochastic) orbits, respectively; black dots give the totals. Lower panels show solutions in which the stochastic orbits were excluded. Solid and dashed lines show the distributions for isotropic spherical models with the same radial density profile, with and without the SBH, respectively.

Figure 3.18 shows the orbital composition of self-consistent models of triaxial nuclei containing SBHs [434]. The density profile in these models is given by equation (3.126), with $c/a = 0.5$ and $\gamma = (1, 2)$. The models extend outward in radius to $\sim 10 r_{\rm m}$, that is, well into the "zone of chaos" discussed above. Self-consistent solutions were constructed by superposition of time-averaged orbits; in addition, the solutions were realized as N-body models and advanced forward in time to verify that the numerical self-consistent solutions corresponded to bona fide steady states. Self-consistency was achieved for maximally triaxial ($T = 0.5$) and nearly oblate ($T = 0.25$) geometries. In both these geometries, solutions exist that include only regular orbits—mostly tube orbits about the short axis. But solutions including fully mixed stochastic orbits are also allowed. Perhaps surprisingly, a large fraction of the mass, exceeding 50%, was found to be assignable to the stochastic orbits without violating self-consistency or inducing noticeable evolution. By contrast, no self-consistent solutions could be found for $T = 0.75$, a nearly prolate model.

The results illustrated in figure 3.18 are from the only published study in which the self-consistency problem for triaxial nuclei on scales $r \lesssim r_m$ was addressed. There is, however, a large body of work addressing the *large-scale* effects of central mass concentrations on triaxial spheroids or bars [11, 125, 229, 266, 267, 372, 399]. These studies generally proceed by first constructing an N-body model for the bar or triaxial spheroid, then a compact mass is inserted or grown at the center and the model is integrated forward. Typically the model isophotes evolve toward rounder and/or more axisymmetric shapes on scales $\gtrsim r_h$, as the preexisting box orbits are converted to stochastic orbits. The evolution can be very striking when the mass of the central object exceeds ∼1% of the total galaxy mass (compared with ∼0.1% for real SBHs) and when the figure is elongated. When the mass of the SBH is smaller, $M_\bullet \lesssim 10^{-3} M_{gal}$, evolution is often still observed but the final shape can remain appreciably nonaxisymmetric. It is currently unclear whether Nature would select stable triaxial configurations for galactic nuclei like those in figure 3.18, or whether the presence of an SBH would mitigate against such equilibria, as it seems to do on larger scales.

Chapter Four

Motion Near Supermassive Black Holes

A supermassive black hole (SBH) is a compact object, and from the point of view of a star that orbits far outside the event horizon, its gravitational field should be nearly indistinguishable from that of a Newtonian point mass. Indeed this assumption was the basis for the entire treatment of nuclear dynamics in the preceding chapter, and estimation of the masses of SBHs from kinematical data almost always assumes Newtonian dynamics as well. But the Newtonian approximation must break down for matter that orbits within a few gravitational radii, r_g, of the SBH:

$$r_g \equiv \frac{G M_\bullet}{c^2} \approx 4.78 \times 10^{-8} \left(\frac{M_\bullet}{10^6 \, M_\odot} \right) \, \text{pc} \tag{4.1}$$

since the orbital velocity at such distances is of order the speed of light.[1]

At first sight, one is struck by the enormous difference between r_g and the gravitational influence radii r_h or r_m that were defined in chapter 2. For instance, in the case of r_h,

$$\frac{r_g}{r_h} = \frac{G M_\bullet}{c^2} \Big/ \frac{G M_\bullet}{\sigma^2} = \frac{\sigma^2}{c^2} \approx 10^{-7} \left(\frac{\sigma}{100 \, \text{km s}^{-1}} \right)^2 \tag{4.2}$$

(recall that σ is the stellar velocity dispersion near the galaxy's center). Equation (4.2) seems to suggest that for the vast majority of stars orbiting within the influence sphere of an SBH, relativity is unimportant.

This conclusion turns out to be misleading, for several reasons. First, the effects of relativity depend less on the size of an orbit than on its distance of closest approach to the SBH. The lowest-order corrections to the Newtonian equations of motion have amplitudes that are of order $\mathcal{P}^{-1}$ where $\mathcal{P}$ is the "penetration parameter":

$$\mathcal{P} \equiv (1 - e^2) \frac{a}{r_g} = (1 + e) \frac{r_p}{r_g}.$$

Here, a and e are the semimajor axis and eccentricity of the orbit, and $r_p = (1 - e)a$ is the distance of closest approach to the SBH. It turns out that the feeding of stars and compact objects to SBHs occurs predominantly from very eccentric orbits. For instance, capture of stellar-mass black holes (BHs) by SBHs—a so-called "extreme-mass-ratio inspiral" (EMRI)—is believed to take place from orbits with semimajor axes of $a \lesssim 0.01$ pc and eccentricities in the range 0.99–0.9999. For such orbits, $\mathcal{P}$ can be of order unity even though the orbits extend outward to a large fraction of r_h.

---

[1] Some authors (e.g., [400]) define the gravitational radius as $2 G M_\bullet / c^2$, equal to the radius of the event horizon for a nonrotating hole.

A second reason why the effects of relativity cannot be ignored has to do with the way in which stars get placed onto orbits of such high eccentricity. One such mechanism is torques (i.e., nonradial forces) that arise from the slightly aspherical distribution of matter near an SBH. These torques remain effective as long as orbits near the SBH—both the orbit of the star being torqued, and the orbits of the torquing stars—maintain their orientations; any mechanism that causes orbits to precess (for instance) tends to randomize the torques. Relativistic precession of the periastron—or, as it is called in the solar system, precession of the perihelion—is such a mechanism. If the timescale for relativistic precession is shorter than the timescale for the torques to do their work, feeding of objects to the SBH will be greatly inhibited. This relativistic quenching effect turns out to be of major importance in the EMRI problem.

A third reason for considering relativistic effects is the prospect of measuring those effects, via careful observation of the orbits of stars at the center of the Milky Way. For the Galactic center SBH, $r_g \approx 1.9 \times 10^{-7}$ pc. The star S2, whose orbital period is 15.8 yr, has a semimajor axis $a \approx 0.005$ pc and an eccentricity $e \approx 0.88$, yielding a penetration parameter $\mathcal{P} \approx 5.9 \times 10^3$. The time required for a star orbiting in the Schwarzschild metric of a nonrotating SBH to precess by an angle π works out to be

$$t_S = \frac{1}{6}\mathcal{P}P,$$ (4.3)

where P is the Keplerian period. For S2, it follows that the argument of periapsis—the angle that defines the orientation of the long axis of the orbit in its plane—should increase with time due to relativity as

$$\frac{\Delta\omega}{\Delta t} = \frac{6\pi}{\mathcal{P}P} \approx 0\overset{''}{.}7 \, \text{yr}^{-1},$$ (4.4)

or roughly $11'$ over its 16 yr period. For comparison, the current measurement uncertainty in this angle is about one degree (table 4.1), suggesting that a detection of the relativistic precession may be quite feasible within the next few years.

But there are other forces acting on S2 which also cause deviations from its otherwise Keplerian trajectory. The largest of these is likely to be the so-called "mass precession," apsidal precession due to the distributed mass (stars, stellar remnants, dark matter, ...) inside its orbit. It turns out that this Newtonian precession is predicted to be of similar magnitude (though opposite in direction) to the relativistic precession, at least if standard, "collisionally relaxed," models of the mass distribution are adopted for the Galaxy.[2]

Roughly speaking, this chapter deals with motion at distances from an SBH that are large compared with r_g and small compared with r_h. In this regime, relativity is (sometimes) important, but it is never dominant, and the appropriate way to include it is via the post-Newtonian expansion. As the example of S2 shows, Newtonian and relativistic perturbations in this regime can be of comparable magnitude. Fortunately, it is straightforward to include both sorts of perturbation into

---

[2]The likelihood of such models is discussed in chapter 7.

Table 4.1 Orbital parameters of the S-stars [192, 193]. The reference plane is the plane of the sky; the nodal line is the north celestial pole. A distance of 8.33 kpc to Sgr A* is assumed, i.e., 1″ corresponds to 40.4 mpc (milliparsecs). The assumed mass of the SBH is $M_\bullet = 4.3 \times 10^6 \, M_\odot$.

Star	a[mpc]	e	i [°]	Ω [°]	ω [°]	P[yr]
S1	20.5 ± 1.13	0.496 ± 0.028	120.82 ± 0.46	341.61 ± 0.51	115.3 ± 2.5	132 ± 11
S2	5.03 ± 0.04	0.883 ± 0.003	134.87 ± 0.78	226.53 ± 0.72	64.98 ± 0.81	15.8 ± 0.11
S4	12.0 ± 0.77	0.406 ± 0.022	77.83 ± 0.32	258.11 ± 0.30	316.4 ± 2.9	59.5 ± 2.6
S5	10.1 ± 1.70	0.842 ± 0.017	143.7 ± 4.7	109 ± 10	236.3 ± 8.2	45.7 ± 6.9
S6	17.6 ± 6.18	0.886 ± 0.026	86.44 ± 0.59	83.46 ± 0.69	129.5 ± 3.1	105 ± 34
S8	16.6 ± 0.16	0.824 ± 0.014	74.01 ± 0.73	315.90 ± 0.50	345.2 ± 1.1	96.1 ± 1.6
S9	11.8 ± 2.10	0.825 ± 0.020	81.00 ± 0.70	147.58 ± 0.44	225.2 ± 2.3	58 ± 9.5
S12	12.4 ± 0.32	0.900 ± 0.003	31.61 ± 0.76	240.4 ± 4.6	308.8 ± 3.8	62.5 ± 2.3
S13	12.0 ± 0.48	0.490 ± 0.023	25.5 ± 1.6	73.1 ± 4.1	248.2 ± 5.4	59.2 ± 3.8
S14	10.3 ± 0.40	0.963 ± 0.006	99.4 ± 1.0	227.74 ± 0.70	339.0 ± 1.6	47.3 ± 2.9
S17	12.6 ± 0.16	0.364 ± 0.015	96.44 ± 0.18	188.06 ± 0.32	319.45 ± 3.2	63.2 ± 2.0
S18	10.7 ± 3.23	0.759 ± 0.052	116.0 ± 2.7	215.2 ± 3.6	151.7 ± 2.9	50 ± 16
S19	32.2 ± 2.59	0.844 ± 0.062	73.58 ± 0.61	342.9 ± 1.2	153.3 ± 3.0	260 ± 31
S21	8.61 ± 1.66	0.784 ± 0.028	54.8 ± 2.7	252.7 ± 4.2	182.6 ± 8.2	35.8 ± 6.9
S24	42.8 ± 7.19	0.933 ± 0.010	106.30 ± 0.93	4.2 ± 1.3	291.5 ± 1.5	398 ± 73
S27	18.3 ± 3.15	0.952 ± 0.006	92.91 ± 0.73	191.90 ± 0.92	308.2 ± 1.8	112 ± 18
S29	16.0 ± 13.53	0.916 ± 0.048	122 ± 11	157.2 ± 2.5	343.3 ± 5.7	91 ± 79
S31	12.0 ± 1.78	0.934 ± 0.007	153.8 ± 5.8	103 ± 11	314 ± 10	59.4 ± 9.2
S33	16.6 ± 3.56	0.731 ± 0.039	42.9 ± 4.5	82.9 ± 5.9	328.1 ± 4.5	96 ± 21
S38	5.62 ± 1.66	0.802 ± 0.041	166 ± 22	286 ± 68	203 ± 68	18.9 ± 5.8
S66	48.9 ± 5.09	0.178 ± 0.039	135.4 ± 2.6	96.8 ± 2.9	106 ± 6.3	486 ± 41
S67	44.2 ± 4.12	0.368 ± 0.041	139.9 ± 2.3	106.0 ± 6.1	215.2 ± 4.8	419 ± 19
S71	42.9 ± 30.9	0.844 ± 0.075	76.3 ± 3.6	34.6 ± 1.5	331.4 ± 7.1	399 ± 283
S83	113. ± 9.45	0.657 ± 0.096	123.8 ± 1.3	73.6 ± 2.1	197.2 ± 3.5	1700 ± 205
S87	50.9 ± 6.50	0.423 ± 0.036	142.7 ± 4.4	109.9 ± 2.9	41.5 ± 3.7	516 ± 44

the machinery that was developed by celestial mechanicians over the last three centuries for treating solar-system motion, and that is the approach adopted in this chapter.

One problem to which this machinery has often been applied is the classical three-body problem. In the context of galactic nuclei, the three-body problem appears whenever there is a binary SBH, or an intermediate-mass black hole (IBH) orbiting around an SBH; the third body might be a star in the nucleus. In the hierarchical three-body problem, the separation of two of the bodies is assumed to be much less than the distance of either body to the third object, and attention in this chapter is restricted to that special case.

There is one context where a fully relativistic treatment cannot be avoided: when deriving the critical orbital parameters corresponding to capture by an SBH. The conditions are well known in the case of capture from circular orbits. As discussed above, in galactic nuclei, capture is more likely to occur from highly elliptical orbits. The relevant question becomes, what are the Keplerian elements (as measured near apoapsis, say) of an eccentric orbit that just avoids continuing into the SBH? The answer to that question will be useful when considering the loss-cone problem in chapter 6.

4.1 KEPLERIAN ORBITS

4.1.1 Description of the motion

Consider two point particles moving in response to their mutual gravitational attraction. Newton's equations of motion are

$$\ddot{x}_1 = -Gm_2 \frac{x_1 - x_2}{r^3},$$

$$\ddot{x}_2 = -Gm_1 \frac{x_2 - x_1}{r^3}, \tag{4.5}$$

where $r \equiv |x_1 - x_2|$ is the separation between the bodies whose masses are m_1 and m_2.

Since

$$m_1 \ddot{x}_1 + m_2 \ddot{x}_2 = 0, \tag{4.6}$$

the center of mass $(m_1 x_1 + m_2 x_2)/(m_1 + m_2)$ moves with constant velocity. Transforming to the frame in which that velocity is zero,

$$0 = m_1 v_1 + m_2 v_2, \tag{4.7}$$

allows the individual positions and velocities to be expressed in terms of the relative position and velocity vectors:

$$r = x_1 - x_2, \quad v = v_1 - v_2 \tag{4.8}$$

since

$$x_1 = \frac{\mu r}{m_1}, \quad x_2 = -\frac{\mu r}{m_2}, \tag{4.9a}$$

$$v_1 = \frac{\mu v}{m_1}, \quad v_2 = -\frac{\mu v}{m_2}, \tag{4.9b}$$

where $\mu \equiv m_1 m_2 / (m_1 + m_2)$ is the **reduced mass**. The equation of relative motion is then easily shown to be

$$\frac{dv}{dt} = -\frac{Gmr}{r^3} \tag{4.10}$$

where $m \equiv m_1 + m_2$ is the total mass.

The conserved quantities are the reduced energy and angular momentum,

$$E = \frac{1}{2} v^2 - \frac{Gm}{r}, \tag{4.11a}$$

$$L = r \times v, \tag{4.11b}$$

which are the center-of-mass energy and angular momentum divided by μ. Conservation of L implies that the orbit is confined to a plane. Writing

$$v^2 = \left(\frac{dr}{dt}\right)^2 + r^2 \left(\frac{d\phi}{dt}\right)^2, \tag{4.12a}$$

$$|r \times v| = r^2 \frac{d\phi}{dt}, \tag{4.12b}$$

where (r, ϕ) are polar coordinates in that plane, conservation of E and L imply

$$\left(\frac{dr}{dt}\right)^2 = A + 2\frac{B}{r} + \frac{C}{r^2}, \tag{4.13a}$$

$$\frac{d\phi}{dt} = \frac{H}{r^2}, \tag{4.13b}$$

where

$$A = 2E, \quad B = Gm, \quad C = -L^2, \quad H = L. \tag{4.14}$$

In the case of bound ($E < 0$) motion, the solution of equations (4.13) can be expressed parametrically in terms of the **eccentric anomaly** E (not to be confused with the energy, E). The radial motion satisfies

$$n(t - t_0) = \mathrm{E} - e \sin \mathrm{E}, \tag{4.15a}$$

$$r = a\,(1 - e \cos \mathrm{E}). \tag{4.15b}$$

Here n is the **mean motion** (i.e., the mean angular velocity) and the **mean anomaly**, M, is defined as

$$\mathrm{M} = n(t - t_0). \tag{4.16}$$

The mean motion is given by

$$n = \frac{(-A)^{3/2}}{B} = \frac{(-2E)^{3/2}}{Gm} \tag{4.17}$$

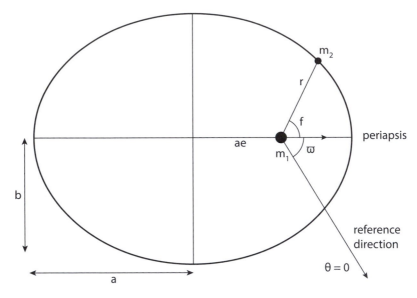

Figure 4.1 The geometry of a Keplerian ellipse, showing the motion of body m_2 with respect to m_1. The semimajor axis of the relative orbit is a, the semiminor axis is b, the eccentricity is e, the longitude of periapsis is ϖ, and the true anomaly is f.

and is equal to the orbital frequency, or $2\pi/P$, where P is the **orbital period**:

$$P = \frac{2\pi}{n} = \frac{2\pi a^{3/2}}{\sqrt{Gm}}. \tag{4.18}$$

The usual symbol for the mean motion, n, is easily confused with the symbol for number density, and throughout most of this book the symbol ν_r will be used instead; the subscript r indicates that this frequency is associated with radial motion (as opposed to frequencies of angular precession, for instance). The constants a and e are the **semimajor axis** and **eccentricity** (figure 4.1):

$$a = -\frac{B}{A} = -\frac{Gm}{2E}, \tag{4.19a}$$

$$e = \left(1 - \frac{AC}{B^2}\right)^{1/2} = \left(1 + \frac{2EL^2}{G^2m^2}\right)^{1/2}, \tag{4.19b}$$

with inverse relations

$$E = -\frac{Gm}{2a}, \tag{4.20a}$$

$$L^2 = Gma\left(1 - e^2\right). \tag{4.20b}$$

The minimum and maximum values of r occur, respectively, at the **periapsis**, r_p, and **apoapsis**, r_a, where

$$r_p = (1 - e)\,a, \quad r_a = (1 + e)\,a. \tag{4.21}$$

The period can be expressed in terms of E as

$$P(E) = \frac{\pi}{\sqrt{2}} \frac{Gm}{(-E)^{3/2}}. \tag{4.22}$$

The solution to the angular equation is

$$\phi - \phi_0 = \frac{H}{(-C)^{1/2}} f = f, \tag{4.23a}$$

$$\tan \frac{f}{2} = \left(\frac{1+e}{1-e}\right)^{1/2} \tan \frac{E}{2}, \tag{4.23b}$$

where f is the **true anomaly** (figure 4.1). The angle ϕ_0 is sometimes defined as the **longitude of pericenter**, ϖ; what is important in what follows is that $f = 0$ at periapsis, the point of closest approach. The orbit in space is obtained by eliminating E between equations (4.15) and (4.23):

$$r(\phi) = \frac{a(1-e^2)}{1 + e\cos(\phi - \phi_0)} = \frac{a(1-e^2)}{1 + e\cos f}, \tag{4.24}$$

an ellipse. Other useful relations are

$$r\cos f = a(\cos E - e), \quad r\sin f = a(1-e^2)^{1/2}\sin E,$$

$$r = a(1 - e\cos E). \tag{4.25}$$

Let $E_{bin} \equiv \mu E$ and $L_{bin} \equiv \mu L$ be the energy and angular momentum of the binary (rather than the quantities per unit of reduced mass) in the center-of-mass frame. The relation between these quantities, and the Keplerian elements a and e, is given by equations (4.19):

$$a = -\frac{Gm_1 m_2}{2E_{bin}}, \quad e = \left(1 + \frac{2E_{bin}L_{bin}^2}{G^2\mu^3 m^2}\right)^{1/2}, \tag{4.26}$$

and the inverse relations are

$$E_{bin} = -\frac{Gm_1 m_2}{2a}, \quad L_{bin} = \mu\sqrt{Gma(1-e^2)}. \tag{4.27}$$

So far we have assumed a bound orbit, $E < 0$. We will have occasion later in this book also to consider the unbound case, when discussing the dynamics of encounters between passing stars. We give the necessary expressions here without derivation. The relative orbit in space is

$$r(\phi) = \frac{L^2}{Gm} \frac{1}{1 + e\cos(\phi - \phi_0)}, \tag{4.28}$$

with $e > 1$ a quantity that plays the role of eccentricity:

$$e^2 = 1 + \frac{2EL^2}{G^2 m^2}, \tag{4.29}$$

identical to the expression (4.19b), except of course that now $E > 0$. The solution to the angular equation is

$$\tan \frac{f}{2} = \left(\frac{e+1}{e-1}\right)^{1/2} \tanh \frac{F}{2}, \tag{4.30}$$

with $f = \phi - \phi_0$ and F playing the role of eccentric anomaly; the analogue to Kepler's equation is

$$e \sinh F - F = M = n (t - t_0) . \qquad (4.31)$$

Throughout this book, we will often be concerned with the motion of a star around an SBH. In this case, we can set $m_1 = M_\bullet \gg m_2$ and $\mu = m_2$, and identify the center of mass of the binary with the SBH, located at the origin with zero velocity (say). The quantities E and L defined in equation (4.20) then become the specific energy and angular momentum of the second body, the same quantities that were identified in chapter 3 as integrals of motion in a general spherical potential.

4.1.2 Orbital distributions

It is useful to rewrite some of the expressions derived in section 3.2 in terms of the alternate integrals of motion a and e, after setting $\Phi(r) = -G M_\bullet / r$. We assume, as in that section, that $f = f(E, L^2)$, that is, that the spatial distribution is spherical and the velocity distribution exhibits no mean motion. The velocity components parallel and perpendicular to a radius vector, v_r and v_t, are

$$v_r^2 = \frac{G M_\bullet}{a} \left[\frac{2a}{r} - 1 - \frac{a^2}{r^2} \left(1 - e^2 \right) \right] = \frac{G M_\bullet}{a} \left(\frac{r_a}{r} - 1 \right) \left(1 - \frac{r_p}{r} \right) ,$$

$$v_t^2 = \frac{G M_\bullet a}{r^2} \left(1 - e^2 \right) = \frac{G M_\bullet}{a} \frac{r_a r_p}{r^2} . \qquad (4.32)$$

The velocity-space volume element, $dv = 2\pi \, v_t \, dv_t \, dv_r$, becomes

$$\begin{aligned} dv &= 2\pi \, v_t \, \frac{\partial (v_r, v_t)}{\partial (a, e)} \, da \, de \\ &= \pi \, \frac{(G M_\bullet)^{3/2}}{r a^{1/2}} \left[a^2 e^2 - (r - a)^2 \right]^{-1/2} \, da \, e \, de \\ &= \pi \, \frac{(G M_\bullet)^{3/2}}{r a^{1/2}} \left[(r_a - r) \left(r - r_p \right) \right]^{-1/2} \, da \, e \, de. \qquad (4.33) \end{aligned}$$

The configuration-space density, $n(r)$, of an ensemble of orbits that are uniformly populated with respect to mean anomaly (thus satisfying Jeans's condition) is

$$n(r) = \int f (x, v) \, dv \qquad (4.34a)$$

$$= 2\pi \, \frac{(G M_\bullet)^{3/2}}{r} \int_{r/2}^{\infty} \frac{da}{a^{1/2}} \int_{(1-r/a)}^{1} \frac{f(a, e^2) \, e \, de}{\sqrt{a^2 e^2 - (r - a)^2}} , \qquad (4.34b)$$

with f the phase-space density. Finally, consider the function $N(a, e)$, where $N(a, e) \, da \, de$ is the number of stars with orbital elements in the intervals da and *de* centered on *a* and *e*. By definition,

$$N(a, e) \, da \, de = 2 \int_{r_p}^{r_a} f(a, e^2) \times 4\pi r^2 dr \times dv \tag{4.35}$$

$$= \frac{8\pi^2}{a^{1/2}} (GM_\bullet)^{3/2} f(a, e^2) \int_{r_p}^{r_a} \frac{r \, dr}{\sqrt{a^2 e^2 - (r-a)^2}} \, da \, e \, de$$

$$= 8\pi^3 (GM_\bullet)^{3/2} a^{1/2} f(a, e^2) \, da \, e \, de.$$

Suppose that the velocity distribution is isotropic, $f = f(a)$. It was shown in chapter 3 that near an SBH, a configuration-space density $n(r) \propto r^{-\gamma}$ is reproduced by an isotropic f of the form $f(E) \propto |E|^{\gamma - 3/2} \propto a^{3/2 - \gamma}$. The corresponding $N(a, e)$ is

$$N(a, e) \, da \, de = N_0 a^{2-\gamma} \, da \, e \, de. \tag{4.36}$$

The normalized eccentricity distribution, for stars with some specified range of a (say), is simply

$$\frac{dN}{de} = 2e. \tag{4.37}$$

This is sometimes referred to as a "thermal" distribution of eccentricities. For such a distribution,

$$\langle e \rangle = \int_0^1 N(e) \, e \, de = \frac{2}{3}, \tag{4.38a}$$

$$\langle e^2 \rangle = \int_0^1 N(e) \, e^2 \, de = \frac{1}{2}. \tag{4.38b}$$

4.2 PERTURBED ORBITS

The remainder of this chapter deals with the effects of small perturbing forces on the otherwise Keplerian orbit of a star around an SBH. The perturbing forces may be due to distributed mass in the form of other stars, stellar remnants, dark matter, etc., or to the effects of relativity. In such problems, it is useful to represent the motion as a slow (compared with orbital frequencies) variation of the constants, or **elements**, that define the shape and orientation of the otherwise-Keplerian orbit. Figure 4.2 illustrates the traditional orbital elements. The semimajor axis a and eccentricity e were defined above. Two more elements define the orientation of the orbital plane: the **inclination** i, defined as the tilt of the ellipse with respect to a reference plane, and the **longitude of the ascending node**, Ω, the angle between the reference direction and the line of intersection of the orbit with the reference plane; the latter is called the **line of nodes**. Two final elements specify the location in the orbital plane: the **argument of periapsis** (or **pericenter**, or **periastron**) is the angle between the line of nodes and the periapsis, and the mean anomaly, defined above, gives the phase along the orbit. In the unperturbed two-body problem, all of the Keplerian elements are conserved except for the mean anomaly, which increases linearly with time.

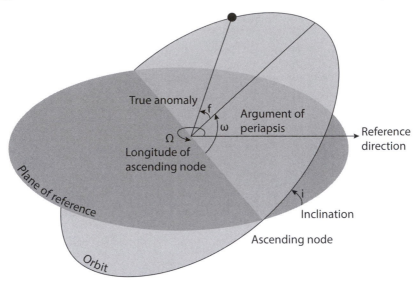

Figure 4.2 The Keplerian orbit in three-dimensional space, showing the angles that define its orientation with respect to a reference plane and a reference direction: the inclination i, the longitude of the ascending node Ω, and the argument of periapsis ω.

When describing observed orbits, for example, the orbits of stars around the Galactic center SBH, the reference plane is often chosen to be the plane of the sky. In other situations, the choice of reference plane is often motivated by symmetries in the problem. For instance, in the case of motion around a spinning SBH, associating the reference plane with the plane perpendicular to the SBH spin vector allows precession due to frame dragging to be described simply in terms of changes in Ω.

The Keplerian elements are the basis for a commonly used set of action-angle variables, the **Delaunay variables** or **elements** [201]. The Delaunay momentum variables are the three actions (J_1, J_2, J_3), where $J_1 = L = |\mathbf{x} \times \mathbf{v}|$, the magnitude of the angular momentum; $J_2 = I = J_r + L$, where J_r is the radial action, defined as

$$J_r = \frac{1}{\pi} \int_{r_{min}}^{r_{max}} v_r \, dr = \frac{1}{\pi} \int_{r_{min}}^{r_{max}} dr \sqrt{2E - 2\Phi(r) - \frac{L^2}{r^2}} \; ; \qquad (4.39)$$

and $J_3 = L_z$, the projection of $\mathbf{L}$ onto the z-axis. The Delaunay actions are related to the elements of a Kepler orbit via

$$L = \sqrt{Gma(1 - e^2)}, \quad I = \sqrt{Gma}, \quad L_z = L \cos i. \qquad (4.40)$$

The variables conjugate to (L, I, L_z)—the "angles"—are, respectively, the argument of periapsis ω, the mean anomaly M, and the longitude of the ascending node Ω.

The Delaunay variables are particularly useful for representing perturbed motion, since in the absence of perturbations, all of the variables (excepting M) are conserved. Expressed in terms of the Delaunay variables, the Hamiltonian describing relative motion in the nonrelativistic two-body problem is simply

$$\mathcal{H}_{\text{Kep}} = -\frac{1}{2}\left(\frac{Gm}{I}\right)^2. \tag{4.41}$$

The Hamiltonian equations for the rates of change of the generalized coordinates q_k and momenta p_k,

$$\dot{q}_k = \frac{\partial \mathcal{H}}{\partial p_k}, \quad \dot{p}_k = -\frac{\partial \mathcal{H}}{\partial q_k}, \tag{4.42}$$

become, in terms of the Delaunay variables,

$$\frac{dI}{dt} = -\frac{\partial \mathcal{H}}{\partial M}, \qquad \frac{dL}{dt} = -\frac{\partial \mathcal{H}}{\partial \omega}, \qquad \frac{dL_z}{dt} = -\frac{\partial \mathcal{H}}{\partial \Omega},$$
$$\frac{dM}{dt} = \frac{\partial \mathcal{H}}{\partial I}, \qquad \frac{d\omega}{dt} = \frac{\partial \mathcal{H}}{\partial L}, \qquad \frac{d\Omega}{dt} = \frac{\partial \mathcal{H}}{\partial L_z}. \tag{4.43}$$

In the unperturbed problem, $\mathcal{H}$ depends only on I and the only nontrivial equation of motion is

$$\frac{dM}{dt} = \frac{\partial \mathcal{H}_{\text{Kep}}}{\partial I} = \frac{G^2 m^2}{I^3} = \frac{\sqrt{Gm}}{a^{3/2}} = \frac{2\pi}{P} = \nu_r. \tag{4.44}$$

In the case of perturbed motion, $\mathcal{H}$ will have an additional piece,

$$\mathcal{H} = \mathcal{H}_{\text{Kep}} + \mathcal{H}_{\text{p}}, \tag{4.45}$$

that describes the perturbation. If the dependence of $\mathcal{H}_{\text{p}}$ on the Cartesian coordinates is known, its dependence on the Delaunay variables can be obtained from the transformations

$$\boldsymbol{r} = r\left[\boldsymbol{u}_1 \cos(f+\omega) + \boldsymbol{u}_2 \sin(f+\omega)\right], \tag{4.46}$$

$$\dot{\boldsymbol{r}} = \left(\frac{Gm}{p}\right)^{1/2}\left\{-\boldsymbol{u}_1\left[e\sin\omega + \sin(f+\omega)\right] + \boldsymbol{u}_2\left[e\cos\omega + \cos(f+\omega)\right]\right\},$$

where $p \equiv (1-e^2)a$ is the **semilatus rectum**, $r = p/(1+e\cos f)$, and the unit vectors $(\boldsymbol{u}_1, \boldsymbol{u}_2, \boldsymbol{u}_3)$ are directed along the line of nodes toward the ascending node, perpendicular to the line of nodes in the plane of the orbit, and perpendicular to the orbital plane, respectively:

$$\boldsymbol{u}_1 = \begin{pmatrix} \cos\Omega \\ \sin\Omega \\ 0 \end{pmatrix}, \quad \boldsymbol{u}_2 = \begin{pmatrix} -\cos i \sin\Omega \\ \cos i \cos\Omega \\ \sin i \end{pmatrix}, \quad \boldsymbol{u}_3 = \begin{pmatrix} \sin i \sin\Omega \\ -\sin i \cos\Omega \\ \cos i \end{pmatrix}. \tag{4.47}$$

In equations (4.46) and (4.47), the reference plane is the x–y plane and the reference line is the x-axis.

Throughout this chapter we assume that the perturbing forces are small compared with the inverse-square force from the SBH. This implies motion that remains close enough to the SBH that the enclosed mass in stars is small compared with $M_\bullet$; but at the same time, not *so* close that relativistic effects begin to dominate the motion: in other words, $r_g \ll r \ll r_h$. In this regime, one expects the motion to be essentially Keplerian on timescales comparable with the period P.

A useful way to deal with such motion is via the technique of **averaging** [55]. Suppose that the equation of motion for an orbital element, say Ω, is

$$\frac{d\Omega}{dt} = \frac{\partial \mathcal{H}_p}{\partial L_z} = g(x, y, z) = g(I, L, L_z, \omega, \Omega, M). \tag{4.48}$$

In many problems of interest, the time dependence of g as seen by the orbiting body will consist of a set of periodic terms with frequencies that are integer multiples of ν_r. This is the case, for instance, if the perturbation is due to the force from a fixed, distributed mass around the SBH. Furthermore, since the perturbation is small, many revolutions will be required before appreciable changes take place in any of the otherwise-conserved elements ($I, L, L_z, \omega, \Omega$). It makes intuitive sense in this case to average the equations of motion over the short timescale associated with radial motion in the unperturbed problem. Thus one replaces the exact equation (4.48) by

$$\left. \frac{d\Omega}{dt} \right|_{Av} = \left\langle \frac{d\Omega}{dt} \right\rangle = \frac{1}{P} \int_0^P g(I, L, L_z, \omega, \Omega, M) dt, \tag{4.49}$$

where it is understood that the orbital elements in the integrand are to be regarded as constants—the **osculating elements**. The result of the averaging is a set of equations describing the gradual evolution of the elements (L, L_z, Ω, ω) due to the perturbing forces. The more rapid, periodic variations that occur over a single orbit typically have amplitudes relative to the gradual changes of $\sim a|\Phi_p|/Gm \ll 1$.

If the perturbed motion is derivable from a velocity-independent potential,

$$\Phi(x) = -\frac{GM_\bullet}{r} + \Phi_p(x), \tag{4.50}$$

with Φ_p the (small) perturbation, the averaged equations of motion can be expressed most simply in terms of the **orbit-averaged Hamiltonian** $\overline{\mathcal{H}}$:

$$\overline{\mathcal{H}} = -\frac{1}{2}\left(\frac{GM_\bullet}{I}\right)^2 + \overline{\Phi}_p(I, L, L_z, \omega, \Omega), \tag{4.51a}$$

$$\overline{\Phi}_p \equiv \oint \frac{dM}{2\pi} \Phi_p = \frac{1}{2\pi} \int_0^{2\pi} dE\,(1 - e\cos E)\,\Phi_p(x). \tag{4.51b}$$

In the final integral, the mean anomaly M has been replaced by the eccentric anomaly E; alternatively, the true anomaly f can be used via $dM/df = (1-e^2)^{3/2}/(1+e\cos f)^2$. In the remainder of this book, averaging will often be carried out using the eccentric anomaly as dependent variable, and it is useful to express the transformation equations (4.46) in terms of E rather than f:

$$r = r\left[w_1(\cos E - e) + w_2\sqrt{1 - e^2}\sin E\right] \tag{4.52}$$

where $r = a(1 - e \cos E)$ and

$$w_1 = \begin{pmatrix} \cos \Omega \cos \omega - \cos i \sin \Omega \sin \omega \\ \sin \Omega \cos \omega + \cos i \cos \Omega \sin \omega \\ \sin i \sin \omega \end{pmatrix}, \tag{4.53a}$$

$$w_2 = \begin{pmatrix} - \cos \Omega \sin \omega - \cos i \sin \Omega \cos \omega \\ - \sin \Omega \sin \omega + \cos i \cos \Omega \cos \omega \\ \sin i \cos \omega \end{pmatrix}. \tag{4.53b}$$

The orbit-averaged equations of motion for the variables (L, L_z, Ω, ω) are then given by Hamilton's equations (4.43), after replacing $\mathcal{H}$ in those equations by $\overline{\mathcal{H}}$.

After the averaging, $\overline{\mathcal{H}}$ is independent of M, implying that a (i.e., I) is conserved. This result can be justified ab initio using the concept of adiabatic invariance; alternatively, it can be shown to be a consequence of the more general equations of perturbed motion to be presented now.

In the post-Newtonian description, the equations of relative motion in the two-body problem have the form

$$\ddot{r} = -\frac{Gm}{r^3} r + a_{\mathrm{p}}, \tag{4.54}$$

where $a_{\mathrm{p}} = a_{\mathrm{PN}}$ depends on r, $\dot{r}$ and (in the higher-order PN equations) on $\ddot{r}$. Over the last three centuries, many techniques have been worked out for obtaining approximate solutions to equation (4.54) in the context of the solar system. The starting point for many of these approximate methods is **Lagrange's planetary equations**, which give the instantaneous rates of change of the osculating elements in terms of the a_p [466]. In writing Lagrange's equations, it is convenient (following Gauss) to first resolve the perturbing acceleration into components (S, T, W):

$$a_p = Sn + Tm + Wk, \tag{4.55}$$

with

$$n = \begin{pmatrix} \cos(f + \omega) \cos \Omega - \sin(f + \omega) \sin \Omega \cos i \\ \cos(f + \omega) \sin \Omega + \sin(f + \omega) \cos \Omega \cos i \\ \sin(f + \omega) \sin i \end{pmatrix}$$

and

$$m = \frac{\partial n}{\partial (f + \omega)}, \quad k = \frac{1}{\sin(f + \omega)} \frac{\partial n}{\partial i}. \tag{4.56}$$

S is the component parallel to the separation vector r, T is the component perpendicular to r in the orbital plane, in the direction such that it makes an angle less than 90° with the velocity vector, and W is the component perpendicular to the orbital plane, in the direction of the orbital angular momentum vector. In terms of

(S, T, W), Lagrange's equations are

$$\frac{da}{dt} = \frac{2}{n\sqrt{1-e^2}}\left(Se\sin f + T\frac{p}{r}\right), \tag{4.57a}$$

$$\frac{de}{dt} = \frac{\sqrt{1-e^2}}{na}\Big(S\sin f + T(\cos f + \cos E)\Big), \tag{4.57b}$$

$$\frac{di}{dt} = \frac{r\cos(\omega + f)}{na^2\sqrt{1-e^2}}W, \tag{4.57c}$$

$$\frac{d\omega}{dt} = -\cos i\frac{d\Omega}{dt} + \frac{\sqrt{1-e^2}}{nae}\left[-S\cos f + T\left(1+\frac{r}{p}\right)\sin f\right], \tag{4.57d}$$

$$\frac{d\Omega}{dt} = \frac{r\sin(\omega + f)}{na^2\sin i\sqrt{1-e^2}}W, \tag{4.57e}$$

$$\frac{dM}{dt} = n - \sqrt{1-e^2}\left(\frac{d\omega}{dt} + \cos i\frac{d\Omega}{dt}\right) - S\frac{2r}{na^2}. \tag{4.57f}$$

Lagrange's equations (4.57) provide a complete description of the motion and in principle could be solved without further approximation. However, if the perturbing force is small compared with the two-body force, the changes in the orbital elements will be slow, and to a first approximation, the elements (with the exception of f) can be set to constant values on the right-hand sides of equations (4.57). Integrating those equations with respect to time then gives the first-order changes in the elements, for example,

$$\Delta\Omega = \int_{t_0}^{t}\frac{d\Omega}{dt}dt = \frac{(1-e^2)^2}{n^2a\sin i}\int_{f(t_0)}^{f(t)}\frac{\sin(\omega+f)}{(1+e\cos f)^3}W(a,e,i,\Omega,\omega;f)df, \tag{4.58}$$

where

$$\frac{df}{dt} = \frac{na^2}{r^2}\sqrt{1-e^2}, \quad r = \frac{a(1-e^2)}{1+e\cos f}$$

have been used. The orbit-averaged rate of change, $\langle d\Omega/dt\rangle$, is then given by $\Delta\Omega/\Delta t$ after setting $(t_0, t) = (0, P)$ and $\Delta t = t - t_0$.

In the remainder of this chapter, the focus will be on changes that take place on timescales long compared with orbital periods, and the orbit-averaged expressions are the most relevant. For this reason, expressions like $d\Omega/dt$ will be understood to mean $\langle d\Omega/dt\rangle$, the orbit-averaged time derivative, unless otherwise noted.

The Delaunay variables are useful for many problems, but they sometimes become awkward to use. For instance, as the inclination goes to zero, the nodal longitude becomes ill defined. Many alternative sets of variables have been proposed. One especially useful set are the **vectorial elements**. They are defined in terms of a set of orthogonal unit vectors $\{e_\ell, e_e, e_n\}$ whose directions are determined by the orientation of the orbit in space. The vector e_ℓ is colinear with the orbital angular momentum. The vector e_e points in the direction of orbital periapsis, or equivalently, along the Laplace–Runge–Lenz vector. The third vector is $e_n = e_\ell \times e_e$.

These vectors are related to Delaunay variables by

$$
\boldsymbol{e}_\ell = \begin{pmatrix} \sin i \sin \Omega \\ -\sin i \cos \Omega \\ \cos i \end{pmatrix}, \quad \boldsymbol{e}_e = \begin{pmatrix} \cos \omega \cos \Omega - \cos i \sin \omega \sin \Omega \\ \cos \omega \sin \Omega + \cos i \sin \omega \cos \Omega \\ \sin i \sin \omega \end{pmatrix},
$$

$$
\boldsymbol{e}_n = \begin{pmatrix} -\sin \omega \cos \Omega - \cos i \cos \omega \sin \Omega \\ -\sin \omega \sin \Omega + \cos i \cos \omega \cos \Omega \\ \sin i \cos \omega \end{pmatrix}. \tag{4.59}
$$

The vectorial elements are parallel to $\{\boldsymbol{e}_\ell, \boldsymbol{e}_e, \boldsymbol{e}_n\}$ but are normalized in different ways by different authors. One choice is

$$
\boldsymbol{L} = L\,\boldsymbol{e}_\ell, \quad \boldsymbol{E} = \sqrt{Gma}\,\boldsymbol{e}_e. \tag{4.60}
$$

The orbit-averaged components of the Cartesian position vector can be written very simply in terms of $\boldsymbol{E}$ as [382]

$$
\overline{x}_i = -\frac{3}{2}\sqrt{\frac{a}{Gm}}\,E_i, \tag{4.61}
$$

and the time-averaged product of two coordinates as

$$
\overline{x_i x_j} = \frac{1}{2}\frac{a}{Gm}\left[L^2\delta_{ij} - J_i J_j + 5E_i E_j\right]. \tag{4.62}
$$

The set of vectors (4.59) will be useful in describing pyramid orbits in section 4.4.3. They also provide the most compact and elegant way for expressing the three-body Hamiltonian in section 4.8, and are useful when describing the changes in $\boldsymbol{L}$ due to "resonant relaxation" (section 5.6).

4.3 THE POST-NEWTONIAN APPROXIMATION

As discussed in the introduction to this chapter, the effects of relativity on the motion of stars orbiting within the influence sphere of an SBH can be large enough to produce significant departures from the predictions of Newtonian theory. In some cases of physical interest, for example, the S-stars at the center of the Milky Way, these departures can be significant even on timescales comparable with orbital periods—which is to say, human lifetimes! But Einstein's equations are notoriously difficult to solve, even in the case $N = 2$, and there is essentially no prospect of obtaining exact solutions in the case of many-body systems like galactic nuclei. Fortunately, we would be satisfied with something far less ambitious—say, a computational framework, preferably "Newtonian-like," that allows us to treat the effects of relativistic perturbations in an approximate way, with an error that is smaller (say) than the amplitudes of the relativistic effects themselves. Such a framework exists: it is called the **post-Newtonian**, or **PN**, **approximation**.

The PN approximation method was developed by Einstein, Droste and De Sitter within one year of the 1916 publication of the general theory of relativity. The initial motivation was to make predictions that could be tested via observations of the solar system, in which the Newtonian effects of the planets' gravity were known to be much greater than the lowest-order corrections due to general relativity. Quantitative predictions were calculated for the relativistic precession of Mercury's perihelion, the deflection of light by the Sun, and the gravitational redshift [104].

The "small parameters" that are the basis for the PN approximation are

$$\beta = \frac{v}{c}, \quad \gamma = \frac{Gm}{c^2 r}, \tag{4.63}$$

where r, v are typical separations and relative velocities of bodies of mass m. In other words, one assumes that objects are moving slowly compared with the speed of light, and that their motion never brings them very near to the gravitational radius of another body. An additional assumption is

$$v^2 \sim \frac{Gm}{r}, \tag{4.64}$$

that is, that characteristic velocities are of the order that would be expected in a system that is bound together by the mutual gravitational attraction of its component bodies. This assumption allows us to write

$$\beta^2 \approx \gamma \ll 1 \tag{4.65}$$

and to express the order of the PN approximation in terms of just one parameter, for example, β. Thus the lowest-order, or 1PN, approximation yields corrections to the Newtonian accelerations of order $O(\beta^2) = O(v^2/c^2)$, the 2PN approximation to order $O\left[(v/c)^4\right]$, etc.

A conceptually difficult point in implementing the post-Newtonian approximation is how to treat the internal structures of bodies. A point-mass approximation might seem natural, but turns out to be surprisingly subtle in the context of general relativity. Furthermore, SBHs are macroscopic bodies, and the effects of their spin and flattening on the motion of stars is often significant. But allowing bodies to have internal degrees of freedom is also complicated, whether their internal structure is Newtonian or relativistic. In some developments of the PN formalism, additional small parameters are introduced that express the assumption that the N bodies are widely separated, for example,

$$\alpha = \frac{R}{r}, \tag{4.66}$$

where R is a typical linear dimension.

Detailed treatments of the PN approximation are given in a number of excellent texts [562, 566] and will not be repeated here. For the purposes of this book, only certain key results are needed.

Consider first a set of N point particles that move in response to their mutual attraction. In the absence of relativity, the acceleration of body a is given by the summed gravitational force from all the other bodies:

$$\left(\frac{d\boldsymbol{v}_a}{dt}\right)_{\text{N}} = \sum_{b \neq a} \frac{Gm_b \boldsymbol{x}_{ab}}{r_{ab}^3}, \tag{4.67}$$

where $\boldsymbol{x}_i, \boldsymbol{v}_i$ are the position and velocity of the ith particle of mass m_i, $\boldsymbol{x}_{ab} = \boldsymbol{x}_b - \boldsymbol{x}_a$, and $r_{ab} = |\boldsymbol{x}_{ab}|$.

The order $O(v^2/c^2)$, or 1PN, corrections to equation (4.67) are called the **Einstein–Infeld–Hoffman**, or **EIH**, **equations of motion** [136]. They are

$$\frac{d\boldsymbol{v}_a}{dt} = \left(\frac{d\boldsymbol{v}_a}{dt}\right)_N + \left(\frac{d\boldsymbol{v}_a}{dt}\right)_{PN}$$

and

$$c^2\left(\frac{d\boldsymbol{v}_a}{dt}\right)_{PN} = \sum_{b\neq a}\frac{Gm_b\boldsymbol{x}_{ab}}{r_{ab}^3}$$

$$\times\left[-4\sum_{c\neq a}\frac{Gm_c}{r_{ac}} + \sum_{c\neq a,b}Gm_c\left(-\frac{1}{r_{bc}} + \frac{\boldsymbol{x}_{ab}\cdot\boldsymbol{x}_{bc}}{2r_{bc}^3}\right)\right.$$

$$\left. - 5\frac{Gm_a}{r_{ab}} + v_a^2 + 2v_b^2 - 4\boldsymbol{v}_a\cdot\boldsymbol{v}_b - \frac{3}{2}\left(\frac{\boldsymbol{v}_b\cdot\boldsymbol{x}_{ab}}{r_{ab}}\right)^2\right]$$

$$+ \sum_{b\neq a}Gm_b\left(\boldsymbol{v}_b - \boldsymbol{v}_a\right)\left[\frac{\boldsymbol{x}_{ab}}{r_{ab}^3}\cdot(4\boldsymbol{v}_a - 3\boldsymbol{v}_b)\right]$$

$$+ \frac{7}{2}\sum_{b\neq a}\sum_{c\neq a,b}\frac{G^2m_b\,m_c\boldsymbol{x}_{bc}}{r_{ab}\ r_{bc}^3}. \tag{4.68}$$

The EIH equations of motion can be derived from Lagrange's equations,

$$\frac{d}{dt}\left(\frac{\partial\mathcal{L}}{\partial v_i}\right) = \frac{\partial\mathcal{L}}{\partial x_i}, \tag{4.69}$$

if the Lagrangian is taken to be

$$\mathcal{L}_{EIH} = \mathcal{L}_N + \mathcal{L}_{PN},$$

$$\mathcal{L}_N = \frac{1}{2}\sum_a m_a v_a^2 + \frac{1}{2}\sum_{b\neq a}\frac{m_a m_b}{r_{ab}},$$

$$c^2\mathcal{L}_{PN} = \frac{1}{8}\sum_a m_a v_a^4$$

$$+ \frac{1}{2}\sum_{b\neq a}\frac{Gm_a m_b}{r_{ab}}\left[3v_a^2 - \frac{7}{2}\boldsymbol{v}_a\cdot\boldsymbol{v}_b - \frac{1}{2}\left(\boldsymbol{v}_a\cdot\boldsymbol{n}_{ab}\right)\left(\boldsymbol{v}_b\cdot\boldsymbol{n}_{ab}\right)\right]$$

$$- \frac{1}{2}\sum_{b\neq a}\sum_{c\neq a}\frac{G^2m_a m_b\,m_c}{r_{ab}\ r_{ac}}, \tag{4.70}$$

where $\boldsymbol{n}_{ab} = (\boldsymbol{x}_a - \boldsymbol{x}_b)/x_{ab}$.

In the solar system, computation of accurate ephemerides for the planets requires the inclusion of relativistic effects, and the EIH equations are the basis for much work in this area. In the case of galactic nuclei, predicting the detailed motion of individual stars is typically not the goal,[3] and the EIH equations, in the form of

---

[3]This situation is likely to change in the near future, as increasingly detailed data become available for stars at the Galactic center.

equation (4.68), have seen relatively little application. However, setting $N = 2$ or $N = 3$ in these equations is a natural starting point for talking about the relativistic two- and three-body problems, and this is the approach we will take below.

In the two-body problem, the lowest-order relativistic corrections predict a precession of the relative orbit, the **relativistic precession of the periastron** (or **periapsis**). As almost everyone knows, the amplitude of this precession was computed by Einstein and others in the case of Mercury's orbit around the Sun as one of the first successful tests of the general theory of relativity.

The EIH equations assume that each body is a point mass. Effects due to the finite sizes of bodies—for example, tidal perturbations, spin-orbit torques, multipole corrections, etc.—are ignored. This **effacement principle** is justified in many problems of interest, where finite size effects can be shown to appear only at high PN order. But it breaks down in the case of motion sufficiently close to a rotating SBH, where even approximate descriptions of the motion must take into account the effects of the hole's spin and of its nonspherical shape.

The spin angular momentum of a black hole can have any value between zero and the maximum value allowed by the Kerr solution,

$$S_{\text{max}} = \frac{G M_\bullet^2}{c}. \tag{4.71}$$

Expressed as a vector,

$$\boldsymbol{S} = \chi S_{\text{max}}, \quad 0 \le \chi \le 1. \tag{4.72}$$

While the spin of no SBH has been convincingly measured, growth of an SBH by accretion from matter in a fixed plane should result in a spin that is close to maximal [22], and so $\chi \lesssim 1$ is a natural assumption to make.

In the weak-field and low-velocity limit, the spin of an SBH has two important consequences for the motion of an orbiting body:

1. The spin introduces a Lorentz-like, velocity-dependent force into the equations of motion. This **spin-orbit acceleration** causes the orbit to precess: both an in-plane precession, which contributes additively to the precession derived in the previous section, and also a precession of the orbital angular momentum vector about the spin axis, which causes the plane of the orbit to change. These spin-related precessions are collectively referred to as **Lense–Thirring**, or **frame-dragging**, precession.

2. A Kerr black hole (like most rotating objects) is not spherically symmetric. This nonsphericity results in an additional nonradial acceleration, which to lowest order in v/c is describable entirely in terms of the **relativistic quadrupole moment** Q of the spinning hole, given by

$$Q = -\frac{1}{c^2} \frac{S^2}{M_\bullet}. \tag{4.73}$$

The negative sign in equation (4.73) indicates that the distortion is oblate in character, that is, that the hole is flattened in the direction parallel to its spin. The quadrupole moment contains the lowest-order information about the flattening of space-time around the spinning hole. It is interesting to note that the acceleration due to the quadrupole moment does not come out of the post-Newtonian expansion, at *any* order, since those equations are derived under the assumption that the bodies are point masses.

The spin-orbit terms have amplitudes of order

$$\frac{GM_\bullet}{r^2} \times \left(\frac{GM_\bullet}{c^2 r} \times \frac{v}{c} \right), \tag{4.74}$$

that is, $O(v^3/c^3)$ relative to the Newtonian acceleration if $GM_\bullet/c^2 r \sim O(v^2/c^2)$. For this reason, spin-orbit effects are most often described as being of 1.5PN order (and higher), even though they formally appear first at 1PN order.

As discussed in chapter 2, one of the motivations for considering motion near SBHs is the prospect of detecting the gravitational radiation that would be produced. Gravitational waves are not discussed in any detail in this book, but it turns out that energy loss due to gravitational radiation appears in the PN expansion at order v^5/c^5. Orbits that evolve into a regime where gravitational-wave emission would become important can therefore be handled via the PN formalism by including terms up to order 2.5PN. However, the PN formalism is not adequate for establishing the capture conditions, that is, the critical values of the orbital elements for which a trajectory continues inside the event horizon of the SBH, since by definition such orbits must come very close to the SBH.

4.4 NEWTONIAN PERTURBATIONS

4.4.1 Distributed mass: Spherical case

We start with the simple case of a spherical star cluster around an SBH of mass $M_\bullet$. Assume that the mass m_2 of the orbiting body is much smaller than both $M_\bullet$ and the mass of the cluster. In this case, the SBH can be assumed to be fixed at the origin, and the gravitational potential is

$$\Phi(r) = -\frac{GM_\bullet}{r} + \Phi_s(r). \tag{4.75}$$

Because the potential is spherically symmetric, an orbiting body will conserve angular momentum and the motion will take place in a fixed plane, just as in the unperturbed Kepler problem. However, the frequencies associated with the radial and angular motions in this plane will no longer be equal, and the orbit will **precess**: the trajectory will trace out a rosette as the orbit's orientation—as defined by the direction of its major axis, say—gradually changes with time. Relating the rate of this "mass precession" to the form of the extended mass distribution is important in many contexts; for instance, it can complicate the detection of precession due to relativistic effects, as discussed later in this chapter, and it will turn out to play a crucial role in "resonant relaxation" (chapter 5).

Assuming that the force from the distributed mass is small compared with the force from the SBH, we can apply the technique of averaging. The averaged stellar potential is

$$\overline{\Phi}_p = \frac{1}{2\pi} \int_0^{2\pi} d\mathrm{E}\,(1 - e\cos\mathrm{E})\,\Phi_s\left[a\,(1 - e\cos\mathrm{E})\right]. \tag{4.76}$$

In terms of the Delaunay variables, precession is defined as the rate of change of ω, the argument of periapsis. Using equations (4.43) and the definition of the angular momentum, the precession rate can be written in the following equivalent forms:

$$\frac{d\omega}{dt} = \frac{\partial\overline{\Phi}_p}{\partial L} = -\frac{1}{I}\frac{\sqrt{1 - e^2}}{e}\frac{\partial\overline{\Phi}_p}{\partial e} \tag{4.77a}$$

$$= \frac{1}{2\pi I}\frac{\sqrt{1 - e^2}}{e}\int_0^{2\pi} d\mathrm{E}\cos\mathrm{E}\left[\Phi_s(r) + a\,(1 - e\cos\mathrm{E})\frac{d\Phi_s}{dr}\right] \tag{4.77b}$$

$$= -\frac{1}{\pi I}\frac{\sqrt{1 - e^2}}{e^2}\int_{r_p}^{r_a} dr\,\frac{e^2 - (1 - r/a)}{\sqrt{e^2 - (1 - r/a)^2}}\frac{d\Phi_s}{dr} \tag{4.77c}$$

$$= -\frac{1}{\pi}\frac{1}{\sqrt{GM_{\bullet}a}}\frac{\sqrt{1 - e^2}}{e^2}\int_{r_p}^{r_a} dr\,\frac{r - a\,(1 - e^2)}{\sqrt{(r - r_p)(r_a - r)}}\frac{d\Phi_s}{dr}. \tag{4.77d}$$

Precession is *retrograde*, that is, opposite in sense to the direction of orbital circulation. In the limit of large eccentricity, the integral in equation (4.77) simplifies:

$$e \to 1, \quad \frac{d\omega}{dt} \to -\frac{\sqrt{1 - e^2}}{\pi\sqrt{GM_{\bullet}a}}\int_0^{2a} dr\,\sqrt{\frac{r}{2a - r}}\frac{d\Phi_s}{dr}. \tag{4.78}$$

Eccentric orbits precess slowly, regardless of the form of Φ_s.

It is often useful to parametrize the mass distribution in a spherical nucleus as a power law:

$$\rho(r) = \rho_0\left(\frac{r}{r_0}\right)^{-\gamma} \quad (\gamma < 3). \tag{4.79}$$

For $\gamma \neq 2$, the corresponding potential is

$$\Phi_s(r) = \frac{4\pi}{(2 - \gamma)(3 - \gamma)}G\rho_0 r_0^2\left(\frac{r}{r_0}\right)^{2-\gamma} + \text{constant}. \tag{4.80}$$

Ignoring the constant term (which does not appear in the equations of motion) and averaging over the unperturbed motion,

$$\overline{\Phi}_p = \frac{4\pi}{(2 - \gamma)(3 - \gamma)}G\rho_0 r_0^2\left(\frac{a}{r_0}\right)^{2-\gamma}F_\gamma(e), \tag{4.81a}$$

$$F_\gamma(e) = \frac{1}{2\pi}\int_0^{2\pi} dx\,(1 - e\cos x)^{3-\gamma} \tag{4.81b}$$

$$= {}_2F_1\left[-\frac{2 - \gamma}{2}, -\frac{3 - \gamma}{2}, 1, e^2\right], \tag{4.81c}$$

where $_2F_1$ is the ordinary hypergeometric function. A good approximation to $F_\gamma(e)$ is

$$F_\gamma(e) \approx 1 + \alpha_1 e^2, \quad \alpha_1 = \frac{2^{3-\gamma} \Gamma\left(\frac{7}{2} - \gamma\right)}{\sqrt{\pi}\,\Gamma\left(4 - \gamma\right)} - 1, \tag{4.82}$$

which is exact for $\gamma = 0$ and $\gamma = 1$; for $0 \le \gamma < 2$, $0 < \alpha_1 \le 3/2$. When $\gamma > 1$ and e is close to 1, a better approximation is

$$F_\gamma(e) \approx 1 + \alpha_1 + \alpha_2\left(e^2 - 1\right), \quad \alpha_2 = \frac{2^{1-\gamma}\left(2 - \gamma\right)\Gamma\left(\frac{5}{2} - \gamma\right)}{\sqrt{\pi}} \frac{\Gamma\left(\frac{5}{2} - \gamma\right)}{\Gamma\left(3 - \gamma\right)}. \tag{4.83}$$

The parameters α_1 and α_2 can themselves be approximated as

$$\alpha_1 \approx \frac{3}{2} - \frac{79}{60}\gamma + \frac{7}{20}\gamma^2 - \frac{1}{30}\gamma^3, \tag{4.84a}$$

$$\alpha_2 \approx \frac{3}{2} - \frac{29}{20}\gamma + \frac{11}{20}\gamma^2 - \frac{1}{10}\gamma^3. \tag{4.84b}$$

The precession rate becomes

$$\frac{d\omega}{dt} = -\frac{4\pi}{(2-\gamma)(3-\gamma)} \frac{G\rho_0 r_0^2}{\sqrt{GM_\bullet a}} \left(\frac{a}{r_0}\right)^{2-\gamma} \frac{\sqrt{1-e^2}}{e} \frac{\partial F_\gamma}{\partial e} \tag{4.85a}$$

$$\approx -\nu_r \frac{2\alpha}{2-\gamma}\sqrt{1-e^2}\left[\frac{M_\star(a)}{M_\bullet}\right] \quad (\gamma \ne 2), \tag{4.85b}$$

where ν_r is the unperturbed (Kepler) frequency,

$$\nu_r = \frac{2\pi}{P} = \frac{\sqrt{GM_\bullet}}{a^{3/2}}.$$

In equation (4.85b), α is identified either with α_1 or α_2 and $M_\star(a)$ is the distributed mass within radius $r = a$.

Setting $\gamma = 2$ gives the "singular isothermal sphere." In this case the potential is

$$\Phi_s(r) = 4\pi G\rho_0 r_0^2 \log\left(\frac{r}{r_0}\right) + \text{constant} \tag{4.86}$$

and the precession rate is given exactly, in the orbit-averaged approximation, by

$$\frac{d\omega}{dt} = -\nu_r \frac{\sqrt{1-e^2}}{1+\sqrt{1-e^2}} \frac{M_\star(a)}{M_\bullet}. \tag{4.87}$$

We can summarize these results by writing

$$\frac{d\omega}{dt} = -\nu_r G_M(e, \gamma)\sqrt{1-e^2}\left[\frac{M_\star(a)}{M_\bullet}\right], \tag{4.88}$$

where

$$G_M(e, \gamma) = -\frac{1}{2\pi e} \frac{3-\gamma}{2-\gamma} \int_0^{2\pi} dx \cos x \left(1 - e\cos x\right)^{2-\gamma} \tag{4.89a}$$

$$= \frac{3-\gamma}{2}\, _2F_1\left[\frac{\gamma}{2} - \frac{1}{2}, \frac{\gamma}{2}, 2, e^2\right] \tag{4.89b}$$

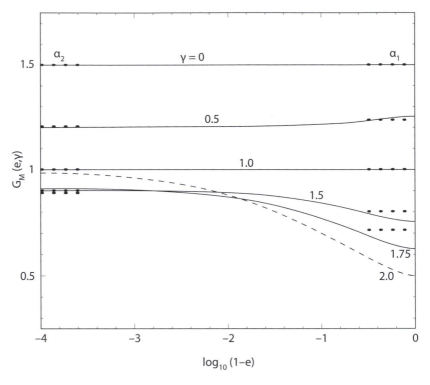

Figure 4.3 The dimensionless function $G_M(e, \gamma)$, equation (4.89), that appears in the expression (4.88) for the precession rate of a nonrelativistic orbit of eccentricity e in a spherical star cluster of density $\rho \propto r^{-\gamma}$ around an SBH. The continuous curves are the exact relations; the dashed curve is for $\gamma = 2$. The dots show the approximation of equation (4.91) (for $\gamma < 2$ only), using the approximate expressions for α_1 and α_2 given by equations (4.84).

for $\gamma \neq 2$, and

$$G_M(e, 2) = \left(1 + \sqrt{1 - e^2}\right)^{-1} \qquad (4.90)$$

for $\gamma = 2$. Figure 4.3 plots $G_M(e, \gamma)$, as well as the two approximations to G_M implied by equation (4.85b) when $\gamma \neq 2$:

$$G_M(e, \gamma) \approx \frac{2}{2 - \gamma} \alpha(\gamma) \qquad (4.91)$$

with $\alpha = \{\alpha_1, \alpha_2\}$. The latter approximation is seen to be quite good; when describing eccentric orbits, α_2 is the better choice, while α_1 is better for $e \lesssim 0.9$.

For orbits of a given eccentricity, the precession rate scales with radius as

$$\frac{d\omega}{dt} \propto v_r(a) M_\star(a) \propto a^{3/2 - \gamma}. \qquad (4.92)$$

For $\gamma < 3/2$, inner orbits precess slower than outer orbits, and vice versa for $\gamma > 3/2$.

Finally, some remarks on terminology. Precession of the kind just discussed, in which only the argument of periapsis changes, is variously called "apsidal precession" or "precession of the periapsis" or "precession of the periastron." Apsidal precession also results when the lowest-order relativistic corrections are added to the equations of motion, as discussed later in this chapter. In the context of galactic nuclei, apsidal precession due to *Newtonian* perturbations is often called **mass precession** to distinguish it from relativistic precession. The quantity $|d\omega/dt| \equiv \nu_M$ is sometimes called the **mass precession rate**. Alternatively, some authors define the mass precession rate more simply as $\nu_r(M_\star/M_\bullet)$, that is, as the precession rate of an orbit with small e.

4.4.2 Distributed mass: Axisymmetric case

4.4.2.1 Tubes and saucers

The precession that occurs in a spherically symmetric star cluster leaves the eccentricity of the precessing orbit unchanged. By contrast, if the nuclear cluster that surrounds the SBH is flattened or elongated, some component of the force exerted on a test star will be perpendicular to the star's radius vector: in other words, there will be a torque, and the star's angular momentum, both magnitude and direction, can change with time. If the eccentricity should become large enough that the orbital periapsis $r = r_p = a(1-e)$ lies inside the SBH event horizon,[4] the star can be captured. This is one way that stars can find their way into the central SBH. "Coherent resonant relaxation," discussed in chapters 5 and 6, operates via essentially the same mechanism: in this case, the nonradial forces are due to irregularities in the mass distribution resulting from the finite number of stars, whose orbits remain nearly fixed in orientation over some limited time.

The gravitational potential of a nonspherical, steady-state nucleus can be expressed quite generally as

$$\Phi(x) = -\frac{GM_\bullet}{r} + \Phi_s(r) + \Phi_t(x, y, z), \tag{4.93}$$

where Φ_s and Φ_t represent the spherical and nonspherical mass components, respectively. Suppose that the spherical component of the mass density is given by equation (4.79). Expressions for the potential of realistic, nonspherical mass distributions can be complicated. Here, we set the nonspherical component in equation (4.93) to

$$\rho_t(r, \theta) = \rho_{t,0} \left(\frac{r}{r_0}\right)^{-\gamma} P_2(\cos\theta), \tag{4.94}$$

with $P_2(x) = (3/2)x^2 - 1/2$, a Legendre polynomial. This density is symmetric about the $\theta = 0$-axis, which we identify with the z-axis; setting $\rho_{t,0} > 0$ then yields a crude representation of a prolate stellar bar, while $\rho_{t,0} < 0$ corresponds to

---

[4]The more exact condition for capture is given in section 4.6.

an oblate nucleus. For the potential corresponding to ρ_t, Poisson's equation gives

$$\Phi_t(r,\theta) = -\Phi_{t,0}\left(\frac{r}{r_0}\right)^{2-\gamma} P_2(\cos\theta),\tag{4.95a}$$

$$\Phi_{t,0} = \frac{4\pi}{\gamma(5-\gamma)}G\rho_{t,0}r_0^2 \quad (\gamma \neq 0).\tag{4.95b}$$

If we assume that the distributed mass is small compared with $M_\bullet$, we can replace the exact equations of motion by the equations derived from the orbit-averaged Hamiltonian:

$$\overline{\mathcal{H}} = -\frac{1}{2}\left(\frac{GM_\bullet}{I}\right)^2 + \overline{\Phi}_s + \overline{\Phi}_t.\tag{4.96}$$

The averaged, spherical part of the potential is given by equation (4.81). The averaged, nonspherical part is

$$2\pi\overline{\Phi}_t = -\Phi_{t,0}\int_0^{2\pi} d\mathrm{E}\left(\frac{r}{a}\right)\left(\frac{r}{r_0}\right)^{2-\gamma} P_2(\cos\theta)\tag{4.97a}$$

$$= -\frac{\Phi_{t,0}}{2}\left(\frac{a}{r_0}\right)^{2-\gamma}\int_0^{2\pi} d\mathrm{E}\left[3\left(\frac{z}{a}\right)^2\left(\frac{r}{a}\right)^{1-\gamma} - \left(\frac{r}{a}\right)^{3-\gamma}\right],$$

$$\tag{4.97b}$$

where r and z are understood to be functions of the unperturbed elements and of the eccentric anomaly via equations (4.52)–(4.53):

$$\frac{r}{a} = 1 - e\cos\mathrm{E},\tag{4.98a}$$

$$\frac{z}{a} = (\cos\mathrm{E} - e)\sin i \sin\omega + \sqrt{1-e^2}\sin i \cos\omega \sin\mathrm{E}.\tag{4.98b}$$

Setting $\gamma = 1$ is both physically reasonable—it corresponds approximately to the nuclear density profile of a giant elliptical galaxy—and it also allows the results of the averaging to be expressed in terms of simple functions:

$$\overline{\Phi}_s = 2\pi\rho_0 r_0 a\left(1 + \alpha_1 e^2\right),\tag{4.99a}$$

$$\overline{\Phi}_t = -\frac{\pi}{4}G\rho_{t,0}r_0 a\left[3\sin^2 i\left(1 - e^2 + 3e^2\sin^2\omega\right) - 2 - e^2\right].\tag{4.99b}$$

Here, $\alpha = \alpha_1$ as defined in equation (4.82); for $\gamma = 1$, α_1 is exactly equal to $1/2$.

Because $\overline{\Phi}_t$ depends on ω and on $\sin^2 i = 1 - (L_z/L)^2$, the equations of motion (4.43) imply that L and Ω will change with time: the line of nodes will precess, and the eccentricity will change due to the torque from the flattened potential. In addition, the instantaneous rate of in-plane precession, $d\omega/dt$, will differ from the precession rate due to the spherical mass component alone: directly because of the torques, and indirectly because the rate of precession due to the spherical component depends on the eccentricity.

The equations of motion can be simplified further by defining dimensionless variables. Let the dimensionless time be $\tau = \nu_0 t$, with ν_0 left unspecified for the moment. A natural choice for the dimensionless, orbit-averaged Hamiltonian is then $H = \overline{\mathcal{H}}/v_0 I$. We can also define a dimensionless angular momentum variable $\ell \equiv L/I = L/L_c(E)$, and $\cos i = \ell_z/\ell$. Note that, to this level of approximation, we can also write

$$\ell = \left(1 - e^2\right)^{1/2} . \tag{4.100}$$

Expressed in terms of these dimensionless quantities, the equations of motion (4.43) are

$$\frac{d\omega}{d\tau} = \frac{\partial H}{\partial \ell}, \quad \frac{d\ell}{d\tau} = -\frac{\partial H}{\partial \omega}, \quad \frac{d\Omega}{d\tau} = -\frac{\partial H}{\partial \ell_z}, \quad \frac{d\ell_z}{d\tau} = \frac{\partial H}{\partial \Omega} = 0. \tag{4.101}$$

These equations become especially simple if we choose

$$v_0 = 2\pi(1 + \Delta)G\rho_0 r_0 \left(\frac{GM_\bullet}{a}\right)^{-1/2}, \quad \Delta = \frac{1}{4}\frac{\rho_{t,0}}{\rho_0} . \tag{4.102}$$

Since $\Delta \ll 1$, $v_0 \approx v_r M_\star(a)/M_\bullet$, or approximately the "mass precession rate" defined in equation (4.88).

Constant terms in H—including terms that depend only on a—do not appear in the equations of motion, and we are free to drop them. The result is

$$H \equiv \frac{\overline{\mathcal{H}}}{v_0 I} = -\frac{\ell^2}{2} + \frac{3}{2}\epsilon \sin^2 i \left[\frac{\ell^2}{3} + (1 - \ell^2)\sin^2 \omega\right]. \tag{4.103}$$

The parameter ϵ, defined as

$$\epsilon \equiv -\frac{3\Delta}{1 + \Delta} \approx -\frac{3}{4}\frac{\rho_{t,0}}{\rho_0}, \tag{4.104}$$

specifies the degree of nuclear flattening or elongation. It is easy to show that the axis ratio q of the isodensity contours, evaluated on the principal axes, is

$$q \approx 1 - 2\epsilon. \tag{4.105}$$

Because ℓ_z is conserved, $H = H(\omega, \ell)$. Solving this expression for $\ell = \ell(H, \omega)$ and substituting the result into the right-hand side of the first of the equations of motion (4.101) yields $d\omega/d\tau = f(\omega)$, which can be integrated numerically. A similar procedure yields an integrable expression for $\ell(\tau)$; and when the solutions $\omega(\tau)$ and $\ell(\tau)$ are substituted into the third of the equations of motion, the time dependence of Ω is likewise determined. It follows that the (orbit-averaged) motion is completely regular [472].

Figure 4.4 shows the results of numerical integrations of the averaged equations of motion in an oblate potential ($\epsilon = 0.1$, or axis ratio ~ 0.8). Neglecting resonances, orbits fall into one of just two families:

- **Tube orbits**. Tube orbits circulate in ω and Ω. As long as ℓ_z is not too small, tube orbits approximately conserve the total angular momentum as well; thus the eccentricity and inclination are approximately constant. In configuration space, the orbit fills an annular region and is symmetric with respect to the symmetry plane.

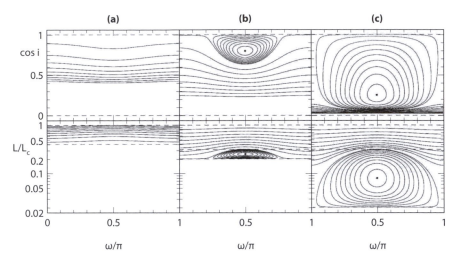

Figure 4.4 Motion in an axisymmetric nucleus around an SBH. These are solutions to the
equations of motion derived from the averaged Hamiltonian (4.103), with $\epsilon =$
0.03. The three panels correspond to different values of ℓ_z, the component of the
(dimensionless) angular momentum parallel to the symmetry axis: (a) $\ell_z = 0.4$,
(b) $\ell_z = 0.2$, (c) $\ell_z = 0.02$. Within each panel, the different phase curves have
different values of the "third integral" H. For large ℓ_z, as in panel (a), all orbits
are tubes; saucer orbits appear in this potential when $\ell_z \lesssim 0.3$, in panels (b) and
(c). In these panels, the fixed-point orbit that is the "generator" of the saucers is
indicated by the dot, and the maximum angular momentum reached by saucers is
indicated with a horizontal dashed line; the lower dashed line is ℓ_z. Saucer orbits
are important because their large angular momentum variations allow stars on
such orbits to come close to the SBH.

- **Saucer orbits**. Below some ℓ_z, a second family of orbits appear in which ω
  librates. The parent of these orbits is an orbit of fixed ℓ and $\omega = \pi/2$, which
  precesses in Ω at constant inclination, tracing out a saucer- or cone-shaped
  region in configuration space. The saucer family includes orbits for which
  L is not even approximately conserved, and for which the inclination varies
  substantially.

Tube orbits were discussed in chapter 3, in the context of motion in general
axisymmetric potentials. The "third integral" associated with tube orbits was expressed there in terms of the "shape invariant" $S_m = 1 - \cos\theta_0$, equation (3.122);
θ_0 was defined as the minimum angular separation with respect to the symmetry
axis reached by the orbit at apoapsis. In the current case, tube orbits reach their
maximum inclination (minimum θ_0) for $\omega = \pi/2$, and for the relation between H
and θ_0 equation (4.103) gives

$$\cos^2\theta_0 \approx \frac{2H}{3\epsilon}, \tag{4.106}$$

showing that H plays the role of the "third integral" here.

The saucer orbits are restricted to those parts of phase space with small L_z, but they are important because of their eccentricity variations: a star on such an orbit can come much closer to the SBH than would be predicted on the basis of its *mean* eccentricity.[5] To understand these orbits in more detail, we begin by noting that in spite of their eccentricity variations, most of the saucer orbits in figure 4.4 never become very circular: $\ell_z^2 \lesssim \ell^2(\tau) \ll 1$. It is therefore reasonable to try simplifying the Hamiltonian (4.103) by ignoring the two terms of order $\epsilon \ell^2$. The result is

$$H \approx H' \equiv -\frac{\ell^2}{2} + \frac{3}{2}\epsilon \sin^2 i \sin^2 \omega, \qquad (4.107)$$

with equations of motion

$$\frac{d\omega}{d\tau} = -\ell + \frac{3\epsilon}{\ell}\cos^2 i \sin^2 \omega, \quad \frac{d\ell}{d\tau} = -\frac{3}{2}\epsilon \sin^2 i \sin(2\omega). \qquad (4.108)$$

The fixed point—the orbit of constant (ω, ℓ, i) that is the generator of the saucers—can be found by setting $d\omega/d\tau = 0$, yielding

$$\ell^2 = \ell_{\rm fp}^2 \equiv \sqrt{3\epsilon}\,\ell_z, \quad \cos^2 i = \cos^2 i_{\rm fp} \equiv \ell_z/\sqrt{3\epsilon}. \qquad (4.109)$$

Evidently, a fixed point will only exist if $\ell_{\rm fp} > \ell_z$, or

$$\ell_z < \sqrt{3\epsilon} \equiv \ell_{\rm sep} \qquad (4.110)$$

and this is the (approximate) condition for saucers to exist. Note that the requirement of small ℓ_z implies that saucer orbits are likely to represent a small fraction of the orbital population of an axisymmetric nucleus.

The separatrix dividing the saucers from the tubes has $H' = -\ell_z^2/2$, and the angular momentum of this orbit spans the maximal range,

$$\ell_{\rm min} \le \ell \le \ell_{\rm max}, \; \ell_{\rm min} = \ell_z, \; \ell_{\rm max} = \sqrt{3\epsilon} = \ell_{\rm sep}. \qquad (4.111)$$

It is interesting that the maximum angular momentum attained by saucers is independent of L_z; it depends only on the degree of elongation of the nucleus. The inclination of the separatrix orbit likewise exhibits the maximal variation:

$$\frac{\ell_z}{\sqrt{3\epsilon}} \le \cos^2 i \le 1. \qquad (4.112)$$

Saucer orbits reach their maximum eccentricity near the equatorial plane; as the orbit tilts up toward the z-axis, conservation of ℓ_z implies that ℓ must increase. Conservation of ℓ_z and H' likewise imply a simple relation between the maximum and minimum values of ℓ reached by any saucer orbit:

$$\ell_- = \frac{\ell_{\rm fp}^2}{\ell_+} = \frac{\ell_{\rm sep}\ell_z}{\ell_+}, \quad \ell_z < \sqrt{3\epsilon}. \qquad (4.113)$$

As ℓ_z tends to zero, figure 4.4 shows that saucer orbits "crowd out" tube orbits; the only tubes that remain are highly inclined and nearly circular.

Many readers will notice the similarities between the behavior of saucer orbits, and the "Lidov–Kozai cycles" that occur in the hierarchical three-body problem [321, 299]. Lidov–Kozai cycles are discussed in detail in section 4.8.

---

[5]Similar orbits can exist in oblate potentials lacking central SBHs; they have been called "pipe orbits" [456] and "reflected banana orbits" [314] among other names.

4.4.2.2 Motion in axisymmetric nuclei

Having established the basic character of the motion in axisymmetric nuclei, a more general treatment will now be presented,[6] relaxing two of the assumptions made above: the index, γ, of the density power law will be allowed to have any value; and it will no longer be assumed that $\epsilon \ell^2 \ll 1$.

Consider then a model of the nucleus similar to the one considered in section 4.4.2.1, consisting of a spherical and a flattened component in addition to the SBH. Let the stellar density and its corresponding potential be

$$\rho(\boldsymbol{x}) = \rho_0 \left(\frac{r}{r_0} \right)^{-\gamma} \left(1 + \epsilon_d \left[\frac{z^2}{r^2} - \frac{1}{3} \right] \right), \tag{4.114a}$$

$$\Phi(\boldsymbol{x}) = \Phi_0 \left(\frac{r}{r_0} \right)^{2-\gamma} \left(1 + \epsilon_p \left[\frac{z^2}{r^2} - \frac{1}{3} \right] \right), \tag{4.114b}$$

where the flattening of the isodensity surfaces is specified via the parameter ϵ_d and the axis of rotational symmetry is the z-axis. Poisson's equation gives

$$\Phi_0 = \frac{4\pi G \rho_0 r_0^2}{(3 - \gamma)(2 - \gamma)}, \tag{4.115a}$$

$$\epsilon_d = \frac{3(1 - q^{-\gamma})}{1 + 2q^{-\gamma}}, \tag{4.115b}$$

$$\epsilon_p = \epsilon_d \frac{(3 - \gamma)(2 - \gamma)}{\gamma(\gamma - 5)}, \quad 0 \le \gamma < 2. \tag{4.115c}$$

Here, $q \le 1$ is the axis ratio of an isodensity surface; since these surfaces are not ellipsoids, q is defined in terms of the points of intersection of an isodensity surface with the z- and x-axes. The parameter ϵ defined in the previous section is approximately equal to ϵ_p.

As in the previous section, we replace the exact equations of motion by the equations derived from the Hamiltonian averaged over the unperturbed motion:

$$\overline{\mathcal{H}} = -\frac{1}{2} \left(\frac{GM_\bullet}{I} \right)^2 + \overline{\Phi}. \tag{4.116}$$

After the averaging, $\overline{\mathcal{H}}$ is independent of the mean anomaly M and I is conserved, as is the semimajor axis a. Since $\overline{\mathcal{H}}$ is an integral of the motion, so is $\overline{\Phi}$. The third integral, of course, is L_z due to the axial symmetry of the potential.

---

[6]This section contains unpublished work carried out in collaboration with E. Vasiliev.

A good approximation to $\overline{\Phi}$ is

$$\overline{\Phi} = \Phi_0 \left(\frac{a}{r_0}\right)^{2-\gamma} H(\ell, \ell_z), \tag{4.117a}$$

$$H = \left(1 - \frac{\epsilon_p}{3}\right)\left[N - (N-1)\ell^2\right] + \tag{4.117b}$$

$$+ \epsilon_p \left(1 - \frac{\ell_z^2}{\ell^2}\right)\left[N(1-\ell^2)\sin^2\omega + \frac{1}{2}\ell^2\right],$$

$$N \equiv \frac{2^{3-\gamma}\,\Gamma\left(\frac{7}{2} - \gamma\right)}{\sqrt{\pi}\,\Gamma(4-\gamma)}. \tag{4.117c}$$

These expressions are exact for $\gamma = 0, 1$ (for which $N = 5/2, 3/2$) and approximate the true value to within few percent in other cases. The special case $\gamma = 1$ was considered in detail in the previous section; the analysis that follows generalizes that treatment to arbitrary γ.

Define a dimensionless time $\tau = \nu_0 t$ where

$$\nu_0 = \frac{\Phi_0}{I}\left(\frac{r_0}{a}\right)^{\gamma-2} \approx \frac{2\pi}{2-\gamma}\frac{1}{P}\frac{\tilde{M}}{M_\bullet}, \tag{4.118}$$

with $\tilde{M} \equiv 4\pi a^3 \rho(a)/(3-\gamma)$, an approximation to the mass in stars within $r = a$. For $\gamma = 1$ and $\epsilon_p \ll 1$, this is the same quantity defined in equation (4.102). In terms of the dimensionless H and τ, the equations of motion are then given by equations (4.101).

It is convenient to replace H by $\mathcal{H}$, a linear combination of the integrals H and ℓ_z:

$$\mathcal{H} \equiv \frac{N(1 - \epsilon_p/3) - H - (N-1)(1 - \epsilon_p/3)\mathcal{R}_z}{(N-1)(1 - \epsilon_p/3) - \epsilon_p/2}$$

$$= (\mathcal{R} - \mathcal{R}_z)\left(1 - \frac{\mathcal{R}_{\text{sep}}}{1 - \mathcal{R}_{\text{sep}}}\frac{1-\mathcal{R}}{\mathcal{R}}\sin^2\omega\right), \tag{4.119}$$

where

$$\mathcal{R} \equiv L^2/L_c^2 = \ell^2, \quad \mathcal{R}_z \equiv \ell_z^2, \tag{4.120}$$

and $\mathcal{R}_{\text{sep}}$ is given by

$$\mathcal{R}_{\text{sep}} \equiv \frac{N\epsilon_p}{(N-1)(1 - \epsilon_p/3) - \epsilon_p/2 + N\epsilon_p}. \tag{4.121}$$

It is shown below that $\mathcal{R}_{\text{sep}}$ is the maximum value of $\mathcal{R}_z$ for which saucers exist. Equation (4.121) therefore generalizes equation (4.110), the approximate condition when $\gamma = 1$.

The equation of motion for $\mathcal{R}$ is obtained by eliminating ω from the first of equations (4.101) using equation (4.119). The result is

$$\frac{\partial \mathcal{R}}{\partial \tau} = -k\sqrt{(\mathcal{R}_1 - \mathcal{R})(\mathcal{R} - \mathcal{R}_2)(\mathcal{R} - \mathcal{R}_3)}, \tag{4.122}$$

where

$$k \equiv \frac{4N\epsilon_p\sqrt{1 - \mathcal{R}_{\text{sep}}}}{\mathcal{R}_{\text{sep}}} \approx 4(N - 1) \text{ for } \epsilon_p \ll 1, \qquad (4.123a)$$

$$\mathcal{R}_{1,2} \equiv \mathcal{R}_\star \pm \sqrt{\mathcal{R}_\star^2 - \mathcal{R}_{\text{sep}}\mathcal{R}_z}, \qquad (4.123b)$$

$$\mathcal{R}_3 \equiv \mathcal{H} + \mathcal{R}_z, \qquad (4.123c)$$

$$\mathcal{R}_\star \equiv \frac{1}{2}\left[\mathcal{R}_{\text{sep}}(1 + \mathcal{R}_z) + (1 - \mathcal{R}_{\text{sep}})(\mathcal{H} + \mathcal{R}_z)\right]. \qquad (4.123d)$$

It is clear from the form of equation (4.122) that $\mathcal{R}$ will oscillate between $\mathcal{R}_+ \equiv \mathcal{R}_1$ and $\mathcal{R}_- \equiv \max(\mathcal{R}_2, \mathcal{R}_3)$. We thus have two classes of orbit, depending on the relation between $\mathcal{R}_2$ and $\mathcal{R}_3$, which are separated by $\mathcal{H} = 0$. For $\mathcal{H} > 0$, $\mathcal{R}_3 > \mathcal{R}_2$ and the orbit is a tube orbit, while for $\mathcal{H} < 0$ the orbit is a saucer. It is easy to show from equations (4.119) and (4.121) that saucer orbits appear only for $\mathcal{R}_z \leq \mathcal{R}_{\text{sep}}$.

Defining one additional variable as $\mathcal{R}_{\text{low}} \equiv \min(\mathcal{R}_2, \mathcal{R}_3)$, the solution to equation (4.122) can be expressed in terms of the elliptic cosine function, cn, as

$$\mathcal{R}(\tau) = \mathcal{R}_- + (\mathcal{R}_+ - \mathcal{R}_-) \qquad (4.124)$$

$$\times \text{cn}^2\left(\frac{k\sqrt{\mathcal{R}_+ - \mathcal{R}_{\text{low}}}}{2}\tau, \sqrt{\frac{\mathcal{R}_+ - \mathcal{R}_-}{\mathcal{R}_+ - \mathcal{R}_{\text{low}}}}\right).$$

The period of oscillation in ω or $\mathcal{R}$ is given, in physical (not dimensionless) units, by

$$T_{\text{prec}} = v_0^{-1}\frac{4}{k\sqrt{\mathcal{R}_+ - \mathcal{R}_{\text{low}}}}K\left(\sqrt{\frac{\mathcal{R}_+ - \mathcal{R}_-}{\mathcal{R}_+ - \mathcal{R}_{\text{low}}}}\right), \qquad (4.125)$$

where $K(k)$ is the complete elliptic integral,

$$K(k) = \int_0^1 \frac{dt}{\sqrt{(1 - t^2)(1 - k^2t^2)}}. \qquad (4.126)$$

For orbits not too close to the separatrix, $K \approx \pi/2$. Note that, in a time T_{prec}, ω varies by π for tube orbits and by an amount $\leq \pi$ for saucers.

Simple expressions can also be found that relate $\mathcal{R}_-$ and $\mathcal{R}_+$. For tube orbits, the relation is

$$\mathcal{R}_- = \mathcal{R}_+ - \frac{\mathcal{R}_{\text{sep}}}{1 - \mathcal{R}_{\text{sep}}}\frac{(1 - \mathcal{R}_+)(\mathcal{R}_+ - \mathcal{R}_z)}{\mathcal{R}_+}. \qquad (4.127)$$

It is easy to see that—if $\epsilon_p \ll 1$ and $\mathcal{R}_{\text{sep}} \ll \mathcal{R} \lesssim 1$—these two values are quite close to each other, verifying that L^2 is approximately conserved (figure 4.4). For saucer orbits the relation is even simpler:

$$\mathcal{R}_-\mathcal{R}_+ = \mathcal{R}_{\text{sep}}\mathcal{R}_z \qquad (4.128)$$

which generalizes equation (4.113). The condition $\mathcal{R}_- = \mathcal{R}_+$ gives the fixed-point orbit

$$\mathcal{R}_{\text{fp}} = (\mathcal{R}_{\text{sep}}\mathcal{R}_z)^{1/2} \approx \sqrt{\frac{N}{N - 1}}\epsilon_p^{1/2}\ell_z \qquad (4.129)$$

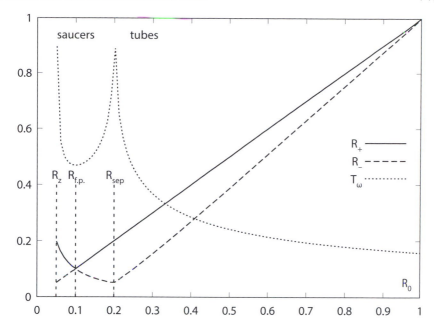

Figure 4.5 Properties of orbits in the axisymmetric Hamiltonian (4.117). The solid and dashed curves show the maximum and minimum values of the dimensionless angular momentum, $\mathcal{R} = L^2/L_c^2$, as a function of $\mathcal{R}_0$, the value of $\mathcal{R}$ when $\omega = \pi/2$. Each saucer orbit appears twice, to the left and right of the fixed point (f.p.). The dotted curve is the period, equation (4.125). This figure assumes $\mathcal{R}_z = 0.05$ and $\mathcal{R}_{sep} = 0.2$.

which generalizes equation (4.109). Figure 4.5 plots some properties of orbits for $\mathcal{R}_z = 0.05$, $\mathcal{R}_{sep} = 0.2$.

By inverting equation (4.119), we can express $\mathcal{R}$ in terms of ω:

$$\mathcal{R}(\omega) = \frac{\mathcal{R}_a \pm \sqrt{\mathcal{R}_a^2 - 4\mathcal{R}_z \mathcal{R}_b(1 + \mathcal{R}_b)}}{2(1 + \mathcal{R}_b)}, \tag{4.130a}$$

$$\mathcal{R}_b \equiv \frac{\mathcal{R}_{sep}}{1 - \mathcal{R}_{sep}} \sin^2 \omega, \tag{4.130b}$$

$$\mathcal{R}_a \equiv \mathcal{H} + \mathcal{R}_z + \mathcal{R}_b(1 + \mathcal{R}_z). \tag{4.130c}$$

In the case of tube orbits only the upper root in equation (4.130a) is relevant, while for saucers both roots are valid, as long as $\sin^2 \omega$ exceeds

$$\sin^2 \omega_{min} = \frac{1 - \mathcal{R}_{sep}}{\mathcal{R}_{sep}} \frac{\mathcal{R}_z(1 - \mathcal{H} - \mathcal{R}_z) - \mathcal{H} + 2\sqrt{-\mathcal{H}\mathcal{R}_z(1 - \mathcal{H} - \mathcal{R}_z)}}{(1 - \mathcal{R}_z)^2}. \tag{4.131}$$

Near the fixed point, the angular momentum of a saucer orbit varies sinusoidally with time. Away from the fixed point, the period increases, becoming infinite on

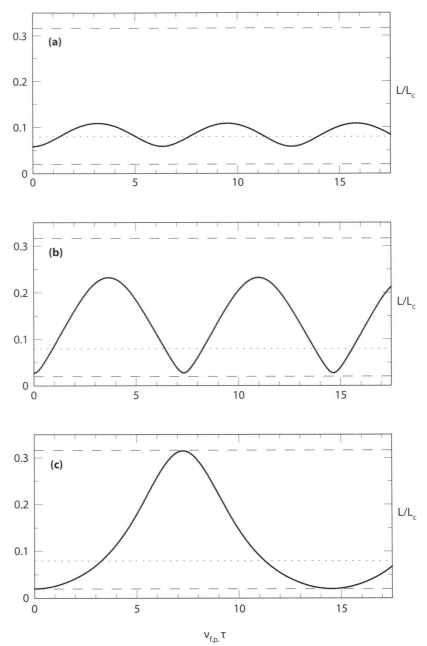

Figure 4.6 Time dependence of the angular momentum for three saucer orbits drawn from figure 4.4c. Orbit (a) is near the fixed-point orbit, while orbit (c) is near the separatrix orbit that separates saucers from tubes. Dotted lines show ℓ_{fp}, the angular momentum of the fixed point; dashed lines are plotted at $\ell = \ell_z$ and $\ell = \epsilon^{1/2}$, the minimum and maximum ℓ reached by saucer orbits. The frequency of oscillations near the fixed point, ν_{fp}, is used to scale the time.

the saucer/tube separatrix. Figure 4.6 illustrates the time dependence of the angular momentum for three saucer orbits from figure 4.4. The sinusoidal behavior of $\ell(\tau)$ near the fixed point changes to a very different form near the separatrix, in which ℓ lingers near its minimum value, $\ell \approx \ell_z$, for a larger and larger fraction of the period.

Finally, we delineate the regions in $(\mathcal{H}, \mathcal{R}_z)$ space corresponding to the orbital types. As noted above, the boundary between tube and saucer orbits is $\mathcal{H} = 0$. The other important boundaries are

$$\mathcal{H} = 1 - \mathcal{R}_z \qquad \text{(circular orbit, i.e., } \mathcal{R} = 1\text{),}$$

$$\mathcal{H} = -\frac{\mathcal{R}_{\text{sep}}}{1 - \mathcal{R}_{\text{sep}}} \left(1 - \sqrt{\frac{\mathcal{R}_z}{\mathcal{R}_{\text{sep}}}} \right)^2 \qquad \text{(fixed-point saucer).} \tag{4.132}$$

Looking ahead to the loss-cone problem (chapter 6), we can ask, what are the conditions on an orbit such that its minimum angular momentum falls below some interesting value ℓ_{lc}—allowing capture by the SBH? Evidently we require $\ell_z < \ell_{\text{lc}}$. If in addition $\ell_z \gtrsim \sqrt{3\epsilon}$, only tubes will exist, and the near constancy of angular momentum for tube orbits allows us to write the capture condition simply as $\ell \lesssim \ell_{\text{lc}}$. If $\ell_z \lesssim \sqrt{3\epsilon}$, saucers will exist as well; the capture condition for saucers is $\ell_- < \ell_{\text{lc}}$, or

$$\ell_+ > \frac{\ell_{\text{sep}}}{\ell_{\text{lc}}} \ell_z = \frac{\sqrt{3\epsilon}}{\ell_{\text{lc}}} \ell_z. \tag{4.133}$$

Because ℓ_+ can greatly exceed ℓ_z, the fraction of stars in an axisymmetric galaxy that are available to feed the SBH[7] can greatly exceed the fraction in an equivalent spherical galaxy. We can estimate the fraction in an axisymmetric galaxy, as follows. Recall that in an isotropic, spherical galaxy, $N(L; E)dL \propto L\, dL$ (equation 3.44); suppose that the same is true in the axisymmetric galaxy. For an orbit drawn randomly from a uniform distribution in L^2 and $\cos i = L_z/L$, the probability that $\ell_- < \ell_{\text{lc}}$ is the product of two factors: the first, $\sim \ell_{\text{sep}}^2$, asserts that the orbit is a saucer (since only very few tube orbits near the separatrix have low ℓ_-); and the second, $\sim \ell_{\text{lc}}/\ell_{\text{sep}}$, demands that $\ell_- = (\ell_z/\ell_+)\ell_{\text{sep}} < \ell_{\text{lc}}$. The fraction of orbits with $\ell_- < \ell_{\text{lc}}$ is then $\sim \ell_{\text{sep}}\ell_{\text{lc}} \approx \sqrt{\epsilon}\ell_{\text{lc}}$. By contrast, in a spherical galaxy, this fraction would be $\sim \ell_{\text{lc}}^2$, that is, smaller by a factor $\sim \ell_{\text{lc}}/\sqrt{\epsilon}$. At least until such a time as the saucer orbits have been "drained" by the SBH, feeding rates in axisymmetric galaxies can be much higher than in spherical galaxies [334].

4.4.3 Distributed mass: Triaxial case

4.4.3.1 Pyramids

The tube and saucer orbits that characterize motion near an SBH in an axisymmetric nucleus are still present in nonaxisymmetric, or triaxial, nuclei. In fact, two families of tube orbits exist, circulating about both the short and the long axes of the triaxial figure, as well as saucers that circulate about the short axis. Like tube

---

[7]The possibility of scattering of stars onto low angular momentum orbits is being ignored here.

orbits in the axisymmetric geometry, tube orbits in triaxial potentials respect a third integral that is similar to the total angular momentum, and they avoid the very center. But as discussed in chapter 3, triaxial potentials can also support orbits that are qualitatively different from tubes and saucers: "centrophilic" orbits that pass arbitrarily close to the SBH. Such orbits are potentially very important for getting stars and stellar remnants into the SBH [371].

Centrophilic orbits exist even in axisymmetric nuclei, but they are restricted to a meridional plane, that is, to a plane that contains the z- (symmetry) axis. Orbits in the meridional plane have $L_z = 0$, and so conservation of L_z does not impose any additional restriction on the motion. Perturbing such an orbit *out* of the meridional plane implies a nonzero L_z: the orbit is converted into a saucer or a tube and again avoids the center. But in the triaxial geometry, L_z is not conserved, and it turns out that a substantial fraction of such "perturbed" planar orbits will maintain their centrophilic character, becoming **pyramid orbits**.

To illustrate how pyramid orbits arise near an SBH in a triaxial nucleus, we return temporarily to the axisymmetric model discussed in the previous section. Consider motion that is restricted to a meridional plane. Setting $i = \pi/2$ in the averaged Hamiltonian (4.103) yields a simple relation between $\ell = (1 - e^2)^{1/2}$ and ω for motion in this plane:

$$\frac{e^2}{e_0^2} = \frac{3 + 2\epsilon}{3 + 2\epsilon - 3\epsilon \cos^2 \omega} \approx 1 + 3\epsilon \sin^2 \omega \quad (\epsilon \ll 1). \tag{4.134}$$

Here, e_0 is the eccentricity when $\omega = \pi/2$, that is, when the orbit is oriented with its major axis parallel to the short (z-) axis. As the orbit precesses away from the symmetry axis, its angular momentum decreases due to the torques from the flattened potential. If

$$e_0 > e_{min} \approx \left(1 + \frac{\epsilon}{3}\right)^{-1/2}, \tag{4.135}$$

the eccentricity reaches unity before the circulation in ω has brought the orbit to the x-axis. In this case, circulation in ω changes to **libration**: the orbit—which is highly eccentric if ϵ is small—librates about the short axis, reaching a maximum angular displacement θ_{max} given by

$$\sin^2 \theta_{max} = \frac{1 - e_0^2}{1 - e_{min}^2} \tag{4.136}$$

(figure 4.7a). When $\theta = \pm\theta_{max}$, the direction of the angular momentum instantaneously flips, and the precession (which is due almost entirely to the spherical mass component if ϵ is small) reverses direction as well. As discussed in more detail below, relativistic effects would necessarily dominate the motion when the eccentricity becomes so large; nevertheless, the essential character of the motion predicted by this model—libration about the short axis, with the eccentricity reaching a maximum value near the turning points—turns out to be robust.

If on the other hand $e_0 < e_{min}$, the orbit never attains unit eccentricity. The motion then consists of circulation in ω and the eccentricity oscillates between a minimum value of $e = e_0$ when $\omega = 0$ and a maximum value of $e \approx (1 + 3\epsilon/2)e_0$

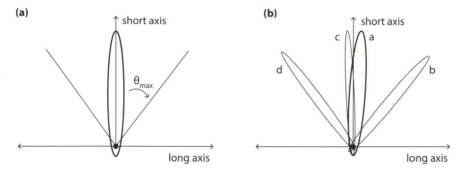

Figure 4.7 (a) The origin of pyramid orbits around an SBH in a triaxial nucleus. The motion
is Newtonian: the orbit precesses due to the spherical component of the distributed mass, and at the same time, its eccentricity changes due to the torque from
the flattened potential. When $\theta = \theta_{max}$, the angular momentum drops to zero (in
the orbit-averaged approximation) and the direction of circulation about the SBH
reverses, as does the sense of the precession. When perturbed out of the plane,
this orbit would become a three-dimensional pyramid orbit. (b) A windshield-
wiper orbit. Here, the motion includes the effects of the relativistic (Schwarzschild) precession. When the eccentricity reaches a critical value (at b), the rate
of Schwarzschild precession matches the rate of mass precession; because the
two act in opposite directions, the precession halts momentarily, then reverses
($b \rightarrow c$). The torques continue to increase the eccentricity, until the orbit crosses
the short axis (c), after which the eccentricity starts to decrease. When e is small
enough that mass precession again dominates, the direction of precession reverses again (d). This orbit's name derives from the fact that motion from $b \rightarrow d$
occurs more quickly than motion from $a \rightarrow b$.

when $\omega \approx 0, \pi, \ldots$, that is, when the orbit is elongated in the direction of the long
axis.

The case of large eccentricity is of particular interest because stars on these orbits
are able to come closest to the SBH. We now consider this case in detail [378]. As
we will see, libration about the short axis now occurs in two directions, with different frequencies, and the orbit only attains unit eccentricity when both oscillations
reach their respective extrema at the same time.

4.4.3.2 Motion in triaxial nuclei

Consider a nucleus in which the gravitational potential is given by equation (4.93)
with

$$\Phi_t(x, y, z) = 2\pi \, G\rho_t \left(T_x x^2 + T_y y^2 + T_z z^2 \right), \qquad (4.137)$$

with ρ_t a constant. Φ_t can be interpreted as the potential of a homogeneous triaxial ellipsoid of density ρ_t—for instance, a large-scale stellar bar. Alternatively,
Φ_t could be taken as a first approximation to the potential of an inhomogeneous
triaxial component. In the former case, the dimensionless coefficients (T_x, T_y, T_z),
of order unity, are expressible in terms of the axis ratios of the ellipsoid via elliptic

functions [80]. As always in this book, the x- and z-axis are assumed to be the long and short axis, respectively, of the triaxial figure; that is, $\{T_x, T_y\} < T_z$.

If the perturbing forces are small compared with the force from the SBH, we can average the Hamiltonian over the fast angle associated with radial motion. Assuming again a power-law mass distribution, equation (4.80), for the averaged, spherical part of the perturbing potential, we find

$$\overline{\Phi}_s = \frac{4\pi}{(2-\gamma)(3-\gamma)} G\rho_0 r_0^2 \left(\frac{a}{r_0}\right)^{2-\gamma} \left(1 + \alpha_1 - \alpha_2 \ell^2\right). \qquad (4.138)$$

Here we have adopted the second of our two approximations to $F_\gamma(e)$, equation (4.83), appropriate when the eccentricity is large. The orbit-averaged triaxial component is given, after some algebra, by

$$\overline{\Phi}_t = 2\pi G\rho_t T_x \, a^2 \left[\frac{5}{2} - \frac{3}{2}\ell^2 + \epsilon_b^{(t)} H_b(\ell, \ell_z, \omega, \Omega) + \epsilon_c^{(t)} H_c(\ell, \ell_z, \omega)\right],$$

$$(4.139a)$$

$$H_b = \frac{1}{2} \left[(5 - 4\ell^2)(c_\omega s_\Omega + c_i c_\Omega s_\omega)^2 + \ell^2(s_\omega s_\Omega - c_i c_\Omega c_\omega)^2\right], \qquad (4.139b)$$

$$H_c = \frac{1}{4}(1 - c_i^2)\left[5 - 3\ell^2 - 5(1 - \ell^2)c_{2\omega}\right], \qquad (4.139c)$$

$$\epsilon_b^{(t)} \equiv T_y/T_x - 1, \quad \epsilon_c^{(t)} \equiv T_z/T_x - 1. \qquad (4.139d)$$

The shorthand s_x, c_x has been used for $\sin x$, $\cos x$. As in the previous section, we have defined dimensionless angular momentum variables $\ell = L/I$ and $\ell_z = L_z/I$; the eccentricity is $e = \sqrt{1 - \ell^2}$ and the orbital inclination i is given by $\cos i = \ell_z/\ell$.

We can simplify the Hamiltonian by dropping constant terms, including terms that depend only on a, and defining a dimensionless time $\tau = v_0 t$, where

$$I v_0 \equiv 2\pi G\rho_0 a^2 \left(\frac{a}{r_0}\right)^{-\gamma} \left[\frac{4}{3}\frac{\alpha}{(3-\gamma)(2-\gamma)} + \frac{\rho_t}{\rho_0} T_x\right]; \qquad (4.140)$$

henceforth, $\alpha \equiv \alpha_2$. Since the orbit-averaged spherical potential, equation (4.138), has the same dependence on ℓ^2 as the first nonconstant term in the averaged triaxial potential, the coefficients at ℓ^2 have been summed when defining v_0. Expressing v_0 in terms of the radial frequency v_r,

$$v_0 = v_r \left[\frac{M_t(a)}{M_\bullet}\frac{3T_x}{2} + \frac{M_s(a)}{M_\bullet}\frac{2\alpha}{3(2-\gamma)}\right], \qquad (4.141a)$$

$$M_t(a) \equiv \frac{4\pi}{3}a^3\rho_t, \quad M_s(a) \equiv \frac{4\pi}{3-\gamma}a^3\rho_0\left(\frac{a}{r_0}\right)^{-\gamma}, \qquad (4.141b)$$

where $M(a)$ denotes the mass enclosed within radius $r = a$. Since $M_t \ll M_s$, v_0 is related to the precession frequency due to the spherical part of the cluster, $v_M = |d\omega/dt|$, by $v_M \approx 3\ell v_0$. The dimensionless Hamiltonian $H \equiv \overline{\Phi}_p/v_p I$ and the equations of motion for the osculating elements are then

$$H = -\frac{3}{2}\ell^2 + \epsilon_b H_b + \epsilon_c H_c, \tag{4.142a}$$

$$\frac{d\ell}{d\tau} = -\frac{\partial H}{\partial \omega}, \quad \frac{d\omega}{d\tau} = \frac{\partial H}{\partial \ell}, \quad \frac{d\ell_z}{d\tau} = -\frac{\partial H}{\partial \Omega}, \quad \frac{d\Omega}{d\tau} = \frac{\partial H}{\partial \ell_z}. \tag{4.142b}$$

Aside from factors of order unity, the renormalized triaxiality coefficients that appear in equation (4.142) are

$$\epsilon_{b,c} \simeq (T_{y,z} - T_x) \frac{\rho_t}{\rho_0} \left(\frac{a}{r_0}\right)^\gamma. \tag{4.143}$$

Solutions to the equations of motion (4.142b) that are characterized by circulation in both ω and Ω correspond to tube orbits about the short axis—similar to the tube orbits in the axisymmetric geometry. Motion that circulates in ω but librates in Ω corresponds to tube orbits about the long axis. Motion that circulates in Ω and librates in ω corresponds to saucer orbits.

Our primary interest here is in orbits that librate in both ω and Ω. Since the rate of precession in ω is proportional to ℓ, for sufficiently low ℓ the triaxial torques can produce substantial changes in ℓ on a precession timescale via the first term in (4.142b). As a result, the circulation in ω can change to libration and the orbital eccentricity can approach arbitrarily close to one. As discussed above, this is the origin of the pyramid orbits. An example of a pyramid orbit, obtained via direct numerical integration of the equations of motion, was shown in figure 3.14. The apex of the pyramid is defined by the orbital periapsis and so lies close to the SBH; the pyramid's base is traced out by the apoapsis as it oscillates in two directions about the short axis of the triaxial figure.

Pyramid orbits can be treated analytically if the following two additional approximations are made [378]:

1. The angular momentum is assumed to be small, $\ell^2 \ll 1$.
2. The density of the triaxial component is assumed to be small compared with that of the spherical component; that is, $\epsilon_b, \epsilon_c \ll 1$.

These two assumptions allow us to omit the second-order terms in ϵ_b, ϵ_c and ℓ^2 from the Hamiltonian (4.142), yielding

$$H \approx H' \equiv -\frac{3}{2}\ell^2 + \frac{5}{2}\left[\epsilon_c(1 - c_i^2)s_\omega^2 + \epsilon_b(c_\omega s_\Omega + c_i s_\omega c_\Omega)^2\right]. \tag{4.144}$$

Pyramid orbits resemble precessing rods. Intuitively, one expects the important variables to be the two angles that describe the orientation of the rod, and its eccentricity. In equation (4.59) we defined a unit vector, e_e, that points along the major axis of the orbit, toward periapsis. Dropping the subscript n, the components of that vector are

$$e_x = \cos\omega \cos\Omega - \sin\omega \cos i \sin\Omega, \tag{4.145}$$
$$e_y = \sin\omega \cos i \cos\Omega + \cos\omega \sin\Omega,$$
$$e_z = \sin\omega \sin i,$$

and $e_x^2 + e_y^2 + e_z^2 = 1$. In terms of these variables, the Hamiltonian (4.144) takes on a particularly simple form:

$$H' = -\frac{3}{2}\ell^2 + \frac{5}{2}\left[\epsilon_c - \epsilon_c e_x^2 - (\epsilon_c - \epsilon_b)e_y^2\right].\qquad(4.146)$$

As expected, H' depends on only three variables: e_x and e_y, which describe the orientation of the orbit's major axis; and the angular momentum ℓ.

To find the equations of motion, we must switch to a Lagrangian formulation. Taking the first time derivatives of equations (4.145), and using equations (4.142b), we find

$$\dot{e}_x = 3\ell(\sin\omega\,\cos\Omega + \cos\omega\,\sin\Omega\,\cos i),$$

$$\dot{e}_y = 3\ell(\sin\omega\,\sin\Omega - \cos\omega\,\cos\Omega\,\cos i),\qquad(4.147)$$

where $\dot{e}_x \equiv de_x/d\tau$ etc. Taking second time derivatives, the variables (Ω, i, ω) drop out, and the equations of motion for e_x and e_y turn out to be expressible purely in terms of e_x and e_y:

$$\ddot{e}_x = -e_x\,6(H' + 3\ell^2)$$

$$= -e_x\left[30\epsilon_c - 6H' - 30\epsilon_c e_x^2 - 30(\epsilon_c - \epsilon_b)e_y^2\right],\qquad(4.148a)$$

$$\ddot{e}_y = -e_y\,6(H' + 3\ell^2 - 5\epsilon_b)$$

$$= -e_y\left[30\epsilon_c - 6H' - 15\epsilon_b - 30\epsilon_c e_x^2 - 30(\epsilon_c - \epsilon_b)e_y^2\right].\qquad(4.148b)$$

Given solutions to these equations, the additional elements $(\ell, \ell_z, \omega, \Omega)$ follow from equations (4.145) and (4.147); in particular, the angular momentum is

$$\ell^2 = \frac{\dot{e}_x^2 + \dot{e}_y^2 - (\dot{e}_x e_y - e_x \dot{e}_y)^2}{9(1 - e_x^2 - e_y^2)} = \frac{1}{9}(\dot{e}_x^2 + \dot{e}_y^2 + \dot{e}_z^2).\qquad(4.149)$$

Ignoring for a moment the nonlinearity of the oscillations (i.e., assuming $e_x, e_y \ll 1$), the oscillations are harmonic and uncoupled, with dimensionless frequencies

$$\nu_x^{(0)} = \sqrt{15\epsilon_c}\,,\quad \nu_y^{(0)} = \sqrt{15(\epsilon_c - \epsilon_b)}.\qquad(4.150)$$

The corners of the pyramid's base are defined by $\dot{e}_x = \dot{e}_y = 0$. From equation (4.147), $(\dot{e}_x, \dot{e}_y) = 0$ implies $\ell = 0$; that is, the eccentricity reaches 1 at the corners. The full solution in this limiting case is

$$e_x(\tau) = e_{x0}\cos(\nu_x^{(0)}\tau + \phi_x),$$

$$e_y(\tau) = e_{y0}\cos(\nu_y^{(0)}\tau + \phi_y),\qquad(4.151a)$$

$$\ell^2(\tau) = \ell_{x0}^2\sin^2(\nu_x^{(0)}\tau + \phi_x) + \ell_{y0}^2\sin^2(\nu_y^{(0)}\tau + \phi_y),$$

$$r_p(\tau) = r_{px0}\sin^2(\nu_x^{(0)}\tau + \phi_x) + r_{py0}\sin^2(\nu_y^{(0)}\tau + \phi_y),\qquad(4.151b)$$

where $r_p(\tau) \approx \ell^2(\tau)/2$ is the periapsis distance and

$$\ell_{x0} = v_x^{(0)} e_{x0}/3, \qquad\qquad \ell_{y0} = v_y^{(0)} e_{y0}/3, \qquad (4.152a)$$

$$r_{px0} = \frac{1}{18} \left(v_x^{(0)} e_{x0}\right)^2, \qquad r_{py0} = \frac{1}{18} \left(v_y^{(0)} e_{y0}\right)^2. \qquad (4.152b)$$

These solutions comprise a two-parameter family: e_{x0} and e_{y0} determine the extent of the pyramid's base, as well as the eccentricity e_0 when the orbit precesses past the z-axis; that is,

$$e_0^2 = 1 - \ell_{x0}^2 - \ell_{y0}^2 \qquad\qquad (4.153a)$$

$$= 1 - \frac{1}{9}\left(v_x^{(0)} e_{x0}\right)^2 - \frac{1}{9}\left(v_y^{(0)} e_{y0}\right)^2 \qquad (4.153b)$$

$$= 1 - \frac{5}{3}\epsilon_c e_{x0}^2 - \frac{5}{3}(\epsilon_c - \epsilon_b)e_{y0}^2. \qquad (4.153c)$$

Equations (4.151) manifestly describe integrable motion. Remarkably, the more general equations of motion (4.148) are integrable as well [378]. The first integral is H', equation (4.142); an equivalent, but nonnegative, integral is U where

$$U \equiv 15\epsilon_c - 6H' = 15\epsilon_c e_x^2 + 15(\epsilon_c - \epsilon_b)e_y^2 + (\dot{e}_x^2 + \dot{e}_y^2 + \dot{e}_z^2). \qquad (4.154)$$

The second integral is obtained after multiplying the first of equations (4.148) by $\epsilon_c \dot{e}_x$, the second by $(\epsilon_c - \epsilon_b)\dot{e}_y$, and adding them to obtain a complete differential. The integral W is then

$$W = \epsilon_c(\dot{e}_x^2 + v_x^2 e_x^2 - 15\epsilon_c e_x^4)$$
$$+ (\epsilon_c - \epsilon_b)\left(\dot{e}_y^2 + \omega_y^2 e_y^2 - 15(\epsilon_c - \epsilon_b)e_y^4\right)$$
$$- 30\epsilon_c(\epsilon_c - \epsilon_b)e_x^2 e_y^2, \qquad (4.155a)$$

$$v_x^2 \equiv U + 15\epsilon_c, \quad v_y^2 \equiv U + 15(\epsilon_c - \epsilon_b). \qquad (4.155b)$$

The existence of two integrals (U, W), for a 2 d.o.f. system, demonstrates regularity of the motion. It is perhaps surprising that completely regular motion can result in a star coming arbitrarily close to the central singularity!

The (dimensionless) periods of oscillation of the planar orbits ($e_x = 0$ or $e_y = 0$) are easily shown to be

$$\tau_{\text{pyr}}(e_{x0}) = \frac{4}{v_x^{(0)}} K(e_{x0}) \quad (e_y = 0),$$

$$\tau_{\text{pyr}}(e_{y0}) = \frac{4}{v_y^{(0)}} K(e_{y0}) \quad (e_x = 0), \qquad (4.156)$$

where $K(k)$ is the complete elliptic integral (equation 4.126). For small k, that is, for a pyramid with a narrow opening angle, $K \approx \pi/2$ and $\tau_{\text{pyr}} \approx 2\pi/v_x^{(0)}$ or $2\pi/v_y^{(0)}$. As $k \to 1$, $K \to \infty$; this corresponds to a pyramid that precesses from the z-axis all the way to the x–y plane.

Figure 4.9 shows how U and W determine the type of orbit in one triaxial model. In the case of pyramid orbits, the two integrals can be expressed more simply by using the fact that $\dot{e}_x = \dot{e}_y = \dot{e}_z = 0$ at the "corners." Recalling that (e_{x0}, e_{y0}) are defined as the values at the corners, we have, for pyramids,

$$U = \left[v_x^{(0)}\right]^2 e_{x0}^2 + \left[v_y^{(0)}\right]^2 e_{y0}^2 , \qquad (4.157a)$$

$$W = \left[v_x^{(0)}\right]^4 e_{x0}^2 + \left[v_y^{(0)}\right]^4 e_{y0}^2 . \qquad (4.157b)$$

The two points in figure 4.9 that demarcate the rightmost boundary of the pyramids are

$$U = \left[v_y^{(0)}\right]^2 , \quad W = \left[v_y^{(0)}\right]^4 , \qquad (4.158a)$$

$$U = \left[v_x^{(0)}\right]^2 , \quad W = \left[v_x^{(0)}\right]^4 , \qquad (4.158b)$$

corresponding, respectively, to pyramids that precess to $\omega = 0$ in the y–z or x–z planes. These extremal pyramids also have the largest angular momenta when precessing past the z-axis; that is,

$$\ell_0 = \ell_{y0} = \sqrt{5\,(\epsilon_c - \epsilon_b)}, \qquad (4.159a)$$

$$\ell_0 = \ell_{x0} = \sqrt{5\epsilon_c}. \qquad (4.159b)$$

   A star on a pyramid orbit comes close to the SBH whenever the two variables (e_x, e_y) are simultaneously close to 1—that is, near the corners of the pyramid. As long as the frequencies of oscillation in e_x and e_y are incommensurable, the vector (e_x, e_y) densely fills the whole available area, which has the form of a distorted rectangle (figure 4.8). According to equation (4.146), $\ell^2 \approx (1/3)\left[5\epsilon_c - 2H - \epsilon_c e_x^2 - (\epsilon_c - \epsilon_b)e_y^2\right]$, where e_x and e_y are close to sine functions. Since $r_p \propto \ell^2$, it follows that the probability of having $r_p < X$ is almost proportional to X for small X. This idea will be developed in more detail in chapter 6.

   Finally, it is interesting to ask whether the four orbit families found here still persist at radii $r \gtrsim r_{\rm m}$, where the approximation of nearly Keplerian motion breaks down. In chapter 3, orbits like the pyramids were called "centrophilic" orbits, and the loss of integrability of centrophilic orbits when their apoapsides exceeded $\sim r_{\rm m}$ was described in section 3.5.1. Figure 4.10 summarizes the results of numerical integrations of the equations of motion derived, without approximation, from the triaxial potential of equation (4.93). As expected, chaotic orbits begin to appear when $a \gtrsim r_{\rm m}$. The fraction of phase space associated with each of the orbit families can be estimated by assuming that the orbital "integrals" (L, L_z) are distributed as they would be in an isotropic, spherical galaxy, $N(L, L_z; E) \propto dL^2 dL_z$, and choosing $\{\varpi, \Omega\}$ uniformly in $\left[0, \frac{\pi}{2}\right]$. Figure 4.10 shows that—whereas the fraction of pyramid orbits gradually declines with increasing a—the fraction of *centrophilic* (pyramid + chaotic) orbits increases again when $a \gtrsim r_{\rm m}$. As discussed in chapter 6, the two types of centrophilic orbit behave similarly in terms of their frequency of close approaches to the SBH.

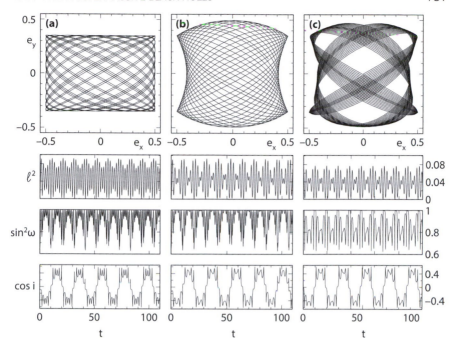

Figure 4.8 A pyramid orbit, in three approximations [378]. Each orbit has the same $(e_{x0}, e_{y0}) = (0.5, 0.35)$. (a) The simple harmonic oscillator (SHO) approximation, equations (4.151), valid for small ℓ^2, (ϵ_b, ϵ_c) and (e_{x0}, e_{y0}); (b) from equations (4.148), which do not assume small (e_{x0}, e_{y0}); (c) from the full orbit-averaged equations (4.142b), which do not assume small ℓ^2, ϵ or (e_{x0}, e_{y0}). Aside from the fact that the latter orbit is fairly close to a $5:2$ resonance, the correspondence between the physically important properties of the approximate orbits is good. The triaxiality parameters are $(\epsilon_b, \epsilon_c) = (0.0578, 0.168)$, corresponding to a pyramid orbit with $a = 0.1r_0$ in a nucleus with triaxial axis ratios $(0.5, 0.75)$, density ratio $\rho_t(r_0)/\rho_s(r_0) = 0.1$, and $\gamma = 1$. The frequencies for the SHO case are $\nu_x^{(0)} = 1.59$, $\nu_y^{(0)} = 1.28$ (equation 4.150); frequencies for planar orbits with the same e_x and e_y amplitudes are 1.48 and 1.24, respectively.

4.5 RELATIVISTIC ORBITS

4.5.1 First post-Newtonian order

In the case of two point bodies, the EIH N-body Lagrangian (4.70) simplifies to

$$\mathcal{L} = \mathcal{L}_N + \mathcal{L}_{PN},$$

$$\mathcal{L}_N = \frac{1}{2}m_1 v_1^2 + \frac{1}{2}m_2 v_2^2 + \frac{Gm_1 m_2}{r},$$

$$c^2 \mathcal{L}_{PN} = \frac{1}{8}m_1 v_1^4 + \frac{1}{8}m_2 v_2^4 + \frac{Gm_1 m_2}{2r}$$

$$\times \left[3\left(v_1^2 + v_2^2\right) - 7\boldsymbol{v}_1 \cdot \boldsymbol{v}_2 - \left(\boldsymbol{v}_1 \cdot \hat{\boldsymbol{n}}\right)\left(\boldsymbol{v}_2 \cdot \hat{\boldsymbol{n}}\right) - \frac{G\left(m_1 + m_2\right)}{r} \right], \quad (4.160)$$

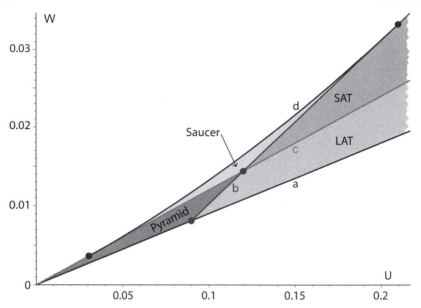

Figure 4.9 Regions in the U–W plane occupied by the different orbit families, in a triaxial nucleus with $\epsilon_b = 0.002$, $\epsilon_c = 0.008$ [378]. LAT = long-axis tubes; SAT = short-axis tubes. This figure is based on the expressions (4.154) and (4.155), which are valid in the case of highly eccentric orbits.

with

$$\hat{\boldsymbol{n}} = \frac{\boldsymbol{x}_1 - \boldsymbol{x}_2}{r}, \quad r = |\boldsymbol{x}_1 - \boldsymbol{x}_2|. \tag{4.161}$$

As in the Newtonian case, we seek the equations that describe the relative motion in the (post-Newtonian) center-of-mass frame. The total momentum is

$$\begin{aligned}
\boldsymbol{P} &= \frac{\partial \mathcal{L}}{\partial \boldsymbol{v}_1} + \frac{\partial \mathcal{L}}{\partial \boldsymbol{v}_2} \\
&= m_1 \boldsymbol{v}_1 + m_2 \boldsymbol{v}_2 + \frac{1}{2} m_1 \frac{v_1^2}{c^2} \boldsymbol{v}_1 + \frac{1}{2} m_2 \frac{v_2^2}{c^2} \boldsymbol{v}_2 \\
&\quad + \frac{G m_1 m_2}{2 c^2 r} \Big[-(\boldsymbol{v}_1 + \boldsymbol{v}_2) - \hat{\boldsymbol{n}} \big(\hat{\boldsymbol{n}} \cdot (\boldsymbol{v}_1 + \boldsymbol{v}_2) \big) \Big],
\end{aligned} \tag{4.162}$$

which is conserved to 1PN order, and the center of mass is

$$\boldsymbol{X} = \frac{m_1^* \boldsymbol{v}_1 + m_2^* \boldsymbol{v}_2}{m_1^* + m_2^*}, \quad m_i^* \equiv m_i + \frac{1}{2} m_i \frac{v_i^2}{c^2} - \frac{1}{2} \frac{G m_1 m_2}{r}. \tag{4.163}$$

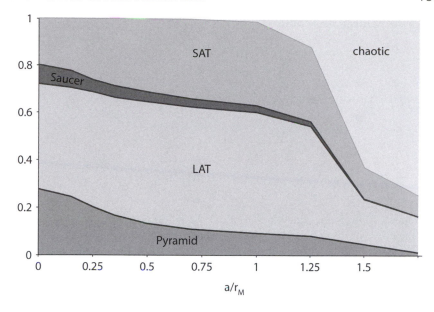

Figure 4.10 Phase-space fractions associated with the various orbit families in a triaxial nucleus, as a function of orbital semimajor axis [378]. This figure is based on numerical integration of the equations of motion corresponding to the potential (4.93), with $\gamma = 1$ and $\epsilon_c = 0.1$ at $a = r_m$. For $a \gtrsim r_m$, most of the low angular-momentum orbits are chaotic.

Transferring[8] to a frame where $\boldsymbol{P} = \boldsymbol{X} = 0$, the individual positions are related to the relative positions by

$$\boldsymbol{x}_1 = \left[\frac{m_2}{m} + \frac{\mu \delta m}{2m^2} \left(v^2 - \frac{Gm}{r} \right) \right] \boldsymbol{x},$$

$$\boldsymbol{x}_2 = \left[-\frac{m_1}{m} + \frac{\mu \delta m}{2m^2} \left(v^2 - \frac{Gm}{r} \right) \right] \boldsymbol{x}, \tag{4.164}$$

where

$$\boldsymbol{x} = \boldsymbol{x}_1 - \boldsymbol{x}_2, \qquad\qquad \boldsymbol{v} = \boldsymbol{v}_1 - \boldsymbol{v}_2,$$

$$m = m_1 + m_2, \qquad\qquad \delta m = m_1 - m_2. \tag{4.165}$$

The relative motion obeys

$$\frac{d\boldsymbol{v}}{dt} = -\frac{Gm\hat{\boldsymbol{n}}}{r^2} + 2(2 - v)\frac{Gm}{c^2 r^2} \boldsymbol{v} \left(\hat{\boldsymbol{n}} \cdot \boldsymbol{v} \right)$$

$$+ \frac{Gm\hat{\boldsymbol{n}}}{c^2 r^2} \left[2(2 + v)\frac{Gm}{r} - (1 + 3v)v^2 + \frac{3}{2}v \left(\hat{\boldsymbol{n}} \cdot \boldsymbol{v} \right)^2 \right], \tag{4.166}$$

[8]The same result is obtained to this PN order if the nonrelativistic center of mass is used [99].

with ν the **reduced mass ratio**,

$$\nu \equiv \frac{\mu}{m} = \frac{m_1 m_2}{m^2}. \tag{4.167}$$

The Lagrangian, expressed in terms of the relative coordinates, becomes [252]

$$\mathcal{L} = \frac{1}{2} v^2 + \frac{Gm}{r} + \frac{1}{8} (1 - 3\nu) \frac{v^4}{c^2} \\ + \frac{Gm}{2c^2 r} \left[(3 + \nu) v^2 + \nu (\hat{\boldsymbol{n}} \cdot \boldsymbol{v})^2 - \frac{Gm}{r} \right]. \tag{4.168}$$

Exact solutions to the equation of motion (4.166) do not exist. One approach is to treat the post-Newtonian accelerations as small perturbations and apply Lagrange's planetary equations (4.69) [552, 52, 142]. We will in fact adopt this approach in most of the remainder of this chapter. However, it turns out that—if one consistently ignores terms of second post-Newtonian order $O(v^4/c^4)$—a "post-Keplerian" description of the motion can be found that is exact to this order and that looks very similar mathematically to the nonrelativistic solution. We present that solution below, following closely the original analysis of Damour and Deruelle [99].

As a first step, one uses the time invariance and the rotational invariance of the Lagrangian (4.168) to write the first integrals of the motion: the specific energy,

$$E = \boldsymbol{v} \cdot \frac{\partial \mathcal{L}}{\partial \boldsymbol{v}} = \frac{1}{2} v^2 - \frac{Gm}{r} + \frac{3}{8} (1 - 3\nu) \frac{v^4}{c^2} \\ + \frac{Gm}{2c^2 r} \left[(3 + \nu) v^2 + \nu (\hat{\boldsymbol{n}} \cdot \boldsymbol{v})^2 + \frac{Gm}{r} \right], \tag{4.169}$$

and the specific angular momentum,

$$L = \left| \boldsymbol{x} \times \frac{\partial \mathcal{L}}{\partial \boldsymbol{v}} \right| = |\boldsymbol{x} \times \boldsymbol{v}| \left[1 + \frac{1}{2} (1 - 3\nu) \frac{v^2}{c^2} + (3 + \nu) \frac{Gm}{c^2 r} \right]. \tag{4.170}$$

As in the nonrelativistic case, motion takes place in a plane. Writing

$$v^2 = \left(\frac{dr}{dt} \right)^2 + r^2 \left(\frac{d\phi}{dt} \right)^2, \quad |\boldsymbol{r} \times \boldsymbol{v}| = r^2 \frac{d\phi}{dt}, \tag{4.171}$$

one finds, to order $O(v^2/c^2)$,

$$\left(\frac{dr}{dt} \right)^2 = A + \frac{2B}{r} + \frac{C}{r^2} + \frac{D}{r^3}, \tag{4.172a}$$

$$\frac{d\phi}{dt} = \frac{H}{r^2} + \frac{I}{r^3}, \tag{4.172b}$$

where

$$A = 2E \left[1 + \frac{3}{2}(3v - 1)\frac{E}{c^2} \right],$$ (4.173a)

$$B = Gm \left[1 + (7v - 6)\frac{E}{c^2} \right],$$ (4.173b)

$$C = -L^2 \left[1 + 2(3v - 1)\frac{E}{c^2} \right] + 5(v - 2)\frac{G^2 m^2}{c^2},$$ (4.173c)

$$D = (8 - 3v)\frac{Gm L^2}{c^2},$$ (4.173d)

$$H = L \left[1 + (3v - 1)\frac{E}{c^2} \right],$$ (4.173e)

$$I = 2(v - 2)\frac{Gm L}{c^2}.$$ (4.173f)

Equations (4.172) differ in functional form from the nonrelativistic equations for the radial and angular motion, equations (4.13), due to the r^{-3} terms. But—again neglecting terms of order $O(v^4/c^4)$—they can be put into the same form as the nonrelativistic expressions by two simple changes of variable. Consider first the radial equation. Write

$$r = \bar{r} + \frac{D}{2C_0},$$ (4.174)

where $C_0 = -L^2$ is the limit of C when $c \rightarrow \infty$ (equation (4.174) is called a "conchoidal transformation"). Substituting (4.174) into (4.172a), and neglecting terms of order $O(v^4/c^4)$, one finds

$$\left(\frac{d\bar{r}}{dt} \right)^2 = A + \frac{2B}{\bar{r}} + \frac{\bar{C}}{\bar{r}^2},$$ (4.175)

where

$$\bar{C} = C - \frac{BD}{C_0}.$$ (4.176)

The right-hand side of equation (4.175) has the same functional form as the nonrelativistic equation (4.13a). By analogy, the solution can be expressed parametrically as

$$n_r(t - t_0) = E - e_t \sin E,$$ (4.177a)
$$\bar{r} = \bar{a}(1 - e_t \cos E),$$ (4.177b)

where

$$\bar{a} = -\frac{B}{A}, \quad e_t^2 = 1 - \frac{A\bar{C}}{B^2}$$ (4.178)

and E plays the role of eccentric anomaly. Equations (4.174) and (4.177b) imply

$$r = a_r(1 - e_r \cos E)$$ (4.179)

with

$$a_r = \frac{D}{2C_0} - \frac{B}{A}, \quad e_r = \left(1 + \frac{AD}{2C_0B}\right)e_t \tag{4.180}$$

(both expressions correct to $O(v^2/c^2)$). The main difference between the relativistic and nonrelativistic solutions is the appearance of two different eccentricities: a **time eccentricity** e_t, and a **radial eccentricity** e_r. Using equations (4.173), we can express a_r, e_r, e_t and n_r in terms of the relativistic E and L:

$$a_r = -\frac{Gm}{2E}\left[1 - \frac{1}{2}(\nu - 7)\frac{E}{c^2}\right], \tag{4.181a}$$

$$e_r = \left\{1 + \frac{2E}{G^2m^2}\left[1 + \frac{5}{2}(\nu - 3)\frac{E}{c^2}\right]\left[L^2 + (\nu - 6)\frac{G^2m^2}{c^2}\right]\right\}^{1/2}, \tag{4.181b}$$

$$e_t = \left\{1 + \frac{2E}{G^2m^2}\left[1 + \frac{1}{2}(-7\nu + 17)\frac{E}{c^2}\right]\left[L^2 + 2(1 - \nu)\frac{G^2m^2}{c^2}\right]\right\}^{1/2}, \tag{4.181c}$$

$$n_r = \frac{(-2E)^{3/2}}{Gm}\left[1 - \frac{1}{4}(\nu - 15)\frac{E}{c^2}\right]. \tag{4.181d}$$

Again neglecting terms of order $O(v^4/c^4)$, the mean motion can be expressed in terms of a_r as

$$n_r = \left(\frac{Gm}{a_r^3}\right)^{1/2}\left[1 + \frac{Gm}{2a_rc^2}(\nu - 9)\right]. \tag{4.182}$$

Thus, just as in the nonrelativistic case, the semimajor axis and mean motion depend only on the energy. We can also write

$$\frac{e_r}{e_t} = 1 + (3\nu - 8)\frac{E}{c^2} \tag{4.183}$$

$$= 1 + \frac{Gm}{a_rc^2}\left(4 - \frac{3}{2}\nu\right). \tag{4.184}$$

The time between periapsis passages is

$$\frac{2\pi}{n_r} = \frac{2\pi a_r^{3/2}}{\sqrt{Gm}}\left[1 + \frac{1}{2}(9 - \nu)\frac{Gm}{c^2a_r}\right]. \tag{4.185}$$

A second conchoidal transformation,

$$r = \tilde{r} + \frac{I}{2H}, \tag{4.186}$$

puts the relativistic angular equation (4.172b) into the same form as the nonrelativistic equation (4.13b):

$$\frac{d\phi}{dt} = \frac{H}{\tilde{r}^2}, \tag{4.187}$$

where, as usual, terms of order $O(v^4/c^4)$ have been neglected. Writing

$$\tilde{r} = \tilde{a}\,(1 - \tilde{e}\cos E)\,, \tag{4.188}$$

with

$$\tilde{a} = a_r - \frac{I}{2H}, \quad \tilde{e} = e_r\left(1 - \frac{AI}{2BH}\right), \tag{4.189}$$

and differentiating equation (4.177a) with respect to time,

$$dt = n_r^{-1}\,(1 - e_t\cos E)\,dE \tag{4.190}$$

yields

$$d\phi = \frac{H}{n\tilde{a}^2}\frac{1 - e_t\cos E}{(1 - \tilde{e}\cos E)^2}\,dE. \tag{4.191}$$

Defining a third eccentricity e_ϕ, the **angular eccentricity**, as

$$e_\phi = 2\tilde{e} - e_t, \tag{4.192}$$

we find

$$\frac{1 - e_t\cos E}{(1 - \tilde{e}\cos E)^2} = \frac{1}{1 - e_\phi\cos E} + O(v^4/c^4), \tag{4.193}$$

and

$$d\phi = \frac{H}{n_r\tilde{a}^2}\frac{dE}{1 - e_\phi\cos E}, \tag{4.194}$$

the same form as the nonrelativistic equation. Integrating,

$$\phi - \phi_0 = Kf, \tag{4.195a}$$

$$\tan\frac{f}{2} = \left(\frac{1 + e_\phi}{1 - e_\phi}\right)^{1/2}\tan\frac{E}{2}, \tag{4.195b}$$

$$K = \frac{H}{n_r\tilde{a}^2\left(1 - e_\phi^2\right)^{1/2}}, \tag{4.195c}$$

where f plays the role of true anomaly. In terms of E and L, the constants e_ϕ and K are

$$e_\phi = (1 - vE)\,e_r = \left(1 + \frac{G\mu}{2a_rc^2}\right)e_r \tag{4.196a}$$

$$= \left\{1 + \frac{2E}{G^2m^2}\left[1 + \frac{1}{2}\,(v - 15)\frac{E}{c^2}\right]\left[L^2 - 6\frac{G^2m^2}{c^2}\right]\right\}^{1/2}, \tag{4.196b}$$

$$K = \frac{L}{\left(L^2 - 6G^2m^2/c^2\right)^{1/2}} \approx 1 + 3\frac{G^2m^2}{L^2c^2}. \tag{4.196c}$$

From equation (4.179), periapsis passages occur for $E = 0, 2\pi, 4\pi, \ldots$. The argument of periapsis precesses each revolution by an angle

$$\Delta\phi = 2\pi\,(K - 1) = \frac{6\pi\,Gm}{a_r\left(1 - e_r^2\right)c^2} + O(v^4/c^4). \tag{4.197}$$

This is the **relativistic precession of the periastron** [105] (also called "geodetic precession," "De Sitter precession," or "Schwarzschild precession"; the latter name differentiates this precession from the precession induced by the Kerr metric of a spinning hole). The direction of the precession is prograde, that is, in the same angular sense as the direction of orbital circulation.

The equation for the relative orbit is found by eliminating E from equations (4.179) and (4.195a):

$$r = \left(a_r - \frac{G\mu}{2c^2}\right)\frac{1 - e_\phi^2}{1 + e_\phi\cos f} + \frac{G\mu}{2c^2}. \tag{4.198}$$

Writing

$$f' = f - \frac{1}{2}e_r\frac{G\mu}{c^2 a_r\left(1 - e_r^2\right)}\sin f \tag{4.199}$$

allows the equation for the relative orbit to be written in a form that parallels the nonrelativistic case:

$$r = \frac{a_r\left(1 - e_r^2\right)}{1 + e_r\cos f'}. \tag{4.200}$$

As a final step, we can derive the motion of each body in the center-of-mass frame by inserting the solutions for the relative motion into equations (4.164). The polar angle of body 1 is the same as the relative angle ϕ and the polar angle of body 2 is $\phi + \pi$. The radial motion for body 1 is then

$$r_1 = a_{r,1}\left(1 - e_{r,1}\cos E\right), \tag{4.201}$$

where

$$a_{r,1} = \frac{m_2}{m}a_r, \tag{4.202a}$$

$$e_{r,1} = e_r\left(1 - \frac{Gm_1\delta m}{2ma_rc^2}\right) \tag{4.202b}$$

(replacing the subscript 1 by 2 gives the motion of body 2), and

$$n_r(t - t_0) = E - e_t\sin E, \tag{4.203a}$$

$$\phi - \phi_0 = Kf, \tag{4.203b}$$

as before. The path traced by body 1 in space is

$$r_1(\phi) = \frac{Gm_1^2 m_2}{2m^2c^2} + \left(a_{r,1} - \frac{Gm_1^2 m_2}{2m^2c^2}\right)\frac{1 - e_\phi^2}{1 + e_\phi\cos\left(\frac{\phi - \phi_0}{K}\right)}, \tag{4.204}$$

and similarly for body 2. Equation (4.204) is the conchoid of a precessing ellipse.

As this derivation shows, conserved quantities exist, to first post-Newtonian order, that are analogous to the Keplerian semimajor axis and eccentricity. Because the motion occurs in a fixed plane, the orbital inclination i and the longitude of the ascending node Ω are also conserved. The only conserved element in the nonrelativistic problem that has no fixed counterpart in the relativistic problem is the argument of periapsis, ω. Its variation between consecutive periapsis passages is just the quantity $\Delta\phi$ that is given by equation (4.197). Another way to state this result is in terms of the **orbit-averaged precession rate**:

$$\left\langle \frac{d\omega}{dt} \right\rangle = \frac{n}{2\pi}\Delta\phi = n\frac{3Gm}{a(1-e^2)c^2} = \frac{3(Gm)^{3/2}}{a^{5/2}(1-e^2)c^2}. \tag{4.205}$$

Here use has been made of the fact that—to first post-Newtonian order—a_r, n_r and e_r in equation (4.197) for $\Delta\phi$ can be replaced by n, a and e.

A quasi-Keplerian representation like Damour and Deruelle's is not possible in every case of perturbed motion, and it is instructive to compute the rate of apsidal precession in a more direct (but less elegant) manner, from Lagrange's planetary equations (4.57). The starting point is again the post-Newtonian equation for the relative motion, equation (4.166). We identify the order $O(v^2/c^2)$ terms in that expression with the perturbing acceleration $\boldsymbol{a}_p$ in equation (4.54). The Gaussian components of $\boldsymbol{a}_p$, equation (4.55), are

$$S = \frac{Gm}{c^2 r^2}\left[2(2+v)\frac{Gm}{r} + (4-v/2)(\hat{\boldsymbol{n}} \cdot \boldsymbol{v})^2 - (1+3v)v^2\right],$$

$$T = \frac{2Gm}{c^2 r^2}(2-v)(\hat{\boldsymbol{n}} \cdot \boldsymbol{v})(\hat{\boldsymbol{m}} \cdot \boldsymbol{v}),$$

$$W = 0. \tag{4.206}$$

Using

$$r = \frac{a(1-e^2)}{1+e\cos f},$$

$$v^2 = \frac{Gm}{a(1-e^2)}\left(1+e^2+2e\cos f\right),$$

$$\hat{\boldsymbol{n}} \cdot \boldsymbol{v} = v_r = \left[\frac{Gm}{a(1-e^2)}\right]^{1/2} e\sin f,$$

$$\hat{\boldsymbol{m}} \cdot \boldsymbol{v} = v_t = \left[\frac{Gm}{a(1-e^2)}\right]^{1/2} (1+e\cos f),$$

with f the true anomaly, we can express (S, T, W) in terms of the orbital elements as

$$S = \frac{G^2 m^2}{c^2 a^3 (1 - e^2)^3} (1 + e \cos f)^2 \left[3(e^2 + 1) + 5v \left(1 - \frac{7}{10} e^2 \right) \right.$$

$$\left. + 2(e - 4v) \cos f - 4e^2 \left(1 - \frac{v}{8} \right) \cos^2 f \right],$$

$$T = \frac{G^2 m^2}{c^2 a^3 (1 - e^2)^3} 2(2 - v) (1 + e \cos f)^3 e \sin f,$$

$$W = 0. \tag{4.207}$$

The dependence of the orbital elements on time, to first order, is then given by substituting these expressions into equations (4.57), fixing all the elements except for f on the right-hand sides, and carrying out the integrations with respect to f. In the case of a, e, Ω and i, the resulting expressions contain only oscillatory terms in f that average to zero over a complete radial oscillation. The amplitudes of the sinusoidal variations in these elements are of order $O(v^2/c^2)$, consistent with the fact that the relativistic elements $\{a_r, e_r, e_t, e_\phi\}$ differ from their nonrelativistic counterparts a and e at the same order. In the case of ω, there appears as well a secular term proportional to f:

$$\omega(t) - \omega(t_0) = \frac{Gm}{c^2 a(1 - e^2)} \left\{ 3f + \left[\frac{v - 3}{e} + \left(1 + \frac{21}{8} v \right) e \right] \sin f \right.$$

$$\left. + \left(2v - \frac{5}{2} \right) \sin 2f + \frac{v}{8} e \sin 3f \right\} \Big|_{t_0}^{t}. \tag{4.208}$$

In this expression, t_0 corresponds to $f = 0$, that is, to periapsis passage. Setting $t = P$ (i.e., $f = 2\pi$), the oscillatory terms go to zero, and the secular term yields the correct, orbit-averaged precession rate, equation (4.205). Figure 4.11 plots $\omega(t)$ over one orbital period for orbits with $v = 0$ and various eccentricities. Most of the precession takes place near periapsis passage, particularly when the orbital eccentricity is large.

4.5.2 Second post-Newtonian order

It is possible to generalize equation (4.166), the relative acceleration in the 1PN two-body problem, to higher post-Newtonian orders. The equation of motion for the relative orbit turns out to have the general form

$$\frac{dv}{dt} = -\frac{Gm}{r^2} \left(\mathcal{A} n + \mathcal{B} v \right), \tag{4.209}$$

where, as before, $m = m_1 + m_2$, $x = x_1 - x_2$, $v = v_1 - v_2$, $r = |x_1 - x_2|$, $v = |v_1 - v_2|$, and $n = x/r$. The coefficients $\mathcal{A}, \mathcal{B}$ are series expansions in the small parameter c^{-1}; that is,

$$\mathcal{A} = \mathcal{A}_0 + c^{-2} \mathcal{A}_1 + c^{-4} \mathcal{A}_2 + \cdots,$$

$$\mathcal{B} = \mathcal{B}_0 + c^{-2} \mathcal{B}_1 + c^{-4} \mathcal{B}_2 + \cdots.$$

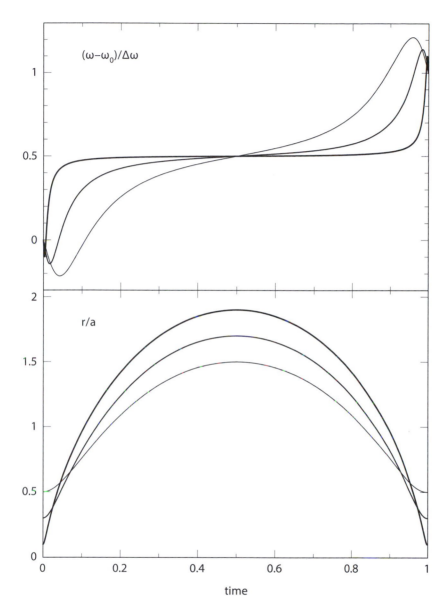

Figure 4.11 Relativistic precession of the argument of periapsis ω at 1PN order. Curves in the top panel are plots of equation (4.208) for test-particle orbits ($\nu = 0$) and for three different eccentricities: $e = 0.5$ (thinnest), $e = 0.7$, $e = 0.9$ (thickest). The bottom panel shows the separation. Time is in units of the radial period. Most of the precession occurs when the star is near periapsis.

(Note that the subscript of $\mathcal{A}_n$ denotes the nth PN term which is of order c^{-2n}.) Up to second post-Newtonian order, the coefficients are [51]

$$\mathcal{A}_0 = 1,$$

$$\mathcal{A}_1 = -\frac{3\dot{r}^2 v}{2} + (1 + 3v)v^2 - 2(2 + 2v)\frac{Gm}{r},$$

$$\mathcal{A}_2 = \frac{15\dot{r}^4 v}{8} - \frac{45\dot{r}^4 v^2}{8} - \frac{9\dot{r}^2 v v^2}{2} + 6\dot{r}^2 v^2 v^2 + 3v v^4 - 4v^2 v^4$$
$$+ \frac{Gm}{r}\left(-2\dot{r}^2 - 25\dot{r}^2 v - 2\dot{r}^2 v^2 - \frac{13 v v^2}{2}\right) + \frac{G^2 m^2}{r^2}\left(9 + \frac{87v}{4}\right),$$

$$\mathcal{B}_0 = 0,$$

$$\mathcal{B}_1 = -2(2 - v)\dot{r},$$

$$\mathcal{B}_2 = \frac{9\dot{r}^3 v}{2} + 3\dot{r}^3 v^2 - \frac{15\dot{r} v v^2}{2} - 2\dot{r} v^2 v^2 + \frac{Gm}{r}\left(2\dot{r} + \frac{41\dot{r} v}{2} + 4\dot{r} v^2\right).$$

$$(4.210)$$

The expressions for $\mathcal{A}_0$, $\mathcal{A}_1$, $\mathcal{B}_0$ and $\mathcal{B}_1$ reproduce the 1PN equation of motion, equation (4.166), after noting that $\dot{r} = \boldsymbol{n} \cdot \boldsymbol{v}$.

In the previous section, Damour and Deruelle's Keplerian-like solution for the 1PN two-body problem was presented. The relative orbit, to 1PN order, was shown to be representable in terms of a parameter E that plays the role of the eccentric anomaly in the nonrelativistic problem

$$n_r(t - t_0) = \mathrm{E} - e_t \sin \mathrm{E}, \tag{4.211a}$$

$$r = a_r (1 - e_r \cos \mathrm{E}), \tag{4.211b}$$

$$\phi - \phi_0 = 2K \tan^{-1}\left(\sqrt{\frac{1 + e_\phi}{1 - e_\phi}} \tan\frac{\mathrm{E}}{2}\right) \tag{4.211c}$$

(cf. equations 4.177a, 4.179, and 4.195a). The quantities $\{n_r, a_r, e_r, e_t, e_\phi, K\}$ differ from their nonrelativistic counterparts $\{n, a, e, 1\}$ by terms of order $O(v^2/c^2)$; precession of the periapsis is a consequence of the fact that $K \neq 1$. It turns out [100] that a similar result holds at second post-Newtonian order: the relative orbit, in the relativistic center-of-mass system, is given by

$$n_r(t - t_0) = \mathrm{E} - e_t \sin \mathrm{E} + f_P \sin v + g_P(v - \mathrm{E}), \tag{4.212a}$$

$$r = a_r (1 - e_r \cos \mathrm{E}), \tag{4.212b}$$

$$\phi - \phi_0 = 2K \left[v + f_\Phi \sin(2v) + g_\Phi \sin(3v)\right], \tag{4.212c}$$

where

$$v = 2 \tan^{-1}\left(\sqrt{\frac{1 + e_\phi}{1 - e_\phi}} \tan\frac{\mathrm{E}}{2}\right) \tag{4.213}$$

plays the role of ϕ in the nonrelativistic problem. The quantities $(K, n_r, a_r, e_t, e_r, e_\phi, f_P, g_P, f_\Phi, g_\Phi)$ can be expressed, to order $O(v^4/c^4)$ accuracy, in terms of E and L, the 2PN generalizations of the energy and angular momentum integrals [478].

The character of the motion is essentially the same as in the 1PN description; the rate of periapsis advance differs by a term of order $O(v^4/c^4)$ [100].

4.5.3 Spin-orbit effects

The spin angular momentum, $\boldsymbol{S}$, of a black hole of mass $M_\bullet$ is expressible in terms of the dimensionless spin vector $\boldsymbol{\chi}$ defined in equation (4.72):

$$\boldsymbol{S} = \boldsymbol{\chi}\,\frac{GM_\bullet^2}{c},$$

where $\chi = 1$ for a maximally spinning hole. The hole's quadrupole moment, $\mathcal{Q}$, was defined in equation (4.73):

$$\mathcal{Q} = -\frac{1}{c^2}\frac{\mathcal{S}^2}{M_\bullet}.$$

In full general relativity, motion of a test particle around a spinning black hole is integrable if expressed in suitable coordinates (section 4.6). There are three conserved quantities: two of these are analogous to the energy and angular momentum per unit mass in the Newtonian case, while the third, the Carter constant, plays a role similar to that of the component of the angular momentum parallel to the symmetry (spin) axis of the hole. Excepting the case of resonance, all orbits belong to a single, toruslike family, roughly similar in configuration-space structure to the tube orbits that were described in chapter 3.

The post-Newtonian equations of motion in the presence of spin can differ depending on how the "center of mass" of a spinning body is defined [26]. Here we adopt the so-called "covariant spin supplementary condition"; the resulting equations of motion for the two-body problem have the property that they do not depend explicitly on the motion of the center of coordinates [278].

Write the mass of the spinning black hole, m_1, as $M_\bullet$, and $m_2 \ll M_\bullet$ is the mass of the orbiting (and nonspinning) body. Let $\hat{n} = \boldsymbol{x}/r$ be a unit vector pointing from $M_\bullet$ to m_2. To lowest order in v/c, the spin-induced acceleration of the second body is given by any of the three equivalent forms [278]

$$\boldsymbol{a}_{\mathrm{J}} = -\frac{2G^2 M_\bullet^2}{c^3 r^3}\left[2\boldsymbol{v} \times \boldsymbol{\chi} - 3\hat{n}(\hat{n} \cdot \boldsymbol{v}) \times \boldsymbol{\chi} - 3\hat{n}(\hat{n} \times \boldsymbol{v}) \cdot \boldsymbol{\chi}\right]$$

(4.214a)

$$= -\frac{2G^2 M_\bullet^2}{c^3 r^3}\left\{\boldsymbol{v} \times \left[2\boldsymbol{\chi} + 3\hat{n} \times (\hat{n} \times \boldsymbol{\chi})\right]\right\}$$

(4.214b)

$$= -\frac{2G^2 M_\bullet^2}{c^3 r^3}\left\{\boldsymbol{v} \times \left[-\boldsymbol{\chi} + 3(\hat{n} \cdot \boldsymbol{\chi})\hat{n}\right]\right\}.$$

(4.214c)

The "gravitomagnetic" character of these equations becomes apparent if the last one is rewritten as

$$\boldsymbol{a}_{\mathrm{J}} = -\frac{2}{c}\boldsymbol{v} \times \boldsymbol{B},$$

(4.215a)

$$\boldsymbol{B} = \frac{G}{c}\frac{1}{r^3}\left[-\boldsymbol{S} + 3(\boldsymbol{S}\cdot\hat{\boldsymbol{n}})\hat{\boldsymbol{n}}\right] = \nabla\times\boldsymbol{A}, \qquad (4.215\text{b})$$

$$\boldsymbol{A} = \frac{G}{c}\frac{\boldsymbol{S}\times\boldsymbol{r}}{r^3}. \qquad (4.215\text{c})$$

The acceleration of the second body due to the quadrupole moment of the spinning hole is, again to lowest order in v/c [27],

$$\boldsymbol{a}_Q = -\frac{3}{2}\chi^2\frac{G^3 M_\bullet^3}{c^4 r^4}\left[5\hat{\boldsymbol{n}}(\hat{\boldsymbol{n}}\cdot\hat{\boldsymbol{e}}_S)^2 - 2(\hat{\boldsymbol{n}}\cdot\hat{\boldsymbol{e}}_S)\hat{\boldsymbol{e}}_S - \hat{\boldsymbol{n}}\right], \qquad (4.216)$$

where $\hat{\boldsymbol{e}}_S \equiv \boldsymbol{S}/S = \boldsymbol{\chi}/\chi$ is a unit vector in the direction of the spin.

Because $\boldsymbol{a}_J$ is perpendicular to the relative velocity vector $\boldsymbol{v}$, the spin does no work on m_2. However, $\boldsymbol{a}_J$ does contribute to the precession of m_2's orbit about the spinning hole. As in previous sections, we calculate the first-order effects by identifying $\boldsymbol{a}_J$ with the perturbing acceleration in equation (4.54). Without loss of generality, the spin (which for the moment we assume to be fixed in magnitude and direction) can be aligned with the z-axis, $\hat{\boldsymbol{e}}_S = \hat{\boldsymbol{e}}_z$. The Gaussian components (4.55) of $\boldsymbol{a}_J$ are then easily seen to be

$$S = \frac{2G^2 M_\bullet^2}{c^3 r^4}\chi\left(xv_y - yv_x\right),$$

$$T = \frac{2G^2 M_\bullet^2}{c^3 r^3}\chi\left[m_x\left(-2v_y + 3\frac{y}{r}v_r\right) + m_y\left(2v_x - 3\frac{x}{r}v_r\right)\right],$$

$$W = \frac{2G^2 M_\bullet^2}{c^3 r^3}\chi\left[k_x\left(-2v_y + 3\frac{y}{r}v_r\right) + k_y\left(2v_x - 3\frac{x}{r}v_r\right)\right], \qquad (4.217)$$

where $\boldsymbol{m}$ and $\boldsymbol{k}$ are the unit vectors defined in equations (4.56); that is,

$$m_x = -\sin(\omega + f)\cos\Omega - \cos(\omega + f)\sin\Omega\cos i,$$

$$m_y = -\sin(\omega + f)\sin\Omega + \cos(\omega + f)\cos\Omega\cos i,$$

$$k_x = \sin\Omega\sin i,$$

$$k_y = -\cos\Omega\sin i.$$

After some algebra, one finds

$$S = \frac{2(GM_\bullet)^{5/2}}{c^3 r^4}\chi\sqrt{a(1 - e^2)}\cos i,$$

$$T = -\frac{2(GM_\bullet)^{5/2}}{c^3 r^3}\frac{1}{\sqrt{a(1 - e^2)}}e\chi\sin f\cos i,$$

$$W = \frac{2(GM_\bullet)^{5/2}}{c^3 r^4}\sqrt{a(1 - e^2)}\chi\sin i$$

$$\times\left[2\sin(\omega + f) + \frac{er}{a(1 - e^2)}\sin f\cos(\omega + f)\right]. \qquad (4.218)$$

Substituting these expressions into equations (4.57) and integrating with respect to f reveals that a exhibits no variations to first order, and that e and i exhibit no secular variations. However, the expressions for $\omega(f)$ and $\Omega(f)$ both contain secular terms:

$$\Omega(f) = \frac{2(GM_\bullet)^{3/2}\chi}{c^3\left[a(1-e^2)\right]^{3/2}}\left[f - \frac{1}{2}\sin 2u + e\left(\sin f - \frac{1}{2}\sin 2u\cos f\right)\right],$$

(4.219a)

$$\omega(f) = -\frac{2(GM_\bullet)^{3/2}\chi}{c^3\left[a(1-e^2)\right]^{3/2}}\cos i\left[3f - \frac{\sin f}{e} - \frac{1}{2}\sin 2u\,(1+e\cos f)\right],$$

(4.219b)

where $u \equiv \omega + f$. The evolution of $\Omega(t)$ is plotted in figure 4.12 for three values of e and for $\omega = 0$. As in the case of 1PN precession of ω, most of the change in Ω takes place near periapsis.

The changes in these elements over one period are

$$(\Delta\Omega)_J = \frac{4\pi\chi}{c^3}\left[\frac{GM_\bullet}{(1-e^2)a}\right]^{3/2},$$

(4.220a)

$$(\Delta\omega)_J = -\frac{12\pi\chi}{c^3}\left[\frac{GM_\bullet}{(1-e^2)a}\right]^{3/2}\cos i = -3\cos i\,(\Delta\Omega)_J,$$

(4.220b)

and the orbit-averaged precession frequencies are

$$\left\langle\frac{d\Omega}{dt}\right\rangle_J = \frac{2G^2M_\bullet^2\chi}{c^3a^3(1-e^2)^{3/2}} = \frac{2GS}{c^2a^3(1-e^2)^{3/2}},$$

(4.221a)

$$\left\langle\frac{d\omega}{dt}\right\rangle_J = -\frac{6G^2M_\bullet^2\chi}{c^3a^3(1-e^2)^{3/2}}\cos i = -\frac{6GS}{c^2a^3(1-e^2)^{3/2}}\cos i.$$

(4.221b)

These spin-related precessions are referred to collectively as the **Lense–Thirring**, or **frame-dragging**, precession [315].[9]

The time for frame dragging to rotate the line of nodes by an angle π can be written in various ways:

$$t_J = \frac{P}{4\chi}\mathcal{P}^{3/2} = \frac{P}{4\chi}\left[\frac{(1-e^2)c^2a}{GM_\bullet}\right]^{3/2} = \frac{\pi(1-e^2)^{3/2}a^3c^3}{2\chi G^2M_\bullet^2}$$

$$\approx 1.4\times10^5\chi^{-1}(1-e^2)^{3/2}\left(\frac{M_\bullet}{4\times10^6\,M_\odot}\right)^{-2}\left(\frac{a}{\text{mpc}}\right)^3\text{yr},$$

(4.222)

where the penetration parameter $\mathcal{P}$ is defined, as above, to be $(1-e^2)a/r_g$. The last of these relations shows that frame dragging can have appreciable effects on stellar orbits inside a milliparsec from the Milky Way SBH, even on timescales that are shorter than main-sequence lifetimes of massive stars [318].

[9]A historically more accurate name might be Einstein–Thirring–Lense precession [428].

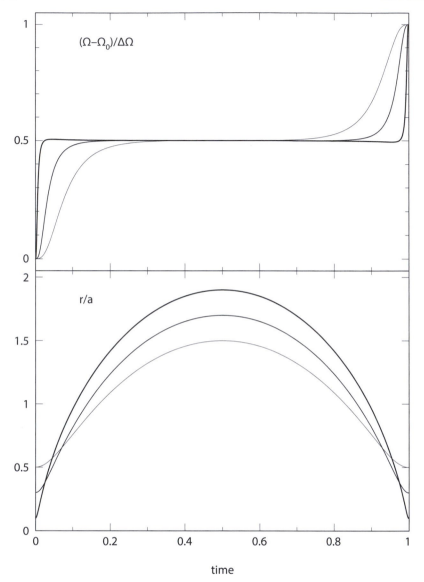

Figure 4.12 Relativistic precession of the line of nodes, Ω, due to frame dragging at 1.5PN order. Curves in the top panel are plots of equation (4.219a) for three different eccentricities: $e = 0.5$ (thinnest), $e = 0.7$, $e = 0.9$ (thickest). The bottom panel shows the separation. The argument of periapsis, ω, was fixed at zero. Time is in units of the radial period.

The precession described by equations (4.220) has two components: precession, at fixed inclination, of the orbit's angular momentum vector about the spin axis of the SBH ($\Delta\Omega$); and precession of the argument of periapsis within the changing orbital plane ($\Delta\omega$). Both precessions cause changes in the spatial location of the orbit's periapsis, and it is natural to define $\Delta\varpi = \Delta\omega + \cos i\,\Delta\Omega$, the precession of the periapsis with respect to a fixed reference direction:[10]

$$(\Delta\varpi)_J = -2\cos i\,(\Delta\Omega)_J. \tag{4.223}$$

Nodal precession induced by frame dragging can also be described in terms of its effect on the direction of the angular momentum vector. From equation (4.59), a unit vector in the direction of L is

$$\frac{L}{L} = (\sin i \sin\Omega)\,e_x - (\sin i \cos\Omega)\,e_y + \cos i\,e_z. \tag{4.224}$$

Since the magnitude of L is unchanged by frame dragging, the orbit-averaged rate of change of L is

$$\left\langle\frac{dL}{dt}\right\rangle_J = L\frac{(\Delta\Omega)_J}{P}\left(\sin i \cos\Omega\,e_x + \sin i \sin\Omega\,e_y\right) \tag{4.225a}$$

$$= \frac{2G^2 M_\bullet^2}{c^3(1-e^2)^{3/2}a^3}\,\chi \times L. \tag{4.225b}$$

The angular momentum precesses about the spin axis of the SBH with a frequency

$$\nu_J = \frac{\pi}{t_J} = \frac{2G^2 M_\bullet^2 \chi}{c^3(1-e^2)^{3/2}a^3}. \tag{4.226}$$

The period of the precession induced by frame dragging is independent of an orbit's inclination with respect to the SBH's equatorial plane. However, over times $\Delta t \ll t_J$, the angular displacement of the direction of L is proportional to δt and does depend on i. Let $\delta\theta$ be the angle between the angular momentum vector at two times, t_i and t_f:

$$\cos(\delta\theta) = \frac{L_i \cdot L_f}{L^2}. \tag{4.227}$$

Since L and i are unchanged by frame dragging,

$$\cos(\delta\theta) = \cos^2 i + \sin^2 i \cos(\delta\Omega). \tag{4.228}$$

For small δt (i.e., small $\delta\Omega$) this becomes

$$\delta\theta \approx \sin i\,\delta\Omega. \tag{4.229}$$

Angular changes in the orientation of a star's orbital plane due to frame dragging are therefore proportional to $\sin i$.

Returning now to the SBH's quadrupole moment: it is straightforward to show that the quadrupole adds an additional secular term to the evolution of both Ω and ω, leading to changes, over one period, of

$$(\Delta\Omega)_Q = \frac{3\pi\chi^2}{c^4}\left[\frac{GM_\bullet}{(1-e^2)a}\right]^2\cos i, \tag{4.230a}$$

$$(\Delta\omega)_Q = \frac{3\pi\chi^2}{2c^4}\left[\frac{GM_\bullet}{(1-e^2)a}\right]^2\left(1-5\cos^2 i\right). \tag{4.230b}$$

---

[10]The symbol ϖ is more commonly used to denote $\omega+\Omega$, the so-called "longitude of pericenter." Except in the case of zero inclination, $\omega + \Omega$ does not correspond to a physical angle.

The time t_Q for the quadrupole torque, by itself, to rotate the line of nodes by an angle π is

$$(\cos i)t_Q = \frac{P}{3\chi^2}\mathcal{P}^2 = \frac{P}{3\chi^2}\left[\frac{(1-e^2)c^2a}{GM_\bullet}\right]^2 = \frac{\pi(1-e^2)^2 a^{7/2}c^4}{6\chi^2(GM_\bullet)^{5/2}}$$

$$\approx 3.3 \times 10^6 \chi^{-2}(1-e^2)^2\left(\frac{M_\bullet}{4 \times 10^6 M_\odot}\right)^{-5/2}\left(\frac{a}{\text{mpc}}\right)^{7/2}\text{yr}.$$

$$(4.231)$$

4.5.4 Post-Newtonian order 2.5, and energy loss

In center-of-mass coordinates, the contribution of the order $O(v^5/c^5)$ terms to the relative acceleration in the two-body problem is

$$\boldsymbol{a}_{2.5} = -\frac{8}{5}\frac{G^2m^2}{c^5r^3}\boldsymbol{v}\left\{\boldsymbol{v}\left[v^2 + 3\frac{Gm}{r}\right] - \boldsymbol{n}(\boldsymbol{n}\cdot\boldsymbol{v})\left[3v^2 + \frac{17}{3}\frac{Gm}{r}\right]\right\}. \quad (4.232)$$

Proceeding as above, we can express $\boldsymbol{a}_{2.5}$ in terms of its (S, T, W) components as

$$S = \frac{8}{5}\frac{G^3m^3n}{c^5a^3(1-e^2)^{9/2}}v\,(1+e\cos f)^3\,e\sin f$$

$$\times\left(\frac{14}{3} + 2e^2 + \frac{20}{3}e\cos f\right),$$

$$T = -\frac{8}{5}\frac{G^3m^3n}{c^5a^3(1-e^2)^{9/2}}v\,(1+e\cos f)^4\,e\,(4+e^2+5e\cos f),$$

$$W = 0. \quad (4.233)$$

Substituting these expressions into Lagrange's equations (4.57), and integrating with respect to true anomaly f while holding the other elements fixed, we find that first-order changes in (Ω, ω, i) are oscillatory in time, while both a and e exhibit secular changes.[11] Averaged over a single period, the latter changes are [423]

$$\left\langle\frac{da}{dt}\right\rangle = -\frac{64}{5}\frac{G^3m_1m_2m}{c^5a^3\left(1-e^2\right)^{7/2}}\left(1 + \frac{73}{24}e^2 + \frac{37}{96}e^4\right), \quad (4.234a)$$

$$\left\langle\frac{de}{dt}\right\rangle = -\frac{304}{15}\frac{G^3m_1m_2me}{c^5a^4\left(1-e^2\right)^{5/2}}\left(1 + \frac{121}{304}e^2\right). \quad (4.234b)$$

In response to loss of energy via gravitational-wave emission, the semimajor axis shrinks, and the eccentricity tends toward zero. If no other processes are acting to change a or e, equations (4.234) imply

$$\frac{da}{de} = \frac{12}{19}\frac{a}{e}\frac{1+(73/24)e^2+(37/96)e^4}{\left(1-e^2\right)\left[1+(121/304)e^2\right]} \quad (4.235)$$

[11]Of course, the 1PN and 2PN terms imply secular changes in ω, and the spin-orbit terms imply secular changes in Ω. The precession induced by these terms has no effect on the secular changes in a and e.

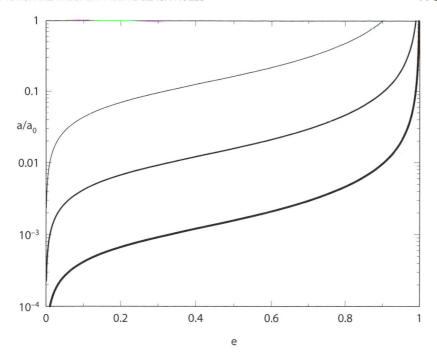

Figure 4.13 Joint evolution of semimajor axis, a, and eccentricity, e, as an orbit evolves in response to gravitational-wave energy loss according to equation (4.235). Initial conditions (upper right) were $e = 0.9$ (thin), $e = 0.99$, and $e = 0.999$ (thick).

with solution

$$a(e) = C(a_0, e_0) \frac{e^{12/19}}{1 - e^2} \left(1 + \frac{121}{304} e^2 \right)^{870/2299} , \qquad (4.236)$$

where $C(a_0, e_0)$ is determined by setting $a(e_0) = a_0$. Given a_0 and e_0, any two of the equations (4.234a)–(4.235) can be numerically integrated to solve for the decay of the orbit in the gravitational-wave-dominated regime, that is, to find $a = a(t)$, $e = e(t)$ (figure 4.13).

Two types of initial condition are particularly interesting. If the eccentricity is initially zero, then equation (4.234b) implies $de/dt = 0$ and the orbit remains circular during the decay. At the other extreme, suppose that the orbit is initially highly eccentric, $e \approx 1$; such initial conditions are relevant to the extreme-mass-ratio inspiral problem discussed in chapter 6 and to the triple-SBH problem discussed in chapter 8. In this limit, equation (4.235) implies

$$\frac{\Delta(1 - e)}{(1 - e)} \approx -\frac{\Delta a}{a}. \qquad (4.237)$$

Since the periapsis distance is $r_p = (1 - e)a$, equation (4.237) implies that r_p remains nearly constant as the orbit decays. In this high-eccentricity regime, loss of energy to gravitational waves acts like a drag force that "turns on" suddenly near periapsis, much like the atmospheric drag force that causes the orbits of artificial satellites around the Earth to decay.

By combining equations (4.234b) and (4.236), we can write the rate of change of eccentricity in terms of e alone:

$$\left\langle \frac{de}{dt} \right\rangle = -\frac{304}{15} \frac{G^3 m_1 m_2 m}{c^5 C(a_0, e_0)^4} \frac{e^{-29/19}(1-e^2)^{3/2}}{\left[1 + (121/304)e^2\right]^{1181/2299}}. \tag{4.238}$$

The time to coalescence, starting from eccentricity e_0, is given by the integral of this expression from $e = 0$ to $e = e_0$. The results are well approximated by

$$t_{\mathrm{GW}} \equiv t(e=0) - t(e=e_0) \tag{4.239}$$

$$\approx 5.8 \times 10^6 \frac{(1+q)^2}{q} \left(\frac{a_0}{10^{-2}\,\mathrm{pc}}\right)^4 \left(\frac{m_1 + m_2}{10^8\,M_\odot}\right)^{-3}$$

$$\times \left(1 - e_0^2\right)^{7/2} f(e_0)\,\mathrm{yr},$$

where $q \equiv m_2/m_1 \le 1$ is the mass ratio of the binary, and $f(e_0)$ is a weak function of the initial eccentricity: $f(0) = 0.979$ and $f(1) = 1.81$. Clearly, gravitational-wave emission only becomes interesting for a massive binary if the separation drops well below one parsec, or if the initial eccentricity is very high.

If the decay has gone on so long that the eccentricity has dropped nearly to zero, the subsequent evolution is obtained by integrating equation (4.234a) after setting $e = 0$:

$$a(t)^4 - a_0^4 = -\frac{256}{5} \frac{G^3 m_1 m_2 m}{c^5} (t - t_0) \tag{4.240}$$

and

$$t_{\mathrm{GW}} \equiv t(a=0) - t(a=a_0) = \frac{5}{256} \frac{c^5 a_0^4}{G^3 m_1 m_2 m} \tag{4.241}$$

$$\approx 5.7 \times 10^6 \frac{(1+q)^2}{q} \left(\frac{a_0}{10^{-2}\,\mathrm{pc}}\right)^4 \left(\frac{m_1 + m_2}{10^8\,M_\odot}\right)^{-3} \mathrm{yr}.$$

4.6 CAPTURE

In the previous section, the calculation of t_{GW} assumed a value of zero for the final radius of the inspiral orbit. Of course, a better choice would have been the radius of the SBH's event horizon; but the orbital decay rate becomes so large toward the end that either choice gives essentially the same result for the total inspiral time. But there are situations in which it is important to know the actual, limiting value of the orbital radius. For instance, solar-type stars are not tidally disrupted by an SBH with $M_\bullet \gtrsim 10^8\,M_\odot$, and their capture rate is determined by the rate at which they are scattered onto **capture orbits**: orbits that progress inevitably inside the event horizon. It turns out that most such captures occur from very eccentric orbits (chapter 6), so we need to derive capture criteria for orbits of arbitrary e, and of course for SBHs of arbitrary spin.

Capture is an inherently relativistic event: it occurs when an object passes within a few gravitational radii of the SBH. It is not possible to derive useful capture criteria in a post-Newtonian framework, and in this section, we cannot avoid presenting some results from the exact theory. The reader who desires a more complete background to these equations is directed toward any of a number of excellent texts on the general theory of relativity [562, 492, 400]. Throughout this section, we assume that the mass of the orbiting object is small compared with the SBH mass.

Start by writing the metric for a spinning black hole:

$$-c^2 d\tau^2 = -\left(1 - \frac{2mr}{\rho^2}\right) c^2 dt^2 + \frac{\rho^2}{\Delta} dr^2 + \rho^2 d\theta^2 - \frac{4mrs}{\rho^2} \sin^2\theta \, c \, dt \, d\phi$$
$$+ \left(r^2 + s^2 + \frac{2mrs^2}{\rho^2} \sin^2\theta\right) \sin^2\theta d\phi^2. \tag{4.242}$$

This is the **Kerr metric** [275] in **Boyer–Lindquist coordinates** [60]. The black-hole mass, $M_\bullet$, has been expressed in terms of $m \equiv GM_\bullet/c^2$; note that m has units of length and in fact is equal to the gravitational radius $r_{\rm g}$. The black hole is assumed to be rotating in the ϕ direction with spin angular momentum

$$\mathcal{S} = cM_\bullet s, \quad \text{i.e.,} \quad s = \frac{GM_\bullet}{c^2}\chi \tag{4.243}$$

with $\chi = s/m$ the dimensionless spin parameter defined in equation (4.72). Since $0 \le \chi \le 1$,

$$0 \le s \le \frac{GM_\bullet}{c^2}.$$

(Relativists, who see no problem with setting $G = c = 1$, prefer to write $0 \le s/M_\bullet \le 1$.) The other quantities in equation (4.242) are

$$\rho^2 \equiv r^2 + s^2 \cos^2\theta, \tag{4.244a}$$
$$\Delta \equiv r^2 + s^2 - 2mr. \tag{4.244b}$$

Boyer–Lindquist coordinates tend, in the large-distance limit, to ordinary, "flat-space" spherical coordinates. The proper time τ becomes the ordinary time in this limit.

The event horizon—the surface from which escape to infinity requires infinite energy—is located at the outer root of the equation $\Delta^2 = 0$, or

$$r = m + \left(m^2 - s^2\right)^{1/2}. \tag{4.245}$$

For $s = 0$, a Schwarzschild black hole, this radius is just $2m$, or $2r_{\rm g}$. Even in the case of a rotating hole, the event horizon is still a sphere, but its radius decreases with increasing s to a limiting value of $r_{\rm g}$ when $s = 1$. When $s \ne 0$, there is a second important surface, the **static limit**, defined by

$$r(\theta) = m + \left(m^2 - s^2 \cos^2\theta\right)^{1/2}. \tag{4.246}$$

Inside this surface, which lies outside the event horizon and is elongated perpendicular to the spin axis, observers are forced to move in the positive ϕ direction by the "dragging of inertial frames."

It turns out that motion of a particle in the Kerr geometry is completely integrable [72]. The constants of the motion are

E_∞ = total energy,

L_z = component of angular momentum parallel to spin axis,

C = Carter constant,

and the equation of motion in r has first integral

$$\left(\frac{dr}{d\tau}\right)^2 - \left(\frac{r^4}{\rho^4}\right) V(r) = 0, \tag{4.247}$$

where

$$\frac{V(r)}{c^2} = \frac{E_\infty^2}{c^2} - \left(1 - \frac{2m}{r} + \frac{s^2}{r^2}\right)\left(1 + \frac{C}{c^2 r^2}\right) - \frac{\beta}{c^2 r^2} + \frac{2m\alpha^2}{c^2 r^3} \tag{4.248}$$

and

$$\alpha = L_z - s E_\infty, \tag{4.249a}$$

$$\beta = L_z^2 - s^2 E_\infty^2. \tag{4.249b}$$

In the large-radius limit, L_z reduces to the Newtonian angular momentum per unit mass, and E_∞ is related to the Newtonian specific energy by

$$E_\infty^2 = 2E + c^2. \tag{4.250}$$

Note that E_∞ includes the rest mass of the orbiting body. In the same limit, the Carter constant becomes

$$C \to L^2 - L_z^2, \quad r \to \infty, \tag{4.251}$$

where L is the total angular momentum. Setting $C = 0$ is equivalent to restricting motion to the hole's equatorial plane.

First consider circular orbits in the equatorial plane: $\theta = \pi/2$, $C = 0$, $\rho^2 = r^2$, and

$$\left(\frac{dr}{d\tau}\right)^2 = V(r). \tag{4.252}$$

Turning points occur where $V(r) = 0$; for a circular orbit, we also require the effective potential to be a minimum, that is,

$$V(r) = 0, \quad dV/dr = 0. \tag{4.253}$$

Solving these equations for E_∞ and L_z,

$$c^{-1} E_\infty = \frac{r^2 - 2mr \pm s\sqrt{mr}}{r\left(r^2 - 3mr \pm 2s\sqrt{mr}\right)^{1/2}}, \tag{4.254a}$$

$$c^{-1} L_z = \frac{\sqrt{mr}\left(r^2 \mp 2s\sqrt{mr} + s^2\right)}{r\left(r^2 - 3mr \pm 2s\sqrt{mr}\right)^{1/2}}. \tag{4.254b}$$

The upper sign refers to corotating (prograde) orbits, that is, orbits with angular momenta parallel to the black-hole spin, and the lower sign to counterrotating orbits.

Not all circular orbits are stable; we are interested in the **innermost stable circular orbit**, or **ISCO**. Stability requires $d^2V/dr^2 \leq 0$. After some algebra, one finds for the ISCO radius [24]

$$r_{\text{ISCO}} = m \left\{ 3 + Z_2 \mp [(3 - Z_1)(3 + Z_1 + 2Z_2)]^{1/2} \right\},$$

$$Z_1 = 1 + \left(1 - \frac{s^2}{m^2}\right)^{1/3} \left[\left(1 + \frac{s}{m}\right)^{1/3} + \left(1 - \frac{s}{m}\right)^{1/3}\right],$$

$$Z_2 = \left(3\frac{s^2}{m^2} + Z_1^2\right)^{1/2}. \tag{4.255}$$

Setting $s = 0$ gives the circular orbit capture radius for a Schwarzschild black hole:

$$r_c = 6m = 6r_g = \frac{6GM_\bullet}{c^2} \quad (s = 0), \tag{4.256}$$

which is three times the radius of the event horizon. The corresponding orbital angular momentum, from equation (4.254b), is

$$L_c = \sqrt{12}r_g c \quad (s = 0) \tag{4.257}$$

and the energy, from equation (4.254a), is

$$E_{\infty,c} = \frac{2\sqrt{2}}{3}c \quad (s = 0). \tag{4.258}$$

A maximally rotating Kerr black hole has $s = m$. There are now two capture radii in the equatorial plane:

$$r_c = \begin{cases} r_g, & \text{prograde,} \\ 9r_g, & \text{retrograde.} \end{cases} \tag{4.259}$$

Figure 4.14 plots capture radii as a function of the dimensionless spin. The expressions given above for the energy and angular momenta are ill defined when $s = m$ and $r = r_g$; alternative forms for these expressions, valid when $s = m$, are

$$c^{-1}E_\infty = \frac{r - m \pm (mr)^{1/2}}{r^{3/4}\left(r^{1/2} \pm 2m^{1/2}\right)^{1/2}} \quad (s = m), \tag{4.260a}$$

$$c^{-1}L_z = \frac{m^{1/2}\left(r^{3/2} \pm m^{1/2}r + mr^{1/2} \mp m^{3/2}\right)}{r^{3/4}\left(r^{1/2} \pm 2m^{1/2}\right)^{1/2}} \quad (s = m). \tag{4.260b}$$

From these expressions, the angular momenta of the capture orbits for $s = m$ are

$$L_c = \begin{cases} (2/\sqrt{3})r_g c, & \text{prograde,} \\ (22/3\sqrt{3})r_g c, & \text{retrograde,} \end{cases} \tag{4.261}$$

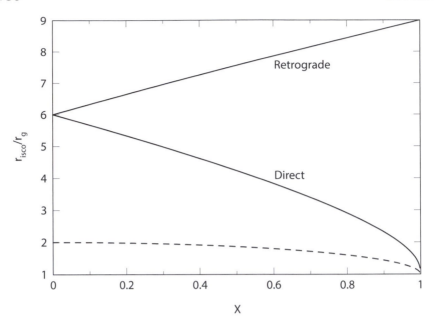

Figure 4.14 Radius of the innermost stable circular orbit as a function of the dimensionless spin parameter χ, for prograde and retrograde orbits. The dashed curve shows the radius of the event horizon.

and the energies are

$$
E_{\infty,c} = \begin{cases} (1/\sqrt{3})c, & \text{prograde}, \\ (5/\sqrt{27})c, & \text{retrograde}. \end{cases} \tag{4.262}
$$

In the discussion of the loss-cone problem later in this book, we will mostly be concerned with objects on orbits that are highly eccentric, and which only come close to the SBH during brief intervals near periapsis. Such objects spend most of their time in the Newtonian or post-Newtonian regime, and the detailed theory of loss-cone repopulation that is developed in chapter 6 is likewise rooted firmly in the Newtonian paradigm. When we insert the "capture conditions" that were just derived into such theory, we are effectively pasting together a fully relativistic description with a Newtonian one. Unfortunately, there is no unique way to do that. Boyer–Lindquist coordinates are the natural generalization of Schwarzschild coordinates and are best for many purposes, but they are not unique, and sufficiently close to a rapidly spinning black hole they behave in a nonintuitive way [24]. It is perhaps for this reason that different authors writing on the astrophysical loss-cone problem adopt different expressions for the "radius" corresponding to capture, particularly in the case of Kerr SBHs.

When considering capture from very eccentric orbits, there is a natural way to proceed [470]. One computes the critical values of the orbital elements corresponding to capture in the adopted metric, then asks for the Keplerian elements that would be inferred when an object on that orbit is far from the SBH, near apoapsis.[12]

Consider then the case of capture from orbits of high eccentricity and (in the case of a spinning black hole) arbitrary orbital inclination. A star on a highly eccentric orbit spends most of its time at radii $r \gg r_g$ where the motion is essentially Newtonian. We can therefore characterize the orbits by the Keplerian elements a and e, defined as the values that would be computed far from the SBH. The energy is $E_\infty^2 = 2E + c^2 \approx c^2(1 - GM_\bullet/c^2 a) \approx c^2$; that is, $E_\infty^2/c^2 \approx 1$. It is useful to define a new Carter constant, given by

$$C' = C + L_z^2. \tag{4.263}$$

In the spherically symmetric Schwarzschild geometry, C' is equal to the square of the total angular momentum, L^2. With this definition, the potential $V(r)$ takes the form

$$\frac{V(r)}{c^2} = \left(1 + \frac{2ms^2}{r^3} + \frac{s^2}{r^2}\right)\frac{E_\infty^2}{c^2} - \frac{\Delta}{r^2}\left(1 + \frac{C'}{c^2 r^2}\right) + \frac{s^2 L_z^2}{c^2 r^4} - \frac{4ms E_\infty L_z}{c^3 r^3}. \tag{4.264}$$

In order to make the connection to the Newtonian part of the motion more natural, we define two new constants of motion: $L_a > 0$ and $\cos i$, related to C' and L_z by

$$L_a^2 \equiv C', \quad \cos i \equiv \frac{L_z}{L_a}. \tag{4.265}$$

In the Newtonian limit, L_a and i correspond to the total angular momentum and the orbital inclination, respectively. For orbits in the equatorial plane, $L_a = |L_z|$, and $i = 0\,(\pi)$ for prograde (retrograde) orbits; for orbits out of the plane, $0 \le i \le \pi/2$ for prograde motion and $\pi/2 \le i \le \pi$ for retrograde motion.

We seek the critical angular momentum such that an orbit will not be "reflected" back to large distances, but instead will continue on to smaller values of r. The turning points of the orbit are still given by the condition $V(r) = 0$. The critical values of E_∞, C', and L_z are those for which the potential has an extremum at that same point; that is, where $d\left[(r^4/\rho^4)V(r)\right]/dr = 0$. But since $V(r) = 0$ this is equivalent to $dV/dr = 0$. This point should also be a minimum of $V(r)$; that is, $d^2V/dr^2 > 0$. The conditions $V(r) = 0$ and $dV/dr = 0$ then yield two quadratic equations for r. Solving one equation for r and substituting into the other yields an algebraic relation for the critical value of C in terms of m, s, and L_z. Scaling out the factors of m by defining the dimensionless variables $\tilde{L}_a = L_a/(mc)$ and using $\chi \equiv s/m$, the equation for the critical value of the angular momentum, $\tilde{L}_c$, has the

---

[12]The remaining material in this section was kindly provided by C. Will in advance of publication.

form

$$0 = Q\left\{ \left(1 - \chi^2 \sin^2 i\right) \tilde{L}_c^8 - 4\chi \cos i \tilde{L}_c^7 - 2\left[8 - \chi^2 \left(3 + 7 \sin^2 i\right)\right] \tilde{L}_c^6 \right.$$
$$+ 4\chi \cos i \left[24 - \chi^2 \left(1 + 9 \sin^2 i\right)\right] \tilde{L}_c^5$$
$$- \chi^2 \left[240 - 192 \sin^2 i - \chi^2 \left(1 + 18 \sin^2 i - 27 \sin^4 i\right)\right] \tilde{L}_c^4$$
$$+ 64\chi^3 \cos i \left(5 - 2 \sin^2 i\right) \tilde{L}_c^3 - 48\chi^4 \left(5 - 4 \sin^2 i\right) \tilde{L}_c^2$$
$$\left. + 96\chi^5 \cos i \tilde{L}_c - 16\chi^6 \right\},$$

(4.266)

where $Q = 1/\left[\tilde{L}_c^2(\tilde{L}_c^4 - 12\tilde{L}_c^2 + 24\chi \tilde{L}_c \cos i - 12\chi^2)^2\right]$. The critical radius corresponding to this unstable orbit is

$$\tilde{r}_c = \frac{\tilde{L}_c^2 \left[\left(\tilde{L}_c - \chi \cos i\right)^2 - 8\chi^2 \sin^2 i\right]}{\tilde{L}_c^4 - 12\left(\tilde{L}_c - \chi \cos i\right)^2 - 12\chi^2 \sin^2 i}.$$

(4.267)

An orbit in the equatorial plane has $i = \{0, \pi\}$, and the condition becomes

$$0 = \frac{\left(\tilde{L}_c^2 \mp 4\tilde{L}_c + 4\chi\right)\left(\tilde{L}_c^2 \pm 4\tilde{L}_c - 4\chi\right)\left(\tilde{L}_c \mp \chi\right)^4}{\tilde{L}_c^2 \left(\tilde{L}_c^4 - 12\tilde{L}_c^2 \pm 24\chi \tilde{L}_c - 12\chi^2\right)^2}.$$

(4.268)

The solutions that correspond to values of r that are outside the event horizon, and to maxima of $V(r)$, are

$$\tilde{L}_{c+} = 2 + 2\sqrt{1 - \chi}, \quad i = 0 \text{ (prograde)},$$
$$\tilde{L}_{c-} = 2 + 2\sqrt{1 + \chi}, \quad i = \pi \text{ (retrograde)}.$$

(4.269)

For $\chi = 0$, the solutions merge to $\tilde{L}_c = 4$.

For arbitrary values of i, equation (4.266) must be solved numerically. Alternatively, an approximate, analytic solution can be sought. Consider

$$\tilde{L}_c = 2 + 2\sqrt{1 - \chi \cos i - \frac{1}{8}\chi^2 \sin^2 i \; F(\chi, \cos i)}$$

(4.270)

which has the correct behavior when $\chi = 0$, and when $i = 0$ or $i = \pi$ for arbitrary χ. Solving equation (4.266) as a power series in χ yields

$$F(\chi, \cos i) = 1 + \frac{1}{2}\chi \cos i + \frac{1}{64}\chi^2 \left(7 + 13 \cos^2 i\right)$$
$$+ \frac{1}{128}\chi^3 \cos i \left(23 + 5 \cos^2 i\right)$$
$$+ \frac{1}{2048}\chi^4 \left(55 + 340 \cos^2 i - 59 \cos^4 i\right) + O(\chi^5).$$

(4.271)

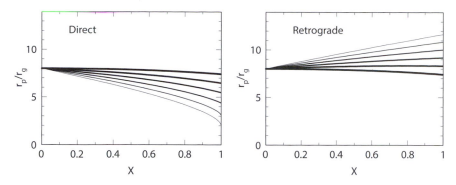

Figure 4.15 Critical, Newtonian periapsis for capture of eccentric orbits ($e \approx 1$) by Kerr black holes as a function of the dimensionless spin parameter χ. Different curves are for different values of $\cos i \equiv L_z/L$ in steps of 0.2, from $\cos i = 1$ (thinnest) to $\cos i = 0$ (thickest).

For $\chi = 1$, the critical value is given to high accuracy by

$$\tilde{L}_c = \begin{cases} 1.575 + 2.276\sqrt{1 - 1.044\cos i}, & -1 \le \cos i \le \sqrt{2/3}, \\ 2/\cos i, & \sqrt{2/3} \le \cos i \le 1. \end{cases} \tag{4.272}$$

Thus for a star in an orbit with a given $\cos i = L_z/L$, capture will occur for $L \le L_c = (GM_\bullet/c)\tilde{L}_c$. When the star is far from the hole, the orbital angular momentum per unit of μc^2 is given in terms of the Newtonian semimajor axis a and eccentricity e by $L^2 = GM_\bullet a(1 - e^2)/c^2 = mr_p(1 + e)$ where r_p is the periapsis distance. For large eccentricity then, capture will occur when

$$r_p \le r_{pc} \equiv \frac{1}{2}r_g\left(\tilde{L}_{c\pm}\right)^2. \tag{4.273}$$

For a Schwarzschild black hole, $r_c = 8r_g$; in the equatorial plane of a maximally rotating Kerr black hole, $r_c = 2(11.6)r_g$ for prograde (retrograde) orbits. Figure 4.15 shows numerically computed values of r_p corresponding to capture orbits for various values of χ and i.

4.7 RELATIVISTIC MOTION IN THE PRESENCE OF A DISTRIBUTED MASS

We are now in a position to consider the effects of relativity on the motion of a star orbiting in a nonspherical nucleus.

As noted above, the relativistic periapsis advance of the Galactic center star S2 should occur at approximately the same, time-averaged rate (although in the opposite sense) as the rate of precession due to the distributed mass near Sgr A*. But there is an even more interesting way in which Newtonian and relativistic perturbations can interact. If torques from the flattened potential of a nonspherical nucleus should cause the angular momentum of a star to decrease to sufficiently low values—as in the case of the saucer orbits described in section 4.4.2, or the pyramid orbits described in section 4.4.3—there will inevitably come a time when precession is dominated by relativistic effects. As we will see, this fact implies an effective upper limit on how eccentric such an orbit can become in response to steady torques [378].

The orbit-averaged rate of relativistic periapsis advance is given by equation (4.205):[13]

$$\left(\frac{d\omega}{dt}\right)_{\mathrm{GR}} = \nu_r \frac{3GM_\bullet}{c^2 a(1-e^2)}. \tag{4.274}$$

The rate of precession due to the distributed (spherical) mass is given by equation (4.88)

$$\left(\frac{d\omega}{dt}\right)_{\mathrm{M}} = -\nu_r G_{\mathrm{M}}(e, \gamma) \frac{M_\star(r < a)}{M_\bullet} \sqrt{1-e^2}, \tag{4.275}$$

where $G_{\mathrm{M}} \approx 1$ (figure 4.3).

We begin by ignoring the (generally smaller) contribution to the precession rate arising from the nonsphericity of the nucleus. The two frequencies (4.274), (4.275) are equal in magnitude (but opposite in sign) when

$$\ell \equiv \sqrt{1-e^2} = \ell_{\mathrm{crit}} \approx \left[\frac{r_{\mathrm{g}}}{a} \frac{M_\bullet}{M_\star}\right]^{1/3}, \tag{4.276}$$

where, as usual, $r_{\mathrm{g}} \equiv GM_\bullet/c^2$. When $\ell \gtrsim \ell_{\mathrm{crit}}$, precession is dominated by the mass term and is retrograde, while for $\ell \lesssim \ell_{\mathrm{crit}}$, relativity dominates and the precession is prograde.

Now consider the effect of the nonspherical component of the potential, and assume that $\ell \ll \ell_{\mathrm{crit}}$. In this regime, the precession rate scales as $\sim \ell^{-2}$, and the sign of the torque as experienced by the rapidly precessing orbit will fluctuate with such a high frequency that its net effect over one precessional period will be negligible: in other words, relativity will "quench" the effects of the torque and the orbit's eccentricity will remain nearly constant.

To estimate the eccentricity at which this occurs, we express the torque as

$$|\mathcal{T}| \approx \epsilon \frac{GM_\star}{a}, \tag{4.277}$$

where $M_\star$ is the stellar mass within $r = a$ and ϵ measures the degree of nuclear elongation. The timescale over which this torque changes a star's angular momentum is

$$\left|\frac{1}{L}\frac{dL}{dt}\right|^{-1} \approx \left|\frac{L}{\mathcal{T}}\right| \approx \frac{M_\bullet}{\epsilon M_\star} \left[\frac{a^3(1-e^2)}{GM_\bullet}\right]^{1/2} \approx \nu_r^{-1} \frac{M_\bullet}{\epsilon M_\star} \sqrt{1-e^2}. \tag{4.278}$$

In order for L to undergo significant variation (i.e., by of order itself), this timescale must be shorter than the timescale associated with relativistic precession,

---

[13]Henceforth in this chapter we ignore the distinction between a and a_r, etc.

$|(1/\pi)(d\omega/dt)|_{\text{GR}}^{-1}$, or

$$\ell \gtrsim \ell_{\min} \approx \frac{r_{\text{g}}}{a} \frac{M_{\bullet}}{\epsilon M_{\star}} \approx \epsilon^{-1} \ell_{\text{crit}}^3. \tag{4.279}$$

We expect that the torque due to a fixed mass distribution will be unable to reduce a star's angular momentum much below this value. Stated differently, orbits which have $\ell \lesssim \ell_{\min}$ will precess so rapidly that their angular momentum hardly changes; orbits which at some moment have $\ell \gtrsim \ell_{\min}$ will experience periodic changes in angular momentum but ℓ will not fall below $\sim \ell_{\min}$.

We can get a more detailed picture of how relativity affects the behavior of orbits by adding the orbit-averaged, 1PN accelerations to the equations of motion that were previously derived for axisymmetric and triaxial nuclei. Consider first the axisymmetric case. Since we are considering orbits of high eccentricity, we adopt equations (4.108), which describe the behavior of low-ℓ orbits in a mildly flattened nucleus with radial density profile $\rho \sim 1/r$. To the equation for $d\omega/dt$ we now add the term (4.274) representing Schwarzschild precession. The results are

$$\frac{d\omega}{d\tau} = -\ell + \frac{3\epsilon}{\ell} \cos^2 i \sin^2 \omega + \frac{\kappa}{\ell^2},$$

$$\frac{d\ell}{d\tau} = -\frac{3}{2}\epsilon \sin^2 i \sin(2\omega),$$

$$\frac{d\ell_z}{d\tau} = \frac{d\Omega}{d\tau} = 0, \tag{4.280}$$

where

$$\kappa \equiv v_0^{-1} \left(\frac{d\omega}{dt}\right)_{\text{GR}} \approx 3\frac{r_{\text{g}}}{a} \frac{M_{\bullet}}{M_{\star}(a)}. \tag{4.281}$$

Recall that the dimensionless time in equations (4.280) is $\tau = v_0 t$ and that v_0, defined in equation (4.102), is approximately equal to $v_r M_{\star}(a)/M_{\bullet}$, the mass precession rate. When the relativistic term is absent ($\kappa = 0$), these equations were found to define two sorts of orbit: tubes and saucers; saucer orbits were present when $\ell_z < \ell_{\text{sep}} \approx \sqrt{3\epsilon}$, where ϵ, defined in equation (4.104), is related to the axis ratio, q, of the nucleus by $\epsilon \approx (1 - q)/2$. Recall also that saucer orbits can reach values of ℓ as low as ℓ_z, which of course is the lowest angular momentum consistent with conservation of ℓ_z. It is interesting to ask: at what value of κ do saucer orbits disappear?

Saucer-like orbits will be present if there is a fixed point of the motion at $\omega = \pi/2$ (figure 4.4). Setting $\dot{\omega} = 0$, $\omega = \pi/2$ in the first of equations (4.280) gives, for the value of ℓ at the fixed point,

$$\ell^4 - \kappa\ell - 3\epsilon\ell_z^2 = 0. \tag{4.282}$$

As κ is increased from zero at fixed ℓ_z, the angular momentum associated with the fixed point increases. When κ exceeds $\kappa_{\max}$, given by

$$\kappa_{\max} = 1 - 3\epsilon\ell_z^2, \tag{4.283}$$

there is no longer a fixed point and the saucer orbits disappear.

In the axisymmetric geometry, a lower limit on ℓ is always set by ℓ_z, even in the absence of the quenching effects of relativity. We now consider the more interesting case of motion in a triaxial nucleus, for which the geometry imposes no lower limit on ℓ.

Consider then the orbit-averaged Hamiltonian (4.142a) that describes motion in a nucleus containing both a spherical, and a triaxial, mass component. We can include the effects of relativistic precession by adding an extra term to H:

$$H = -\frac{3}{2}\ell^2 + \epsilon_b H_b + \epsilon_c H_c - \frac{\kappa}{\ell}, \tag{4.284}$$

which is equivalent to adding the term κ/l^2 to the right-hand side of $d\omega/d\tau = \partial H/\partial \ell$. Here κ is given by

$$\kappa \equiv \frac{3GM_\bullet}{c^2 a}\frac{v_r}{v_0} \approx \frac{r_g}{a}\frac{M_\bullet}{M(a)} \tag{4.285}$$

with v_0 defined in equation (4.141a). In terms of κ, the critical value of ℓ at which relativistic precession cancels precession due to the distributed mass is

$$\ell_{\rm crit} = \left(\frac{\kappa}{3}\right)^{1/3}. \tag{4.286}$$

Consider an orbit for which $\ell \approx \ell_{\rm crit}$. Since κ and $\ell_{\rm crit}$ are typically much less than one, it is appropriate in this case to replace (4.284) by the simplified Hamiltonian (4.146). Adding the relativistic term,

$$\frac{5}{2}\epsilon_c - H = \left[\frac{3}{2}\ell^2 + \frac{\kappa}{\ell}\right] + \left[\frac{5}{2}\epsilon_c e_x^2 + \frac{5}{2}(\epsilon_c - \epsilon_b)e_y^2\right]$$
$$\equiv P(\ell) + Q(e_x, e_y), \tag{4.287}$$

where P and Q denote the expressions in the first and second sets of square brackets. The minimum of $P(\ell)$ occurs at $\ell = \ell_{\rm crit}$:

$$P_{\rm min} = \left(\frac{81\kappa^2}{8}\right)^{1/3}. \tag{4.288}$$

The function Q can vary from 0 to some maximum value $Q_{\rm max}$ due to the limitation that $e_x^2 + e_y^2 \le 1$. For each value of (e_x, e_y) (and therefore Q), there are *two* allowed values of ℓ; one of these is smaller than $\ell_{\rm crit}$ while the other is greater (figure 4.16). In the case of a librating orbit, these two values correspond to precession first in one, then the other sense at a given (e_x, e_y); in the case of a circulating orbit, the two values correspond to orbits that precess in one, or the other sense when they cross, for example, the z-axis. The minimum and maximum values of ℓ—both of which correspond to the maximum value of P (figure 4.16)—are attained when $Q = 0$, that is, when $e_x = e_y = 0$; the maximum of Q corresponds to $\ell = \ell_{\rm crit}$. (In the Newtonian case, the minimum of ℓ corresponds to the maximum of Q.)

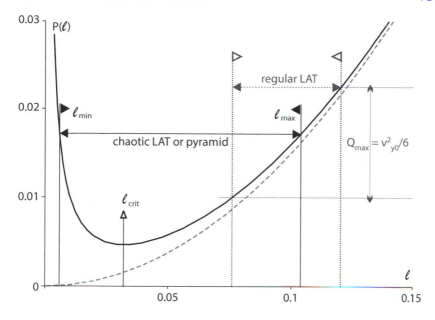

Figure 4.16 The allowed variations in angular momentum ℓ for orbits in a triaxial nucleus and in the presence of general relativistic precession [378]. The solid curve represents the function $P(\ell)$ (equation 4.287); the dashed parabola is the same function in the Newtonian case ($\kappa = 0$). If $\kappa \neq 0$, $P(\ell)$ has a minimum at ℓ_{crit} (equation 4.286). Orbits make excursions along the curve $P(\ell)$ in the range from a certain value $P_{\max}$ to $P_{\max} - Q_{\max}$ (here $Q_{\max}$ is given for the case of planar orbits). If during such an excursion ℓ does not cross ℓ_{crit}, then the orbit resides on one branch of $P(\ell)$; otherwise it "flips" to the other branch, reaching lower values of $\ell_{\min}$ (equation 4.292), becoming a pyramid orbit or a long-axis tube orbit (LAT).

Consider orbits that are confined to the y–z plane. Setting $\Omega = \pi/2$, $\ell_z = 0$ in equation (4.144), the Hamiltonian and equations of motion become

$$\frac{5}{2}\epsilon_c - H = \frac{3}{2}\ell_0^2 + \frac{\kappa}{\ell_0} = \frac{3}{2}\ell^2 + \frac{\kappa}{\ell} + \frac{v_y^{(0)2}}{6}\cos^2\omega, \qquad (4.289a)$$

$$\dot{\ell} = -\frac{v_y^{(0)2}}{6}\sin 2\omega, \qquad \dot{\omega} = -3\ell + \frac{\kappa}{\ell^2}. \qquad (4.289b)$$

Here, $\ell = \ell_0$ is the angular momentum when the orbit coincides with the short (z-) axis; that is, $\omega = \pi/2$, $e_y = 0$. The orbit in the course of its evolution may or may not cross the y-axis ($\omega = 0$). If it does, then the angle ω circulates monotonically, with $\dot{\omega} \neq 0$. In figure 4.16, the condition $\dot{\omega} = 0$ corresponds to reaching the lowest point in the $P(\ell)$ curve, $\ell = \ell_{\text{crit}}$. Whether this happens depends on the value of ℓ_0: since the orbit starts from $Q = 0$ and $P = P(\ell_0)$, it can "descend" the $P(\ell)$ curve at most by $Q_{\max} = v_{y0}^2/6$. The critical values of ℓ_0, $\ell = (\ell_{0+}, \ell_{0-})$, at which circulation changes to libration are the solutions to

$$\frac{3}{2}\ell_{0\pm}^2 + \frac{\kappa}{\ell_{0\pm}} = \frac{3}{2}\ell_{\text{crit}}^2 + \frac{\kappa}{\ell_{\text{crit}}} + \frac{v_{y0}^2}{6}; \qquad (4.290)$$

ℓ_{0+} and ℓ_{0-} are the upper and lower positive roots of this equation. For a librating (i.e., pyramid) orbit, these are the values of ℓ when the orbit precesses first in one sense, then the other, past the short (z-) axis. For such orbits, $\dot{\omega}$ changes sign exactly at $\ell = \ell_{\text{crit}}$, but the angular momentum continues to decrease beyond the point of turnaround, reaching its minimum value only when ω returns again to $\pi/2$, that is, the z-axis. The two semiperiods of oscillation are not equal: the first ($\ell > \ell_{\text{crit}}$ and $\dot{\omega} < 0$) is slower, the other is more abrupt (figure 4.17, cases b, d). In effect, these **windshield-wiper orbits**[14] are "reflected" by "striking" the relativistic angular momentum barrier, and so never reach unit eccentricity as in the case of Newtonian pyramid orbits (figure 4.7). After the orbit precesses past the z-axis in the opposite sense, the angular momentum begins to increase again, reaching its original value after precession in ω has completed a full cycle and the orbit has returned to the z-axis from the other side.

The extreme values of ℓ both occur at $\omega = \pi/2$, when the orbit precesses past the short (z-) axis. Setting $Q = 0$ (i.e., $P(\ell) = P(\ell_0)$) gives

$$\ell_{\text{extr,P}} = \frac{\ell_0}{2}\left(\sqrt{1 + 8\ell_{\text{crit}}^3/\ell_0^3} - 1\right). \qquad (4.291)$$

If $\ell_0 > \ell_{\text{crit}}$, this root corresponds to the minimum ℓ, with ℓ_0 the maximum value; in the opposite case they exchange places. For $\kappa \ll 3\ell_0^3$ this additional root is

$$\ell_{\min} \approx \frac{2\ell_{\text{crit}}^3}{\ell_0^2} = \frac{2}{3}\frac{\kappa}{\ell_0^2}. \qquad (4.292)$$

Evidently, the minimum angular momentum achievable by a pyramid orbit in the presence of relativistic precession scales as ℓ_0^{-2}—the pyramids with the widest bases come closest to the SBH. Combining equation (4.292) with equation (4.159), the maximum value of ℓ_0 for (nonrelativistic) pyramids, we find that the minimum value of ℓ for *any* pyramid is

$$\ell_{\min} \approx \frac{\kappa}{\epsilon} \approx \epsilon^{-1}\ell_{\text{crit}}^3, \qquad (4.293)$$

consistent with the order-of-magnitude estimate (4.279) made at the start of this section.

Figure 4.18 shows the dependence of the maximum and minimum values of ℓ on ℓ_0 for the various orbit families.

In the case of pyramid orbits that are not restricted to a principal plane, numerical solution of the equations of motion derived from the Hamiltonian (4.284) are observed to be generally chaotic, increasingly so as κ is increased (figure 4.19). This may be attributed to the "scattering" effect of the relativistic term κ/l in the Hamiltonian, which causes the vector (e_x, e_y) to be deflected by an almost random

---

[14]The reader may be old enough to remember when automobile windshield wipers behaved in this way.

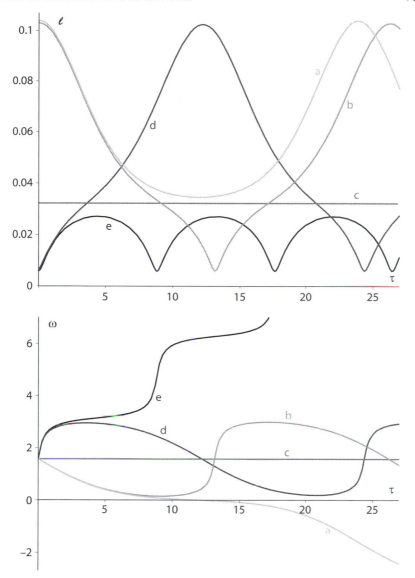

Figure 4.17 Behavior of orbits in the y–z plane of a triaxial nucleus when the 1PN effects of relativity are included [378]. These are solutions to the equations of motion (4.289b) in a potential with $\epsilon_c = 10^{-2}$, $\epsilon_b = \epsilon_c/2$, and with $\kappa = 10^{-4}$. All five orbits were started with $\omega = \pi/2$ (i.e., along the z-axis) but with different ℓ_0: (a) 0.104, (b) 0.103, (c) 0.0322, (d) 0.006, (e) 0.0058. The first two orbits lie close to the separatrix between long-axis tubes (LATs) and pyramids at $\ell_{0+} = 0.10392$ (4.290); the third is the stationary orbit with $\ell_0 = \ell_{\mathrm{crit}}$; and the last two lie near the separatrix between pyramids and GR-precession-dominated LATs, $\ell_{0-} = 0.005845$. For pyramid orbits (b, d), the angle ω librates around $\pi/2$, and ℓ crosses the critical value ℓ_{crit}; tube orbits (a,e) have ω monotonically circulating, and ℓ is always above or below ℓ_{crit}.

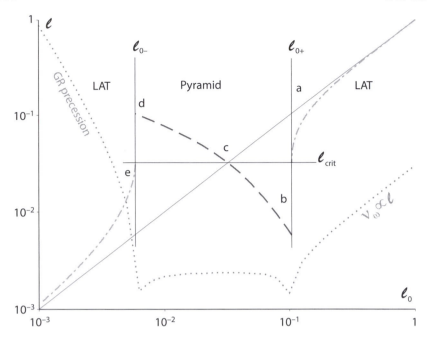

Figure 4.18 Minimum and maximum values of the dimensionless angular momentum ℓ for planar orbits like those in Figure 4.17 [378]. The straight line is $\ell = \ell_0$; the dashed line is the extremum for pyramids, equation (4.291). These two curves intersect at ℓ_{crit} (equation 4.286), where they exchange roles. For $\ell > \ell_{0+}$ and $\ell < \ell_{0-}$ (equation 4.290) the orbit is a tube. The dotted gray line shows the leading frequency of ω oscillations, $\nu_\omega \times 10^{-2}$; for high-$\ell$ orbits, $\dot{\omega} \approx 3\ell$, while for orbits dominated by relativistic precession, $\dot{\omega} \approx 2\kappa/\ell^2$. Letters denote the position of orbits shown in figure 4.17.

angle whenever ℓ approaches zero. In the limit that the motion is fully chaotic, H remains the only integral of the motion, and equation (4.287) implies that the vector (e_x, e_y) can lie anywhere inside an ellipse

$$Q(e_x, e_y) \equiv \frac{5}{2}\left[\epsilon_c e_x^2 + (\epsilon_c - \epsilon_b)e_y^2\right] \leq Q_{\mathrm{max}}, \qquad (4.294)$$

whose boundary is given by

$$Q_{\mathrm{max}} = \frac{5}{2}\epsilon_c - H - P_{\mathrm{min}}. \qquad (4.295)$$

This ellipse defines the base of the "pyramid" (which now rather resembles a cone). As in the planar case, the maximum and minimum values of ℓ are attained—not on the boundary of this ellipse (i.e., the corners in the Newtonian case)—but rather at $e_x = e_y = 0$, where $Q = 0$ and P attains its maximum. The ellipse (4.294) serves as a "reflection boundary" for trajectories that come below $\ell \approx \ell_{\mathrm{crit}}$. If this happens,

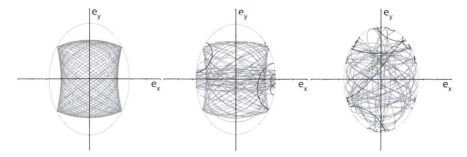

Figure 4.19 The effect of relativistic precession on three-dimensional pyramid orbits [378]. The three orbits were all started with the same initial conditions ($\ell = 0.05$, $\ell_z = 0.02$, $\omega = \Omega = \pi/2$) in a triaxial potential with $\epsilon_c = 0.01$, $\epsilon_b = 0.005$ and in the absence of relativity, each would be a pyramid orbit. The three panels have different values of the coefficient κ (equation 4.285) that determines the relative speed of relativistic and Newtonian precessions. Left: $\kappa = 0$ (regular); middle: $\kappa = 10^{-6}$ (weakly chaotic); right: $\kappa = 10^{-5}$ (strongly chaotic). The ellipse marks the maximal extent of the (e_x, e_y) vector, equation (4.294), i.e. $\ell = \ell_{\text{crit}}$, equation (4.286).

the vector (e_x, e_y) is observed to be quickly "scattered" by an almost random angle (figure 4.19, right, denoted by the red segments), similar to the rapid change in ω that occurs in the planar case (figure 4.17). It turns out [378] that the minimum value of the angular momentum attained in the chaotic case is approximately the same as in equation (4.291). Roughly speaking, all pyramid orbits and some tube orbits (those that may attain $\ell \leq \ell_{\text{crit}}$) are found to be chaotic.

So far in this section, we have considered only the lowest-order relativistic corrections to the equations of motion. The same approach can be used to include progressively higher-order PN terms, representing the effects of frame dragging, of torques due to the SBH's quadrupole moment, etc. Of course, these higher-order corrections become progressively more important at smaller distances from the SBH. But sufficiently close to the SBH, the number of stars enclosed within any orbit is so small that it may no longer make sense to represent the gravitational potential from the stars as a smooth, symmetric function of position. Instead, the nonsphericity of the potential may be due mostly to the fact that at any moment, there are different numbers of stars on one side of the SBH as compared with another. Here we are anticipating the discussion of "resonant relaxation" in chapter 5: the idea is that—at least for some span of time—orbits near the SBH are nearly Keplerian, and maintain their orbital elements, including particularly their orientations. As argued in section 5.6, the magnitude of the torque acting on a test star in this regime is roughly

$$|\mathcal{T}| \approx \sqrt{N} \frac{G m_\star}{a}, \qquad (4.296)$$

where N is the number of stars inside the test-star's orbit, of semimajor axis a, and $m_\star$ is the mass of one star. This "$\sqrt{N}$ torque" will dominate the torque from the large-scale nonsphericity, equation (4.277), if

$$N(a) \lesssim \epsilon^{-2}. \tag{4.297}$$

So, for instance, if the nucleus is only slightly elongated, say $\epsilon \approx 10^{-2}$, then at radii where $N(< a) \lesssim 10^4$, the $\sqrt{N}$ torques will dominate torques from the large-scale distortion. In the Milky Way, the corresponding radius might be $\sim 10^{-2} r_m \sim 10^{-2}$ pc. One can obtain an approximate understanding of the motion in this regime by supposing that the $\sqrt{N}$ torques are representable approximately in terms of an axisymmetric or triaxial distortion (say), with amplitude given by $\epsilon \approx 1/\sqrt{N}$, and applying the orbit-averaged equations derived above. What makes the problem much more interesting, and difficult, is the fact that the orbits generating the torque do not maintain their orientations forever: they precess, causing the direction of the torque to change with time in some complicated way. Understanding the motion in this regime is extremely important but such work is still in its infancy. Elsewhere in this book, two early results from that research are presented. In section 5.7.1 it is shown that there is a "sphere of rotational influence" inside of which dragging of inertial frames by a spinning SBH dominates the $\sqrt{N}$ torques from nearby stars [358]. In section 6.4, interaction of the $\sqrt{N}$ torques with the 1PN precession is shown to create a "barrier" that reflects orbits to lower eccentricities, strongly mediating the capture rate of stars near the SBH [359].

4.8 MOTION IN THE PRESENCE OF A SECOND MASSIVE BODY

So far in this chapter, we have considered several examples in which the two-body equations of motion were modified by an additional, time-independent force term: due to a distributed mass, or to relativity, or both. But there are many interesting problems in nuclear dynamics that involve the presence of a second massive body which itself is orbiting about the SBH. An extreme example would be a binary SBH, formed, for instance, in a galaxy merger. Binary SBHs can radically change the distribution of mass near the center of a galaxy, so much so that we postpone a discussion of their dynamics until chapter 8. But there are less extreme possibilities as well: for instance, an intermediate-mass black hole (IBH) that orbits about the SBH. As discussed in chapter 2, these hypothesized objects would have masses between $\sim 10^2$ and $\sim 10^6 \, M_\odot$. An IBH orbiting within the influence sphere of an SBH would substantially perturb the orbits of stars at $r \lesssim r_h$ [219].

Ignoring the influence of any distributed mass, what we are talking about here is the famous **three-body problem** of classical mechanics. The three-body problem has a long history and a voluminous literature, and there are a number of texts devoted entirely to this subfield of celestial mechanics [336, 532, 31]. In this section, we restrict our attention to a special (but still very useful) case: the **hierarchical three-body problem**, in which the separation of two of the bodies is much less than the distance of either body to the third object. In addition, we will assume that the evolution of interest takes place on timescales long compared with the periods of the inner or outer "binaries," allowing us to apply the averaging techniques that were developed earlier in this chapter.

4.8.1 The hierarchical three-body problem

Consider three bodies of mass M, m_1, and m_2. We define mass M as the "primary," and will later set $M > (m_1, m_2)$, but for the moment the ordering of masses is left unspecified. Let the position vectors of the three bodies with respect to an inertial coordinate system be $\boldsymbol{R}$, $\boldsymbol{R_1}$, $\boldsymbol{R_2}$, and the positions of m_1, m_2 relative to the primary be $\boldsymbol{r_1}$, $\boldsymbol{r_2}$:

$$\boldsymbol{r}_1 = \boldsymbol{R}_1 - \boldsymbol{R}, \quad \boldsymbol{r}_2 = \boldsymbol{R}_2 - \boldsymbol{R} \tag{4.298}$$

(figure 4.20a). The equations of motion of the three masses are

$$M\ddot{\boldsymbol{R}} = GMm_1 \frac{\boldsymbol{r}_1}{r_1^3} + GMm_2 \frac{\boldsymbol{r}_2}{r_2^3}, \tag{4.299a}$$

$$m_1\ddot{\boldsymbol{R}}_1 = Gm_1m_2 \frac{\boldsymbol{r}_2 - \boldsymbol{r}_1}{|\boldsymbol{r}_2 - \boldsymbol{r}_1|^3} - GMm_1 \frac{\boldsymbol{r}_1}{r_1^3}, \tag{4.299b}$$

$$m_2\ddot{\boldsymbol{R}}_2 = Gm_2m_1 \frac{\boldsymbol{r}_1 - \boldsymbol{r}_2}{|\boldsymbol{r}_1 - \boldsymbol{r}_2|^3} - GMm_2 \frac{\boldsymbol{r}_2}{r_2^3}. \tag{4.299c}$$

By differencing these expressions, we derive the accelerations of m_1 and m_2 relative to the primary:

$$\ddot{\boldsymbol{r}}_1 + G(M + m_1) \frac{\boldsymbol{r}_1}{r_1^3} = Gm_2 \left(\frac{\boldsymbol{r}_2 - \boldsymbol{r}_1}{|\boldsymbol{r}_2 - \boldsymbol{r}_1|^3} - \frac{\boldsymbol{r}_2}{r_2^3} \right), \tag{4.300a}$$

$$\ddot{\boldsymbol{r}}_2 + G(M + m_2) \frac{\boldsymbol{r}_2}{r_2^3} = Gm_1 \left(\frac{\boldsymbol{r}_1 - \boldsymbol{r}_2}{|\boldsymbol{r}_1 - \boldsymbol{r}_2|^3} - \frac{\boldsymbol{r}_1}{r_1^3} \right). \tag{4.300b}$$

Equations (4.300) can be written in terms of gradients of scalar functions:

$$\ddot{\boldsymbol{r}}_1 = -\nabla_1(\Phi_1 + R_1), \quad \ddot{\boldsymbol{r}}_2 = -\nabla_2(\Phi_2 + R_2), \tag{4.301}$$

where the subscript i on ∇ denotes differentiation with respect to the coordinates $\boldsymbol{r}_i$. The functions Φ_i are

$$\Phi_1 = -G\frac{M + m_1}{r_1}, \quad \Phi_2 = -G\frac{M + m_2}{r_2}, \tag{4.302}$$

which are just the two-body, or Keplerian, parts of the total potential. The functions R_i are called the **disturbing functions**:

$$R_1 = -\frac{Gm_2}{|\boldsymbol{r}_2 - \boldsymbol{r}_1|} + Gm_2 \frac{\boldsymbol{r}_1 \cdot \boldsymbol{r}_2}{r_2^3}, \tag{4.303a}$$

$$R_2 = -\frac{Gm_1}{|\boldsymbol{r}_1 - \boldsymbol{r}_2|} + Gm_1 \frac{\boldsymbol{r}_1 \cdot \boldsymbol{r}_2}{r_1^3}. \tag{4.303b}$$

The first terms in equations (4.303) are called the **direct terms**. They represent the pairwise interaction potentials of m_1 and m_2 and are symmetric in $\{\boldsymbol{r}_1, \boldsymbol{r}_2\}$. The second terms are called the **indirect terms**, and their particular form depends on the choice of coordinate system. In particular, the indirect terms are not symmetric in $\{\boldsymbol{r}_1, \boldsymbol{r}_2\}$.

(a)

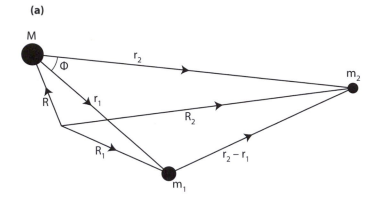

(b)

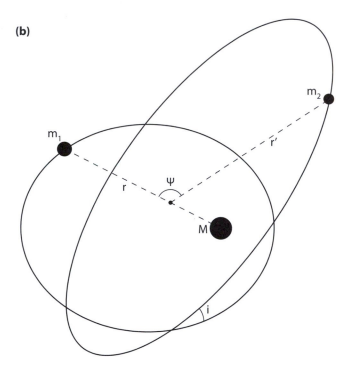

Figure 4.20 Two coordinate systems that are useful when discussing the three-body problem. (a) "Astrocentric" coordinates. The position vectors r_1 and r_2, of masses
m_1 and m_2, are defined with respect to the primary mass M. (b) "Jacobi" coordinates. The position vector r is defined in the same way as r_1. The position
vector r_2 is defined with respect to the center of mass of the (M, m_1) system.
In the averaged equations of motion, the unperturbed outer "binary" is defined
in terms of the motion of m_2 about this center of mass, and not in terms of the
(M, m_2) binary.

Suppose that $r_2 > r_1$. Then the interaction potential $|r_2 - r_1|^{-1}$ can be expanded in a Legendre series as

$$\frac{1}{|r_2 - r_1|} = \frac{1}{r_2} \sum_{l=0}^{\infty} \left(\frac{r_1}{r_2}\right)^l P_l(\cos\phi), \tag{4.304}$$

where ϕ is the angle between r_1 and r_2; that is, $r_1 \cdot r_2 = r_1 r_2 \cos\phi$ (figure 4.20a). The two disturbing functions become

$$R_1 = -\frac{Gm_2}{r_2} \sum_{l=1}^{\infty} \left(\frac{r_1}{r_2}\right)^l P_l(\cos\phi) + Gm_2 \frac{r_1}{r_2^2} \cos\phi, \tag{4.305a}$$

$$R_2 = -\frac{Gm_1}{r_2} \sum_{l=1}^{\infty} \left(\frac{r_1}{r_2}\right)^l P_l(\cos\phi) + Gm_1 \frac{r_2}{r_1^2} \cos\phi. \tag{4.305b}$$

In these expressions, the $l = 0$ (constant) terms have been neglected since they do not contribute to the accelerations. In the disturbing function for the inner body, the $l = 1$ term in the Legendre expansion is exactly canceled by the additional (indirect) term on the right-hand side of equation (4.305a), so that

$$R_1 = -\frac{Gm_2}{r_2} \sum_{l=2}^{\infty} \left(\frac{r_1}{r_2}\right)^l P_l(\cos\phi). \tag{4.306}$$

The lowest-order contribution to the perturbed potential is $l = 2$. In the case of the outer body, this cancellation does not occur.

At this point, it is standard practice to effect a change in variables. **Jacobi coordinates** are defined in the following way (figure 4.20b). The first position vector, r, is equated with r_1: it is the separation vector between M and m_1, the two components of the inner binary. The second position vector, r', connects the center of mass of the inner binary to the location of m_2. Thus

$$r = r_1, \quad r' = r_2 - \frac{m_1}{M + m_1} r. \tag{4.307}$$

In effect, the outer "binary" consists of m_2 orbiting about the center of mass of the inner binary.[15] If we define a new angle ψ such that $r \cdot r' = rr' \cos\psi$ (figure 4.20b), the Hamiltonian of the three-body system becomes, after some algebra [228]:

$$\mathcal{H} = -\frac{GMm_1}{2a_1} - \frac{G(M + m_1)m_2}{2a_2}$$
$$- \frac{G}{a_2} \sum_{l=2}^{\infty} \left(\frac{a_1}{a_2}\right)^l M_l \left(\frac{r}{a_1}\right)^l \left(\frac{a_2}{r'}\right)^{l+1} P_l(\cos\psi), \tag{4.308}$$

where

$$M_l = Mm_1 m_2 \frac{M^{l-1} - (-m_1)^{l-1}}{(M + m_1)^l}. \tag{4.309}$$

---

[15] Jacobi coordinates can be generalized to an arbitrary number of bodies [432].

In equation (4.308), a_1 is the semimajor axis of the (osculating) orbit of the inner binary, and a_2 is the semimajor axis of the outer "binary," that is, the orbit defined by m_2's motion about the center of mass of the inner binary. Note that the lowest-order term in the ψ expansion has $l = 2$: a "quadrupole."

We are now in a position to invoke our assumption that the triple system is hierarchical. If $a_1 \ll a_2$, the series in equation (4.308) will rapidly converge. Taking only the lowest-order, $l = 2$, term yields the **quadrupole Hamiltonian**:

$$\mathcal{H} = \mathcal{H}_{\text{Kep}} + \mathcal{H}_p, \tag{4.310}$$

$$\mathcal{H}_{\text{Kep}} = -\frac{GMm_1}{2a_1} - \frac{G(M + m_1)m_2}{2a_2},$$

$$\mathcal{H}_p = -\frac{G}{2a_2}\frac{Mm_1m_2}{M + m_1}\frac{r^2}{r'^3}\left(3\cos^2\psi - 1\right).$$

Equations (4.310) are the basis for most studies of the hierarchical three-body problem. The quadrupole approximation turns out to be adequate for many problems, but can yield misleading results in certain cases; for instance, if the two orbits are nearly coplanar, or highly eccentric, and of course the approximation breaks down if $a_1 \approx a_2$. Some of these special cases are discussed in more detail below.

If we assume that $\{m_1, m_2\} \ll M$, the motion of the two bodies will remain close to Keplerian for many orbital periods, and it is appropriate to average the Hamiltonian (4.310) over the unperturbed orbits. (No assumption is made yet about the relative sizes of m_1 and m_2.) As before in this chapter, the Hamiltonian is expressed in terms of Delaunay variables and $\mathcal{H}$ is integrated with respect to the mean anomaly. In the current problem, the averaging is carried out twice: once with respect to the outer orbit and once with respect to the inner orbit. Because of the fact that we are dealing with *two* orbits, of arbitrary relative inclination, the algebra can become daunting. It turns out that this is a problem for which the vectorial orbital elements defined in section 4.2 are very useful. In terms of the unit vectors defined in equation (4.59), the doubly averaged $\mathcal{H}_p$ can be written very compactly as [156]

$$\overline{\overline{\mathcal{H}}}_p = -\frac{3GMm_1m_2}{8(M + m_1)(1 - e_2^2)^{3/2}}\frac{a_1^2}{a_2^3} \tag{4.311}$$

$$\times \left[-\frac{1}{3} + 2e_1^2 + \left(1 - e_1^2\right)\left(e_{\ell,1} \cdot e_{\ell,2}\right)^2 - 5e_1^2\left(e_{e,1} \cdot e_{\ell,2}\right)^2\right].$$

Recall that e_ℓ is a unit vector in the direction of the orbital angular momentum and e_e is a unit vector parallel to the Runge–Lenz vector, pointing toward orbital periapsis; the subscripts 1 and 2 refer to the inner and outer "binaries."

The doubly averaged $\mathcal{H}_p$ is independent of ω_2, the argument of periapsis of the outer orbit. Since $dL_2/dt = -\partial\mathcal{H}/\partial\omega_2$, this means that the eccentricity of the outer orbit is constant. Remarkably, this is true regardless of the eccentricity e_1 of the *inner* orbit—in spite of the fact that an eccentric inner orbit produces a nonaxisymmetric force on the outer orbit! Not surprisingly, the constancy of e_2 does not hold at the next higher (octopole) order of expansion unless $e_1 = 0$.

While the motion corresponding to (4.311) is integrable, the solutions are complicated, and it is common to consider two limiting cases. Setting $m_1 \ll m_2$ gives the **inner restricted problem**;[16] setting $m_2 \ll m_1$ gives the **outer restricted problem**. The inner restricted problem has received by far the most attention from celestial mechanicians; it applies, for instance, to the motion of an artificial satellite that is perturbed by the Moon, or to an asteroid perturbed by Jupiter. In the context of galactic nuclei, the inner restricted problem applies, for instance, to a star that orbits about an SBH while being perturbed by an IBH orbiting farther out.

4.8.2 The inner restricted problem

The inner restricted problem is associated with the names M. L. Lidov [321] and Y. Kozai [299], who first established the existence of solutions in which conservation of L_z (in the quadrupolar approximation) allows the inner body to exchange eccentricity with inclination:

$$\left(1 - e_1^2\right)^{1/2} \cos i = \text{constant} \tag{4.312}$$

with $e_1 \to 1$ if the "initial" inclination is favorable. This is the **Lidov–Kozai mechanism**.[17] In the context of triple SBHs, the Lidov–Kozai mechanism is important because it can lead to greatly reduced timescales for gravitational-wave coalescence of the inner binary.

Setting $m_1 = 0$, the doubly averaged perturbing function (per unit of m_1) becomes [253]

$$\overline{\overline{\mathcal{H}}}_p = \frac{Gm_2}{8(1 - e_2^2)^{3/2}} \frac{a^2}{a_2^3} \left[-2 - 3e^2 + 3\sin^2 i \left(5e^2 \sin^2 \omega + 1 - e^2\right)\right]. \tag{4.313}$$

Here, the variables (a, e, ω, e) lacking subscripts refer to the inner orbit, and the inclination is defined with respect to the outer orbital plane, which is fixed in this limit.

We saw a very similar Hamiltonian in section 4.4.2, in our discussion of motion in an axisymmetric star cluster around an SBH. As in that case, we can define a dimensionless, averaged Hamiltonian as $H = \overline{\overline{H}}_p/(\nu_0 I)$, where $I = \sqrt{GMa}$ and a natural choice for ν_0 is

$$\nu_0^{-1} \equiv T_{\text{Kozai}} = \frac{\sqrt{GM}}{Gm_2} \frac{a_2^3}{a^{3/2}} \left(1 - e_2^2\right)^{3/2} = \frac{1}{2\pi} \frac{M}{m_2} \frac{P_2^2}{P} \left(1 - e_2^2\right)^{3/2}. \tag{4.314}$$

Thus

$$H = \frac{1}{8} \left[-5 + 3\ell^2 + 3\sin^2 i \left(\ell^2 + 5e^2 \sin^2 \omega\right)\right]. \tag{4.315}$$

As before, $\ell \equiv L/I = (1 - e^2)^{1/2}$ is the dimensionless angular momentum variable and $\cos i = \ell_z/\ell$. The equations of motion are

$$\frac{d\omega}{d\tau} = \frac{\partial H}{\partial \ell} = \frac{3}{4\ell} \left[2\ell^2 + 5\sin^2 \omega \left(e^2 - \sin^2 i\right)\right], \tag{4.316a}$$

---

[16]"Restricted" here refers to the fact that one of the masses is negligible.
[17]Commonly attributed just to Y. Kozai.

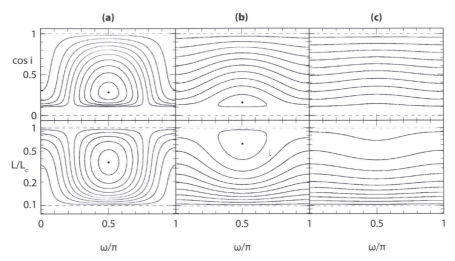

Figure 4.21 Lidov–Kozai oscillations of the inner binary (M, m_1) in the inner restricted hierarchical three-body problem $(m_1 \ll m_2)$. (a) Motion for $\ell_z = 0.1$. These are solutions to the equations of motion (4.316), with no relativistic corrections. The fixed-point orbit that is the "generator" of the librating orbits is indicated by the dot. (b) Motion for $\ell_z = 0.1$ and $\kappa = 0.5$, where κ is defined in equation (4.334) and measures the strength of the relativistic term in the equation of motion (4.333). (c) Motion for $\ell_z = 0.1$ and $\kappa = \kappa_{max}$, the maximum value of κ that allows librating solutions for this value of ℓ_z (equation 4.335). In this case, relativistic precession is so rapid that the eccentricity of the inner binary is hardly affected by torques from m_2.

$$\frac{d\ell}{d\tau} = -\frac{\partial H}{\partial \omega} = -\frac{15}{8} e^2 \sin^2 i \sin(2\omega), \tag{4.316b}$$

$$\frac{d\Omega}{d\tau} = -\frac{\partial H}{\partial \ell_z} = \frac{3}{4} \frac{\cos i}{\ell} \left(\ell^2 + 5e^2 \sin^2 \omega \right), \tag{4.316c}$$

$$\frac{d\ell_z}{d\tau} = 0. \tag{4.316d}$$

The last of these is equivalent to equation (4.312). In addition to ℓ_z, there is a second conserved quantity, H. Using the constancy of $\ell_z^2 = (1 - e^2) \cos^2 i$, the second conserved quantity can also be written as

$$Q = e^2 \left(5 \sin^2 i \, \sin^2 \omega - 2 \right), \tag{4.317}$$

which for a given ℓ_z differs from H only by a constant.

As in the case of the axisymmetric star cluster, we can solve $H = H(\ell, \omega)$ given ℓ_z to find $\ell = \ell(H, \omega)$, substitute the result into the right-hand side of equation (4.316a), and integrate to find $\omega(\tau)$. Inserting the result into equation (4.316b) and integrating then yields $\ell(\tau)$, and $\Omega(\tau)$ follows from equation (4.316c). Figure 4.21a shows numerical solutions for $\ell_z = 0.1$.

Just as in the axisymmetric problem discussed earlier, there are two classes of solution, corresponding to motion that circulates in ω ($0 \leq \omega \leq 2\pi$), or that librates over a limited range in ω. Librating solutions correspond to the "saucer" orbits in the axisymmetric problem. Libration requires the existence of a fixed point, at $\omega = \pm\pi/2$, where

$$\ell^2 = \ell_{\text{fp}}^2 = \sqrt{\frac{5}{3}}\ell_z, \quad \cos^2 i = \cos^2 i_{\text{fp}} = \sqrt{\frac{3}{5}}\ell_z. \tag{4.318}$$

Fixed points only exist for $\ell_z < \sqrt{3/5} \approx 0.775$; at the maximum allowed ℓ_z, the fixed-point inclination is $\cos i = \sqrt{3/5}$, that is, $i = (39.23°, 140.8°)$. If $\ell_z < \sqrt{3/5}$, there will be a separatrix dividing the two sorts of motion. Figure 4.21 shows that the separatrix passes through $e = 0$ when $\omega = 0$, implying

$$H_{\text{sep}} = \frac{1}{8}\left(1 - 3\ell_z^2\right). \tag{4.319}$$

Along the separatrix, $\cos i$ varies between

$$\ell_z \leq \cos i \leq \sqrt{\frac{3}{5}} \tag{4.320}$$

and the angular momentum varies between

$$\frac{5}{3}\ell_z^2 \leq \ell^2 \leq 1. \tag{4.321}$$

For high "initial" inclinations, the eccentricity attains values close to 1 as the orbit tilts down toward the plane of the outer binary. Remarkably, the degree of possible variation in ℓ is independent of the masses involved, which only set the timescale of the oscillations.

We can obtain an estimate of that timescale by considering motion near the fixed points. Linearizing the equations of motion about ($\ell = \ell_{\text{fp}}, \omega = \pi/2$) yields

$$\frac{d\omega}{d\tau} = -9\left(\ell - \ell_{\text{fp}}\right), \tag{4.322a}$$

$$\frac{d\ell}{d\tau} = \frac{1}{4}\left[15 - 3\ell_{\text{fp}}^2\left(8 - 3\ell_{\text{fp}}^2\right)\right](\omega - \pi/2) \tag{4.322b}$$

with (dimensionless) period

$$\frac{4\pi}{9}e_{\text{fp}}^{-1}\left(\frac{5}{3} - \ell_{\text{fp}}^2\right)^{-1/2}. \tag{4.323}$$

Recalling that the unit of time is given by equation (4.314), the (dimensional) period near the fixed point becomes

$$T_{\text{fp}} = \frac{4\pi}{9}\sqrt{\frac{3}{5}}\left[\left(\sqrt{\frac{5}{3}} - \ell_z\right)\left(\sqrt{\frac{3}{5}} - \ell_z\right)\right]^{-1/2}T_{\text{Kozai}}, \tag{4.324}$$

where ℓ_{fp}^2 has been replaced by $\sqrt{5/3}\ell_z$. For $\ell_z = 0$, the numerical factor multiplying T_{Kozai} is 1.082. When $\ell_z = \sqrt{3/5}$, the fixed point is coincident with the separatrix and the period diverges. For intermediate values of ℓ_z, the period near the fixed point lies between these extremes.

It turns out that the period of librating orbits can be expressed simply in terms of special functions, even for orbits that do not lie near the fixed point [544, 287]. The period depends on ℓ_z and on a second parameter that characterizes the amplitude of the oscillations. It is convenient to take as that second parameter the minimum or maximum value of the angular momentum, $\ell_\pm$, reached during the cycle. These are related by

$$\ell_- \ell_+ = \ell_{fp}^2 \tag{4.325}$$

(as in equation 4.113). Let $z_1 = 1 - \ell_-^2$, $z_2 = 1 - \ell_+^2$, and

$$z_3 = -\frac{3}{2}\left(1 - \ell_\pm^2\right)\left(1 - \frac{5}{3}\frac{\ell_z^2}{\ell_\pm^2}\right) \tag{4.326}$$

with $z_3 < 0 < z_2 < z_1$. Then

$$T = \frac{8}{3\sqrt{6}}\frac{K(k^2)}{\sqrt{z_1 - z_3}}T_{\text{Kozai}}, \tag{4.327}$$

where

$$k^2 = \frac{z_1 - z_2}{z_1 - z_3} \tag{4.328}$$

and $K(k^2)$ is the complete elliptic integral of the first kind with modulus k. For motion near the fixed point, $K \approx \pi/2$ and equation (4.324) is recovered. The time dependence of (ω, ℓ, Ω), all of which vary with period T, can likewise be expressed in terms of special functions [544, 287].

In the case of zero inclination, $\sin i = 0$, equations (4.316) predict that the eccentricity is constant, and conservation of ℓ_z then implies that the motion is restricted to the $i = 0$ plane. This result seems counterintuitive, since two coplanar, eccentric orbits will clearly exert torques on one another! The reason for this nonphysical result is the quadrupolar approximation, which implies an effective potential that is axisymmetric, even when the orbits involved are eccentric. Clearly in this case, we need to take the expansion in equation (4.308) to one higher (octopole) order. If we do so, then average the perturbing function over the unperturbed motion, and define a dimensionless Hamiltonian as in equation (4.314), we find [313]

$$H = \frac{1}{8}\left[-5 + 3\ell^2 + \frac{15}{8}\frac{a}{a_2}\frac{ee_2}{(1 - e_2^2)}\left(7 - 3\ell^2\right)\cos\left(\omega - \omega_2\right)\right]. \tag{4.329}$$

Ignoring changes in the outer binary, appropriate if $m_2 \gg m$, the equations of motion of the inner binary are

$$\frac{d\omega}{d\tau} = \frac{\partial H}{\partial \ell} = \frac{3}{4}\ell - \frac{C}{8}\frac{\ell}{e}\left(13 - 9\ell^2\right)\cos\omega, \tag{4.330a}$$

$$\frac{d\ell}{d\tau} = -\frac{\partial H}{\partial \omega} = -\frac{C}{8}e\left(7 - 3\ell^2\right)\sin\omega, \tag{4.330b}$$

where ω_2 has been set to 0 and

$$C = \frac{15}{8}\frac{a}{a_2}\frac{e_2}{(1 - e_2^2)}. \tag{4.331}$$

The second of these equations states the reasonable result that the torque acting on the inner binary varies as the sine of the angle between it and the perturbing (outer) orbit. Note also that the torque depends linearly on both eccentricities and tends to 0 if either e or e_2 is 0.

These results suggest that for eccentric, and/or nearly coplanar, orbits, the quadrupole approximation can give misleading results. Indeed, direct numerical integration of the three-body equations of motion reveals that qualitatively—and in some cases, strikingly—different behavior can appear if the dominant binary is sufficiently eccentric; for instance, orbits can "flip," changing the sign of their angular momentum [367]. Some examples of motion in this "eccentric Kozai problem" are presented in section 8.6.4.

We also expect the model presented so far to break down if the inner binary is so tight that relativistic precession begins to be important [243]. Roughly speaking, this will be the case when the Kozai period, equation (4.314), is longer than the time required for ω of the inner binary to advance by 2π due to relativity. Using equation (4.205), this condition is

$$
\left(1 - e^2\right) \frac{a}{r_g} \lesssim \frac{3}{2\pi} \frac{M_\bullet}{m_2} \frac{a_2^3}{a^3} \left(1 - e_2^2\right)^{3/2}, \tag{4.332}
$$

where $r_g \equiv GM_\bullet/c^2$ is the gravitational radius of the dominant SBH. As in section 4.7, we can include the lowest-order (1PN) effects of relativity by adding a term $\dot\omega_{GR}$, the orbit-averaged rate of periapsis advance, to the orbit-averaged equation of motion for ω. In terms of the dimensionless time $\tau = v_0 t$, the result is [50]

$$
\frac{d\omega}{d\tau} = \frac{3}{4\ell} \left[2\ell^2 + 5\sin^2\omega \left(e^2 - \sin^2 i\right)\right] + \frac{\kappa}{\ell^2}, \tag{4.333}
$$

where

$$
\kappa \equiv 3\frac{M}{m_2}\frac{r_g a_2^3}{a^4}\left(1 - e_2^2\right)^{3/2} \tag{4.334a}
$$

$$
\approx 0.3\left(\frac{M}{10m_2}\right)\left(\frac{a}{10^2 r_g}\right)^{-1}\left(\frac{a_2}{a}\right)^3\left(1 - e_2^2\right)^{3/2}. \tag{4.334b}
$$

Note that the motion remains regular. Figure 4.21 shows how increasing κ, at fixed ℓ_z, causes the region associated with libration to shrink, until it disappears at

$$
\kappa = \kappa_{max} = \frac{3}{4}\left(3 - 5\ell_z^2\right) \quad \left(\ell_z \leq \sqrt{3/5}\right). \tag{4.335}
$$

As κ is increased above κ_{max}, the variations in eccentricity over one cycle decrease. For sufficiently large κ, the motion consists of precession in ω at a nearly constant (orbit-averaged) rate $\dot\omega_{GR} \approx \kappa/\langle\ell^2\rangle$, with small-amplitude variations in eccentricity:

$$
\omega(\tau) \approx \omega_0 + \dot\omega_{GR}t, \tag{4.336a}
$$

$$
\ell(\tau) \approx \frac{15}{8\kappa}\left(1 - \langle\ell^2\rangle\right)\left(\langle\ell^2\rangle - \ell_z^2\right)\cos\left[2\left(\omega_0 + \dot\omega_{GR}\tau\right)\right]. \tag{4.336b}
$$

4.8.3 The outer restricted problem

In the opposite limit of $m_2 \ll m_1$, the doubly averaged perturbing function becomes [156]

$$\overline{\overline{\mathcal{H}}}_p = -\frac{3}{16} \frac{GMm_1}{M+m_1} \frac{a^2}{a_2^3} \frac{1}{(1-e_2^2)^{3/2}}$$

$$\times \left[2\cos^2 i_2 - e^2 \sin^2 i_2 (3 - 5\cos 2\Omega_2)\right]. \tag{4.337}$$

Once again, we define a dimensionless averaged Hamiltonian as $H = \overline{\overline{\mathcal{H}}}_p/(\nu_0 I)$, now setting $I = \sqrt{G(M+m_1)a_2}$ and

$$\nu_0^{-1} \equiv T_{\text{outer}} = \frac{[G(M+m_1)]^{3/2} \cdot a_2^{7/2}}{G^2 Mm_1} \frac{}{a^2} \left(1 - e_2^2\right)^2. \tag{4.338}$$

The result is

$$H = -\frac{\ell_2}{4} \left[2\cos^2 i_2 - e^2 \sin^2 i_2 (3 - 5\cos 2\Omega_2)\right]. \tag{4.339}$$

Since ω_2 does not appear in H, the conjugate momentum ℓ_2 is constant: the eccentricity of the outer binary does not change (in the quadrupole approximation). However, its orientation does evolve, due to changes in each of the elements (ω, Ω, ℓ_z).

Consider first the case of a circular inner binary, $e = 0$. Then

$$H = -\frac{1}{2}\ell_2 \cos^2 i_2. \tag{4.340}$$

The inclination is fixed, and the only motion consists of a uniform precession of the line of nodes:

$$\frac{d\Omega_2}{dt} = -\frac{1}{T_0} \frac{\cos i_2}{\sqrt{1-e_2^2}}. \tag{4.341}$$

If the inner binary is eccentric, the equations of motion of the outer binary are more complicated. Those equations are greatly simplified if expressed in terms of the vectorial elements. In particular, identify the principal plane with the (fixed) plane of the inner orbit. Then the changes in the direction of the outer orbit's angular momentum can be expressed in terms of the components of the unit vector $\boldsymbol{e}_\ell$:

$$x \equiv \boldsymbol{e}_\ell \cdot \boldsymbol{e}_x = \sin i_2 \sin \Omega_2,$$

$$y \equiv \boldsymbol{e}_\ell \cdot \boldsymbol{e}_y = -\sin i_2 \cos \Omega_2,$$

$$z \equiv \boldsymbol{e}_\ell \cdot \boldsymbol{e}_z = \cos i_2, \tag{4.342}$$

satisfying $x^2 + y^2 + z^2 = 1$. Using these variables, H can be written

$$H = -\frac{1}{2}\left[z^2 - e^2\left(4x^2 - y^2\right)\right] \tag{4.343}$$

and $-(1/2) \leq H \leq 2e^2$. The equations of motion for (x, y, z) can be found by the chain rule; the results are

$$\dot{x} = \left(1 - e^2\right) yx,$$
$$\dot{y} = -\left(1 + 4e^2\right) xz,$$
$$\dot{z} = 5e^2 xy. \tag{4.344}$$

The region accessible to the motion can be found by writing the two equations

$$\left(1 + 4e^2\right) x^2 + \left(1 - e^2\right) y^2 = 1 - h \geq 0, \tag{4.345a}$$
$$x^2 + y^2 + z^2 = 1. \tag{4.345b}$$

The first of these equations is just the energy integral after expressing z^2 in terms of x^2 and y^2, and $h \equiv -2H$. The tip of the angular momentum vector moves along the intersection of these two surfaces: the first is an elliptic cylinder and the second is a sphere. There are two distinct kinds of trajectory [156]. On the one hand, motion can consist of closed trajectories around one of the poles of the sphere, $(x, y, z) = (0, 0, \pm 1)$. The angular momentum precesses about the angular momentum of the binary with an inclination that is always less than, or always greater than, $90°$. On the other hand, the angular momentum vector can circulate about $(x, y, z) = (\pm 1, 0, 0)$, corresponding to precession about the periapsis of the inner binary or its opposite.

The period is [156]

$$T = \frac{16}{3} \frac{K(\kappa^2)}{\sqrt{h + 4e^2}} T_{\text{outer}}, \tag{4.346}$$

where

$$T_{\text{outer}} = \frac{1}{2\pi} \frac{(M + m)^2}{M + m} \frac{P_2^{7/3}}{P^{4/3}} \frac{\left(1 - e_2^2\right)^2}{\sqrt{1 - e^2}}; \tag{4.347}$$

as before, variables lacking subscripts refer to the inner binary. The argument of the elliptic integral is now

$$\kappa^2 = \frac{5e^2}{1 - e^2} \frac{1 - h}{h + 4e^2}. \tag{4.348}$$

If $\kappa^2 < 1$ ($h > e^2$), motion is of the first kind defined above, that is, the angular momentum precesses about that of the inner binary; in the opposite case, the angular momentum precesses about the Runge–Lenz vector.

4.9 STELLAR MOTIONS AT THE CENTER OF THE MILKY WAY

A remarkable cluster of about 20 bright stars—the so-called **S-stars**[18]—is observed in the central arcsecond (roughly 0.05 pc) of the Milky Way, centered on Sgr A*, the presumed location of the SBH [300]. These appear to be main-sequence stars, mostly of spectral type B [189, 137]; such stars would have masses in the range

---

[18]Apparently, "S" stands simply for "(infrared) source" [127].

3–20 $M_\odot$ and main-sequence lifetimes from 10–100 Myr, implying that the S-stars are quite young, astronomically speaking. The S-stars are not the only population of bright, young stars found within the influence radius of the SBH: there are also two partially overlapping disklike structures consisting mostly of blue supergiant stars, with ages of 10 Myr or less, extending from ~ 0.05 to ~ 0.4 pc [183, 415], as well as a more uniformly distributed population of main-sequence B stars. The provenance of these young populations—the so-called paradox of youth at the Galactic center—is an important outstanding problem.

Whatever their origin, the S-stars have played a crucial role in constraining the mass of the central dark object. Their periods,

$$P = \frac{2\pi a^{3/2}}{\sqrt{GM_\bullet}} \approx 1.48 \left(\frac{M_\bullet}{4 \times 10^6 \, M_\odot} \right)^{-1/2} \left(\frac{a}{\mathrm{mpc}} \right)^{3/2} \mathrm{yr}, \qquad (4.349)$$

(mpc $\equiv$ milliparsec) are measured in years, and several of the S-stars have completed a significant fraction of one full orbit since astrometric monitoring began around 1992. The brightest S-star, called S2, also happens to have the shortest period: just 15.8 yr. This star appears to be a normal, B0–B2.5 star with a mass that is estimated between 15 and 20 $M_\odot$ [340]. The astrometric data for S2, and the fitted orbit, are shown in figure 4.22. All of the Keplerian elements are well determined [191, 192]; the semimajor axis and eccentricity are

$$a = 5.03 \pm 0.04 \, \mathrm{mpc}, \quad e = 0.883 \pm 0.003 \, ; \qquad (4.350)$$

these numbers assume $M_\bullet = 4.3 \times 10^6 \, M_\odot$. Periapsis was reached in 2002; a radial velocity of about 1600 km s^{-1}, or $\sim 0.005c$, was measured shortly after that time.

The orbital parameters of all the S-stars are listed in table 4.1. Unfortunately, two similar, but slightly different, labeling conventions are in use for these stars. A group centered in the Max-Planck-Institut für extraterrestrische Physik near Munich, Germany uses Sn, with n a numeral; we adopt that convention here. A group centered in Los Angeles, California prefers S0-n. The following is a partial conversion table.

MPE:	S1	S2	S4	S8	S9	S11	S12	S13	S14
UCLA:	S0-1	S0-2	S0-3	S0-4	S0-5	S0-9	S0-19	S0-20	S0-16

The fitted orbits are plotted, as they would be seen in projection on the plane of the sky, in figure 4.23.

In fitting Keplerian orbits to the astrometric and radial velocity data, the mass, position, distance, and velocity of the SBH are all free parameters. While combined fits to all the S-star data can be and have been carried out [191, 193], it turns out that essentially all the constraints on the SBH-related parameters come from just one star: S2. The most recent study, which combines data from both groups, finds [192]

$$M_\bullet = 4.30 \pm 0.20|_{\mathrm{stat}} \pm 0.30|_{\mathrm{sys}} \times 10^6 \, M_\odot, \qquad (4.351\mathrm{a})$$

$$R_0 = 8.28 \pm 0.15|_{\mathrm{stat}} \pm 0.29|_{\mathrm{sys}} \, \mathrm{kpc}, \qquad (4.351\mathrm{b})$$

with R_0 the distance to the SBH.

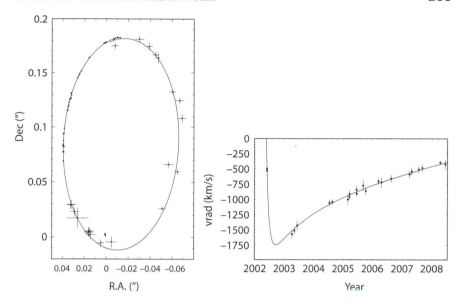

Figure 4.22 Orbit of the star S2 (S0-2) at the center of the Milky Way [193]. The left panel
shows the apparent position on the plane of the sky during the time period 1992–
2010; the motion is clockwise on this plot, starting from near apoapsis. The
switch in 2002 from astrometry based on speckle interferometry to adaptive
objects is reflected in a sudden decrease in the size of the error bars. The ellipse
is from a Keplerian model in which the velocity of the central point mass and
its location were free parameters; because the best-fit velocity of the SBH is
not zero, the orbit does not close when it reaches apoapsis the second time. The
elongated dot near orbital periapsis is the fitted location of the SBH; its extent
and shape reflect the uncertainty in the fitted position. The right panel shows
measured radial velocities, compared with predictions from the orbital fit.

In addition to their usefulness in determining the mass and distance of Sgr A*,
the S-stars also hold the promise of revealing relativistic deviations from Keplerian
motion. Table 4.2 lists the penetration parameter, $\mathcal{P}$, for each of the S-stars, as
defined in the introduction to this chapter:

$$\mathcal{P} \equiv (1 - e^2)\frac{ac^2}{GM_\bullet} = \frac{(1+e)r_p}{r_g}, \tag{4.352}$$

where $r_p = (1 - e)a$ is the (Keplerian) distance of closest approach and $r_g \equiv
GM_\bullet/c^2$. Using the results obtained in this chapter, we can express the lowest-
order relativistic changes in the orbital elements in terms of $\mathcal{P}$ as follows: Let $\Delta\omega$
and $\Delta\Omega$ be the advance in one period, due to the lowest-order relativistic corrections to the equations of motion, of the argument of periapsis and the nodal angle,
respectively.

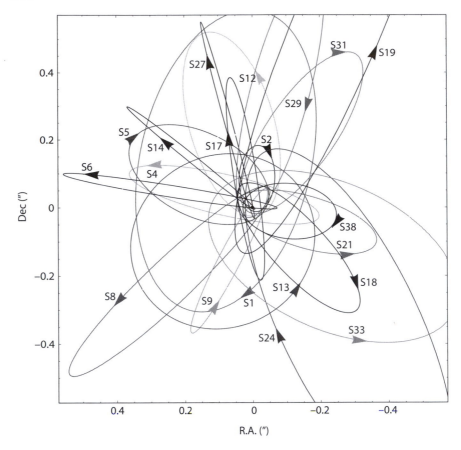

Figure 4.23 Projection onto the plane of the sky of the orbits of the S-stars [193]. Shown here are all the orbits for which the astrometric data were deemed good enough to carry out a fit.

Then

$$\Delta\omega = A_S - 3A_J \cos i - \frac{1}{2} A_Q \left(1 - 5\cos^2 i\right), \qquad (4.353a)$$

$$\Delta\Omega = A_J - A_Q \cos i, \qquad (4.353b)$$

where

$$A_S = 6\pi \mathcal{P}^{-1}, \quad A_J = 4\pi \chi \mathcal{P}^{-3/2}, \quad A_Q = -3\pi \chi^2 \mathcal{P}^{-2}. \qquad (4.354)$$

The subscripts S, J and Q refer to the geodetic (Schwarzschild), frame-dragging (Kerr or Lense–Thirring), and quadrupole contributions, respectively. In equations (4.353), the inclination i is defined with respect to the equatorial plane of the SBH. Since $\mathcal{P} > 1$ and $\chi < 1$, these expressions show clearly the ranking of the various contributions to the orbital precession. In the case of ω, geodetic precession always dominates, followed by Lense–Thirring precession and

Table 4.2 Relativistic parameters of the S-stars, computed from the orbital elements in table 4.1. $\mathcal{P}$ is the penetration parameter, equation (4.352). A_S, A_J, and A_Q are the angular precessions per orbital period due to the Schwarzschild, Kerr and quadrupole contributions to the metric, equation (4.354). The values of A_J and A_Q assume $\chi = 1$, that is, maximal spin of the SBH; they scale, respectively, as χ and χ^2.

Star	$\mathcal{P}$	$A_S('')$	$A_J('')$	$A_Q('')$
S1	8.27×10^4	0.784	0.109	2.84×10^{-4}
S2	5.99×10^3	10.82	5.590	5.40×10^{-2}
S4	5.37×10^4	1.206	0.208	6.73×10^{-4}
S5	1.57×10^4	4.125	1.317	7.88×10^{-3}
S6	2.02×10^4	3.202	0.900	4.75×10^{-3}
S8	2.85×10^4	2.275	0.539	2.40×10^{-3}
S9	2.02×10^4	3.208	0.903	4.76×10^{-3}
S12	1.26×10^4	5.129	1.825	1.22×10^{-2}
S13	4.87×10^4	1.330	0.241	8.19×10^{-4}
S14	4.01×10^3	16.14	10.19	1.21×10^{-1}
S17	5.83×10^4	1.113	0.184	5.73×10^{-4}
S18	2.43×10^4	2.672	0.686	3.31×10^{-3}
S19	4.96×10^4	1.308	0.235	7.92×10^{-4}
S21	1.77×10^4	3.657	1.099	6.19×10^{-3}
S24	2.96×10^4	2.186	0.508	2.21×10^{-3}
S27	9.18×10^3	7.056	2.945	2.03×10^{-2}
S29	1.38×10^4	4.698	1.600	1.02×10^{-2}
S31	8.21×10^3	7.891	3.483	2.88×10^{-2}
S33	4.12×10^4	1.572	0.310	1.14×10^{-3}
S38	1.07×10^4	6.052	2.340	1.70×10^{-2}
S66	2.53×10^5	0.256	0.020	3.04×10^{-5}
S67	2.04×10^5	0.317	0.028	4.65×10^{-5}
S71	6.59×10^4	0.983	0.153	4.48×10^{-4}
S83	3.42×10^5	0.190	0.013	1.67×10^{-5}
S87	2.23×10^5	0.290	0.025	3.90×10^{-5}

precession due to the quadrupole moment; in the case of Ω, Lense–Thirring precession dominates quadrupolar precession. In table 4.2, the values given for A_J and A_Q assume $\chi = 1$, that is, maximal spin. Currently, the constraints on the spin of the Milky Way SBH are weak, but astrophysicists often assume that SBHs are rapidly spinning.

Not surprisingly, some of the largest relativistic deviations are predicted for S2. The predicted amplitude of advance of the periapsis is about $11'$ per revolution. By comparison, table 4.1 shows that the current measurement uncertainty in ω, the argument of periapsis, is just under one degree—roughly five times larger.

Fits to the astrometric data for S2 that include the possibility of relativistic precession find no significant effect [193].

Even if a significant precession were detected, it would not necessarily be due to relativity. There is also likely to be a purely Newtonian contribution due to the distributed mass (stars, stellar remnants) within S2's orbit, as discussed in section 4.4.1. For the Newtonian contribution to the precession, equation (4.88) gives

$$\Delta\omega = -2\pi\, G_{\mathrm{M}}(e, \gamma)\sqrt{1-e^2}\,\frac{M_\star(r < a)}{M_\bullet}. \tag{4.355}$$

(Recall from section 4.4.1 that the distributed mass density is assumed here to follow $\rho \propto r^{-\gamma}$; $M_\star(< a)$ is the distributed mass within a sphere of radius $r = a$; and G_{M} is defined in equation 4.89.) The minus sign means that this contribution to the precession is retrograde—opposite in sense to the relativistic (geodetic) precession. Unfortunately, essentially nothing is known about the amount or distribution of mass at these distances from the SBH, and the best we can do for S2 is to write

$$\Delta\omega \approx -23\rlap{.}{'}4\, G_{\mathrm{M}}\,\frac{M_\star(r < 5.0\,\mathrm{mpc})}{10^4\, M_\odot}, \tag{4.356}$$

where $G_{\mathrm{M}} = (1.5, 1.0, 0.68)$ for $\gamma = (0, 1, 2)$ and $e = 0.883$. Apparently, if the distributed mass within S2's orbit is $\sim 1\%$ of the mass of the SBH, the shift in ω during one orbit is roughly one degree—substantially greater than the relativistic precession. Turning the problem around, and placing limits on the amount of distributed mass from the measured positions and velocities, one finds that $M_\star(r < a)$ must be less than about $10^4\, M_\odot$ [193].

The prospects for detecting the effects of relativity are not as hopeless as this comparison might suggest. As shown in figure 4.11, almost all of the relativistic precession occurs near periapsis, and observations made during the next periapsis passage (in 2018) might allow the Newtonian and relativistic contributions to be sorted out, particularly if accurate radial velocities are available [584, 6].

If the stellar cluster around Sgr A* is spherically symmetric, the Newtonian contribution to the precession leaves the plane of a star's orbit unchanged. The same is true of the relativistic precession if the SBH is not rotating. But orbits around a rotating SBH also experience precession of the nodal angle Ω (defined with respect to a reference plane that is perpendicular to the SBH's spin axis). One way to separate relativistic from Newtonian effects might be to find stars orbiting close enough to Sgr A* that the out-of-plane precession due to the SBH's spin is measurable [561].

Figure 4.24 plots the timescales associated with precession of orbital planes due to frame dragging (t_J) and due to the SBH's quadrupole moment (t_Q), where

$$t_J \equiv \left[\frac{A_J(a, e)}{\pi P(a)}\right]^{-1} = \frac{P}{4\chi}\left[\frac{c^2 a(1 - e^2)}{G M_\bullet}\right]^{3/2}$$

$$\approx 1.39 \times 10^5 \left(1 - e^2\right)^{3/2} \chi^{-1} a_{\mathrm{mpc}}^3\ \mathrm{yr}, \tag{4.357a}$$

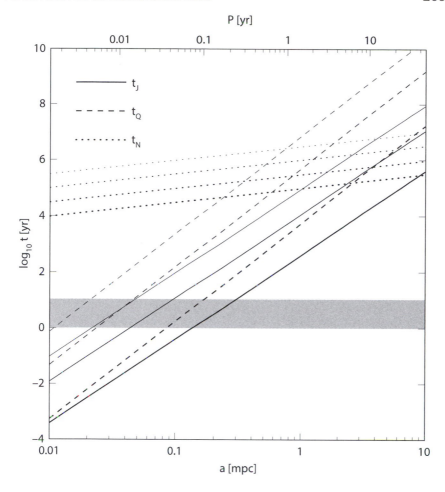

Figure 4.24 Timescales associated with precession of orbital planes about the Milky Way SBH [358]. The quantities t_J and t_Q are the precession timescales due to frame dragging and to the quadrupole torque from a maximally spinning ($\chi = 1$) SBH. Line thickness denotes orbital eccentricity, from $e = 0.99$ (thickest) to $e = 0.9$ and $e = 0.5$ (thinnest). The quantity t_N is an estimate of the timescale for torquing of orbital planes due to Newtonian perturbations from other stars, assumed to have one solar mass (equation 4.362a). The line thickness denotes the total distributed mass within 1 mpc from the SBH, from $10^3 \, M_\odot$ (thickest) to $1 \, M_\odot$ (thinnest), assuming that density falls off as r^{-1}. The shaded region shows the range of interesting time intervals for observation, $1 \, \mathrm{yr} \leq \Delta t \leq 10 \, \mathrm{yr}$.

$$t_Q \equiv \left[\frac{A_Q(a, e)}{\pi P(a)} \right]^{-1} = \frac{P}{3\chi^2} \left[\frac{c^2 a(1 - e^2)}{GM_\bullet} \right]^2$$

$$\approx 1.34 \times 10^7 \left(1 - e^2\right)^2 \chi^{-2} a_{\mathrm{mpc}}^{7/2} \, \mathrm{yr}, \qquad (4.357b)$$

as functions of a and e for stars at the Galactic center, assuming a maximally spinning SBH ($\chi = 1$); a_{mpc} is the semimajor axis in units of milliparsecs. Observing changes of $\sim \pi$ in the nodal angle in a time less than 10 yr would require finding stars that orbit well inside 1 mpc (recall that $a \approx 5$ mpc for S2) and, if possible, that have high eccentricities.

Determining the orbits of such stars allows one to do more than detect relativistic effects. In principle, it becomes possible to test theories of gravity [567]. According to uniqueness, or **no-hair theorems**, of general relativity, an electrically neutral black hole is completely characterized by its mass M and its spin angular momentum $\mathcal{S}$. As a consequence, all the multipole moments of its external space-time are functions of M and $\mathcal{S}$. This is true, in particular, of the quadrupole moment Q, and equation (4.353) makes a unique prediction about the relation between the nodal precession amplitudes due to quadrupole torques and frame dragging:

$$\frac{A_Q}{A_J} = \frac{3}{4} \frac{\chi}{\mathcal{P}^{1/2}}. \tag{4.358}$$

The orbital angular momentum vector of a star is predicted to precess according to

$$\frac{d\boldsymbol{L}}{dt} = P^{-1} \left(A_J - A_Q \cos i \right) (\boldsymbol{\mathcal{S}} \times \boldsymbol{L}) \tag{4.359a}$$

$$= 4\pi P^{-1} \mathcal{P}^{-3/2} \left[1 + \frac{3}{4} \mathcal{P}^{-1/2} \boldsymbol{\chi} \cdot \frac{\boldsymbol{L}}{L} \right] (\boldsymbol{\chi} \times \boldsymbol{L}), \tag{4.359b}$$

where i is the inclination with respect to the SBH's equatorial plane. Keplerian fits to the astrometric data for a single star yield P, $\mathcal{P}$, and $\boldsymbol{L}$ for that star. Measurement of the change in the direction of $\boldsymbol{L}$ for *two* stars—that is, four numbers—yields enough information to determine $\boldsymbol{\chi}$ (three numbers) and, independently, an estimate of the ratio of the first (frame-dragging) and second (quadrupole) terms in the square brackets. If this ratio does not have the "correct" value, a violation of the uniqueness theorems has been detected [567].

Here again, a caveat is in order. If the distributed mass within the observed orbits is spherically symmetric, there will be no Newtonian contribution to the out-of-plane precession. But this assumption is certain to be violated at some level: either because the nucleus is inherently aspherical, or due simply to the fact that the number of stars within the orbit at any given time is finite. Since we are considering orbits within a milliparsec or so from the SBH, the number of stars at smaller radii could be small.

To estimate the timescale for the orientation of an orbit to change due to these Newtonian perturbations, we can use results from section 4.4 on motion in nonspherical nuclei. For instance, for the orbit-averaged rate of change of the nodal angle Ω, equations (4.108), (4.102) and (4.104) give

$$\frac{d\Omega}{dt} = -2v_0 \frac{\rho_t}{\rho} \times f(e, i, \omega), \tag{4.360}$$

where $v_0 \approx v_r M_\star(r < a)/M_\bullet$, ρ_t/ρ is the fractional amplitude of the nonspherical part of the density, and the function f depends on the orbital elements. Ignoring f,

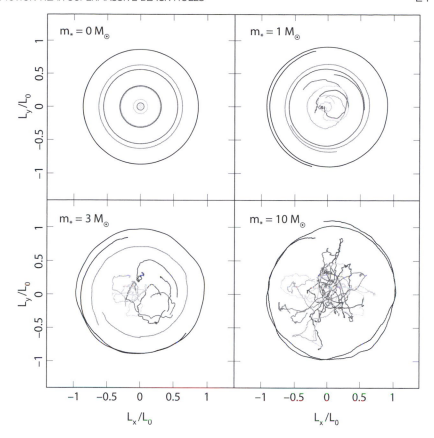

Figure 4.25 Evolution of orbital planes in a cluster of eight stars orbiting about the Galactic
center SBH for an elapsed time of 2×10^6 yr [358]. The SBH rotates about
the z-axis with maximal spin. Four different values were assumed for the stellar
masses $m_\star$, as indicated. Stars were placed initially on orbits with semimajor
axis $a = 2$ mpc and eccentricity 0.5 and with random orientations. In a nucleus
containing stars of a given mass, the transition between motion like that in the
first and last panels occurs at the "rotational influence radius" defined in section
5.7.1.

we can rewrite this approximately as

$$\frac{d\Omega}{dt} \approx P^{-1} \frac{M_\star}{M_\bullet} \frac{\rho}{\rho_t}. \tag{4.361}$$

Suppose we ascribe the nonspherical part of the potential to the finite number, N, of
stars at radii $r \lesssim a$. We are concerned here with timescales that are long compared
with orbital periods, and so these "root-N fluctuations" are due to the fact that the
orbit-averaged density of the N stars is not precisely spherical. Then we can write
$\rho_t/\rho \sim N^{-1/2}$, and the timescale for changes in the orbit's plane due to Newtonian

perturbations becomes

$$t_N \approx \frac{P}{\sqrt{N}} \frac{M_\bullet}{m_\star} \tag{4.362a}$$

$$\approx 6 \times 10^5 \left(\frac{M_\bullet}{4 \times 10^6 \, M_\odot} \right)^{1/2} \left(\frac{N}{10^2} \right)^{-1/2} \left(\frac{m_\star}{1 \, M_\odot} \right)^{-1} \text{yr}, \tag{4.362b}$$

where $m_\star = M_\star/N$.[19] This estimate, which is plotted in figure 4.24, can only be considered very approximate; among other things, it ignores any dependence on orbital eccentricity.[20] Nevertheless, figure 4.24 suggests that Newtonian torquing of orbital planes at the Galactic center is likely to overwhelm relativistic changes beyond roughly a milliparsec from Sgr A* [358].

Figure 4.25 shows how the orbital planes of stars evolve in response to the combined effects of frame dragging, the SBH's quadrupole moment, and perturbations from a small number of other stars. As the magnitude of the stellar perturbations increases, the orderly precession about the SBH spin axis is converted into a more chaotic evolution. A comprehensive N-body study [358] suggests thet detection of frame-dragging precession may be feasible after a few years' monitoring with the next generation of astrometric telescopes of stars in the radial range $0.2 \, \text{mpc} \lesssim a \lesssim 1 \, \text{mpc}$; at smaller radii, the number of detectable stars is likely to be too small, while at larger radii, Newtonian $\sqrt{N}$ perturbations become too large. Quadrupole-induced precession stands out from stellar perturbations only in a narrow class of assumed models for the nuclear cluster, and even then, only at radii well inside 1 mpc.

---

[19]In chapter 5, changes in orbital planes due to torquing by $N^{1/2}$ fluctuations in the background potential will be called "coherent resonant relaxation."

[20]A more precise criterion is given in section 5.7.1.

Chapter Five

Theory of Gravitational Encounters

5.1 BASIC CONCEPTS AND TIME OF RELAXATION

It is convenient to divide the forces acting on a star into two kinds. The gravitational potential of the galaxy as a whole generates a force that can be approximated as a smoothly varying function of position and (if need be) of time. This smooth force determines the orbit of the star, as discussed in chapters 3 and 4. Each star is also influenced by other forces arising from the fact that the mass making up a galaxy is discrete: it is composed of individual stars and stellar remnants, as well as more massive objects like star clusters and giant molecular clouds. As a star moves along its orbit, it experiences fluctuations in the total force as its distance from these objects changes—due both to the star's own motion, and to the motion of the other bodies. These **gravitational encounters** (so called to distinguish them from the much rarer **physical collisions**, in which two bodies collide) alter the kinetic energy and the direction of motion of each star, causing its orbit to deviate gradually from the zero-order solutions described in chapters 3 and 4. The **relaxation time**, T_r, can be defined as the time over which the cumulative effect of gravitational encounters becomes significant for a typical star.

A more precise definition of the relaxation time is possible in terms of the integrals of motion, the quantities that are conserved in the unperturbed motion. For instance, if the smooth potential is time independent, the energy E is conserved for each orbit,[1] and the relaxation time can be defined as the mean time for E to change by of order itself as a result of encounters. If $\Phi(x, t)$ is spherically symmetric, a second relaxation time can be defined in terms of changes in L; and so forth.

These different ways of defining T_r can sometimes yield very different estimates for the time of relaxation. For instance, in the case of stars orbiting near the center of a galaxy containing a supermassive black hole (SBH), it turns out that changes in L due to gravitational encounters can occur much more rapidly than changes in E; this is a consequence of the fact that the unperturbed orbits are nearly Keplerian and the encounters are correlated over time.

But before dealing with such subtleties, it is useful to consider the simplest possible case: the effect of random encounters on a star that is following a rectilinear, unaccelerated orbit in an infinite, homogeneous galaxy.

Since we will be satisfied here with an approximate calculation of T_r (more careful treatments will follow later in this chapter), it is sufficient to consider encounters

---

[1] Unless otherwise noted, the symbols E and L in this chapter refer to the energy and angular momentum per unit mass.

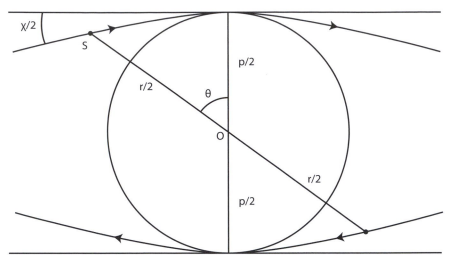

Figure 5.1 The geometry of an encounter between two stars of equal mass, in the limit of large impact parameter.

with a single velocity V, defined as the relative speed between the test and perturbing stars at infinity. We also assume for the moment that the two stars have equal masses, m, and that their impact parameter p is so large that the change in velocity due to the encounter is slight. Figure 5.1 illustrates the geometry of such an encounter in the center-of-mass frame.

When star 1 is at point S, at a distance $r/2$ from the center of mass and a distance r from the second star, the acceleration along the separation vector OS is

$$\frac{Gm}{r^2} = \frac{Gm}{p^2} \cos^2 \theta. \tag{5.1}$$

The component of this acceleration along the *fixed* direction perpendicular to the trajectory is

$$\frac{Gm}{p^2} \cos^3 \theta. \tag{5.2}$$

The total change in velocity of star 1 in this direction is then

$$\Delta v_\perp = \int_{-\infty}^{+\infty} \frac{Gm \cos^3 \theta}{p^2} dt = \frac{Gm}{pV} \int_{-\pi/2}^{\pi/2} \cos \theta \, d\theta = \frac{2Gm}{pV}. \tag{5.3}$$

We are interested in the cumulative effect of many encounters. If the encounters are random, the expectation value of $\Delta v_\perp$ will tend to zero. However, if we imagine a set of stars, each of which experiences its own encounters, the distribution of their velocities will gradually broaden over time, even if the mean velocity remains unchanged. This argument suggests that we look at the mean value of the *squared* velocity change of our single star.

Accordingly, define $\langle (\Delta v)^2 \rangle$ as the mean value of $(\Delta v_\perp)^2$ per unit interval of time. Invoking the assumption of an infinite homogeneous medium, with n the number density of stars, we can compute $\langle (\Delta v)^2 \rangle$ by multiplying $(\Delta v_\perp)^2$ by $2\pi p n V\, dp$ and integrating over p. The result is

$$\langle (\Delta v)^2 \rangle = \frac{8\pi G^2 m^2 n}{V} \int \frac{dp}{p} = \frac{8\pi G^2 m^2 n}{V} \ln \left(\frac{p_{max}}{p_{min}} \right). \tag{5.4}$$

A problem is now apparent: the integral over impact parameters diverges, both at small and large p!

The divergence at small p is not necessarily worrisome, since our expression for $\Delta v_\perp$ *assumed* that p was large. Encounters with small p are known to result in finite velocity changes, and their number is small; hence we expect this divergence to disappear in a more exact treatment. For now, we can use equation (5.3) to identify the impact parameter p_0 corresponding to $\Delta v_\perp \approx V$, that is,

$$p_0 = \frac{2Gm}{V^2}, \tag{5.5}$$

and adopt this as an effective lower limit for p in equation (5.4). Encounters with $p \leq p_0$ will be called **close encounters**, and those with $p > p_0$ called **distant encounters**. For conditions characteristic of galactic nuclei,

$$p_0 \approx 9 \times 10^{-7} \left(\frac{m}{M_\odot} \right) \left(\frac{V}{10^2\,\mathrm{km\,s^{-1}}} \right)^{-2} \mathrm{pc}. \tag{5.6}$$

If we assume that there is also an effective *upper* limit to the impact parameter, $p = p_{max}$, we can write equation (5.4) as:

$$\langle (\Delta v)^2 \rangle = \frac{8\pi G^2 m\rho}{V} \ln \left(\frac{p_{max}}{p_0} \right), \tag{5.7}$$

where $\rho = nm$ is the mass density of perturbers.

What should we choose for p_{max}? An absolute upper limit to p is set by the physical size of the stellar system. In the case of a galaxy, this argument suggests that $p_{max} \lesssim 10\,\mathrm{kpc}$ and $\ln(p_{max}/p_0) \lesssim 20$. But this is almost certainly an overestimate, since real stellar systems are inhomogeneous; if we are interested in the relaxation time near the center of a galaxy, the density of stars with large impact parameters (i.e., at large distances) will be much lower than the local density and their contribution to the velocity change will be much less than implied by equation (5.7) with fixed ρ. A more reasonable guess for p_{max} in this case might be the linear scale over which the stellar density changes by of order itself. Near the center of a galaxy like the Milky Way, this length is of order one parsec, implying $\ln(p_{max}/p_0) \approx 10$.

Whatever the proper definition of p_{max}, it is clear that $p_{max} \gg p_0$, and hence that distant encounters are responsible for the dominant contribution to $\langle (\Delta v)^2 \rangle$. Changes in velocity can therefore be assumed to result from the cumulative sum of many, small velocity changes, rather than from the rare close encounters that produce large $\Delta \boldsymbol{v}$'s.

It remains only to convert equation (5.7) into an estimate of the relaxation time T_r. We can do this by setting

$$\langle (\Delta v)^2 \rangle \times T_r = V^2. \tag{5.8}$$

In other words, in a time T_r, the mean value of $\sum(\Delta v_\perp)^2$ is equal to V^2. The result is

$$T_r = \frac{V^3}{8\pi G^2 m\rho \ln \Lambda},$$ (5.9)

where $\ln \Lambda \equiv \ln(p_{max}/p_0)$, the **Coulomb logarithm**. Equation (5.9) is the same, to within a factor of order unity, as the relaxation time defined earlier in this book (equation 3.2), if V^2 is replaced by the mean square relative velocity between stars having a Maxwellian velocity distribution.

5.2 DIFFUSION COEFFICIENTS

The quantity $\langle(\Delta v)^2\rangle$ defined above is an example of a **diffusion coefficient**. In order to more completely describe the effect of gravitational encounters, we need to define a number of additional diffusion coefficients, corresponding to the various ways in which velocities can change.

Accordingly, let $\langle\Delta\boldsymbol{v}\rangle$ be the vector sum of the velocity changes $\Delta\boldsymbol{v}$ experienced by a test star per unit interval of time, in interactions with all the other stars ("field stars"). If the components of $\boldsymbol{v}$ in some orthogonal coordinate system are v_i, the diffusion coefficient corresponding to the ith coordinate is $\langle\Delta v_i\rangle$. In the same way, if we imagine summing $\Delta\boldsymbol{v}\Delta\boldsymbol{v}$ over all encounters, we can derive the second-order diffusion coefficients $\langle\Delta v_i\Delta v_j\rangle$.

A complete description of the velocity changes would require an infinite set of diffusion coefficients, extending to all orders in $\Delta\boldsymbol{v}$. In practice it is almost never necessary to go beyond second order, for reasons that are set out in more detail below.

We assume that the distribution of field-star velocities is isotropic in the frame in which the galaxy is at rest. But a little thought makes it clear that velocity changes of the test star might be different, on average, in directions parallel and perpendicular to its instantaneous motion, since the field-star velocity distribution as seen from a moving frame is *not* isotropic. In fact there are three independent diffusion coefficients in this case. If the instantaneous velocity of the test star is taken to be parallel to the x-axis, then $\langle(\Delta v_x)^2\rangle$ can differ from $\langle(\Delta v_y)^2\rangle$ and $\langle(\Delta v_z)^2\rangle$, while the latter two coefficients are equal due to symmetry. We write

$$\langle(\Delta v_\parallel)^2\rangle = \langle(\Delta v_x)^2\rangle$$ (5.10)

to describe the mean squared changes in the component of the velocity parallel to the motion, and

$$\langle(\Delta v_\perp)^2\rangle = \langle(\Delta v_y)^2\rangle + \langle(\Delta v_z)^2\rangle$$ (5.11)

for the diffusion coefficient that describes changes perpendicular to the motion.

The only other diffusion coefficient that differs from zero is

$$\langle\Delta v_\parallel\rangle = \langle\Delta v_x\rangle.$$ (5.12)

We anticipate a later result by calling $\langle\Delta v_\parallel\rangle$ the **coefficient of dynamical friction**: as its name suggests, this first-order coefficient describes a slowing down of

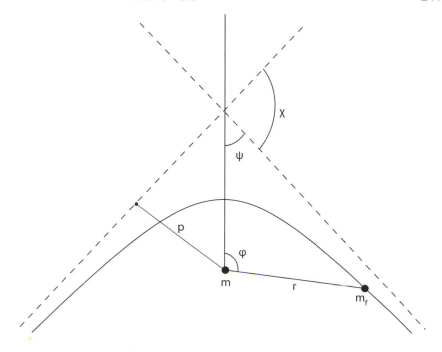

Figure 5.2   A hyperbolic encounter, shown in a frame in which the test star is at rest.

the test particle. The second-order diffusion coefficients are sometimes referred to collectively as **coefficients of scattering**.

5.2.1 Coefficient of dynamical friction

Consider an encounter between a test star of mass m and a field star of mass m_f. We assume that the two stars approach each other on an unbound orbit, with impact parameter p (no longer assumed large) and relative velocity at infinity V. Figure 5.2 plots the relative orbit, the equations for which were presented in section 4.1 in terms of E and L, the energy and angular momentum divided by the reduced mass.

It is useful to rewrite equations (4.28), (4.29) in terms of p and V using $E = V^2/2$, $L = pV$:

$$\frac{1}{r} = \frac{G(m + m_f)}{p^2 V^2} (1 + e \cos \phi), \qquad (5.13a)$$

$$e^2 = 1 + \frac{p^2 V^4}{G^2 (m + m_f)^2}, \qquad (5.13b)$$

where $\phi = 0$ corresponds to the point of closest approach. To derive the diffusion coefficients, we require the vector change between the initial and final velocity of the test star during the time that the relative velocity rotates through the angle χ; evidently the magnitude of the relative velocity, V, is unchanged by the encounter.

When r is infinite, we see from figure 5.2 that $\phi = \pi \pm \psi$, and equation (5.13a) implies $\cos \phi = -1/e$; using $\chi = \pi - 2\psi$ we find

$$\sin \chi = \frac{2p/p_0}{1 + (p/p_0)^2}, \quad \cos \chi = 1 - \frac{2}{1 + (p/p_0)^2}, \tag{5.14}$$

where

$$p_0 = \frac{G(m_f + m)}{V^2}. \tag{5.15}$$

The factor $m_f/(m_f + m)$ converts the relative velocity to the velocity of the test star in the center of mass frame, yielding for the changes in the test-star velocity

$$\Delta v_\parallel = -\frac{m_f}{m + m_f} V (1 - \cos \chi) = -2V \frac{m_f}{m + m_f} \frac{1}{1 + (p/p_0)^2}, \tag{5.16a}$$

$$\Delta v_\perp = \frac{m_f}{m + m_f} V \sin \chi = 2V \frac{m_f}{m + m_f} \frac{p/p_0}{1 + (p/p_0)^2}; \tag{5.16b}$$

these are the changes parallel, and perpendicular, to V, the initial relative velocity.

We are interested here in $\Delta v_\parallel$. In order to derive the coefficient of dynamical friction, we need to sum the velocity changes, per unit interval of time, over all impact parameters and over all values for the relative velocity at infinity. The first operation proceeds as before: multiplying the velocity changes by $2\pi p n_f V$, with V assumed fixed, and integrating over p yields

$$\overline{(\Delta v_\parallel)} = -\frac{2\pi G^2 m_f (m_f + m) n_f}{V^2} \ln \left(1 + p_{\max}^2/p_0^2\right). \tag{5.17}$$

The divergence at small p that appeared previously is no longer present due to our proper treatment of close encounters. However, the integral dp still diverges at large values of p, and so an upper limit $p_{\max}$ will still be required.

The final step is the integration over field-star velocities. The relative velocity is $V = v - v_f$, where v is the velocity of the test star. Since equation (5.17) gives the velocity change in the direction of the initial *relative* motion, we must multiply it by

$$\frac{V \cdot v}{V v} = \frac{v - v_{fx}}{V} \tag{5.18}$$

to convert it into a velocity change in the direction of the test star's motion, assumed here to be along the x-axis. We then replace n_f by $\int f(v_f) dv_f$, where $f(v_f)$ is the phase-space number density of field stars, and integrate over v_f. We find for the dynamical friction coefficient

$$\langle (\Delta v_\parallel) \rangle = \int f(v_f) \overline{(\Delta v_\parallel)} \frac{v - v_{fx}}{V} dv_f \tag{5.19}$$

$$= -2\pi G^2 (m_f + m) m_f$$

$$\times \int f(v_f) \frac{v - v_{fx}}{V^3} \ln \left[1 + \frac{p_{\max}^2 V^4}{G^2 (m_f + m)^2}\right] dv_f.$$

We now invoke our assumption that the field-star distribution is isotropic in velocity space. Having made this assumption, there are various ways to simplify the integral over velocities in equation (5.19). Here we follow Chandrasekhar [79] and represent the velocity-space volume element in terms of v_f and V, using

$$v - v_{f_x} = \frac{V^2 + v^2 - v_f^2}{2v}.$$

The result is

$$\langle (\Delta v_\parallel) \rangle = -\frac{16\pi^2 G^2 (m_f + m) m_f}{v^2} \int_0^\infty dv_f \, v_f^2 \, f(v_f) \mathcal{H}\left(v, v_f, p_{\max}\right),$$

(5.20a)

$$\mathcal{H}(v, v_f, p_{\max}) = \frac{1}{8 v_f} \int_{|v-v_f|}^{v+v_f} dV \left(1 + \frac{v^2 - v_f^2}{V^2}\right) \ln\left[1 + \frac{p_{\max}^2 V^4}{G^2 (m_f + m)^2}\right].$$

(5.20b)

The integral that defines the weighting function $\mathcal{H}$ has an analytic solution which the reader is invited to derive.[2]

The coefficient of dynamical friction is always negative: in other words, it describes a decrease in the test body's velocity along the direction of its instantaneous motion. This is a consequence of the fact that any deflection of the motion from its otherwise linear path results in a decrease in $v_\parallel$. Those same deflections can result in an *increase* in the kinetic energy in directions perpendicular to the motion, as reflected in the diffusion coefficients $\langle (\Delta v_\parallel)^2 \rangle$, $\langle (\Delta v_\perp)^2 \rangle$ (which we have yet to derive). Furthermore, when the mass m of the test body greatly exceeds the mass of a field star, equation (5.20) states that the deceleration is proportional to m. It will turn out that the other diffusion coefficients do not have this property; thus the effect of encounters on a massive body is almost exclusively to slow down its motion. This is the reason that $\langle \Delta v_\parallel \rangle$ is called the coefficient of "dynamical friction."

In his 1943 paper [79], Chandrasekhar noticed a natural approximation that greatly simplifies the expressions just derived. Suppose that one simply ignores the velocity dependence of the logarithmic term in equation (5.20). Instead, write this term as

$$\ln\left[1 + \frac{p_{\max}^2 V^4}{G^2 (m_f + m)^2}\right] \equiv 2 \ln \Lambda,$$

(5.21)

where $\ln \Lambda$, the Coulomb logarithm (which we saw before, in a slightly different form) is assumed constant. The weighting function then becomes simply

$$\mathcal{H} = \begin{cases} \ln \Lambda & \text{if } v > v_f, \\ 0 & \text{if } v < v_f \end{cases}$$

(5.22)

[2]The quantity $\mathcal{H}$ is related to the quantity J defined by Chandrasekhar in his 1943 paper [79] by $\mathcal{H} = J/(8v_f)$.

—stars whose velocity at infinity is greater than the test star's velocity do not contribute at all to the frictional force! The dynamical friction coefficient becomes

$$\langle \Delta v_\parallel \rangle = -\frac{16\pi^2 G^2 (m_f + m) m_f \ln \Lambda}{v^2} \int_0^v dv_f v_f^2 f(v_f) \tag{5.23a}$$

$$= -\frac{4\pi G^2 (m_f + m) \ln \Lambda}{v^2} \rho \left(v_f < v \right), \tag{5.23b}$$

where $\rho \left(v_f < v \right)$ is the mass density contributed by stars moving more slowly than the test star.[3]

Equation (5.23) is very commonly used to define the coefficient of dynamical friction [48]. We will follow that practice for the next few paragraphs, before returning to the more exact form.

Suppose that $f(v_f)$ is Maxwellian;[4] that is,

$$f(v_f) = \frac{n_f}{(2\pi \sigma_f^2)^{3/2}} e^{-v_f^2/(2\sigma_f^2)} \tag{5.24}$$

with σ_f the mean square velocity of the field stars in one direction. Writing $x = v/(\sqrt{2}\sigma_f)$, equation (5.23) becomes

$$\langle \Delta v_\parallel \rangle = - \left(1 + \frac{m}{m_f} \right) \frac{n_f \Gamma}{\sigma_f^2} G(x), \tag{5.25a}$$

$$G(x) = \frac{\operatorname{erf}(x) - x \operatorname{erf}'(x)}{2x^2}. \tag{5.25b}$$

Here

$$\Gamma \equiv 4\pi G^2 m_f^2 \ln \Lambda \tag{5.26}$$

and erf is the error function,

$$\operatorname{erf}(x) \equiv \frac{2}{\sqrt{\pi}} \int_0^x e^{-y^2} dy, \tag{5.27}$$

with $\operatorname{erf}'(x) \equiv (d/dx)\operatorname{erf}(x)$.

The function $G(x)$ has the following forms in the limits of large and small x:

$$G(x) = \begin{cases} \frac{2x}{3\sqrt{\pi}} & \text{if } x \ll 1, \\ \frac{1}{2x^2} & \text{if } x \gg 1. \end{cases} \tag{5.28}$$

Thus, the dynamical friction force increases linearly with the test body's velocity at low speeds ("Hooke's law") and declines as v^{-2} at large v. The peak value of G, 0.214, occurs for $x = 1$; in other words, the dynamical friction force is maximized when $v = \sqrt{2}\sigma_f$.

We can define the **dynamical friction time** as

$$T_{\text{df}} \equiv \left| \frac{\langle \Delta v_\parallel \rangle}{v} \right|^{-1}, \tag{5.29}$$

---

[3] Note that v_f is the velocity at infinity, in the frame in which the test star's velocity is v. Stars with $v_f < v$ may be moving faster than v as they approach the test body.
[4] As we saw in chapter 3, this is probably not a good approximation near an SBH.

the time for v to decrease by of order itself. This time depends on v, except in the limit that v is small. In that limit,

$$T_{df} = \frac{3}{8}\sqrt{\frac{2}{\pi}}\frac{\sigma_f^3}{G^2\rho m \ln \Lambda}, \tag{5.30}$$

where $m \gg m_f$ has been assumed. At higher test-body velocities, T_{df} is increased relative to this expression by the factor Q^{-1} where

$$Q(x) \equiv \frac{3\sqrt{\pi}}{2x}G(x)$$

$$= \frac{3}{2x^3}\left(\int_0^x e^{-y^2}dy - xe^{-x^2}\right), \quad x \equiv v/(\sqrt{2}\sigma_f); \tag{5.31}$$

$Q \to 1$ for small x. Expressing T_{df} in terms of Q, and in terms of physical scales relevant to galactic nuclei, yields

$$T_{df} \approx \frac{5 \times 10^9}{Q}$$

$$\times \left(\frac{\sigma_f}{100\,\text{km s}^{-1}}\right)^3 \left(\frac{\rho}{10^5\,M_\odot\,\text{pc}^{-3}}\right)^{-1} \left(\frac{m}{10\,M_\odot}\right)^{-1} \left(\frac{\ln \Lambda}{10}\right)^{-1}\,\text{yr}. \tag{5.32}$$

This expression implies, for instance, that a $10\,M_\odot$ BH near the center of a galaxy like the Milky Way would lose most of its kinetic energy over the age of the universe. As a result, it would spiral into the center—a process discussed in more detail in chapters 7 and 8.

The simplicity of the dynamical friction coefficient that we have been using until now, equation (5.23), is a consequence of the brute-force way in which the logarithmic term was "taken out of the integral." Unfortunately, there is a price to pay for this brutality: we really have no idea how to define $\ln \Lambda$. The definition given in equation (5.21) includes the *variable V*, the relative velocity between test and field stars. Somehow, we need to decide what number to associate with that V.

A myriad of different recipes have been proposed for doing this. Most authors have accepted Chandrasekhar's [79] suggested form:

$$\Lambda = \frac{p_{max}v_{rms}^2}{G(m_f + m)}, \tag{5.33}$$

where $v_{rms} = \sqrt{3}\sigma_f$, and recast the discussion in terms of the choice of p_{max}. In a famous paper from 1942 [81], Chandrasekhar and von Neumann argued that the effective limiting distance for encounters should be the interparticle distance. But most subsequent authors (e.g., [88]) have advocated larger values for p_{max}: the half-mass radius of the stellar system, the core radius (if there is a core), etc.

For instance, consider a test mass orbiting near the center of a galaxy that contains an SBH, and assume $m \approx m_f \ll M_\bullet$. As discussed in chapter 2, many galaxies, including the Milky Way, are observed to contain cores with radii $r_c \approx r_h$, with the density falling off rapidly beyond $r = r_c$. If the test mass is inside the core, a reasonable guess for p_{max} is $\sim r_c \approx r_h$. With this ansatz, equation (5.21) becomes

$$2 \ln \Lambda \approx \ln \left[1 + \frac{r_h^2 V^4}{G^2 (m_f + m)^2} \right] \approx \ln \left[1 + \frac{1}{4} \frac{M_\bullet^2}{m^2} \frac{V^4}{\sigma^4} \right] \tag{5.34}$$

or, replacing V^2 again by $3\sigma^2$,

$$\ln \Lambda \approx \ln \left(\frac{3 M_\bullet}{2m} \right) \approx \ln N_h, \tag{5.35}$$

where N_h is the number of stars whose combined mass equals $M_\bullet$. For $m \approx M_\odot$ and $M_\bullet \approx 10^6 M_\odot$, this prescription gives $\ln \Lambda \approx 13$, while for $M_\bullet \approx 10^9 M_\odot$, $\ln \Lambda \approx 20$. These values should be considered very approximate and subject to verification by numerical simulation (section 5.2.3.3).

Given that there will always be some ambiguity in the choice of p_{max}, is there anything to be gained by replacing the simple expression for $\langle \Delta v_\parallel \rangle$, equation (5.23), by the more exact expression (5.20)? In fact, it is easy to think of circumstances where "taking the logarithm out of the integrand" is unjustified. One example, considered later in this book (section 7.2), is a star orbiting near the center of a galaxy containing an SBH and a low-density core. If the density profile inside the SBH sphere of influence is flat, $\rho \sim r^{-1/2}$, there will be *no* stars moving more slowly than the local circular velocity, and the standard expression for the coefficient of dynamical friction would predict zero frictional force. Numerical integrations demonstrate that the frictional force in this case is, in fact, nonzero and is much better fit by including the contribution from the fast-moving stars [8].

A second example is the frictional force experienced by a massive object near the center of a galaxy without a central SBH; for instance, the force on an SBH that has been displaced from the center. Such an object would normally move much more slowly than the surrounding stars. But if $v \ll v_{rms}$, and if (as equation 5.23 tells us) the only field stars contributing to the frictional force are those with $v_f < v$, then the logarithmic term in equation (5.20b) will be close to zero for *all* field stars that contribute to the frictional force. Either the frictional force in this case is much less than implied by equation (5.23), or else most of the friction must come from stars with $v_f > v$.

To sort this out, we must return to the more exact expression for $\langle \Delta v_\parallel \rangle$, equation (5.20). Setting $m = M \gg m_f$ in that equation and expanding about $v = 0$, we find $\langle \Delta v_\parallel \rangle = -Av + Bv^3 - \cdots$, where

$$A = \frac{32}{3} \pi^2 G^2 M \rho \int_0^\infty f(v_f) \frac{p_{max}^2 v_f^4 / G^2 M^2}{1 + p_{max}^2 v_f^4 / G^2 M^2} \frac{dv_f}{v_f}. \tag{5.36}$$

It is evident from this expression that field stars of every velocity contribute to the dynamical friction force; there is no sharp cutoff at $v_f \approx v$. If $f(v_f)$ is a Maxwellian, the result of the integration is

$$\langle \Delta v_\parallel \rangle = -\frac{4\sqrt{2\pi}}{3} \frac{G^2 M \rho \ln \Lambda'}{\sigma_f^3} v, \tag{5.37}$$

where

$$\ln \Lambda' \equiv \frac{1}{2} \int_0^\infty dz \, e^{-z} \ln \left(1 + \frac{4 p_{max}^2 \sigma_f^4}{G^2 M^2} z^2 \right) \approx \ln \sqrt{1 + \frac{2 p_{max}^2 \sigma_f^4}{G^2 M^2}}. \quad (5.38)$$

This result for $\langle \Delta v_\parallel \rangle$ turns out to be identical to (5.25) in the limit that v is small, but only if $\ln \Lambda$ in that equation is identified with $\ln \Lambda'$ as defined in equation (5.38). This example suggests that the approximate formula for $\langle \Delta v_\parallel \rangle$, equation (5.23), may be adequate even in cases where it is unjustified to "remove $\ln \Lambda$ from the integral," as long as one is willing to adjust the definition of $\ln \Lambda$. Given the indeterminacy associated with the choice of p_{max}, this may or may not be deemed important. But regardless of its implications for the magnitude of the frictional force, it is clear that a substantial fraction of that force can come from stars that move faster than the test star.[5]

Chandrasekhar [79] was aware of the shortcomings of his approximate formula, and derived two, less approximate forms for $\langle \Delta v_\parallel \rangle$ by assuming a large, but finite, value for the argument of the logarithm in equation (5.20). The second of these is more commonly used: rather than setting $\mathcal{H}$ to a step function, as in equation (5.22), one writes instead

$$\mathcal{H}(v, v_f, p_{max}) \approx \begin{cases} \ln \left[\Theta(v^2 - v_f^2) \right] & \text{if } v > v_f, \\ \frac{1}{2} \ln \left(4 \Theta v_f^2 \right) - 1 & \text{if } v = v_f, \\ \ln \left(\frac{v_f + v}{v_f - v} \right) - 2 \frac{v}{v_f} & \text{if } v < v_f, \end{cases} \quad (5.39)$$

where

$$\Theta \equiv \frac{p_{max}}{G(m + m_f)}. \quad (5.40)$$

The last of equations (5.39) illustrates what Chandrasekhar called the "nondominant terms": terms appearing in the diffusion coefficients that do not contain $\ln(\Theta V^2)$. The "standard" approximation to $\langle \Delta v_\parallel \rangle$, equation (5.23), is lacking the nondominant terms (as well as being approximate in other ways).

In general, the smaller Θ, or p_{max}, the greater the contribution of the fast-moving stars to the dynamical friction force. This can be important in the case of small-N systems [563]. Equating p_{max} with the size R of the system and V with the rms velocity, and using the virial relation $GM/R \approx v_{rms}^2$ with M the total mass, it follows that $\Theta \sim M/m \sim N$.

5.2.2 Scattering

We now wish to derive the two, second-order diffusion coefficients defined above, $\langle (\Delta v_\parallel)^2 \rangle$ and $\langle (\Delta v_\perp)^2 \rangle$.

[5]Figure 5.3 presents a concrete example.

The squared velocity changes of the test star in one encounter with a field star are given by equations (5.16):

$$(\Delta v_\parallel)^2 = 4V^2 \frac{m_f^2}{(m + m_f)^2} \frac{1}{(1 + p^2/p_0^2)^2}, \tag{5.41a}$$

$$(\Delta v_\perp)^2 = 4V^2 \frac{m_f^2}{(m + m_f)^2} \frac{p^2/p_0^2}{(1 + p^2/p_0^2)^2}. \tag{5.41b}$$

Multiplying by $2\pi p n_f V \, dp$ and integrating over p as before,

$$\overline{(\Delta v_\parallel)^2} = \frac{4\pi G^2 n_f m_f^2}{V} \left(\frac{p_{max}^2/p_0^2}{1 + p_{max}^2/p_0^2} \right), \tag{5.42a}$$

$$\overline{(\Delta v_\perp)^2} = \frac{4\pi G^2 n_f m_f^2}{V} \left[\ln \left(1 + \frac{p_{max}^2}{p_0^2} \right) - \frac{p_{max}^2/p_0^2}{1 + p_{max}^2/p_0^2} \right]. \tag{5.42b}$$

These are velocity changes with respect to the direction of the initial relative velocity vector V. We need to transform to a fixed frame in which the test particle has velocity v. Let the fixed coordinate system be defined by the Cartesian unit vectors (e_1, e_2, e_3). A second coordinate system is defined in terms of the unit vectors (e_1', e_2', e_3'), with e_1' parallel to the initial relative velocity V, and e_2', e_3' perpendicular to e_1'. The velocity changes in the fixed (e) frame are

$$\langle \Delta v_i \Delta v_j \rangle = (e_i \cdot e_1')(e_j \cdot e_1') \overline{(\Delta v_\parallel)^2} \tag{5.43}$$
$$+ \frac{1}{2} \left[(e_i \cdot e_2')(e_j \cdot e_2') + (e_i \cdot e_3')(e_j \cdot e_3') \right] \overline{(\Delta v_\perp)^2}.$$

It is clear that $e_i \cdot e_1' = V_i/V$ where V_i is the component of V along the e_i-axis. Hence

$$(e_i \cdot e_1')(e_j \cdot e_1') = \frac{V_i V_j}{V^2}. \tag{5.44}$$

Furthermore,

$$e_i = \sum_{k=1}^{3} (e_i \cdot e_k') e_k'. \tag{5.45}$$

Writing this equation again for e_j and using $e_i \cdot e_j = \delta_{ij}$, we find

$$(e_i \cdot e_2')(e_j \cdot e_2') + (e_i \cdot e_3')(e_j \cdot e_3') = \delta_{ij} - \frac{V_i V_j}{V^2}. \tag{5.46}$$

Combining equations (5.43), (5.44) and (5.46),

$$\langle \Delta v_i \Delta v_j \rangle = \frac{V_i V_j}{V^2} \left[\overline{(\Delta v_\parallel)^2} - \frac{1}{2} \overline{(\Delta v_\perp)^2} \right] + \frac{1}{2} \delta_{ij} \overline{(\Delta v_\perp)^2} \tag{5.47}$$

or

$$\langle \Delta v_1 \Delta v_1 \rangle = \frac{V_x^2}{V^2} \overline{(\Delta v_\parallel)^2} + \frac{1}{2}\left(1 - \frac{V_x^2}{V^2}\right)\overline{(\Delta v_\perp)^2}, \qquad (5.48a)$$

$$\langle \Delta v_2 \Delta v_2 \rangle + \langle \Delta v_3 \Delta v_3 \rangle = \frac{V_y^2 + V_z^2}{V^2}\overline{(\Delta v_\parallel)^2} + \left(1 - \frac{1}{2}\frac{V_y^2 + V_z^2}{V^2}\right)\overline{(\Delta v_\perp)^2}.$$

$$(5.48b)$$

The final step is the integration over field-star velocities. Recalling that the e_1-axis is oriented parallel to v, the two diffusion coefficients are given by

$$\langle \Delta v_\parallel^2 \rangle = \frac{2\pi}{n_f v}\int_0^\infty dv_f \, v_f f(v_f)\int_{|v-v_f|}^{v+v_f} dV \, V \, \langle \Delta v_1 \Delta v_1 \rangle, \qquad (5.49a)$$

$$\langle \Delta v_\perp^2 \rangle = \frac{2\pi}{n_f v}\int_0^\infty dv_f \, v_f f(v_f)\int_{|v-v_f|}^{v+v_f} dV \, V \, [\langle \Delta v_2 \Delta v_2 \rangle + \langle \Delta v_3 \Delta v_3 \rangle]. \qquad (5.49b)$$

Writing

$$V_x = \frac{V^2 + v^2 - v_f^2}{2v}, \qquad V_y^2 + V_z^2 = 1 - V_x^2, \qquad (5.50)$$

we find

$$\langle \Delta v_\parallel^2 \rangle = \frac{8\pi}{3}(4\pi G^2 m_f^2)v\int_0^\infty dv_f \left(\frac{v_f}{v}\right)^2 f(v_f)\mathcal{H}_2(v, v_f, p_{max}), \qquad (5.51a)$$

$$\langle \Delta v_\perp^2 \rangle = \frac{8\pi}{3}(4\pi G^2 m_f^2)v\int_0^\infty dv_f \left(\frac{v_f}{v}\right)^2 f(v_f)\mathcal{H}_3(v, v_f, p_{max}), \qquad (5.51b)$$

where $\mathcal{H}_2$ and $\mathcal{H}_3$ are weighting functions given by

$$\mathcal{H}_2(v, v_f, p_{max})$$

$$= \frac{3}{8v_f}\int_{|v-v_f|}^{v+v_f} dV \left[1 - \frac{V^2}{4v^2}\left(1 + \frac{v^2 - v_f^2}{V^2}\right)^2\right]\ln\left(1 + \Theta^2 V^4\right)$$

$$+ \left[\frac{3}{4}\frac{V^2}{v^2}\left(1 + \frac{v^2 - v_f^2}{v^2}\right) - 1\right]\frac{\Theta^2 V^4}{1 + \Theta^2 V^4}, \qquad (5.52a)$$

$$\mathcal{H}_3(v, v_f, p_{max})$$

$$= \frac{3}{8v_f}\int_{|v-v_f|}^{v+v_f} dV \left[1 + \frac{V^2}{4v^2}\left(1 + \frac{v^2 - v_f^2}{V^2}\right)^2\right]\ln\left(1 + \Theta^2 V^4\right)$$

$$+ \left[1 - \frac{3}{4}\frac{V^2}{v^2}\left(1 + \frac{v^2 - v_f^2}{v^2}\right)\right]\frac{\Theta^2 V^4}{1 + \Theta^2 V^4}. \qquad (5.52b)$$

We could stop at this point, but it is natural to explore the same sort of approximation that was used to simplify the expression for the dynamical friction coefficient.
We first note that—unlike in the case of $\mathcal{H}_1$, the weighting function in the integral
for $\langle \Delta v_\parallel \rangle$—the weighting functions $\{\mathcal{H}_2, \mathcal{H}_3\}$ both contain nondominant terms. Ignoring those terms, and replacing $\ln(1 + \Theta V^4)$ by the constant $2 \ln \Lambda$ as before, we
find

$$\mathcal{H}_2 = \begin{cases} \ln \Lambda \left(\frac{v_f}{v}\right)^2 & \text{if } v > v_f, \\ \ln \Lambda \left(\frac{v}{v_f}\right) & \text{if } v < v_f, \end{cases} \tag{5.53}$$

$$\mathcal{H}_3 = \begin{cases} \ln \Lambda \left(3 - \frac{v_f^2}{v^2}\right) & \text{if } v > v_f, \\ 2 \ln \Lambda \left(\frac{v}{v_f}\right) & \text{if } v < v_f. \end{cases} \tag{5.54}$$

Even under this approximation, both fast- and slow-moving field stars make "dominant" contributions to the diffusion coefficients. The latter are

$$\langle (\Delta v_\parallel)^2 \rangle = \frac{32\pi^2}{3} G^2 m_f^2 \ln \Lambda \, v \, [F_4(v) + E_1(v)], \tag{5.55a}$$

$$\langle (\Delta v_\perp)^2 \rangle = \frac{32\pi^2}{3} G^2 m_f^2 \ln \Lambda \, v \, [3F_2(v) - F_4(v) + 2E_1(v)], \tag{5.55b}$$

where the functions $E_n(v)$ and $F_n(v)$ are given by

$$E_n(v) = \int_v^\infty \left(\frac{v_f}{v}\right)^n f(v_f) \, dv_f, \tag{5.56a}$$

$$F_n(v) = \int_0^v \left(\frac{v_f}{v}\right)^n f(v_f) \, dv_f. \tag{5.56b}$$

Equations (5.23) and (5.55), which assume an isotropic $f(v_f)$ and ignore the nondominant terms, are very commonly used to define the diffusion coefficients [48].
Henceforth in this book, we refer to these expressions as the "standard forms" of
the diffusion coefficients.

In the case that $f(v_f)$ is Maxwellian, the standard diffusion coefficients become

$$\langle (\Delta v_\parallel)^2 \rangle = \sqrt{2} \frac{n_f \Gamma}{\sigma_f} \frac{G(x)}{x}, \tag{5.57a}$$

$$\langle (\Delta v_\perp)^2 \rangle = \sqrt{2} \frac{n_f \Gamma}{\sigma_f} \left(\frac{\text{erf}(x) - G(x)}{x}\right) \tag{5.57b}$$

with $\Gamma \equiv 4\pi G^2 m_f^2 \ln \Lambda$ and $x = v/(\sqrt{2}\sigma_f)$ as above, and $G(x)$ is defined as in
equation (5.25b).

In our discussion of the dynamical friction coefficient, we showed that the frictional force acting on the test mass is proportional to its velocity v when v is low.
We can also examine the low-velocity limits of the two scattering coefficients. Returning to the exact expressions (5.51) and expanding to lowest order in v/σ_f gives,

after some algebra,

$$\langle (\Delta v_\parallel)^2 \rangle = C + Dv^2 + \cdots , \tag{5.58a}$$

$$\langle (\Delta v_\perp)^2 \rangle = 2(E + Fv^2) + \cdots , \tag{5.58b}$$

with

$$C = E = \frac{8\sqrt{2\pi}}{3} \frac{G^2 m_f \rho_f}{\sigma_f} \ln \Lambda'; \tag{5.59}$$

$\ln \Lambda'$ is again given by equation (5.38). Equations (5.58) express the reasonable results that the rate of diffusion in velocity tends toward a constant nonzero value when the test body moves slowly, and that the diffusion rate is the same in all directions.

As noted above, diffusion coefficients of even higher order, for example, $\langle \Delta v^\alpha \Delta v^\beta \Delta v^\gamma \rangle$, can also be computed. However, it can be shown [465] that these higher-order coefficients contain only nondominant terms. To the extent that $\ln \Lambda$ is large, it is therefore justifiable to ignore them compared with the first- and second-order coefficients. A roughly equivalent statement is that the higher-order coefficients are only important in situations where close encounters dominate the velocity changes.

We are now in a position to compute a more precise expression for the time of relaxation. One common definition is [503]

$$T_r \equiv \frac{1}{3} \frac{v_{\text{rms}}^2}{\langle (\Delta v_\parallel)^2 \rangle} \tag{5.60}$$

with $\langle (\Delta v_\parallel)^2 \rangle$ evaluated at $v = v_{\text{rms}}$. In a time T_r, the mean value of $\sum (\Delta v_\parallel)^2$, for stars moving at velocity v_{rms}, is $v_{\text{rms}}^2/3 = \sigma_f^2$. Adopting equation (5.57a) for $\langle (\Delta v_\parallel)^2 \rangle$, which assumes a Maxwellian velocity distribution with one-dimensional velocity dispersion σ_f, we find

$$T_r = \frac{0.34 \sigma_f^3}{G^2 m \rho \ln \Lambda} \tag{5.61}$$

$$\approx 0.95 \times 10^{10} \left(\frac{\sigma_f}{200 \, \text{km s}^{-1}} \right)^3 \left(\frac{\rho}{10^6 \, M_\odot \, \text{pc}^{-3}} \right)^{-1} \left(\frac{m}{M_\odot} \right)^{-1} \left(\frac{\ln \Lambda}{15} \right)^{-1} \text{yr}.$$

This is the same expression for T_r that was given at the start of chapter 3.

Equation (5.61) describes the relaxation time for a system containing stars of a single mass, m. But galaxies contain stars with a range of masses, as well as other objects—stellar remnants, star clusters, etc.—with masses that may be much greater than typical stellar masses.[6] A little thought shows that there is no unique way to define a single relaxation time for a system with a spectrum of masses, since the time for gravitational encounters to change the velocity of one species may be different from that of another species. But if we are willing to assume that each species instantaneously obeys the same velocity distribution—for instance, a

---

[6]The expected form of the stellar mass distribution in galactic nuclei is discussed in section 7.1.2.

Maxwellian distribution with dispersion σ—then there is a natural way to generalize T_r. The second-order scattering coefficients, $\langle (\Delta v_\parallel)^2 \rangle$ and $\langle (\Delta v_\perp)^2 \rangle$, do not depend on the mass of the test star; this is because the acceleration produced by a given gravitational force is equal for all masses. From the expressions given above, for example, equation (5.57), it is clear that the rate of scattering due to field stars with a spectrum of masses is proportional to

$$\sum_i m_i^2 n_i \rightarrow \int n(m) m^2 dm = \tilde{m} \rho, \tag{5.62a}$$

$$\tilde{m} \equiv \frac{\int n(m) m^2 dm}{\int n(m) m \, dm}. \tag{5.62b}$$

Here $n(m) dm$ is the number of (field) stars with masses m to $m + dm$, and ρ has the same meaning as in equation (5.61), that is, the total mass density. Replacing m by $\tilde{m}$ in equation (5.61) then gives the desired generalization of T_r. As discussed in chapter 7, "standard" forms of $n(m)$ describing an old stellar population imply that $\tilde{m}$ should be roughly equal to one solar mass. But because $\tilde{m}$ is proportional to the second moment of the mass distribution, even a small number of "massive perturbers" can substantially reduce the relaxation time (section 7.4).

In the case of the Galactic center, the peak stellar density near Sgr A* is measured to be $\sim 10^5 \, M_\odot \, \text{pc}^{-3}$ at a distance of $\sim 0.5 \, \text{pc}$ from the SBH [481]; at this radius, $\sigma \approx 100 \, \text{km s}^{-1}$. Assuming $\tilde{m} = 1 \, M_\odot$, the relaxation time[7] works out to be $\sim 10^{10} \, \text{yr}$. Roughly speaking, this is a minimum value, in the sense that σ^3 / ρ increases both toward larger and smaller r. Since T_r is a measure of the time for the stellar velocity distribution to reach a steady state under the influence of encounters, it would evidently be unwise to assume that the Galactic center is collisionally relaxed.

5.2.3 Alternative approaches

It is worthwhile at this point to review the assumptions that have been made up to now in deriving the velocity diffusion coefficients:

1. The field-star distribution is infinite, homogenous and isotropic.
2. The unperturbed orbit of the test star is a straight line.
3. The effects of each encounter are transmitted instantaneously to the test star.
4. Gravitational encounters are independent and uncorrelated.
5. The distribution of field stars as seen by the test star is unchanging; that is, there is no "memory" of initial conditions.

Some of the expressions derived above, for example, equations (5.25) and (5.55), embody additional approximations:

6. The nondominant terms can be ignored.
7. $f(v_f)$ is isotropic, or Maxwellian.

---

[7]In chapter 7 it will be argued that $\tilde{m} \approx 0.5 \, M_\odot$, implying an even longer relaxation time.

Of course, none of these assumptions is strictly correct, and it is not hard to think of physical situations in which each is seriously violated. Assumptions 1 and 2 are invalid in any system that is inhomogeneous and finite—which, of course, includes every stellar system. In the case of motion near an SBH, some stars will be moving in bound Keplerian orbits; such motion violates not only assumption 2, but also assumption 4, since stars on similar orbits will encounter each other at intervals of time roughly equal to the radial period. Condition 6 was shown above to fail in the case of Brownian motion of a massive body. And there are many cases where assumption 5 of a steady state is violated.

A common practice is to accept the expressions derived above for the diffusion coefficients, but to adjust the constants ρ, σ_f, $\ln \Lambda$, etc. in such a manner as to compensate for the limitations of the theory. For instance, inhomogeneity in a stellar system can be accounted for, at least approximately, by setting p_{max} equal to the density scale length, and by averaging the diffusion coefficients over the field-star velocity distribution as seen by the test star as it moves along its trajectory ("orbit averaging"). Whether these ad hoc adjustments are adequate can only be verified by a detailed comparison with exact N-body integrations.

An alternative approach is to relax one or more of the simplifying assumptions that were made in deriving the diffusion coefficients. In this section we discuss some of these alternative approaches. One special case—the effect of encounters on stars moving on bound orbits very near an SBH—deserves a more extended treatment and will be dealt with in section 5.6.

5.2.3.1 Dynamical friction as the force from a density wake

In the limit that the mass of the test body, M, is much greater than that of a field star, the dynamical friction coefficient (5.23) obeys

$$\langle \Delta v_\| \rangle \propto -M\rho(< v)\, v^{-2}. \tag{5.63}$$

The deceleration depends only on the mass density of field stars, not on their individual masses. The second-order diffusion coefficients, on the other hand, have magnitudes that scale as $n_f m_f^2 \propto m_f \rho$, and in the limit $M \gg m_f$, the changes they predict in v are smaller by a factor $\sim m_f/M$. In this limit, changes in the direction of the test mass's motion due to encounters can be ignored.

These arguments suggest an alternative way of computing the dynamical friction acceleration [101, 394]: transfer to a frame moving with the massive body's velocity v_t (assumed constant) and find the steady-state distribution function $f(x, v)$ of the field stars in this frame, with the condition that $f(x, v - v_t) \to f(v_f)$ far from the test mass. In this approach, dynamical friction is interpreted as the net force due to the asymmetric distribution of field stars, $\rho(x) = \int f(x, v)dv$, around the test mass—the **density wake**. Rather confusingly, dynamical friction in this limit is describable using the formalism of chapter 3—which shows that there is not always a clear distinction to be made between "collisional" and "collisionless" dynamics.

This way of computing the dynamical friction force is based on essentially the same physical picture that was the basis for Chandrasekhar's calculation, and it should give the same result for the acceleration in the $M \gg m_f$ limit (with one qualification, related to the choice of p_{max}). We nevertheless include the method here because it suggests a more general way of looking at the origin of the frictional force.

Consider then a test mass that is moving at constant velocity $\boldsymbol{v}_t$ relative to a fixed frame. Let $\boldsymbol{v}_f$ be the velocity of a field star in that frame, and $\boldsymbol{v} = \boldsymbol{v}_f - \boldsymbol{v}_t$ its velocity in a frame moving with the test mass. Assume that the field-star velocity distribution far from the test mass is a known function: $f(\boldsymbol{v}_f)$ in the fixed frame and $f(\boldsymbol{v} + \boldsymbol{v}_t)$ in the moving frame. Finally, assume that $f(\boldsymbol{v}_f)$ is isotropic. To satisfy Jeans's theorem in the moving frame, we need to find an expression, in terms of the integrals of motion in the two-body problem, that reduces to

$$(\boldsymbol{v} + \boldsymbol{v}_t)^2 = v^2 + v_t^2 + 2\boldsymbol{v} \cdot \boldsymbol{v}_t = v^2 + v_t^2 - vv_t \cos\psi \qquad (5.64)$$

far from the test mass; here, ψ is the angle between $\boldsymbol{v}_t$ and the asymptote of the hyperbolic orbit of the field star with respect to the test mass. At infinity, $2E = v^2 - GM/r \to v^2$. The angle ψ can similarly be expressed in terms of the components of the angular momentum of the relative orbit [394]. Replacing v and ψ in equation (5.64) by these expressions then gives, via Jeans's theorem, the steady-state $f(\boldsymbol{x}, \boldsymbol{v})$ from which the response density can be computed.

Figure 5.3 shows the density computed in this way, assuming a field-star velocity distribution at infinity of

$$f(v_f) = f_0 \left(2v_c^2 - v_f^2\right)^{\gamma - 3/2}, \qquad v_f \le 2^{1/2} v_c, \qquad (5.65)$$

with $\gamma = 5/4$. This is the velocity distribution of stars in a power-law cusp around an SBH, equation (3.49), at a distance from the SBH such that the circular velocity is v_c. In figure 5.3, the velocity v_t of the test mass has been set equal to v_c; in other words, that figure represents the dynamical friction wake around a body in a circular orbit around an SBH.[8] The contributions to the density wake of stars with $v_f < v_t$ and $v_f > v_t$ were computed separately. For this f and v_t, most of the field stars are moving faster than the test mass, and they dominate the density in the wake as well.

Given $\rho(\boldsymbol{x}, \boldsymbol{v})$ computed in this way, the dynamical friction force can be computed by integrating the force exerted by the perturbed medium on the test mass. At first sight, this operation would seem to be independent of p_{max} and $\ln\Lambda$, since the density plotted in figure 5.3 includes the contributions from field stars that come from arbitrarily large distances. But while the response density is uniquely determined everywhere, the net force on the test mass depends on the size of the region over which the density is integrated: roughly speaking, the linear size of this region corresponds to p_{max} [394]. In the example plotted in figure 5.3, one can get a sense of how the relative contribution to the net force, of stars with $v_f > v_t$ and $v_f < v_t$, changes with distance: the fast stars are most important close to the test mass, while at larger distances, it is the asymmetry in the distribution of the slower stars that

---

[8]Note that we are identifying $f(v_f)$, the velocity distribution at infinity, with $f(v; r)$, the *local* velocity distribution in the assumed model if the test mass were not present. This may seem an unreasonable approximation, but it is precisely what is done when computing orbit-averaged diffusion coefficients for the Fokker–Planck equation (section 5.5).

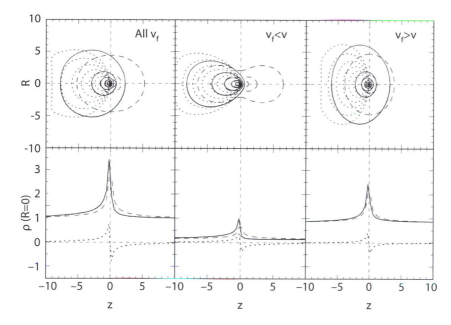

Figure 5.3 Dynamical friction wake around a massive object, assuming the velocity distribution at infinity of equation (5.65) [8]. The panels at the top show contours of the
density, in a frame that follows the massive body (located at the origin); motion
is to the right at constant speed v. The left panel shows the total density of the
wake, the middle panel shows the density contributed only by stars with velocity at infinity less than that of the massive body, and the right panel shows the
contribution from the complement of stars that move faster than v. Thick (solid)
curves show the total response from the indicated stars; medium (dashed) curves
show the part of the response that is symmetric with respect to z; thin (dotted)
curves show the antisymmetric part (only on one side), which is responsible for
the frictional force. The lower panels show the density along a line through the
moving body in the direction of its motion.

dominates the net force. This corresponds, roughly, to Chandrasekhar's result that
field stars with $v_f > v_t$ contribute negligibly to the force if p_{max} is very large.

5.2.3.2 Perturbative approaches; dynamical friction in inhomogeneous systems

In treatments like Chandrasekhar's, the unperturbed field-star trajectories consist
of straight lines that approach and recede from infinity. In reality, both test and
field stars follow unperturbed orbits about the center of the galaxy. Chandrasekhar's
theory might be expected to give approximately correct results for the dynamical
friction force even in this case, as long as $p_{max} \gg p_0$, since over many decades
in scale the orbits of the field stars will appear nearly rectilinear as seen by the
test body. But if the dynamical friction time is long compared with orbital periods,
individual field stars will encounter the test mass again and again.

The consequences are most easily seen by considering a test body that moves along a circular orbit in a spherical galaxy. We can transform to a frame in which the test mass is stationary, that is, a frame that rotates with frequency $\Omega_t = 2\pi/v_{\text{circ}}$. In this rotating frame, the apsides of field stars will be seen to precess, either forward or backward, at a rate determined by the star's energy and angular momentum and by Ω_t. Over some limited span of time, a field star will exert a torque on the test mass, causing it to either accelerate or decelerate. But over a time long compared with both the radial and precessional periods in the rotating frame, the time-averaged torques will vanish, since the field star will spend equal amounts of time ahead of and behind the test mass. However, there will always be some orbits that are nearly closed in this frame; for instance, orbits that precess in the inertial frame with a frequency close to Ω_t. Such **resonant orbits** can exert a net torque; in the case of an orbit that is nearly, but not quite, resonant, a very long time will be required for its time-averaged torque to tend to zero.

One can carry out a perturbative analysis by assuming that the perturbing potential, and the changes induced by it in the orbits of field stars, are small [331]. It turns out that essentially all of the torque acting on the test body comes from orbits near resonance. The net torque induced by a particular resonance is proportional to gradients in the unperturbed f near the resonance, that is, by the relative numbers of stars on one or the other "side" of the resonance. Furthermore, the acceleration induced by the resonant orbits depends on how quickly the orbit of the test mass is evolving [521]. If orbital decay is very slow, the influence of a single resonance can build up, invalidating the perturbative assumption. But if Ω_t changes rapidly enough, field stars will "sweep through" a given resonance before the changes induced in their orbits are appreciable. In this case, the response of the field stars to forcing by the test body is calculable. The equation describing the rate of change of the specific angular momentum of the test mass takes the form [559]

$$\frac{dL}{dt} = \pi^3 m_T \sum_l \int \int \int dE \, dL \, d(\cos i) P(E, L) \left(l_3 \Omega_t \frac{\partial f}{\partial E} + l_2 \frac{\partial f}{\partial L} \right)$$

$$\times |\Psi_l|^2 \, \delta(l_1 v_1 + l_2 v_2 - l_3 v_t). \qquad (5.66)$$

In this equation, $f(E, L)$ is the unperturbed field-star phase-space density, $v_i = \partial H/\partial J_i$ is the frequency associated with the ith action J_i in the unperturbed galaxy potential, and the Ψ_l are coefficients that appear when the potential of the test mass is expanded as a Fourier series in $(\boldsymbol{J}, \boldsymbol{\theta})$:

$$\psi_t(\boldsymbol{r}, t) = m_T \text{Re} \left\{ \sum_l \Psi_l(\boldsymbol{J}) \exp \left[i \left(\boldsymbol{l} \cdot \boldsymbol{\theta} - l_3 \Omega_t t \right) \right] \right\}, \qquad (5.67)$$

The same analysis leading to equation (5.66) also yields expressions for the density wake [559]. While it may seem strange that only resonant orbits contribute to the torque, a given resonance (i.e., a given set of integers (l_1, l_2)) corresponds to a manifold of orbits, with different (E, L), that satisfy $l_1 v_1(E, L) + l_2 v_2(E, L) - l_3 \Omega_t = 0$. Furthermore, in a real galaxy, the frequency spectrum of the perturbing potential is not made up of sharp lines, but rather is broadened by the time dependence of the decaying orbit and by the finite age of the galaxy.

Formally, the advantage of the perturbative treatment is that it relaxes the assumption of locality: the diffusion coefficients so derived reflect the inhomogeneous and bounded nature of real galaxies. On the negative side, evaluating expressions like (5.66) entails a considerable computational effort, which is probably why only a few cases have been worked out in detail [559, 560]. It is also difficult to be certain whether a real system sits in a regime where the assumptions of the perturbative treatment are justified: the test body's orbit must decay quickly enough that the response of the resonant orbits remains linear, but not so quickly that the steady-state approximation is invalidated. So far, these studies have tended to validate Chandrasekhar's results; the main element they add is a quantitative estimate of the Coulomb logarithm. Another important application is to cases where the assumption of locality is clearly violated; for instance, a satellite that orbits just outside a galaxy where the local density is zero [412].

5.2.3.3 Numerical N-body treatments

The number of stars in a galaxy far exceeds the number that can be directly integrated on a computer, at least with high accuracy and for interestingly long times. But an N-body algorithm can still be useful for estimating certain parameters that appear in Chandrasekhar's theory; for instance, $\ln \Lambda$. One well-studied example is the inspiral of a massive body into the center of a galaxy: in the limit $m \gg m_f$, equations like (5.63) tell us that the frictional force depends on the mass density of the field stars, not on their individual masses. The only N dependence occurs implicitly though the Coulomb logarithm. The value of $\ln \Lambda$, and its dependence on N, can be evaluated numerically by fitting the computed trajectories to the predictions of Chandrasekhar's formulas.

The biggest potential pitfall associated with direct numerical integration is the need for accurate treatment of close encounters. Most N-body codes deal with close encounters via some sort of numerical "softening": the forces between particles are modified to avoid the singularities that occur when separations tend to zero. For instance, one common scheme is to replace the force between m_1 and m_2 by

$$F_{12} = Gm_1 m_2 \frac{(x_1 - x_2)}{\left(|x_1 - x_2|^2 + \epsilon^2\right)^{3/2}}, \qquad (5.68)$$

where ϵ is the **softening length** [1]. When simulating dynamical friction on a computer, the softening length must be smaller than p_0 given by equation (5.5), or

$$\epsilon \lesssim 2Gm/v_{\rm rms}^2. \qquad (5.69)$$

In so-called "standard N-body units" [232], the total mass of a galaxy and the gravitational constant are unity, and the total energy (potential plus kinetic) is $-1/4$. The virial theorem then gives $v_{\rm rms} = 1/\sqrt{2}$. In these units, equation (5.69) becomes

$$\epsilon \lesssim 4/N. \qquad (5.70)$$

Large values for N are also desirable; for instance, in simulating dynamical friction, the mass of the test body, M, should be much greater than the field-star mass, where $m_f = N^{-1}$. Together, these requirements are difficult to satisfy without sacrificing

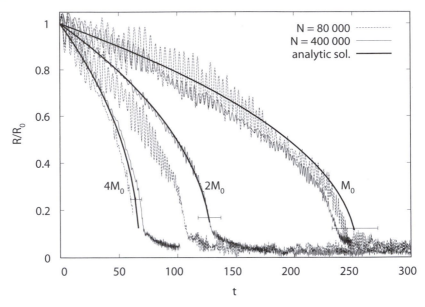

Figure 5.4 Numerical experiments designed to evaluate the Coulomb logarithm [500]. The
inspiral of a test mass was followed with an N-body code starting from a circular
orbit at the half-mass radius of the galaxy model. Results are shown for three
different values of the test-particle mass; $M_0 \approx 5 \times 10^{-4}$ in N-body units. The
solid curves show theoretical trajectories computed using the best-fit parameters
in equation (5.71).

accuracy, and in practice, the effect of the finite softening length must be taken into
account when evaluating the effects of encounters.

When fitting a numerically computed inspiral to equations like (5.23), it is reasonable to replace the Coulomb logarithm by something like

$$\ln \Lambda = \ln \frac{p_{\max}}{p_{\min} + \epsilon}$$
$$= \ln p_{\max} - \ln (p_{\min} + \epsilon) \tag{5.71}$$

since encounters closer than $\sim \epsilon$ do not feel the full deflection. The two free parameters in equation (5.71) can be estimated by integrating the motion of a test mass
on an initially circular orbit (say) for different values of ϵ. Results from one such
study [500] are illustrated in figure 5.4, based on a model for the N-body galaxy
motivated by observations of the Milky Way nuclear star cluster. The authors found

$$p_{\min} \approx 8.0 \times 10^{-4}, \quad p_{\max} \approx 0.39 \tag{5.72}$$

in N-body units. The former value is consistent with equation (5.5); the value for
$p_{\max}$ implies that encounters with "impact parameters" greater than about $1/4$ the
linear size of the system are ineffective.

Another interesting application of N-body methods is to the motion of a massive
object, like an SBH, as its orbit decays near the center of a galaxy. This case was

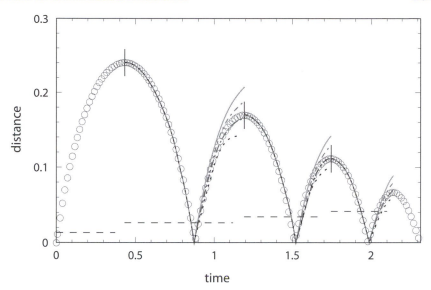

Figure 5.5 Numerical evaluation of the Coulomb logarithm in N-body integrations of a
          massive body (SBH) that has been ejected from the center of a galaxy [218].
          The open circles show the numerically computed trajectory and the lines are fits
          of Chandrasekhar's formula, equation (5.23). The latter were computed in a
          piecewise manner, starting from extrema in the SBH's trajectory (vertical solid
          lines) and continuing until the next extremum. For the purposes of computing the
          theoretical trajectories, the properties (e.g., density) of the N-body model were
          extracted at the time the SBH passed through the center and were assumed to
          remain fixed until the next central passage. Estimated core radii are shown by
          the horizontal dashed lines. Line styles correspond to different values of $\ln \Lambda$:
          1 (solid), 2 (dashed), 3 (dot-dashed), 4 (dotted).

considered in the context of Chandrasekhar's theory in section 5.2.1; while the role
of the fast-moving stars was clarified, that treatment still contained an undetermined
parameter p_{max} in the argument of the logarithm, equation (5.38):

$$\ln \Lambda \approx \ln \sqrt{1 + \frac{2 p_{\mathrm{max}}^2}{r_{\mathrm{h}}^2}}, \tag{5.73}$$

where $G M / \sigma_f^2$ has been replaced by r_{h}. Furthermore, the mass of an SBH is comparable with that of a galaxy core and its displacement from the center can be
expected to substantially affect the local distribution of stars, violating another of
Chandrasekhar's approximations.

It was argued above that p_{max} for a test mass near the center of a galaxy containing an SBH should be of order r_{h}. If so, equation (5.73) implies that $\ln \Lambda$ should be
small, of order unity.

Figure 5.5 shows an N-body simulation of this situation [218]: an SBH, with
mass 10^{-3} in N-body units, was given an impulsive kick at $t = 0$ equal to 70% of

the escape velocity from the galaxy center. As its (nearly radial) orbit decayed, the trajectory was fit to equation (5.23) with various values of $\ln \Lambda$. Since the structure of the nucleus changed each time the SBH passed through, the parameters defining the local density were determined after each central passage and were assumed to remain fixed until the next passage. The best-fit values of $\ln \Lambda$ were indeed found to be small: $2 \lesssim \ln \Lambda \lesssim 3$.

Yet another case in which N-body methods are useful is the inspiral of a massive body into the center of a galaxy containing a central SBH. A simple estimate was given in equation (5.35) for $\ln \Lambda$ in this situation:

$$\ln \Lambda \approx \ln N_h \qquad (5.74)$$

with N_h the number of stars whose mass equals $M_\bullet$, the mass of the SBH. Of course, in an N-body simulation, N_h will be much smaller than its value in a real galaxy, but the prediction can nevertheless be tested. Figure 8.6 shows an example; the theoretical curve in that figure was computed assuming $\ln \Lambda = 5.7$. Since $M_\bullet = 0.01$ in this simulation and $N = 2 \times 10^5$, $N_h \approx 2000$ and equation (5.74) predicts $\ln \Lambda \approx 7.6$.

Numerous other N-body studies have been carried out to evaluate Chandrasekhar's formula in the case of a massive particle inspiraling toward the center of a galaxy [564, 57, 94, 8, 264]. Early work was typically based on approximate N-body schemes and the results were often discrepant from study to study [580]; these differences appear to have been resolved in the last few years through the use of higher-accuracy N-body algorithms.

5.3 FOKKER–PLANCK EQUATION

So far we have discussed gravitational encounters in terms of their effect on the motion of a single test star. When dealing with a stellar system, we are also interested in the evolution of the phase-space density $f(x, v, t)$ describing all the stars. An important use of the diffusion coefficients is to compute this evolution.[9]

Formally, the effect of encounters on f can be described by adding a term to the right-hand side of equation (3.7):

$$\frac{Df}{Dt} = \frac{\partial f}{\partial t} + \sum_i v_i \frac{\partial f}{\partial x_i} + \sum_i a_i \frac{\partial f}{\partial v_i} = \left(\frac{\partial f}{\partial t} \right)_c. \qquad (5.75)$$

In the general case, $(\partial f/\partial t)_c$ is very complicated. For instance, in a close encounter, the velocity change Δv can be comparable to v, causing a star to jump suddenly from one point in phase space to another. Such changes must be described by an integral expression, as in the Boltzmann equation. The encounter term can be greatly simplified if only the effects of distant encounters are included, that is, if gravitational deflections are assumed to be small; in this approximation, the encounter

---

[9]To keep the notation as simple as possible, f will be used to denote both the test-star and field-star velocity distributions; in the latter case it will always be written as $f(v_f)$. This notation is a natural one in cases where $f(v)$ and $f(v_f)$ describe the same set of stars, as they often do.

term becomes a differential operator. The result is the **Fokker–Planck equation** [167, 431].

Two basic forms of the Fokker–Planck equation are commonly used in galactic dynamics. The **local Fokker–Planck equation** expresses $(\partial f/\partial t)_c$ in terms of the phase-space variables (x, v), without regard to the fact that the unperturbed motion consists of orbits in the smooth galactic potential. The **orbit-averaged Fokker–Planck equation** invokes the approximation that the timescale for collisional effects to change f is long compared with orbital periods; the diffusion coefficients are replaced by time averages over the unperturbed trajectories and the phase-space variables are replaced by the integrals of motion in the unperturbed problem. In this section the local Fokker–Planck equation is derived, and subsequent sections deal with the orbit-averaged equation.

Let Δt denote an interval of time that is short compared with the time over which the velocity of a star changes due to encounters, but still long enough that many encounters occur. Define the **transition probability** $\Psi(v; \Delta v)$ that v changes by Δv in time Δt. Then

$$f(v, t + \Delta t) = \int f(v - \Delta v, t) \Psi(v - \Delta v; \Delta v) d\Delta v. \qquad (5.76)$$

This equation assumes in addition that the evolution of f depends only on its instantaneous value, that is, that its previous history can be ignored. This is the definition of a **Markov process**.

We now expand $f(v, t + \Delta t)$ on the left-hand side as a Taylor series in Δt, and $f(v - \Delta v, t)$ and $\Psi(v - \Delta v; \Delta v)$ on the right-hand side as Taylor series in Δv_i, where i runs from 1 to 3. The result is

$$f(v, t) + \frac{\partial f}{\partial t}\Delta t + O\left[(\Delta t)^2\right] = \qquad (5.77)$$

$$\int \left[f(v, t) - \sum_i \frac{\partial f}{\partial v_i}\Delta v_i + \frac{1}{2}\sum_i \frac{\partial^2 f}{\partial v_i^2}\Delta v_i^2 + \sum_{i<j} \frac{\partial^2 f}{\partial v_i \partial v_j}\Delta v_i \Delta v_j + \cdots \right]$$

$$\times \left[\Psi(v; \Delta v) - \sum_i \frac{\partial \Psi}{\partial v_i}\Delta v_i + \frac{1}{2}\sum_i \frac{\partial^2 \Psi}{\partial v_i^2}\Delta v_i^2 + \sum_{i<j} \frac{\partial^2 \Psi}{\partial v_i \partial v_j}\Delta v_i \Delta v_j + \cdots \right]$$

$$\times d\Delta v.$$

Writing

$$\langle \Delta v_i \rangle = \frac{1}{\Delta t}\int \Psi(v; \Delta v)\Delta v_i \, d\Delta v, \qquad (5.78a)$$

$$\langle \Delta v_i \Delta v_j \rangle = \frac{1}{\Delta t}\int \Psi(v; \Delta v)\Delta v_i \Delta v_j \, d\Delta v, \qquad (5.78b)$$

this becomes

$$\frac{\partial f}{\partial t} + O\left[(\Delta t)\right] = -\sum_i \frac{\partial f}{\partial v_i}\langle \Delta v_i\rangle + \frac{1}{2}\sum_i \frac{\partial^2 f}{\partial v_i^2}\langle \Delta v_i^2\rangle$$

$$+ \sum_{i<j} \frac{\partial^2 f}{\partial v_i \partial v_j}\langle \Delta v_i \Delta v_j\rangle - \sum_i f\frac{\partial}{\partial v_i}\langle \Delta v_i\rangle$$

$$+ \sum_i \frac{\partial}{\partial v_i}\langle \Delta v_i^2\rangle\frac{\partial f}{\partial v_i} + \sum_{i\neq j} \frac{\partial f}{\partial v_i}\frac{\partial}{\partial v_j}\langle \Delta v_i \Delta v_j\rangle$$

$$+ \frac{1}{2}\sum_i f\frac{\partial^2}{\partial v_i^2}\langle \Delta v_i^2\rangle + \sum_{i<j} f\frac{\partial^2}{\partial v_i}\partial v_j\langle \Delta v_i \Delta v_j\rangle$$

$$+ O\left[\langle \Delta v_i \Delta v_j \Delta v_k\rangle\right], \tag{5.79}$$

where the last term includes averages of quantities like $\Delta v_i \Delta v_j \Delta v_k$. Retaining only the lowest-order terms, equation (5.79) gives for $\partial f/\partial t \equiv (\partial f/\partial t)_c$:

$$\left(\frac{\partial f}{\partial t}\right)_c = -\sum_i \frac{\partial}{\partial v_i}\left(f\langle \Delta v_i\rangle\right) + \frac{1}{2}\sum_{i,j} \frac{\partial^2}{\partial v_i \partial v_j}\left(f\langle \Delta v_i \Delta v_j\rangle\right). \tag{5.80}$$

Substituting equation (5.80) into equation (5.75) then yields the Fokker–Planck equation. The quantities defined in equations (5.78) are the same diffusion coefficients that were computed above.

Including only the first- and second-order terms in the Taylor expansions of f and Ψ may seem ad hoc. If the source of the velocity diffusion is random perturbations with a Gaussian white-noise spectrum, it can be shown [458] that the expansion truncates after the first two terms. This is not the case for gravitational encounters; however, as noted above, the higher-order diffusion coefficients contain only nondominant terms and in many circumstances are small compared with $\langle \Delta v_i\rangle$, $\langle \Delta v_i \Delta v_j\rangle$. For consistency, neglect of the higher-order diffusion coefficients in the Fokker–Planck equation implies that the nondominant terms in $\langle \Delta v_i\rangle$ and $\langle \Delta v_i \Delta v_j\rangle$ should also be omitted. It can also be argued that the nondominant terms describe predominantly close encounters and should be excluded from the Fokker–Planck equation for this reason [502].

Although it is much simpler than the Boltzmann equation, equation (5.80) is still not very useful for calculations. The diffusion coefficients are expressed in terms of Cartesian coordinates, which are not related in an obvious way to the quantities $\langle \Delta v_{\parallel}\rangle$, $\langle (\Delta v_{\perp})^2\rangle$, $\langle (\Delta v_{\parallel})^2\rangle$ whose forms have already been derived.

A scheme for expressing the Fokker–Planck equation in any coordinate system will be presented below. An alternative approach, which is simpler in many cases of interest, is to rederive $(\partial f/\partial t)_c$ from scratch.

Consider, for instance, a stellar system in which the velocity distribution is isotropic, $f = f(v)$. Let $\Psi(v, \Delta v)$ be the probability that v changes by Δv in time Δt. Then

$$N(v, t + \Delta t) = \int N(v - \Delta v)\Psi(v - \Delta v; \Delta v)d\Delta v, \tag{5.81}$$

where $N(v)dv$ is the number of stars in velocity interval v to $v + dv$; evidently

$$N(v)dv = 4\pi v^2 f(v)dv. \qquad (5.82)$$

Carrying out the same series expansions as before, we arrive at

$$\left(\frac{\partial N}{\partial t}\right)_c = -\frac{\partial}{\partial v}\left[N(v)\langle \Delta v \rangle\right] + \frac{1}{2}\frac{\partial^2}{\partial v^2}\left[N(v)\langle (\Delta v)^2 \rangle\right]. \qquad (5.83)$$

The coefficients $\langle \Delta v \rangle$ and $\langle (\Delta v)^2 \rangle$ are defined in the usual way as sums, over a unit interval of time, of Δv and $(\Delta v)^2$ due to encounters with field stars. We would like to express these in terms of the coefficients $\langle \Delta v_\parallel \rangle$ and $\langle \Delta v_\perp \rangle$. Clearly,

$$\Delta v = \left[(v + \Delta v_\parallel)^2 + (\Delta v_\perp)^2\right]^{1/2} - v. \qquad (5.84)$$

Expanding to second order in the small quantities $\Delta v_\parallel$ and $\Delta v_\perp$ and taking means,

$$\langle \Delta v \rangle = \langle \Delta v_\parallel \rangle + \frac{1}{2}\frac{\langle (\Delta v_\perp)^2 \rangle}{v}, \qquad \langle (\Delta v)^2 \rangle = \langle (\Delta v_\parallel)^2 \rangle. \qquad (5.85)$$

Substituting equations (5.82) and (5.85) into equation (5.83) then yields

$$\left(\frac{\partial f}{\partial t}\right)_c =$$

$$\frac{1}{v^2}\frac{\partial}{\partial v}\left[-vf\left(v\langle \Delta v_\parallel \rangle + \frac{1}{2}\langle (\Delta v_\perp)^2 \rangle\right) + \frac{1}{2}\frac{\partial}{\partial v}\left(v^2 f\langle (\Delta v_\parallel)^2 \rangle\right)\right]. \quad (5.86)$$

Adopting the standard forms of the diffusion coefficients, this becomes

$$\left(\frac{\partial f}{\partial t}\right)_c = \frac{4\pi \Gamma}{v^2}\frac{\partial}{\partial v}\left[\frac{m}{m_f}fv^2 F_2 + \frac{v^3}{3}(E_1 + F_4)\frac{\partial f}{\partial v}\right]. \qquad (5.87)$$

Finally, if $f(v_f)$ is Maxwellian,

$$\left(\frac{\partial f}{\partial \tau}\right)_c = \frac{1}{x^2}\frac{\partial}{\partial x}\left[2xG(x)\left(2x\frac{m}{m_f}f + \frac{\partial f}{\partial x}\right)\right], \qquad (5.88)$$

where $x = v/(\sqrt{2}\sigma_f)$ as before and

$$\tau \equiv \frac{t}{t_0}, \quad t_0 = 2^{5/2}\frac{\sigma_f^3}{n\Gamma} = \frac{\sqrt{2}\sigma_f^3}{\pi G^2 m_f^2 n \ln \Lambda}; \qquad (5.89)$$

t_0 is 4/3 times the relaxation time defined in equation (5.61).

Evolution of stellar systems due to encounters is properly the subject of chapter 7, but it is natural to ask here what equation (5.88) implies about the steady-state form of f. A "zero-flux" solution is obtained by setting

$$2x\frac{m}{m_f}f + \frac{\partial f}{\partial x}$$

to zero. Upon integration,

$$f(v) = f_0 e^{-v^2/2\sigma^2}, \quad \sigma^2 = \frac{m_f}{m}\sigma_f^2. \qquad (5.90)$$

The steady-state velocity distribution is Maxwellian, and the velocity dispersion satisfies the "equipartition" condition $m \langle v^2 \rangle = m_f \langle v_f^2 \rangle$.

Following Rosenbluth, MacDonald and Judd [465], we now present a general method for expressing the local Fokker–Planck equation in any velocity-space coordinates. We ignore the nondominant terms in the diffusion coefficients by "taking the logarithmic term out of the integral." In addition, the diffusion coefficients already derived will be generalized, in an obvious way, to the case that the field stars have a variety of masses. We define $f_b(\boldsymbol{v}_f)$ as the phase-space number density of field stars of mass m_b.

Begin by returning to equation (5.19) for the dynamical friction coefficient. That equation gave the rate of change of $\boldsymbol{v}$ in the direction of the test particle's motion. A straightforward generalization gives the dynamical friction coefficient along any of the three Cartesian directions:

$$\langle \Delta v_i \rangle = -\Gamma \sum_b \left(\frac{m + m_b}{m_b} \right) \int \frac{f_b(\boldsymbol{v}_f)}{V^2} (\boldsymbol{e}_i \cdot \boldsymbol{e}'_1) d\boldsymbol{v}_f, \tag{5.91}$$

where $(\boldsymbol{e}, \boldsymbol{e}')$ are the orthogonal basis vectors defined above, oriented, respectively, along the fixed axes, and such that $\boldsymbol{e}'_1$ is parallel to the relative velocity vector $\boldsymbol{V}$. Equation (5.91) can be written

$$\langle \Delta v_i \rangle = \Gamma' \sum_b m_b (m + m_b) \frac{\partial h_b}{\partial v_i}, \tag{5.92a}$$

$$h_b(\boldsymbol{v}) \equiv \int \frac{f_b(\boldsymbol{v}_f)}{|\boldsymbol{v} - \boldsymbol{v}_f|} d\boldsymbol{v}_f, \tag{5.92b}$$

where $\Gamma' \equiv 4\pi G^2 \ln \Lambda$. A similar representation is possible for the second-order coefficients, if the nondominant terms are ignored:

$$\langle \Delta v_i \Delta v_j \rangle = \Gamma' \sum_b m_b^2 \frac{\partial^2 g_b}{\partial v_i \partial v_j}, \tag{5.93a}$$

$$g_b(\boldsymbol{v}) \equiv \int f_b(\boldsymbol{v}_f) |\boldsymbol{v} - \boldsymbol{v}_f| d\boldsymbol{v}_f. \tag{5.93b}$$

In the special case of an isotropic $f(\boldsymbol{v}_f)$, we can replace $\partial/\partial v_i$ in these expressions by $(v_i/v)(d/dv)$; the three diffusion coefficients already derived are then expressible in the compact forms

$$\langle \Delta v_\parallel \rangle = \Gamma' \sum_b m_b (m + m_b) \frac{dh_b}{dv}, \tag{5.94a}$$

$$\langle (\Delta v_\parallel)^2 \rangle = \Gamma' \sum_b m_b^2 \frac{d^2 g_b}{dv^2}, \tag{5.94b}$$

$$\langle (\Delta v_\perp)^2 \rangle = \frac{2\Gamma'}{v} \sum_b m_b^2 \frac{dg_b}{dv}, \tag{5.94c}$$

where

$$h_b(v) = 4\pi v \left[I_2(v) + J_1(v) \right], \tag{5.95a}$$

$$g_b(v) = \frac{4\pi v^3}{3} \left[3I_2(v) + I_4(v) + 3J_3(v) + J_1(v) \right] \tag{5.95b}$$

and

$$I_n(v) = \int_0^v \left(\frac{v_f}{v} \right)^n f_b(v_f) dv_f, \tag{5.96a}$$

$$J_n(v) = \int_v^\infty \left(\frac{v_f}{v} \right)^n f_b(v_f) dv_f. \tag{5.96b}$$

Returning to the case of a general $f(\boldsymbol{v}_f)$, and writing the encounter term in the Fokker–Planck equation (5.80) in terms of g and h,

$$\frac{1}{\Gamma'} \left(\frac{\partial f}{\partial t} \right)_c = -\frac{\partial}{\partial v_i} \left[f \sum_b m_b (m + m_b) \frac{\partial h_b}{\partial v_i} \right]$$

$$+ \frac{1}{2} \frac{\partial^2}{\partial v_i^2} \left[f \sum_b m_b^2 \frac{\partial^2 g_b}{\partial v_i \partial v_j} \right]. \tag{5.97}$$

Equation 5.97 is expressed in Cartesian coordinates. Let v^i, $i = 1, 2, 3$ be general velocity-space coordinates; writing the index as a superscript indicates that $\boldsymbol{v}$ transforms as a contravariant vector.[10] The velocity-space metric associated with these coordinates is a_{ij} where

$$ds^2 = \sum_{ij} a_{ij} dv^i dv^j \tag{5.98}$$

and ds is the distance between two points whose coordinates differ by dv^1, dv^2, dv^3. For the remainder of this section, we adopt the summation convention: appearance of an index twice in a single term implies a summation over all its possible values. Hence $ds^2 = a_{ij} dv^i dv^j$.

The quantity

$$T^i = \langle \Delta v^i \rangle = \Gamma' \sum_b m_b(m + m_b) \frac{\partial h_b}{\partial v^i} \tag{5.99}$$

clearly transforms as a contravariant vector between different Cartesian coordinate systems, and the quantity

$$S^{ij} = \langle \Delta v^i \Delta v^j \rangle = \Gamma' \sum_b m_b^2 \frac{\partial^2 g_b}{\partial v^i \partial v^j} \tag{5.100}$$

as a contravariant tensor. The invariant form of the Fokker–Planck encounter term is therefore

$$\left(\frac{\partial f}{\partial t} \right)_c = - \left(f T^i \right)_{;i} + \frac{1}{2} \left(f S^{ij} \right)_{;ij}, \tag{5.101}$$

[10]The next few paragraphs make use of standard results from tensor analysis. Readers in need of a refresher are referred to chapter 4 of Weinberg's text [562].

where the semicolons denote covariant differentiation. The covariant divergences can be expressed in terms of a as

$$\left(fT^i\right)_{;i} = a^{-1/2}\frac{\partial}{\partial v^i}\left(a^{1/2}fT^i\right) \tag{5.102}$$

and

$$\left(fS^{ij}\right)_{;ij} = a^{-1/2}\frac{\partial^2}{\partial v^i\partial v^j}\left(a^{1/2}fS^{ij}\right) + a^{-1/2}\frac{\partial}{\partial v^j}\left(a^{1/2}\Gamma^j_{ik}fS^{ik}\right), \tag{5.103}$$

where the affine connection Γ^j_{ik} is

$$\Gamma^j_{ik} = \frac{1}{2}a^{jm}\left(\frac{\partial a_{mk}}{\partial v^i} + \frac{\partial a_{mi}}{\partial v^k} - \frac{\partial a_{ik}}{\partial v^m}\right). \tag{5.104}$$

Substituting these expressions into equation (5.101) yields

$$\left(\frac{\partial f}{\partial t}\right)_c = -a^{1/2}\left[a^{1/2}f\left(T^i - \frac{1}{2}\Gamma^i_{jk}S^{jk}\right)\right]_{,i} + \frac{1}{2}a^{-1/2}\left[a^{1/2}fS^{ij}\right]_{,ij}, \tag{5.105}$$

where the commas indicate simple derivatives, for example, $x_{,i} \equiv \partial x/\partial v^i$. Finally, we express T^i and S^{ij} explicitly in contravariant form as

$$T^i = \Gamma'\sum_b m_b(m + m_b)\left[a^{ij}\frac{\partial h_b}{\partial v^j}\right]$$

$$\equiv \Gamma'a^{ij}\frac{\partial\mathcal{H}}{\partial v^j}, \tag{5.106a}$$

$$S^{ij} = \Gamma'\sum_b m_b^2\left[a^{ik}a^{jl}\left(\frac{\partial^2 g_b}{\partial v^k\partial v^l} - \Gamma^m_{kl}\frac{\partial g_b}{\partial v^m}\right)\right]$$

$$\equiv \Gamma'\left[a^{ik}a^{jl}\left(\frac{\partial^2\mathcal{G}}{\partial v^k\partial v^l} - \Gamma^m_{kl}\frac{\partial\mathcal{G}}{\partial v^m}\right)\right], \tag{5.106b}$$

where

$$\mathcal{G}(v) \equiv \sum_b m_b^2 g_b(v), \qquad \mathcal{H}(v) \equiv \sum_b m_b(m + m_b)h_b(v). \tag{5.107}$$

Equations (5.105)–(5.106), together with equations (5.92b) and (5.93b) for h_b and g_b, are the desired equations.

Note that, in the case of an isotropic field-star distribution, T^i and S^{ij} can be expressed in terms of the three Cartesian diffusion coefficients as

$$T^i = \frac{v^i}{v}\langle\Delta v_\parallel\rangle, \tag{5.108a}$$

$$S^{ij} = \frac{v^iv^j}{v^2}\left(\langle\Delta v_\parallel^2\rangle - \frac{1}{2}\langle\Delta v_\perp^2\rangle\right) + \frac{1}{2}\delta^{ij}\langle\Delta v_\perp^2\rangle. \tag{5.108b}$$

A useful choice for the v^i is spherical polar coordinates:

$$v^1 = v, \qquad v^2 = \mu = \cos\theta, \qquad v^3 = \phi. \tag{5.109}$$

The reader may question the utility of a coordinate system that defines a preferred direction, given that up to now we have been talking about infinite homogeneous systems! However, before long, we will express the Fokker–Planck operator in terms of the angular momentum as a velocity variable, and the preferred axis will be the radius vector from the center of the galaxy to the point in question. In anticipation, we ignore any dependence of f on ϕ, and write $f = f(v, \mu)$.

The nonzero components of the metric tensor, and its inverse, are

$$a_{11} = 1, \qquad a_{22} = v^2(1 - \mu^2)^{-1}, \qquad a_{33} = v^2(1 - \mu^2),$$
$$a^{11} = 1, \qquad a^{22} = v^{-2}(1 - \mu^2), \qquad a^{33} = v^{-2}(1 - \mu^2)^{-1}. \qquad (5.110)$$

The nonzero components of the affine connection are

$$\Gamma_{22}^1 = -v(1 - \mu^2)^{-1}, \qquad \Gamma_{33}^1 = -v(1 - \mu^2), \qquad \Gamma_{12}^2 = v^{-1},$$
$$\Gamma_{22}^2 = \mu \left(1 - \mu^2\right)^{-1}, \qquad \Gamma_{33}^2 = \mu(1 - \mu^2), \qquad \Gamma_{13}^3 = v^{-1},$$
$$\Gamma_{23}^3 = -\mu(1 - \mu^2)^{-1}. \qquad (5.111)$$

The components of T^i are

$$T^1 = \Gamma' \frac{\partial \mathcal{H}}{\partial v}, \qquad T^2 = \Gamma' v^{-2}(1 - \mu^2) \frac{\partial \mathcal{H}}{\partial \mu}, \qquad T^3 = 0, \qquad (5.112)$$

and the components of S^{ij} are

$$S^{11} = \Gamma' \frac{\partial^2 \mathcal{G}}{\partial v^2},$$

$$S^{22} = \Gamma' v^{-4}(1 - \mu^2)^2 \left[\frac{\partial^2 \mathcal{G}}{\partial \mu^2} + v(1 - \mu^2)^{-1} \frac{\partial \mathcal{G}}{\partial v} - \mu(1 - \mu^2)^{-1} \frac{\partial \mathcal{G}}{\partial \mu} \right],$$

$$S^{33} = \Gamma' v^{-4} \left(1 - \mu^2\right)^{-1} \left(v \frac{\partial \mathcal{G}}{\partial v} - \mu \frac{\partial \mathcal{G}}{\partial \mu} \right),$$

$$S^{12} = \Gamma' v^{-2} \left(1 - \mu^2\right) \left(\frac{\partial^2 \mathcal{G}}{\partial v \partial \mu} - v^{-1} \frac{\partial \mathcal{G}}{\partial \mu} \right),$$

$$S^{13} = S^{23} = 0. \qquad (5.113)$$

After just a bit more algebra (which we leave to the reader), equation (5.105) becomes

$$\frac{1}{\Gamma'} \left(\frac{\partial f}{\partial t} \right)_c = -\frac{1}{v^2} \frac{\partial}{\partial v} \left(f v^2 \frac{\partial \mathcal{H}}{\partial v} \right) - \frac{1}{v^2} \frac{\partial}{\partial \mu} \left[f(1 - \mu^2) \frac{\partial \mathcal{H}}{\partial \mu} \right]$$
$$+ \frac{1}{2v^2} \frac{\partial^2}{\partial \mu^2} \left\{ f \left[\frac{1}{v^2} (1 - \mu^2)^2 \frac{\partial^2 \mathcal{G}}{\partial \mu^2} + \frac{1}{v}(1 - \mu^2) \frac{\partial \mathcal{G}}{\partial v} - \frac{1}{v^2} \mu(1 - \mu^2) \frac{\partial \mathcal{G}}{\partial \mu} \right] \right\}$$

$$+ \frac{1}{2v^2} \frac{\partial^2}{\partial v^2} \left(fv^2 \frac{\partial^2 \mathcal{G}}{\partial v^2} \right) + \frac{1}{v^2} \frac{\partial^2}{\partial \mu \partial v} \left\{ f(1 - \mu^2) \left[\frac{\partial^2 \mathcal{G}}{\partial \mu \partial v} - \frac{1}{v} \frac{\partial \mathcal{G}}{\partial \mu} \right] \right\}$$

$$+ \frac{1}{2v^2} \frac{\partial}{\partial v} \left\{ f \left[-\frac{1}{v}(1 - \mu^2) \frac{\partial^2 \mathcal{G}}{\partial \mu^2} - 2 \frac{\partial \mathcal{G}}{\partial v} + 2 \frac{\mu}{v} \frac{\partial \mathcal{G}}{\partial \mu} \right] \right\}$$

$$+ \frac{1}{2v^2} \frac{\partial}{\partial \mu} \left\{ f \left[\frac{\mu}{v^2}(1 - \mu^2) \frac{\partial^2 \mathcal{G}}{\partial \mu^2} + \frac{2\mu}{v} \frac{\partial \mathcal{G}}{\partial v} + \frac{2}{v}(1 - \mu^2) \frac{\partial^2 \mathcal{G}}{\partial \mu \partial v} - \frac{2}{v^2} \frac{\partial \mathcal{G}}{\partial \mu} \right] \right\}.$$
$$(5.114)$$

Earlier, we derived $(\partial f / \partial t)_c$ in the case that $f(\boldsymbol{v})$ and $f(\boldsymbol{v}_f)$ were isotropic, and the field stars had a single mass $m_b = m_f$ (equation 5.86). We can check that equation (5.114) reduces to this form by setting to zero the derivatives with respect to μ:

$$\frac{1}{\Gamma'} \left(\frac{\partial f}{\partial t} \right)_c = \frac{1}{v^2} \frac{\partial}{\partial v} \left[-f \left(v^2 \frac{\partial \mathcal{H}}{\partial v} + \frac{\partial \mathcal{G}}{\partial v} \right) + \frac{1}{2} \frac{\partial}{\partial v} \left(fv^2 \frac{\partial^2 \mathcal{G}}{\partial v^2} \right) \right]. \quad (5.115)$$

Comparison with equations (5.94), (5.106), and (5.107) confirms the agreement; note that equation (5.115) is slightly more general than equation (5.86) in that it allows for a general distribution of field-star masses.

A second special case, not considered so far, is one in which $f(\boldsymbol{v}_f)$ is isotropic but $f(\boldsymbol{v}) = f(v, \mu)$ is not. This may seem unnatural, since the test and field stars are drawn from the same population! However, to simplify numerical calculations, it is often argued that anisotropies in the field-star distribution are less important than those in the test distribution. In this approximation, $f(\boldsymbol{v}_f)$ can be replaced by $\overline{f}(v_f)$, an appropriate average over μ and ϕ, when computing $\mathcal{G}$ and $\mathcal{H}$; the latter then become functions only of v, as in equation (5.95). The encounter term in the Fokker–Planck equation is then

$$\frac{1}{\Gamma'} \left(\frac{\partial f}{\partial t} \right)_c = v^{-2} \frac{\partial}{\partial v} \left[-f \left(v^2 \frac{\partial \mathcal{H}}{\partial v} + \frac{\partial \mathcal{G}}{\partial v} \right) + \frac{1}{2} \frac{\partial}{\partial v} \left(fv^2 \frac{\partial^2 \mathcal{G}}{\partial v^2} \right) \right]$$

$$+ v^{-3} \frac{\partial \mathcal{G}}{\partial v} \frac{\partial}{\partial \mu} \left\{ f\mu + \frac{1}{2} \frac{\partial}{\partial \mu} \left[f \left(1 - \mu^2 \right) \right] \right\}. \quad (5.116)$$

Almost all of the applications of the Fokker–Planck equation in the remainder of this text will be based on equation (5.115) (isotropic f) or equation (5.116) (anisotropic f).

So far in this section, we have ignored the fact that f in a steady-state galaxy must obey Jeans's theorem. More precisely, if changes in f and Φ due to gravitational encounters occur on a timescale much longer than a crossing time, then $f(\boldsymbol{x}, \boldsymbol{v}, t)$ at time $t = t_0$ must be expressible as some function of the integrals of motion in $\Phi(\boldsymbol{x}, t = t_0)$. For instance, in a spherical, nonrotating galaxy, we know from chapter 3 that

$$f(\boldsymbol{x}, \boldsymbol{v}) = f(E, L), \quad (5.117)$$

where $E = (v^2/2) + \Phi(r)$, $L = |\boldsymbol{x} \times \boldsymbol{v}| = rv \sin \theta = rv(1 - \mu^2)^{1/2}$. If we think of equation (5.116) as describing the evolution of f at some fixed radius r in such a galaxy, it makes sense to adopt E and L as velocity variables instead of v and μ.

One way to do this would be to repeat the entire derivation that began with equation (5.109), replacing (v, μ, ϕ) by (E, L, ϕ) as velocity-space variables. A second way would be to carry out a brute-force change of variables in equation (5.105). But it turns out to be simpler, and also more useful, to ask how the terms in equation (5.105) transform when the coordinates v^1, v^2, v^3 are changed to the coordinates V^1, V^2, V^3 [91].

We begin by writing that equation in the form

$$\left(\frac{\partial f}{\partial t}\right)_c = -a^{-1/2}\left[a^{1/2} f\langle\Delta v^i\rangle\right]_{,i} + \frac{1}{2}a^{-1/2}\left[a^{1/2} f\langle\Delta v^i \Delta v^j\rangle\right]_{,ij}, \quad (5.118)$$

where

$$\langle\Delta v^i\rangle \equiv T^i - \frac{1}{2}\Gamma^i_{jk}S^{jk}, \quad \langle\Delta v^i \Delta v^j\rangle \equiv S^{ij}. \quad (5.119)$$

Equations (5.119) define generalized diffusion coefficients. These transform as

$$\langle\Delta V^\lambda\rangle = \frac{\partial V^\lambda}{\partial v^i}T^i$$

$$- \frac{1}{2}\frac{\partial V^\mu}{\partial v^i}\frac{\partial V^\nu}{\partial v^j}\left(\frac{\partial V^\lambda}{\partial v^k}\frac{\partial v^l}{\partial V^\mu}\frac{\partial v^m}{\partial V^\nu}\Gamma^k_{lm} - \frac{\partial v^k}{\partial V^\nu}\frac{\partial v^m}{\partial V^\mu}\frac{\partial^2 V^\lambda}{\partial v^l \partial v^m}\right)S^{ij}$$

$$= \frac{\partial V^\lambda}{\partial v^i}T^i + \frac{1}{2}\left(\frac{\partial^2 V^\lambda}{\partial v^i \partial v^j} - \frac{\partial V^\lambda}{\partial v^k}\Gamma^k_{ij}\right)S^{ij}$$

$$= V^\lambda_{,i}T^i + \frac{1}{2}V^\lambda_{;ij}S^{ij}, \quad (5.120a)$$

$$\langle\Delta V^\mu \Delta V^\nu\rangle = \frac{\partial V^\mu}{\partial v^i}\frac{\partial V^\nu}{\partial v^j}S^{ij} = V^\mu_{,i}V^\nu_{,j}S^{ij}, \quad (5.120b)$$

where the μ, ν, λ indices are associated with the new (V) coordinate system and i, j, k with the old (v) system. Equations (5.120) can be used to express the new diffusion coefficients in terms of the T^i, S^{ij} that were already derived. The transformed encounter term is then

$$\left(\frac{\partial f}{\partial t}\right)_c = -a^{-1/2}\left[a^{1/2} f\langle\Delta V^\lambda\rangle\right]_{,\lambda} + \frac{1}{2}a^{-1/2}\left[a^{1/2} f\langle\Delta V^\mu \Delta V^\nu\rangle\right]_{,\mu\nu}, \quad (5.121)$$

where a is the determinant of the metric tensor $a^{\mu\nu}$ corresponding to the V^μ coordinate system.

We illustrate the transformation by defining new velocity-space coordinates (E, L, ϕ) with ϕ defined as in equation (5.109). The nonzero components of the metric tensor, and its inverse, are

$$a_{11} = v_r^{-2}, \quad a_{22} = \frac{v^2}{r^2 v_r^2}, \quad a_{33} = \frac{L^2}{r^2}, \quad a_{12} = -\frac{L}{r^2 v_r^2},$$

$$a^{11} = v^2, \quad a^{22} = r^2, \quad a^{33} = \frac{r^2}{L^2}, \quad a^{12} = L, \quad (5.122)$$

where $v_r \equiv \left[2(E - \Phi(r)) - L^2/r^2\right]^{1/2}$ is the radial velocity, and $a^{1/2} = L/(r^2 v_r)$. Using equation (5.120a), and the components Γ^k_{ij} of the affine connection given in equation (5.111), we find

$$\langle \Delta E \rangle = E_{,i} T^i + \frac{1}{2} E_{;ij} S^{ij}$$

$$= E_{,i} T^i + \frac{1}{2} \left[E_{,ij} - \Gamma_{ij}^k E_{,k} \right] S^{ij}$$

$$= v T^1 + \frac{1}{2} S^{11} + \frac{v^2}{2(1-\mu^2)} S^{22} + \frac{v^2}{2} (1-\mu^2) S^{33}. \qquad (5.123)$$

If we again assume that the field-star velocity distribution is isotropic, this simplifies to

$$\langle \Delta E \rangle = \Gamma' \left[v \frac{\partial \mathcal{H}}{\partial v} + \frac{1}{2} \frac{\partial^2 \mathcal{G}}{\partial v^2} + \frac{1}{v} \frac{\partial \mathcal{G}}{\partial v} \right]$$

$$= v \langle \Delta v_\parallel \rangle + \frac{1}{2} \langle (\Delta v_\parallel)^2 \rangle + \frac{1}{2} \langle (\Delta v_\perp)^2 \rangle, \qquad (5.124)$$

where $\Gamma' \equiv 4\pi G^2 \ln \Lambda$ as above. The remaining diffusion coefficients are easily shown to be

$$\langle (\Delta E)^2 \rangle = \Gamma' v^2 \frac{\partial^2 \mathcal{G}}{\partial v^2} = v^2 \langle (\Delta v_\parallel)^2 \rangle, \qquad (5.125\text{a})$$

$$L \langle \Delta L \rangle = \Gamma' \left[\frac{L^2}{v} \frac{\partial \mathcal{H}}{\partial v} + \frac{r^2}{2v} \frac{\partial \mathcal{G}}{\partial v} \right]$$

$$= \frac{L^2}{v} \langle \Delta v_\parallel \rangle + \frac{r^2}{4} \langle (\Delta v_\perp)^2 \rangle, \qquad (5.125\text{b})$$

$$\langle (\Delta L)^2 \rangle = \Gamma' \left[\frac{L^2}{v^2} \frac{\partial^2 \mathcal{G}}{\partial v^2} + \frac{1}{v} \left(r^2 - \frac{L^2}{v^2} \right) \frac{\partial \mathcal{G}}{\partial v} \right]$$

$$= \frac{L^2}{v^2} \langle (\Delta v_\parallel)^2 \rangle + \frac{1}{2} \left(r^2 - \frac{L^2}{v^2} \right) \langle (\Delta v_\perp)^2 \rangle, \qquad (5.125\text{c})$$

$$\langle \Delta E \Delta L \rangle = \Gamma' L \frac{\partial^2 \mathcal{G}}{\partial v^2} = L \langle (\Delta v_\parallel)^2 \rangle. \qquad (5.125\text{d})$$

Finally, writing equation (5.121) explicitly in terms of the new variables,

$$a^{1/2} \left(\frac{\partial f}{\partial t} \right)_c = -\frac{\partial}{\partial E} \left(a^{1/2} f \langle \Delta E \rangle \right) + \frac{1}{2} \frac{\partial^2}{\partial E^2} \left(a^{1/2} f \langle (\Delta E)^2 \rangle \right)$$

$$- \frac{\partial}{\partial L} \left(a^{1/2} f \langle \Delta L \rangle \right) + \frac{1}{2} \frac{\partial^2}{\partial L^2} \left(a^{1/2} f \langle (\Delta L)^2 \rangle \right)$$

$$+ \frac{\partial^2}{\partial E \partial L} \left(a^{1/2} f \langle \Delta E \Delta L \rangle \right). \qquad (5.126)$$

5.4 GRAVITATIONAL BROWNIAN MOTION

The Fokker–Planck equation is most often used to describe the evolution of stellar systems, after imposing the additional condition that the relaxation time is long compared with orbital periods. Before considering that approximation in detail, we first discuss a problem for which such a separation of timescales is not required.

Consider a massive body (which we will call the SBH) that is located near the center of a galaxy. The velocity of the SBH will undergo continuous small changes due to perturbations from other stars, and as a consequence, its position will fluctuate about the galaxy's center. This motion is similar in many ways to the Brownian motion of a pollen grain in a fluid—although as we will see, there are some distinguishing features. By analogy with the fluid case, we might expect the rms velocity of the SBH to be of order the "equipartition" value; that is,

$$V_{\rm rms} \approx \left(\frac{m}{M}\right)^{1/2} v_{\rm rms}, \tag{5.127}$$

where M is the SBH's mass, m is a typical stellar mass, and $v_{\rm rms}$ is the rms stellar velocity.

The most natural way to describe Brownian motion is via the **Langevin equation**, the equation of motion of a particle that moves in response to stochastic perturbing forces [306]. It was shown by A. Einstein [133, 134] and M. V. Smoluchowski [550, 551] that given certain reasonable conditions, Brownian motion could equally well be described via the Fokker–Planck equation; the function f that appears in that equation is interpreted as the probability density of the particle's velocity.

Following the approach of Einstein and Smoluchowski, we define $f(\boldsymbol{x}, \boldsymbol{v}, t)$ $d\boldsymbol{x}\, d\boldsymbol{v}$ as the probability that the SBH is located in phase-space volume element $d\boldsymbol{x}\, d\boldsymbol{v}$ near $(\boldsymbol{x}, \boldsymbol{v})$ at time t. The evolution equation for f is equation (5.75),

$$\frac{Df}{Dt} = \frac{\partial f}{\partial t} + \sum_i v_i \frac{\partial f}{\partial x_i} - \sum_i \frac{d\Phi}{dx_i}\frac{\partial f}{\partial v_i} = \left(\frac{\partial f}{\partial t}\right)_c, \tag{5.128}$$

where $\Phi(\boldsymbol{x}, t)$ is the gravitational potential in which the SBH moves.[11]

We are interested in steady-state solutions, $\partial f/\partial t = 0$. Suppose that the steady-state f is a function only of the energy E. By Jeans's theorem, the left-hand side of equation (5.128) is then identically zero, and finding a steady-state solution reduces to finding an $f(E) = f\left[v^2/2 + \Phi(r)\right]$ for which the right-hand side is also zero at every r and v.

When $f = f(r, v)$, the Fokker–Planck encounter term is given by equation (5.86). We force this term to be zero by requiring the velocity-space flux to vanish at each r and v:

$$0 = f(r, v)\left(v\langle\Delta v_\parallel\rangle + \frac{1}{2}\langle(\Delta v_\perp)^2\rangle\right) - \frac{1}{2v}\frac{\partial}{\partial v}\left(v^2 f(r, v)\langle(\Delta v_\parallel)^2\rangle\right). \tag{5.129}$$

The diffusion coefficients appearing in equation (5.129) are complicated functions of the field-star velocity distribution. But we can make use of equation (5.127), which states our expectation that the test-particle velocity is very low compared with σ_f. Expanding the diffusion coefficients about $v = 0$, as in sections 5.2.1

---

[11]When applied to the Brownian motion problem, equation (5.128), containing a fixed external potential, is sometimes called the **Klein–Kramers equation** [289].

and 5.2.2,

$$\langle \Delta v_\parallel \rangle = -Av + Bv^3 + \cdots ,$$
$$\langle (\Delta v_\parallel)^2 \rangle = C + Dv^2 + \cdots ,$$
$$\langle (\Delta v_\perp)^2 \rangle = 2(E + Fv^2) + \cdots ,$$

and inserting these expressions into equation (5.129) gives

$$0 = f(r, v) \left[v^2 (A + 2D - F) - Bv^4 + C - E \right] + (C + Dv^2) \frac{v}{2} \frac{\partial f}{\partial v}. \quad (5.130)$$

Setting $v = 0$ implies $C = E$. Keeping only the lowest-order terms in v,

$$0 = f(r, v)v^2 (A + 2D - F) + v \frac{C}{2} \frac{\partial f}{\partial v}$$
$$\approx Av^2 f + v \frac{C}{2} \frac{\partial f}{\partial v}, \quad (5.131)$$

where the second expression makes use of the fact that A, representing dynamical friction, is larger than D or F by factors of order $M/m \gg 1$. The solution is

$$f(r, v) = g(r)e^{-v^2/2\sigma^2}, \quad \sigma^2 = \frac{C}{2A}. \quad (5.132)$$

Finally, we obtain $f = f(E)$ by a judicious choice of $g(r)$:

$$f(r, v) = f_0 e^{-\Phi(r)/\sigma^2} e^{-v^2/2\sigma^2}$$
$$= f_0 e^{-E/2\sigma^2}. \quad (5.133)$$

Perhaps not surprisingly, the steady-state probability density is just the Boltzmann distribution.

Figure 5.6 shows the results of numerical experiments in which a massive body was placed initially at the center of a galaxy model. The figure verifies that the distribution of instantaneous velocities of the massive particle, $N(v)$, is well described by the Maxwell–Boltzmann formula,

$$N(v)dv = 4\pi v^2 \left(2\pi \langle v^2 \rangle / 3 \right)^{-3/2} \exp\left(-3v^2/2\langle v^2 \rangle \right) dv . \quad (5.134)$$

In figure 5.6, the rms velocity of the massive particle, which appears as an argument of equation (5.134), was computed directly from the measured velocities. We would like to evaluate this quantity theoretically from equation (5.132), which tells us that $\sigma = (C/2A)^{1/2}$. In sections 5.2.1 and 5.2.2 we computed A and C assuming that the field-star velocity distribution was Maxwellian. Using equations (5.37) and (5.59) from that section together with (5.133), we find

$$V_{\text{rms}}^2 = \frac{3C}{2A} = \frac{m}{M} v_{\text{rms}}^2, \quad (5.135)$$

precisely what we might have expected on the basis of "equipartition" [526]. Interestingly, V_{rms} is independent of the form of $\Phi(r)$, and is the same value that would have been obtained in the case of no confining potential.

However, in order to make statements about the SBH's position, we need to assume a form for $\Phi(r)$. At this point we cannot ignore one difference between the

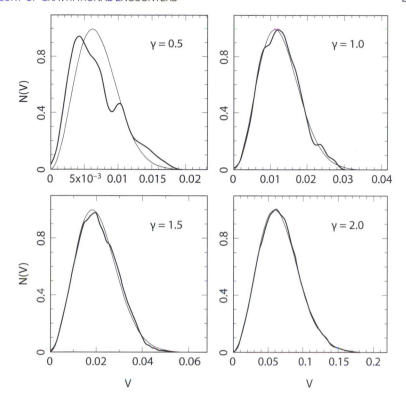

Figure 5.6 The thick curves show the measured distribution of instantaneous velocities, V,
of a massive particle at the center of four N-body models [360]. The galaxy
models differed in terms of the density of stars near the center: $n(r) \propto r^{-\gamma}$.
The thin curves are Maxwell–Boltzmann distributions; the rms velocity in the
Maxwell–Boltzmann formula (5.134) was computed directly from the time series
of measured velocities.

fluid and gravitational cases. An SBH at the center of a galaxy will carry with it a
retinue of bound stars: some fraction of the stars within its gravitational influence
radius $r_{\rm h} = GM_\bullet/\sigma_f^2$. Since $r_{\rm h} \approx r_{\rm m}$ (section 2.2), it follows that the stellar mass
within $r_{\rm h}$ is $\sim 2M_\bullet$, hence the bound mass is comparable to $M_\bullet$. (Exceptions would
be galaxies with a "hole" around the SBH, or some other peculiar distribution.)
As the SBH moves in response to gravitational perturbations, it will carry with
it the bound stars, increasing its effective mass in equation (5.135), and lowering
$V_{\rm rms}$. The effective potential in which the SBH moves will be the original stellar
potential minus the contribution from the bound stars. This argument suggests that
the effective potential is rather shallow; a reasonable form might be

$$\Phi(r) = \Phi(0) + \frac{1}{2}\Omega^2 r^2, \tag{5.136}$$

the potential corresponding to a constant-density core; here $\Omega = \left[4\pi G\rho/3\right]^{1/2}$,
with ρ given roughly by the density at the influence radius. For this $\Phi(r)$, the

distributions in position and velocity, equation (5.133), are both Maxwellian, and the relation between the rms displacements in radius and velocity is

$$\langle V^2 \rangle = \Omega^2 \langle r^2 \rangle, \tag{5.137}$$

a result that can also be derived directly from the differential equation (5.128) after inserting equation (5.136) for $\Phi(r)$ and taking moments over r and v.

The bound stars affect the Brownian motion in a second way [360]. Ignoring perturbations from the unbound stars, the center of mass of the bound subsystem is fixed; that is,

$$M V(t) + m \sum_{\text{bound}} v_i(t) = 0, \tag{5.138}$$

where V and v_i are, respectively, the velocity of the SBH and of a bound star, with respect to the center of mass of the bound system (assumed to be at rest). Assuming no correlations between the motions of the bound stars, equation (5.138) implies

$$\langle V^2 \rangle = \frac{M_b m}{M^2} \langle v^2 \rangle \tag{5.139}$$

with $\langle v^2 \rangle$ the mean square velocity of the bound stars and M_b their total mass. If the latter is comparable to M, as it is likely to be, then the contribution of the bound stars to the SBH's Brownian motion will be comparable to the contribution from unbound stars. Furthermore, since $\langle v^2 \rangle$ for the bound population is determined in large part by the gravitational force from the SBH itself, the dependence of the Brownian velocity on M will no longer be linear: instead one finds [360]

$$\langle V^2 \rangle \propto M^{-1/(3-\gamma)} \tag{5.140}$$

in models where the density of stars around the SBH falls off as $n(r) \propto r^{-\gamma}$.

Yet another difference between the fluid and gravitational cases is the possibility of non-Maxwellian field-star velocity distributions [84]. While the result that the probability density of v for a massive body is Maxwellian (equation 5.132) is robust—it is a consequence of the linear dependence of the dynamical friction coefficient on v for low v (or "Stokes's law" in the fluid case)—the *width* of that Maxwellian depends on the details of the field-star velocity distribution and can differ from the value in equation (5.135) which was derived assuming a Maxwellian $f(v_f)$. The more general result is [354]

$$\sigma^2 = \frac{m}{M} \frac{\int_0^\infty dv_f v_f f(v_f) \ln\left(1 + \frac{p_{\max}^2 v_f^4}{G^2 M^2}\right)}{\int_0^\infty dv_f \left(-df/dv_f\right) f(v_f) \ln\left(1 + \frac{p_{\max}^2 v_f^4}{G^2 M^2}\right)}. \tag{5.141}$$

Reasonable choices for $f(v_f)$ and $p_{\max}$ typically imply larger values of $\langle V^2 \rangle / \langle v^2 \rangle$ than in equation (5.135).

So far we have assumed a single value m for the field-star mass. Suppose instead that there is a distribution of field-star masses. If field stars of all masses are assumed to have the same velocity distribution, it is straightforward to show that

$$\langle V^2 \rangle = \frac{\tilde{m}}{M} \langle v^2 \rangle, \tag{5.142}$$

where $\tilde{m}$ is defined in equation (5.62b). If the field-star mass function contains even a small number of "massive perturbers" [422] with $m \gg \langle m \rangle$, the Brownian displacement of the test mass can be substantially increased.

Would the Brownian motion of an SBH be observable? In the case of the Milky Way SBH, the predicted amplitude of the velocity fluctuations is

$$(\Delta V)_{\text{rms}} \approx 0.2 \left(\frac{\tilde{m}}{1 \, M_\odot} \right)^{1/2} \left(\frac{M}{4 \times 10^6 \, M_\odot} \right)^{-1/2} \text{km s}^{-1} \tag{5.143}$$

and the rms displacement is

$$(\Delta r)_{\text{rms}} \approx 5 \left(\frac{\rho}{10^5 \, M_\odot \, \text{pc}^{-3}} \right)^{-1/2} \left(\frac{\tilde{m}}{1 \, M_\odot} \right)^{1/2} \left(\frac{M}{4 \times 10^6 \, M_\odot} \right)^{-1/2} \text{mpc}. \tag{5.144}$$

These are small numbers. A displacement as small as $\sim 0.01 \, \text{pc}$ would be very difficult to measure given uncertainties about the "center" of the Milky Way. In the case of (ΔV), current upper limits on the peculiar velocity of the radio source Sgr A* are a few kilometers per second [453].

Gravitational Brownian motion becomes much more interesting in the case of putative, intermediate-mass black holes (IBHs) at the centers of globular clusters (section 2.5). Setting $M \approx 10^3 \, M_\odot$, and using the fact that the most massive globular clusters have $\sigma_f \approx 20 \, \text{km s}^{-1}$, core radii $r_c \approx 0.5 \, \text{pc}$, and central densities $\rho \approx 10^5 \, M_\odot \, \text{pc}^{-3}$, we predict

$$(\Delta V)_{\text{rms}} \approx 1 \left(\frac{M}{10^3 \tilde{m}} \right)^{-1/2} \left(\frac{\sigma_f}{20 \, \text{km s}^{-1}} \right) \text{km s}^{-1}, \tag{5.145a}$$

$$(\Delta r)_{\text{rms}} \approx 0.03 \left(\frac{\rho}{10^5 \, M_\odot \, \text{pc}^{-3}} \right)^{-1/2} \left(\frac{M/\tilde{m}}{10^3 \, M_\odot} \right)^{-1/2} \left(\frac{\sigma_f}{20 \, \text{km s}^{-1}} \right) \text{pc}. \tag{5.145b}$$

Thus $(\Delta r)_{\text{rms}} \approx 0.1 r_c$. As discussed in chapter 2, there are currently no *kinematical* data that compel belief in IBHs in globular clusters. An alternative approach is to search for the (possibly very weak) radio and X-ray emission that would result from accretion of ambient gas onto a compact object. Such studies need to examine all sources within a distance $\sim (\Delta r)_{\text{rms}}$ from the cluster center. If and when a candidate source is identified, its measured displacement could be used to construct a probability function for its mass [360].

5.5 ORBIT-AVERAGED FOKKER–PLANCK EQUATION

The Brownian motion problem discussed above is one of only a few in galaxy dynamics for which the Fokker–Planck equation admits an exact solution. Most applications of the Fokker–Planck equation to stellar systems invoke an additional approximation: that the relaxation time is long compared with orbital periods. In this approximation, the phase-space density f is assumed to satisfy Jeans's theorem,

$f(\boldsymbol{x}, \boldsymbol{v}) = f(E, I_2, \ldots)$, at any given time, but the functional form of f is allowed to change gradually with time in response to encounters. Because a star is assumed to complete many orbits during the time that f changes appreciably, the diffusion coefficients that appear in the (local) Fokker–Planck equation can be averaged over each orbit, resulting in an evolution equation that depends only on the integrals of motion and the time as independent variables—the **orbit-averaged Fokker–Planck equation** [323, 91].

We begin with equation (5.118), the local Fokker–Planck encounter term expressed in generalized velocity-space coordinates v^i:

$$a^{1/2} \left(\frac{\partial f}{\partial t} \right)_c = \left[a^{1/2} f \langle \Delta v^i \rangle \right]_{,i} + \frac{1}{2} \left[a^{1/2} f \langle \Delta v^i \Delta v^j \rangle \right]_{,ij} \qquad (5.146)$$

with $a = \det(a_{ij})$, the determinant of the velocity-space metric. Now, from a theorem of integral calculus [258], we also know that $a^{1/2} d\boldsymbol{v}$ is an invariant velocity-space volume element. This means that the number of stars in phase-space volume element $d\boldsymbol{x}\, d\boldsymbol{v}$ is given by

$$dN = f(\boldsymbol{x}, \boldsymbol{v}, t) d\boldsymbol{x}\, a^{1/2} d\boldsymbol{v}. \qquad (5.147)$$

Suppose we choose as velocity-space variables the integrals of motion in the unperturbed problem. Then the unperturbed f is constant over the configuration-space volume filled by an orbit. Integrating over this volume, we obtain

$$N(E, I_2, \ldots) = f(E, I_2, \ldots) \int_{\mathrm{orb}} a^{1/2} d\boldsymbol{x}, \qquad (5.148)$$

where $N(E, I_2, \ldots) dE\, dI_2 \ldots$ is the number of stars in the interval $dE\, dI_2 \ldots$ centered at $(E, I_2, \ldots)$.

We now invoke our assumption that the relaxation time is long compared with orbital periods, and approximate f in equation (5.146) by $f(E, I_2, \ldots)$; in other words we ignore the small variations in f along an orbit due to encounters. The orbit-averaged Fokker–Planck equation is obtained by integrating equation (5.146) over the configuration-space volume filled by an orbit:

$$\int_{\mathrm{orb}} a^{1/2} \left(\frac{\partial f}{\partial t} \right)_c d\boldsymbol{x} = - \int_{\mathrm{orb}} \frac{\partial}{\partial E} \left(a^{1/2} f \langle \Delta E \rangle \right) d\boldsymbol{x} + \cdots. \qquad (5.149)$$

We define the left-hand side as $(\partial N / \partial t)_{\mathrm{enc}}$. Each term on the right-hand side is treated in the same manner and so we consider only the first. Invoking the Leibniz–Reynolds transport theorem [436], we can exchange the order of integration and differentiation:

$$\left(\frac{\partial N}{\partial t} \right)_{\mathrm{enc}} = - \frac{\partial}{\partial E} \left(f \int_{\mathrm{orb}} \langle \Delta E \rangle a^{1/2} d\boldsymbol{x} \right) + \cdots. \qquad (5.150)$$

We then define the **orbit average** of the diffusion coefficient $\langle \Delta E \rangle$ as

$$\langle \Delta E \rangle_t \equiv \frac{\int_{\mathrm{orb}} \langle \Delta E \rangle a^{1/2} d\boldsymbol{x}}{\int_{\mathrm{orb}} a^{1/2} d\boldsymbol{x}}. \qquad (5.151)$$

Evidently, $\langle \Delta E \rangle_t$ is an average of $\langle \Delta E \rangle$ that weights each region of the orbit by the time the star spends in that region. Then

$$\left(\frac{\partial N}{\partial t} \right)_{\text{enc}} = -\frac{\partial}{\partial E} \left[f \left(\int_{\text{orb}} a^{1/2} dx \right) \langle \Delta E \rangle_t \right] + \cdots$$

$$= -\frac{\partial}{\partial E} (N \langle \Delta E \rangle_t) + \cdots . \tag{5.152}$$

Repeating for the other terms on the right-hand side of equation (5.149) gives the orbit-averaged Fokker–Planck equation,

$$\left(\frac{\partial N}{\partial t} \right)_{\text{enc}} = -\frac{\partial}{\partial E} (N \langle \Delta E \rangle_t) + \frac{1}{2} \frac{\partial^2}{\partial E^2} \left(N \langle (\Delta E)^2 \rangle_t \right) - \cdots . \tag{5.153}$$

Before proceeding further, an important caveat is in order. The orbit-averaged Fokker–Planck equation is a good example of what computer scientists call a "kludge."[12] The equation is nonlocal only in a very limited sense. While it is true that the diffusion coefficients are averaged over the volume filled by an orbit, no account is taken of the fact that the perturbations acting on a test star come from stars in regions that may have very different properties than those near the test star. Quantities like $\langle \Delta v_\perp \rangle$, from which the orbit-averaged coefficients are derived, are computed under the assumption that the local properties (density, velocity distribution) apply everywhere, in spite of the fact that those same quantities are assumed to vary along the orbit. Because of this basic inconsistency, there is no sense in which the orbit-averaged Fokker–Planck equation tends (in some large-N limit, say), to a correct description of the evolution of an inhomogeneous system.

5.5.1 Phase-space density a function of E and L

We can apply the formalism derived in the previous section to the case of a spherical galaxy in which $f = f(E, L)$. Let the third velocity-space variable be the angle ϕ defined in equation (5.109). The determinant of the metric tensor is

$$a^{1/2} = \frac{L}{r^2 v_r}, \tag{5.154}$$

and the orbit averages look like

$$\langle \Delta E \rangle_t = \frac{\int_{r_-}^{r_+} 4\pi r^2 dr (L/r^2 v_r) \langle \Delta E \rangle}{\int_{r_-}^{r_+} 4\pi r^2 dr (L/r^2 v_r)}$$

$$= \frac{2}{P} \int_{r_-}^{r_+} \frac{dr}{v_r} \langle \Delta E \rangle. \tag{5.155}$$

The quantities $r_-(E, L)$ and $r_+(E, L)$ are the turning points of the orbit, that is, the roots of

$$0 = 2 [\Phi(r) - E] - \frac{L^2}{r^2}, \tag{5.156}$$

[12]"An ill-assorted collection of poorly matching parts, forming a distressing whole" [212].

and P is the radial period,

$$P(E, L) = 2 \int_{r_-}^{r_+} \frac{dr}{v_r}. \tag{5.157}$$

The relation between $N(E, L)$ and $f(E, L)$ (the dependence on m and t is understood) is

$$N(E, L)dE\, dL\, d\phi = 2f(E, L)\left[\int_{r_-}^{r_+} dr 4\pi r^2 a^{1/2}\right] dE\, dL\, d\phi; \tag{5.158}$$

the factor of two accounts for the fact that orbits of a given E and L can have two signs of the radial velocity at any radius. Thus

$$N(E, L)dE\, dL = 16\pi^2 L f(E, L)\left[\int_{r_-}^{r_+} \frac{dr}{v_r}\right] dE\, dL$$

$$= 8\pi^2 L P(E, L) f(E, L) dE\, dL. \tag{5.159}$$

In terms of these quantities, the orbit-averaged Fokker–Planck equation for $f = f(E, L)$ is

$$\frac{\partial N}{\partial t} = -\frac{\partial}{\partial E}\left(N\langle\Delta E\rangle_t\right) + \frac{1}{2}\frac{\partial^2}{\partial E^2}\left(N\langle(\Delta E)^2\rangle_t\right) - \frac{\partial}{\partial L}\left(N\langle\Delta L\rangle_t\right)$$

$$+ \frac{1}{2}\frac{\partial^2}{\partial L^2}\left(N\langle(\Delta L)^2\rangle_t\right) + \frac{\partial^2}{\partial E \partial L}\left(N\langle\Delta E \Delta L\rangle_t\right). \tag{5.160}$$

Using the formulas (5.125) derived in the previous section for the local (E, L) diffusion coefficients, it is straightforward to derive explicit expressions for the orbit-averaged diffusion coefficients that appear in equation (5.160).

Rather than give those expressions here,[13] we choose to make yet another change of variables [91]. Instead of (E, L), we adopt $(\mathcal{E}, \mathcal{R})$, where

$$\mathcal{E} = -E = -\frac{v^2}{2} + \psi(r), \quad \mathcal{R} = \frac{L^2}{L_c^2(\mathcal{E})}; \tag{5.161}$$

here $\psi(r) = -\Phi(r)$ and $L_c(\mathcal{E})$ is the maximum angular momentum of an orbit of energy $\mathcal{E}$—that is, the angular momentum of a circular orbit. $\mathcal{E}$ is the binding energy and has the desirable property that it is positive for bound orbits. $\mathcal{R}$ is a dimensionless angular momentum variable; for every positive $\mathcal{E}$, the range of $\mathcal{R}$ is $[0,1]$, with the endpoints of this interval corresponding to radial and circular orbits, respectively. Clearly,

$$L_c^2(\mathcal{E}) = -r_c^3 \frac{d\psi}{dr}\bigg|_{r_c}, \tag{5.162}$$

where $r_c(\mathcal{E})$ is the radius of the circular orbit of energy $\mathcal{E}$; $r_c(\mathcal{E})$ is determined by the equation

$$2[\psi(r_c) - \mathcal{E}] + r_c\frac{d\psi}{dr}\bigg|_{r_c} = 0, \tag{5.163}$$

---

[13]They are given by various authors, e.g., [491].

while the two turning points $r_\pm(\mathcal{E})$ are the solutions to

$$2\left[\psi(r_\pm) - \mathcal{E}\right] - \frac{L^2}{r_\pm^2} = 0. \tag{5.164}$$

The velocity-space volume element is

$$2\pi a^{1/2} d\mathcal{E} \, d\mathcal{R} = 2\pi \frac{L_c^2(\mathcal{E})}{r^2 v_r} d\mathcal{E} \, d\mathcal{R} \tag{5.165}$$

and the definition of an orbit-averaged quantity is the same as in equation (5.151). The number density in $(\mathcal{E}, \mathcal{R})$ space is related to f by

$$\begin{aligned}
N(\mathcal{E}, \mathcal{R}) &= 4\pi^2 L_c^2(\mathcal{E}) P(\mathcal{E}, \mathcal{R}) f(\mathcal{E}, \mathcal{R}) \\
&\equiv \mathcal{J}(\mathcal{E}, \mathcal{R}) f(\mathcal{E}, \mathcal{R}).
\end{aligned} \tag{5.166}$$

The orbit-averaged Fokker–Planck equation expressed in terms of $(\mathcal{E}, \mathcal{R})$ is obtained by simply replacing E by $\mathcal{E}$ and L by R in equation (5.160), although of course the new diffusion coefficients need to be derived. Clearly $\langle \Delta \mathcal{E} \rangle = -\langle \Delta E \rangle$, $\langle (\Delta \mathcal{E})^2 \rangle = \langle (\Delta E)^2 \rangle$, and using the transformation equations (5.120),

$$\begin{aligned}
\langle \Delta \mathcal{R} \rangle &= \frac{g'}{g} \mathcal{R} \langle \Delta E \rangle + 2gL \langle \Delta L \rangle + \frac{g''}{2g} \mathcal{R} \langle (\Delta E)^2 \rangle + g \langle (\Delta L)^2 \rangle \\
&\quad + 2g' L \langle \Delta E \Delta L \rangle,
\end{aligned} \tag{5.167a}$$

$$\langle (\Delta \mathcal{R})^2 \rangle = \left(\frac{g'}{g}\right)^2 \mathcal{R}^2 \langle (\Delta E)^2 \rangle + 4g\mathcal{R} \langle (\Delta L)^2 \rangle + 4g' \mathcal{R} L \langle \Delta E \Delta L \rangle, \tag{5.167b}$$

$$\langle \Delta \mathcal{E} \Delta \mathcal{R} \rangle = -\frac{g'}{g} \mathcal{R} \langle (\Delta E)^2 \rangle - 2gL \langle \Delta E \Delta L \rangle. \tag{5.167c}$$

We have defined $g(E) = 1/L_c^2(E)$; a prime denotes differentiation with respect to E.

It is convenient to write the Fokker–Planck equation in "flux-conservation" form:

$$\frac{\partial N}{\partial t} = -\frac{\partial \mathcal{F}_\mathcal{E}}{\partial \mathcal{E}} - \frac{\partial \mathcal{F}_\mathcal{R}}{\partial \mathcal{R}}, \tag{5.168}$$

where the components of the flux vector in $(\mathcal{E}, \mathcal{R})$ space are given by

$$\begin{aligned}
-\mathcal{F}_\mathcal{E} &= D_{\mathcal{E}\mathcal{E}} \frac{\partial f}{\partial \mathcal{E}} + D_{\mathcal{E}\mathcal{R}} \frac{\partial f}{\partial \mathcal{R}} + m D_\mathcal{E} f, \\
-\mathcal{F}_\mathcal{R} &= D_{\mathcal{R}\mathcal{E}} \frac{\partial f}{\partial \mathcal{E}} + D_{\mathcal{R}\mathcal{R}} \frac{\partial f}{\partial \mathcal{R}} + m D_\mathcal{R} f,
\end{aligned} \tag{5.169}$$

and

$$D_{\mathcal{E}} = -\mathcal{J}\langle\Delta\mathcal{E}\rangle_t + \frac{1}{2}\frac{\partial}{\partial\mathcal{E}}\left(\mathcal{J}\langle(\Delta\mathcal{E})^2\rangle_t\right) + \frac{1}{2}\frac{\partial}{\partial\mathcal{R}}\left(\mathcal{J}\langle\Delta\mathcal{E}\Delta\mathcal{R}\rangle_t\right),$$

$$D_{\mathcal{E}\mathcal{E}} = \frac{1}{2}\mathcal{J}\langle(\Delta\mathcal{E})^2\rangle_t,$$

$$D_{\mathcal{R}} = -\mathcal{J}\langle\Delta\mathcal{R}\rangle_t + \frac{1}{2}\frac{\partial}{\partial\mathcal{R}}\left(\mathcal{J}\langle(\Delta\mathcal{R})^2\rangle_t\right) + \frac{1}{2}\frac{\partial}{\partial\mathcal{E}}\left(\mathcal{J}\langle\Delta\mathcal{E}\Delta\mathcal{R}\rangle_t\right),$$

$$D_{\mathcal{R}\mathcal{R}} = \frac{1}{2}\mathcal{J}\langle(\Delta\mathcal{R})^2\rangle_t,$$

$$D_{\mathcal{E}\mathcal{R}} = D_{\mathcal{R}\mathcal{E}} = \frac{1}{2}\mathcal{J}\langle\Delta\mathcal{E}\Delta\mathcal{R}\rangle_t. \tag{5.170}$$

When computing the diffusion coefficients in the anisotropic Fokker–Planck equation, it is a common practice to equate the field-star distribution function with $\overline{f}$, an average over angular momentum of the test-star distribution. There are various ways to do this. One possibility is to average f over L at each (E, r). Using equation (5.165) for the $(\mathcal{E}, \mathcal{R})$ velocity-space volume element, the result is [89]

$$\overline{f}(\mathcal{E}, r) \equiv \frac{1}{2\mathcal{R}_{\max}^{1/2}}\int_0^{\mathcal{R}_{\max}}\frac{f(\mathcal{E}, \mathcal{R})}{(\mathcal{R}_{\max} - \mathcal{R})^{1/2}}d\mathcal{R}, \tag{5.171}$$

where $\mathcal{R}_{\max}(\mathcal{E}, r) = 2r^2\left[\psi(r) - \mathcal{E}\right]/L_c^2(\mathcal{E})$ is the maximum allowed value of $\mathcal{R}$ for an orbit of energy $\mathcal{E}$ which passes through the radius r. This $\overline{f}$ has the computationally undesirable property that it depends on r as well as $\mathcal{E}$. A second, simpler choice is [491]

$$\overline{f}(\mathcal{E}) \equiv \int_0^1 f(\mathcal{E}, \mathcal{R})d\mathcal{R}. \tag{5.172}$$

The second definition is less well justified, but it greatly simplifies the computation of the orbit-averaged diffusion coefficients. In what follows we will allow $\overline{f}$ to have the more general dependence implied by equation (5.171).

One more item remains to be dealt with before writing explicit expressions for the terms in equation (5.170). We would like to allow for a continuous distribution of stellar masses, rather than the discrete distribution that was assumed in writing equations (5.107). Accordingly, let $f(x, v, m)dm$ be the number density in phase space of stars with masses in the range m to $m + dm$. We define the first two moments over mass of f as

$$\nu(E, L, t) \equiv \int f(E, L, m, t)m\, dm, \tag{5.173a}$$

$$\mu(E, L, t) \equiv \int f(E, L, m, t)m^2 dm, \tag{5.173b}$$

and $\overline{\nu}$ and $\overline{\mu}$ are the angular-momentum averages of ν and μ.

The flux coefficients that appear in the $(\mathcal{E}, \mathcal{R})$ orbit-averaged equation are then

$$D_{\mathcal{E}} = -8\pi^2 \Gamma' L_c^2 \int \frac{dr}{v_r} N_1,$$

$$D_{\mathcal{E}\mathcal{E}} = \frac{8\pi^2}{3} \Gamma' L_c^2 \int \frac{dr}{v_r} v^2 \left(M_0 + M_2\right),$$

$$D_{\mathcal{R}} = -16\pi^2 \Gamma' R r_c^2 \int \frac{dr}{v_r} \left(1 - \frac{v_c^2}{v^2}\right) N_1,$$

$$D_{\mathcal{R}\mathcal{R}} = \frac{16\pi^2}{3} \Gamma' R \int \frac{dr}{v_r} \left\{ 2\frac{r^2}{v^2} \left[v_t^2 \left(\frac{v^2}{v_c^2} - 1\right)^2 + v_r^2 \right] M_0 \right.$$
$$\left. + 3r^2 \frac{v_r^2}{v^2} M_1 + \frac{r^2}{v^2} \left[2v_t^2 \left(\frac{v^2}{v_c^2} - 1\right)^2 - v_r^2 \right] M_2 \right\},$$

$$D_{\mathcal{E}\mathcal{R}} = \frac{16\pi^2}{3} \Gamma' L^2 \int \frac{dr}{v_r} \left(\frac{v^2}{v_c^2} - 1\right) \left(M_0 + M_2\right). \tag{5.174}$$

In each of these expressions the radial integral extends from r_- to r_+. We have defined v_c, the velocity of a circular orbit, $v_c^2 = L_c^2/r_c^2$, and v_t, the tangential velocity, $v_t^2 = v^2 - v_r^2$. The functions M_0, N_1, M_1, and M_2 are

$$M_0(\mathcal{E}, r) = 4\pi \int_0^{\mathcal{E}} \overline{\mu}(\mathcal{E}', r) d\mathcal{E}',$$

$$N_1(\mathcal{E}, r) = 4\pi \int_{\mathcal{E}}^{\psi} \left(\frac{\psi - \mathcal{E}'}{\psi - \mathcal{E}}\right)^{1/2} \overline{\nu}(\mathcal{E}', r) d\mathcal{E}',$$

$$M_1(\mathcal{E}, r) = 4\pi \int_{\mathcal{E}}^{\psi} \left(\frac{\psi - \mathcal{E}'}{\psi - \mathcal{E}}\right)^{1/2} \overline{\mu}(\mathcal{E}', r) d\mathcal{E}',$$

$$M_2(\mathcal{E}, r) = 4\pi \int_{\mathcal{E}}^{\psi} \left(\frac{\psi - \mathcal{E}'}{\psi - \mathcal{E}}\right)^{3/2} \overline{\mu}(\mathcal{E}', r) d\mathcal{E}'. \tag{5.175}$$

These expressions will underlie much of the discussion of loss-cone dynamics in chapter 6 and of collisional evolution in chapter 7.

5.5.2 Phase-space density a function of E; the Bahcall–Wolf solution

After a few relaxation times, the distribution of velocities in a galaxy will be approximately isotropic, implying $f = f(E)$. This seemingly innocuous statement begs the question of whether the relaxation time at the center of any galaxy is short enough for isotropization to occur! For instance, in our own galaxy—which contains one of the densest nuclei known—the relaxation time appears to nowhere fall below about 10^{10} yr (figure 7.1b). But the observed trend of $T_r(r_h)$ with galaxy luminosity, figure 3.1, suggests that nuclear relaxation times may be shorter in galaxies with spheroids fainter than that of the Milky Way. And as discussed in chapter 3, it is possible that *collisionless* processes acting during the formation of a galaxy could have resulted in velocity distributions that are close to isotropic.

But unless the relaxation time is much shorter than the age of the galaxy—something which may not be true in any nucleus—there is no strong justification for ignoring the dependence of f on angular momentum. The approximation $f = f(E, t)$ can nevertheless be motivated by the following argument. Roughly speaking, the mean distance of a star from the center of a galaxy is determined by its energy, while its distance of closest approach to the center is determined by both its energy and its angular momentum. If we are primarily concerned with the evolution of a galaxy's density profile due to encounters, then to a first approximation, it is adequate to look at the evolution of $N(E)$, ignoring the dependence of f on L. On the other hand, if our primary concern is the rate of supply of stars to the very center (as in the next chapter), then changes in L are crucial.

There are various ways to derive the isotropic version of equation (5.160), and the reader who has followed the developments of the last few sections should be able to figure out at least one or two. But perhaps the easiest way is to integrate equation (5.160) over angular momenta. The result, expressed in flux-conservation form, is

$$\frac{\partial N}{\partial t} = -\frac{\partial \mathcal{F}_\mathcal{E}}{\partial \mathcal{E}}, \tag{5.176a}$$

$$\mathcal{F}_\mathcal{E} = -D_{\mathcal{E}\mathcal{E}}\frac{\partial f}{\partial \mathcal{E}} - m D_\mathcal{E} f, \tag{5.176b}$$

$$D_\mathcal{E} = -16\pi^3\Gamma' \int_\mathcal{E}^{\psi(0)} d\mathcal{E}' \, p(\mathcal{E}')v(\mathcal{E}', t), \tag{5.176c}$$

$$D_{\mathcal{E}\mathcal{E}} = 16\pi^3\Gamma' \left[q(\mathcal{E}) \int_0^\mathcal{E} d\mathcal{E}' \mu(\mathcal{E}', t) + \int_\mathcal{E}^{\psi(0)} d\mathcal{E}' q(\mathcal{E}')\mu(\mathcal{E}', t) \right]. \tag{5.176d}$$

Here N, the number of stars per unit energy, is related to f via

$$N(\mathcal{E}, m, t) = \int_0^{L_c(\mathcal{E})} N(\mathcal{E}, L, m, t)dL \tag{5.177a}$$

$$= 4\pi^2 p(\mathcal{E}) f(\mathcal{E}, m, t), \tag{5.177b}$$

and the function $p(\mathcal{E})$, which has already been defined in equation (3.52), is

$$p(\mathcal{E}) = \int_0^{L_c^2(\mathcal{E})} P(\mathcal{E}, L)dL^2 = 4\int_0^{\psi^{-1}(\mathcal{E})} v(r, \mathcal{E})r^2dr; \tag{5.178}$$

$\psi^{-1}(\mathcal{E})$ is the inverse of the potential function $\psi(r) = -\Phi(r)$, and $v = [2\psi(r) - 2\mathcal{E}]^{1/2}$. The function $q(\mathcal{E})$ is

$$q(\mathcal{E}) = \int_\mathcal{E}^{\psi(0)} d\mathcal{E}' \, p(\mathcal{E}') = \frac{4}{3}\int_0^{\psi^{-1}(\mathcal{E})} dr \, r^2 v^3(r, \mathcal{E}). \tag{5.179}$$

Orbit averages now take the form

$$\langle \Delta\mathcal{E} \rangle_t = p^{-1}\int_0^{\psi^{-1}(\mathcal{E})} \langle \Delta\mathcal{E} \rangle vr^2dr. \tag{5.180}$$

The functions $\nu(\mathcal{E}, t)$ and $\mu(\mathcal{E}, t)$ are defined as in equation (5.173):

$$\nu(\mathcal{E}, t) \equiv \int f(\mathcal{E}, m, t)m\, dm, \tag{5.181a}$$

$$\mu(\mathcal{E}, t) \equiv \int f(\mathcal{E}, m, t)m^2 dm; \tag{5.181b}$$

thus ν is the mass density in phase space and $\mu = \tilde{m}\nu$, with $\tilde{m}$ defined as in equation (5.62b).

Equations (5.176)–(5.179) are simple enough that one is tempted to look for steady-state solutions, $\partial N/\partial t = 0$. Analytic solutions do not exist for arbitrary $\psi(r)$, but one special case is especially interesting: motion near an SBH, where the gravitational potential is

$$\psi(r) = \frac{GM_\bullet}{r}.$$

In this case, the functions $p(\mathcal{E})$ and $q(\mathcal{E})$ become

$$p(\mathcal{E}) = \frac{\sqrt{2\pi}}{4} G^3 M_\bullet^3 \mathcal{E}^{-5/2}, \tag{5.182a}$$

$$q(\mathcal{E}) = \frac{\sqrt{2\pi}}{6} G^3 M_\bullet^3 \mathcal{E}^{-3/2}. \tag{5.182b}$$

(Since $P(\mathcal{E}, L) = P(\mathcal{E})$ in this case, $p(\mathcal{E})$ can be computed most simply from equation (3.56), $p(\mathcal{E}) = L_c^2(\mathcal{E})P(\mathcal{E})$, using $L_c^2 = GM_\bullet a$, $P = 2\pi a^{3/2}/\sqrt{GM_\bullet}$, and $\mathcal{E} = GM_\bullet/(2a)$.) We further assume a single mass group so that $\nu = mf$ and $\mu = m^2 f$.

Since $\psi(r)$, $p(\mathcal{E})$ and $q(\mathcal{E})$ are all power laws, it is natural to attempt a solution of the form $f(\mathcal{E}) = f_0 \mathcal{E}^p$. With this ansatz, equations (5.176) and (5.182) imply

$$\mathcal{F}_\mathcal{E}(\mathcal{E}) = -4\sqrt{2}\pi^4 \Gamma G^3 M_\bullet^3 f_0^2 \mathcal{E}^{2p-3/2} g(p), \tag{5.183}$$

$$g(p) = \frac{2(4p-1)}{3 - 5p - 4p^2 + 4p^3}, \quad -1 < p < 1/2,$$

with $\Gamma \equiv m^2 \Gamma' = 4\pi^2 G^2 m^2 \ln \Lambda$.

Now it is apparent from equation (5.176a) that two different sorts of steady-state solution might exist:

I. $\dfrac{\partial \mathcal{F}_\mathcal{E}}{\partial \mathcal{E}} = 0.$

II. $\mathcal{F}_\mathcal{E} = 0.$

A solution of the first type is a "constant-flux" solution, while the second is a "zero-flux" solution; of course, the second is a special case of the first. The flow of stars into the SBH implies an outward flux of energy, and so it would seem natural to consider solutions of type I [417]. According to equation (5.183), a constant flux requires $p = 3/4$; but for this value of p, the integrals that define $D_\mathcal{E}$ and $D_{\mathcal{E}\mathcal{E}}$ are divergent: the constant flux is an infinite flux!

It is possible that we were led to this unphysical result by our assumption of a power-law form for $f(\mathcal{E})$. But there is another way out: we can set the flux identically to zero by choosing p such that $g(p) = 0$, that is, by setting $p = 1/4$. This value of p *is* mathematically acceptable.

The configuration-space density corresponding to $f(\mathcal{E}) = f_0 \mathcal{E}^{1/4}$ is

$$n(r) = \int f(v, r)(4\pi v^2 dv) = 4\pi \sqrt{2} f_0 \int_0^{\psi(r)} \mathcal{E}^{1/4} \sqrt{\psi(r) - \mathcal{E}} \, d\mathcal{E}$$

or

$$n(r) \propto r^{-7/4}. \tag{5.184}$$

This solution ($f \propto \mathcal{E}^{1/4}$, $n \propto r^{-7/4}$) is often referred to as a **Bahcall–Wolf cusp** [14]. As we will see in chapter 7, this scale-free solution differs only slightly from solutions that impose more proper boundary conditions near the SBH.

5.5.3 Axisymmetric nuclei

The orbit-averaged Fokker–Planck equation (5.153) was derived from the local equation (5.126) by an integration over the configuration-space volume filled by an orbit. That operation requires, at the very least, some knowledge about the unperturbed orbits—what regions they fill, for instance. From the point of view of computational feasibility, one requires considerably more information: analytic expressions for the isolating integrals of motion in terms of position and velocity. Without such expressions, the elegant formalism presented above for the derivation of the generalized diffusion coefficients is essentially useless.

Unfortunately, spherical potentials are the most general for which one has access to analytic expressions for all the isolating integrals. In axisymmetric potentials, motion always respects two integrals: the energy, and the component L_z of the angular momentum along the symmetry axis. Two isolating integrals are not sufficient to make orbits regular in a three-degrees-of-freedom system, but as discussed in section 3.4, numerical integrations reveal that most orbits in axisymmetric potentials *are* regular, implying the existence of a third integral. However, except in the limiting case of motion near the SBH, as treated in section 4.4.2, simple and general expressions for that integral are not available, nor does the extra integral exist for all orbits.

Because E and L_z are the only analytic integrals in most axisymmetric potentials, Fokker–Planck treatments of axisymmetric galaxies have generally assumed $f = f(E, L_z)$ [205, 132, 162]. At first blush, such an approach might seem no less justified than setting $f = f(E)$ in the Fokker–Planck equation for spherical galaxies. But there is one important difference. Gravitational encounters drive the local velocity distribution toward isotropy. In the spherical geometry, this is equivalent to the statement $f \rightarrow f(E, t)$. No such special role is played by a distribution function of the form $f(E, L_z, t)$, since the implied three-dimensional velocity distribution is *an*isotropic: only the two velocity dispersions in the meridional plane, σ_ϖ and σ_z, are required to be equal when $f = f(E, L_z)$ (section 3.4). Forcing f to remain a function of E and L_z in the face of encounters therefore places an arbitrary and probably unphysical constraint on the manner in which f is allowed to evolve. Nevertheless, it is the best that can be done in general.

As always, there are many ways to proceed. One starting point would be the spherical polar coordinates of equation (5.109). But it is probably simpler to define new coordinates that reflect the axisymmetric geometry [132]. Keeping with the notation defined in chapter 3, let (ϖ, z, φ) be configuration-space coordinates, where $\varpi^2 = x^2 + y^2$ and, as always, the z-axis is the axis of symmetry. As velocity-space variables we take

$$v^1 = v, \quad v^2 = \psi, \quad v^3 = v_\varphi. \tag{5.185}$$

Here $v^2 = v_x^2 + v_y^2 + v_z^2$, $v_\varphi = \boldsymbol{v} \cdot \boldsymbol{e}_\varphi = (y v_x - x v_y)/\varpi$, and $\tan \psi = v_\varpi/v_z$ specifies the orientation of the velocity vector in the meridional ϖ–z plane. In terms of these variables, $L_z = \varpi v_\varphi$. Assuming no integrals aside from E and L_z, the velocity-space volume element is simply

$$d^3v = \frac{2\pi}{\varpi} dE \, dL_z \tag{5.186}$$

and the relation between $N(E, L_z)$ and $f(E, L_z)$ (the dependence on m and t being understood) is

$$
\begin{aligned}
N(E, L_z) \, dE \, dL_z &= \int d\boldsymbol{x} \, (f \, d\boldsymbol{v}) \\
&= \int d\varpi \, 2\pi\varpi \int dz \times \frac{2\pi \, dE \, dL_z}{\varpi} f(E, L_z) \\
&= A(E, L_z) f(E, L_z) dE \, dL_z, \tag{5.187}
\end{aligned}
$$

where

$$A(E, L_z) = 4\pi^2 \int \int_M d\varpi \, dz \tag{5.188}$$

and the subscript M denotes the region in the meridional plane that is accessible to an orbit with specified E and L_z.

The nonzero components of the metric tensor, and its inverse, are

$$
a_{11} = v^2/v_M^2, \quad a_{22} = v_M^2, \quad a_{33} = v^2/v_M^2, \quad a_{13} = a_{31} = -v v_\varphi/v_M^2,
$$
$$
a^{11} = 1, \quad a^{22} = v_M^{-2}, \quad a^{33} = 1, \quad a^{13} = a^{31} = v_\varphi/v, \tag{5.189}
$$

where $v_M = (v_\varpi^2 + v_z^2)^{1/2}$ is the velocity parallel to the meridional plane. The nonzero components of the affine connection are

$$
\Gamma_{11}^1 = -v_\varphi^2/(v v_M^2), \quad \Gamma_{22}^1 = -v_M^2/v, \quad \Gamma_{31}^1 = v_\varphi/v_M^2,
$$
$$
\Gamma_{33}^1 = -v/v_M^2, \quad \Gamma_{12}^2 = v/v_M^2, \quad \Gamma_{32}^2 = -v_\varphi/v_M^2. \tag{5.190}
$$

The components of T^i are

$$
T^1 = \Gamma' \left[\frac{\partial \mathcal{H}}{\partial v} + \frac{v_\varphi}{v} \frac{\partial \mathcal{H}}{\partial v_\varphi} \right], \quad T^2 = \Gamma' \frac{1}{v_M^2} \frac{\partial \mathcal{H}}{\partial \Psi}, \quad T^3 = \Gamma' \left[\frac{\partial \mathcal{H}}{\partial v_\varphi} + \frac{v_\varphi}{v} \frac{\partial \mathcal{H}}{\partial v} \right],
$$
$$
\tag{5.191}
$$

and the nonzero components of S^{ij} are

$$S^{11} = \Gamma' \left(\frac{\partial^2 \mathcal{G}}{\partial v^2} + \frac{v_\varphi^2}{v^2} \frac{\partial^2 \mathcal{G}}{\partial v_\varphi^2} + \frac{2v_\varphi}{v} \frac{\partial^2 \mathcal{G}}{\partial v \partial v_\varphi} \right),$$

$$S^{22} = \Gamma' \frac{1}{v v_M^2} \frac{\partial \mathcal{G}}{\partial \Psi},$$

$$S^{33} = \Gamma' \left(\frac{v_\varphi^2}{v^2} \frac{\partial^2 \mathcal{G}}{\partial v^2} + \frac{\partial^2 \mathcal{G}}{\partial v_\varphi^2} + \frac{v_M^2}{v^3} \frac{\partial \mathcal{G}}{\partial v} + \frac{2v_\varphi}{v} \frac{\partial^2 \mathcal{G}}{\partial v_\varphi^2} \right),$$

$$S^{13} = \Gamma' \left[\frac{v_\varphi}{v} \frac{\partial^2 \mathcal{G}}{\partial v^2} + \left(1 + \frac{v_\varphi^2}{v^2} \right) \frac{\partial^2 \mathcal{G}}{\partial v \partial v_\varphi} + \frac{v_\varphi}{v} \frac{\partial^2 \mathcal{G}}{\partial v_\varphi^2} \right]. \tag{5.192}$$

The diffusion coefficients involving L_z immediately follow by inserting these expressions into equations (5.120):

$$\langle \Delta L_z \rangle = \Gamma' \left[\frac{L_z}{v} \frac{\partial \mathcal{H}}{\partial v} - \varpi \frac{\partial \mathcal{H}}{\partial v_\varphi} \right], \tag{5.193a}$$

$$\langle (\Delta L_z)^2 \rangle = \Gamma' \left[\frac{L_z^2}{v^2} \frac{\partial^2 \mathcal{G}}{\partial v^2} + \frac{\varpi^2 v^2 - L_z^2}{v^3} \frac{\partial \mathcal{G}}{\partial v} + \varpi \left(\varpi + \frac{2L_z}{v} \right) \frac{\partial^2 \mathcal{G}}{\partial v_\varphi^2} \right], \tag{5.193b}$$

$$\langle \Delta E \Delta L_z \rangle = -\Gamma' \left[L_z \frac{\partial^2 \mathcal{G}}{\partial v^2} + v \varpi \left(1 + \frac{v_\varphi^2}{v^2} \right) \frac{\partial^2 \mathcal{G}}{\partial v \partial v_\varphi} + L_z \frac{\partial^2 \mathcal{G}}{\partial v_\varphi^2} \right]. \tag{5.193c}$$

Using equation (5.151), the orbit-averaged diffusion coefficients look like

$$\langle \Delta L_z \rangle_t = \frac{\int \int_M \langle \Delta L_z \rangle (2\pi/\varpi) \varpi \, d\varpi \, dz}{\int \int_M (2\pi/\varpi) \varpi \, d\varpi \, dz} = \frac{\int \int_M \langle \Delta L_z \rangle d\varpi \, dz}{\int \int_M d\varpi \, dz}. \tag{5.194}$$

Note the nonintuitive result that the weighting has no explicit dependence on orbital velocity! In this particular case, the term "orbit averaging" is something of a misnomer. In deriving equation (5.194), we have *assumed* that orbits somehow manage to spread uniformly over the region M accessible to a given E and L_z. If there is a third integral, there is no reason for this to happen; our "orbit average" is really an average over orbits having the same E and L_z but possibly different third integrals. It turns out that, in the axisymmetric geometry, averaging so defined is independent of the gravitational potential, except insofar as the latter determines the integration boundary in equation (5.194).

Having obtained expressions for the orbit-averaged diffusion coefficients, we can write the orbit-averaged Fokker–Planck equation for $N(E, L_z, t)$ as

$$\frac{\partial N}{\partial t} = -\frac{\partial}{\partial E} \left(N \langle \Delta E \rangle_t \right) + \frac{1}{2} \frac{\partial^2}{\partial E^2} \left(N \langle (\Delta E)^2 \rangle_t \right) - \frac{\partial}{\partial L_z} \left(N \langle \Delta L_z \rangle_t \right)$$

$$+ \frac{1}{2} \frac{\partial^2}{\partial L_z^2} \left(N \langle (\Delta L_z)^2 \rangle_t \right) + \frac{\partial^2}{\partial E \partial L_z} \left(N \langle \Delta E \Delta L_z \rangle_t \right). \tag{5.195}$$

Equations (5.193)–(5.195) are valid for any field-star velocity distribution. In practice, $f(\boldsymbol{v}_f)$ is usually replaced by some simpler approximation, as we have already seen in the case of $f(E, L)$. For instance, one could replace $f(\boldsymbol{v}_f)$ by an isotropic approximation $f(v_f)$, yielding for the diffusion coefficients the simpler forms

$$\langle \Delta L_z \rangle = \Gamma' \frac{L_z}{v} \frac{\partial \mathcal{H}}{\partial v} = \frac{L_z}{v} \langle \Delta v_\parallel \rangle, \tag{5.196a}$$

$$\langle (\Delta L_z)^2 \rangle = \Gamma' \left[\frac{L_z^2}{v^2} \frac{\partial^2 \mathcal{G}}{\partial v^2} + \frac{\varpi^2 v^2 - L_z^2}{v^3} \frac{\partial \mathcal{G}}{\partial v} \right]$$

$$= \frac{L_z^2}{v^2} \langle (\Delta v_\parallel)^2 \rangle + \frac{1}{2} \frac{(\varpi^2 v^2 - L_z^2)}{v^2} \langle (\Delta v_\perp)^2 \rangle, \tag{5.196b}$$

$$\langle \Delta E \Delta L_z \rangle = -\Gamma' L_z \frac{\partial^2 \mathcal{G}}{\partial v^2} = -L_z \langle (\Delta v_\parallel)^2 \rangle. \tag{5.196c}$$

But making $f(\boldsymbol{v}_f)$ isotropic is probably too extreme an approximation, especially if we want to use equation (5.195) to follow the evolution of *rotating* systems.

Probably the next simplest approximation is to make $f(\boldsymbol{v}_f)$ isotropic in a frame that rotates with the local, mean velocity of the stars. Writing that mean velocity as $\overline{\boldsymbol{v}}_\varphi = \varpi \, \Omega(\varpi, z)\boldsymbol{e}_\varphi$, the peculiar velocity $\boldsymbol{u}$ is

$$\boldsymbol{u} = \boldsymbol{v} - \varpi \Omega \boldsymbol{e}_\varphi \tag{5.197}$$

and our approximation consists of writing the field-star velocity distribution in the locally corotating frame as $f(\boldsymbol{u})$. It is then straightforward to transform derivatives with respect to v and v_φ into derivatives with respect to u; for instance, $\partial/\partial v_\varphi = -(\varpi \Omega/u)\partial/\partial u$. The resulting, local diffusion coefficients are [132]

$$\langle \Delta E \rangle = \left(u + \frac{L_z \Omega}{u} - \frac{\varpi^2 \Omega^2}{u} \right) \frac{\partial \mathcal{H}}{\partial u} + \frac{1}{2} \frac{\partial^2 \mathcal{G}}{\partial u^2} + \frac{1}{u} \frac{\partial \mathcal{G}}{\partial u}, \tag{5.198a}$$

$$\langle \Delta L_z \rangle = \left(\frac{L_z}{u} - \frac{\varpi^2 \Omega}{u} \right) \frac{\partial \mathcal{H}}{\partial u}, \tag{5.198b}$$

$$\langle (\Delta E)^2 \rangle = \left(u^2 + 2L_z \Omega - 2\varpi^2 \Omega^2 + \frac{L_z^2 \Omega^2}{u^2} - \frac{2L_z \varpi^2 \Omega^3}{u^2} + \frac{\varpi^4 \Omega^4}{u^2} \right) \frac{\partial^2 \mathcal{G}}{\partial u^2}$$

$$+ \left(\frac{\varpi^2 \Omega^2}{u} - \frac{L_z^2 \Omega^2}{u^3} + \frac{2L_z \varpi^2 \Omega^3}{u^3} - \frac{\varpi^4 \Omega^4}{u^3} \right) \frac{\partial \mathcal{G}}{\partial u}, \tag{5.198c}$$

$$\langle (\Delta L_z)^2 \rangle = \left(\frac{L_z^2}{u^2} + \frac{\varpi^4 \Omega^2}{u^2} - \frac{2L_z \varpi^2 \Omega}{u^2} \right) \frac{\partial^2 \mathcal{G}}{\partial u^2}$$

$$+ \left(\frac{-L_z^2}{u^3} - \frac{\varpi^4 \Omega^2}{u^3} + \frac{\varpi^2}{u} + \frac{2L_z \varpi^2 \Omega}{u^3} \right) \frac{\partial \mathcal{G}}{\partial u}, \tag{5.198d}$$

$$\langle \Delta E \Delta L_z \rangle = \left(L_z - \varpi^2 \Omega + \frac{L_z^2 \Omega}{u^2} - \frac{2L_z \varpi^2 \Omega^2}{u^2} + \frac{\varpi^4 \Omega^3}{u^2} \right) \frac{\partial^2 \mathcal{G}}{\partial u^2}$$

$$+ \left(\frac{\varpi^2 \Omega}{u} + \frac{2L_z \varpi^2 \Omega^2}{u^3} - \frac{L_z^2 \Omega}{u^3} - \frac{\varpi^4 \Omega^3}{u^3} \right) \frac{\partial \mathcal{G}}{\partial u}. \tag{5.198e}$$

We leave it as an exercise for the reader to carry out the orbit-averaging of these coefficients.

5.6 GRAVITATIONAL ENCOUNTERS NEAR A SUPERMASSIVE BLACK HOLE

As noted above, the orbit-averaged Fokker–Planck equation accounts for the fact that stars move along orbits, but it does so in a very approximate way. For instance, no allowance is made for the fact that the perturbations acting on a test star come from stars which themselves are moving on orbits, or for the fact that the field-star velocity distribution varies with location in the galaxy. An even more basic criticism can be made of the orbit-averaging procedure itself. Consider a test star on an eccentric orbit. At apoapsis, the star comes nearly to rest, and then its velocity reverses. It is unreasonable to suppose that the dynamical friction wake that accompanies the star (figure 5.3) would be able to switch, instantaneously, from one side of the star to the other; there is undoubtedly a time just after turnaround when the most of the wake lies *in front of* the star. A simple averaging of the diffusion coefficients over the trajectory fails to take account of such orbit-dependent effects.

By and large, these details do not seem to matter very much: comparisons of Fokker–Planck evolutionary models with fully general N-body simulations typically find good agreement, at least in terms of the evolution of macroscopic variables like density and velocity dispersion [279]. But there is one important regime in which the theory of random gravitational encounters worked out in the previous sections breaks down [449]. Sufficiently close to an SBH, both test and field stars move on unperturbed orbits that are Keplerian ellipses. It is straightforward to compute the orbit-averaged diffusion coefficients in this case, using the techniques described in the previous sections, but doing so misses an essential element: the unperturbed orbits are closed, or "resonant": the frequencies of motion in the radial and angular directions are the same.[14] Because of this property of Keplerian motion, the unperturbed orbits are fixed in shape and orientation, and gravitational interactions between stars are highly correlated. These correlations will persist for as long as the stars maintain a fixed relative orientation—roughly speaking, for a time equal to the average field-star precession time. We will call this time the **coherence time**, t_{coh}; a more precise definition of t_{coh} is given below.

Suppose (as is usually the case) that t_{coh} for some set of N stars near an SBH is long compared with orbital periods P. Then for elapsed times Δt such that $P \ll \Delta t \ll t_{coh}$, we can imagine replacing each star by an elliptical ring, or (in the case of eccentric orbits) a rod, of fixed orientation, whose linear density is inversely proportional to the local speed in the Keplerian orbit. The total gravitational force produced by the N rings and rods is stationary, and so it induces no change in the energy of a test star. But because the number of stars is finite, the overall mass distribution will differ slightly from spherical symmetry, and there will be a net

---

[14]This is a somewhat idiosyncratic use of the term "resonant".

torque on the test star, causing its angular momentum to change with time. In this **coherent resonant relaxation** regime, the angular momentum of a test star changes with time at roughly a constant rate.

On timescales longer than t_{coh}, the perturbing potential due to the N stars will have changed its orientation with respect to the test star. The magnitude of the torque will be roughly the same, but its direction will certainly be different. If we assume that the direction of the torque from the field stars is essentially randomized after each t_{coh}, the angular momentum of a test star will undergo a random walk, with step size given by the product of the torque and the coherence time. The evolution of L in this **incoherent resonant relaxation** regime is qualitatively similar to the evolution discussed in the previous sections ("nonresonant relaxation"), but as we will see, resonant relaxation can be much more efficient than nonresonant relaxation at changing L near an SBH.

5.6.1 Coherent resonant relaxation

5.6.1.1 Basic concepts

Consider two stars orbiting around an SBH, with orbits of semimajor axis a and masses m. The stars exert a mutual torque (per unit mass) of order $|\langle F \times r \rangle| \approx (Gm/a^2) \times a \approx Gm/a$. If there are N stars in the region $r \lesssim a$, with randomly oriented orbits, the net torque will be of order $[N(< a)]^{1/2} (Gm/a)$.

We begin by considering the evolution of a single ("test") orbit over times shorter than t_{coh}. By definition, the orientations of the "field" orbits are nearly constant, as is the direction of the net torque acting on the test orbit. As a consequence, its angular momentum will change approximately linearly with time, at a rate $|\dot{L}| \approx \sqrt{N}(Gm/a)$. Expressing this in terms of the orbital period,

$$P(a) = \frac{2\pi a^{3/2}}{\sqrt{GM_\bullet}},$$

and the angular momentum L_c of a circular orbit, $L_c(a) = \sqrt{GM_\bullet a}$, we find for the change in L over times $\Delta t < t_{coh}$

$$\frac{\Delta L}{L_c} \approx \sqrt{N} \frac{Gm}{a} \times \frac{\Delta t}{\sqrt{GM_\bullet a}} \approx 2\pi \frac{m\sqrt{N}}{M_\bullet} \frac{\Delta t}{P}. \tag{5.199}$$

The **coherent resonant relaxation time** can be defined as the Δt for which $\Delta L = L_c$:

$$T_{RR,coh} \equiv \frac{P}{2\pi} \frac{M_\bullet}{m} \frac{1}{\sqrt{N}} \tag{5.200}$$

$$\approx 1.5 \times 10^4 \left(\frac{a}{mpc}\right)^{3/2} \left(\frac{M_\bullet}{10^6 M_\odot}\right)^{-1/2} \left(\frac{q}{10^{-6}}\right)^{-1} \left(\frac{N}{10^3}\right)^{-1/2} \text{yr,}$$

where $q \equiv m/M_\bullet$ and mpc is milliparsecs. Note that N on the right-hand side of this equation is understood to be a function of a.

Resonant relaxation is a local phenomenon, in the sense that most of the torque comes from stars with semimajor axes close to that of the test star. As a consequence, there is no equivalent to the $\ln \Lambda$ term that appears in nonresonant relaxation. On the other hand, the precise value of the dimensionless factors that appear in equations like (5.199) and (5.200) are poorly determined and will differ depending on the details of the orbital distribution.

The coherence time is the time associated with the most rapid source of precession of the field stars. There are three likely sources of precession near an SBH:

1. **Mass precession.** Mass distributed around the SBH breaks the equality between radial and angular periods, leading to a retrograde advance of the argument of periastron ω (section 4.4.1). To a first approximation, the effects of the distributed mass can be computed by assuming that its distribution is spherically symmetric. Equation (4.88) gives the orbit-averaged precession rate in the case that the density follows a power law, $\rho(r) \propto r^{-\gamma}$. The time required for ω to advance by π is

$$t_{\mathrm{M}}(a, e) \approx \frac{1}{2}(1 - e^2)^{-1/2} \frac{M_\bullet}{M_\star(a)} P(a), \qquad (5.201)$$

where $M_\star(a)$ is the distributed mass within radius $r = a$; the dimensionless quantity $G_{\mathrm{M}}(e, \gamma)$ defined in equation (4.89) has been set to 1 (figure 4.3). The precession time defined by equation (5.201) depends on both a and e. The *coherence* time corresponding to this precession is the *average* time for all orbits at $r \approx a$ to precess. Averaging equation (5.201) over eccentricity assuming a "thermal" (isotropic) distribution, $N(e)de = 2e\,de$ (equation 4.37), we find the **mass coherence time** for orbits of semimajor axis a:

$$t_{\mathrm{coh,M}} \approx \frac{M_\bullet}{Nm} P. \qquad (5.202)$$

2. **Relativistic precession.** The lowest-order relativistic corrections to the equations of motion imply precession at a rate given by equation (4.274). Like mass precession, this "Schwarzschild precession" leaves the plane of the orbit unchanged, but it is prograde, that is, opposite in sense to the mass precession. The time required for ω to advance by π is

$$t_{\mathrm{S}}(a, e) = \frac{1}{6}(1 - e^2) \frac{c^2 a}{G M_\bullet} P(a). \qquad (5.203)$$

Again averaging over e assuming a thermal distribution yields the **relativistic coherence time**

$$t_{\mathrm{coh,S}} \approx \frac{1}{12} \frac{a}{r_{\mathrm{g}}} P \qquad (5.204)$$

with $r_{\mathrm{g}} \equiv G M_\bullet / c^2$. Precession due to the spin of the SBH ("Kerr precession") is slower than the Schwarzschild precession unless $a \approx r_{\mathrm{g}}$.

3. **Precession due to resonant relaxation.** By changing L, resonant relaxation causes orbital planes to precess, with a characteristic time given by equation (5.200). Ignoring for the moment the dependence of this time on eccentricity, the **self-coherence time** is roughly

$$t_{\mathrm{coh,N}} \approx T_{\mathrm{RR,coh}} \approx \frac{1}{2} \frac{M_\bullet}{m\sqrt{N}} P. \qquad (5.205)$$

Comparison of equations (5.202) and (5.205) shows that $t_{\text{coh,M}} \approx t_{\text{coh,N}}/\sqrt{N}$: the mass coherence time is always shorter than the self-coherence time. But sufficiently close to the SBH, relativistic precession must dominate: $t_{\text{coh,S}} < t_{\text{coh,M}}$ when

$$\frac{a}{r_{\text{g}}} \lesssim 12 \frac{M_{\bullet}}{m\,N}. \tag{5.206}$$

Because mass precession and Schwarzschild precession are in opposite directions, one might think that equation (5.206) defines a radius around the SBH at which the coherence time goes to infinity. This is not so, since the two sorts of precession depend differently on eccentricity. Precession rates of *individual orbits* are zero if $t_M(a, e) = t_S(a, e)$, that is,

$$(1 - e^2)^{3/2} \approx \frac{r_{\text{g}}}{a} \frac{M_{\bullet}}{M_{\star}(a)}, \tag{5.207}$$

which occurs at different a for orbits of different e. Even if a single orbit satisfies this condition, the orbits of the other stars at similar a will not, implying a finite coherence time.

5.6.1.2 A simple model for coherent resonant relaxation

To gain a deeper understanding of coherent resonant relaxation, we need to know something about the typical form of the perturbing force due to N stars around an SBH. In section 4.8.1, equations of motion were derived for two stars orbiting about a third body. Identifying that third body with the SBH, it is straightforward to generalize the three-body equations of motion to the case of N orbiting bodies. The result, for the ith (test) body, is

$$\ddot{\boldsymbol{r}}_i + G(M_{\bullet} + m_i)\frac{\boldsymbol{r}_i}{r_i^3} = G \sum_{\substack{j=1,\dots,N \\ j \neq i}} m_j \left(\frac{\boldsymbol{r}_j - \boldsymbol{r}_i}{r_{ij}^3} - \frac{\boldsymbol{r}_j}{r_j^3} \right), \tag{5.208}$$

where $r_{ij} \equiv |\boldsymbol{r}_i - \boldsymbol{r}_j|$ and r_j is the distance of the jth star from the SBH. Recall that the first of the two terms of the right-hand side of equation (5.208) represents the direct force exerted by the $N - 1$ bodies on the test mass, while the second term is the "indirect force" resulting from the fact that the SBH is itself tugged to and fro. We begin by noting that if all the r_j are comparable in magnitude, then the indirect term can have a magnitude that is comparable with the direct term, regardless of the value of $(M_{\bullet}/m)$! However, in the context of resonant relaxation, we are concerned not so much with the instantaneous accelerations as with the time-averaged torques acting on a test star. The instantaneous torque is

$$\boldsymbol{\tau}_i = \boldsymbol{r}_i \times \ddot{\boldsymbol{r}}_i$$

$$= G \sum_{\substack{j=1,\dots,N \\ j \neq i}} m_j \left(\frac{\boldsymbol{r}_i \times \boldsymbol{r}_j}{r_{ij}^3} - \frac{\boldsymbol{r}_i \times \boldsymbol{r}_j}{r_j^3} \right)$$

$$= \boldsymbol{r}_i \times \left[\sum_{\substack{j=1,\dots,N \\ j \neq i}} G m_j \boldsymbol{r}_j \left(\frac{1}{r_{ij}^3} - \frac{1}{r_j^3} \right) \right]. \tag{5.209}$$

The "force" responsible for the torque has two components, direct and indirect:

$$F_D = \sum Gm_j \frac{r_j}{r_{ij}^3}, \quad F_I = -\sum Gm_j \frac{r_j}{r_j^3}. \qquad (5.210)$$

Without loss of generality, we can place star j in the x–z plane. Then its unperturbed motion satisfies

$$x_j(t) = -a_j\sqrt{1 - e_j^2} \sin E, \quad z_j(t) = a_j(\cos E - e_j), \qquad (5.211)$$

where E = eccentric anomaly and (a_j, e_j) are constants. The contribution of this star to the indirect force depends on

$$\frac{r_j}{r_j^3} = a_j^{-2}(R_x \hat{e}_x + R_z \hat{e}_z), \qquad (5.212)$$

where

$$R_x = -\frac{\sqrt{1 - e_j^2} \sin E}{\left(1 - e_j \cos E\right)^3}, \quad R_z = -\frac{\cos E - e_j}{(1 - e_j \cos E)^3}. \qquad (5.213)$$

Time averages are computed as

$$\langle Q \rangle = \frac{1}{2\pi} \int_0^{2\pi} dE \, (1 - e \cos E) \, Q(E) \qquad (5.214)$$

and it is easy to show that $\langle R_x \rangle = \langle R_z \rangle = 0$. In other words, the "rods" or "rings" corresponding to the indirect force exert no torque on the test mass.

This result suggests a straightforward way to evaluate the torquing force: distribute N stars, or rather N time-averaged orbital rings, about the SBH and compute the time-averaged torque on a test body that is moving on one of the orbits. The results of such a calculation will of course be different for each different "realization" of the N stars—that is, for each choice of the $5N$ variables that describe the Keplerian elements $\{a, e, i, \Omega, \omega\}$ for each star.

One way to describe the results of such experiments is in terms of the "average" form of the field-star potential corresponding to N stars. For instance, one could express the torquing potential as seen by each of the N stars in terms of multipoles, and ask which angular term is typically dominant. Not surprisingly, such experiments reveal that most of the torque can be ascribed to the lowest-order multipoles: the dipole and quadrupole.

These results suggest the following simple form for the potential seen by a test star:

$$\Phi = -\frac{GM_\bullet}{r} + \Phi_s(r) - S(a)a \cos\theta. \qquad (5.215)$$

$\Phi_s(r)$ represents the spherically distributed mass; we adopt equation (4.80) for this term, which assumes $\rho(r) \propto r^{-\gamma}$. The final term represents the nonspherical part of the potential due to the $\sqrt{N}$ fluctuations if we write

$$S(a) \approx \frac{Gm\sqrt{N}}{a^2} = \frac{GM_\star(a)}{a^2\sqrt{N}}; \qquad (5.216)$$

the elongation is assumed to be along the z-axis and $\cos\theta = z/r$. The evolution of a test star's orbit in the potential (5.215), on timescales long compared with P, can be found using the averaging procedure that was applied to a similar problem in section 4.4.3. We first define a dimensionless time $\tau = \nu_0 t$ where

$$\nu_0 \equiv \nu_r \frac{M_\star(a)}{M_\bullet}, \quad \frac{2\pi}{\nu_0} = t_{\text{coh,M}}. \tag{5.217}$$

Following the averaging procedure, the semimajor axis is removed from the Hamiltonian, and the equations of motion for the remaining elements are

$$\frac{d\omega}{d\tau} = -\ell + \frac{1}{\sqrt{N}} \left[-\frac{\ell}{e} \sin i + \frac{e \cos^2 i}{\ell \sin i} \right] \sin\omega, \tag{5.218a}$$

$$\frac{d\ell}{d\tau} = -\frac{1}{\sqrt{N}} e \sin i \cos\omega, \tag{5.218b}$$

$$\frac{d\Omega}{d\tau} = \frac{1}{\sqrt{N}} e \frac{\ell_z \cos i}{\ell^2 \sin\omega}, \tag{5.218c}$$

$$\frac{d\ell_z}{d\tau} = 0. \tag{5.218d}$$

As in section 4.4.3, we have defined a dimensionless angular momentum $\ell \equiv L/L_c(a) = (1 - e^2)^{1/2}$, and $\cos i = \ell_z/\ell$. The plane of reference has been taken to be the x–y plane and the reference direction is the x-axis; thus an orbit in the x–z plane has $\sin i = 1$, and for such an orbit, $\omega = \pi/2$ corresponds to an orientation parallel to the z-axis, the assumed direction of the lopsided distortion.

Consider first the case $\sin i = 1$. Equations (5.218) simplify to

$$\frac{d\omega}{d\tau} = -\ell - \frac{1}{\sqrt{N}} \frac{\ell}{e} \sin\omega, \tag{5.219a}$$

$$\frac{d\ell}{d\tau} = -\frac{1}{\sqrt{N}} e \cos\omega \tag{5.219b}$$

and $d\Omega/d\tau = d\ell_z/d\tau = 0$: the orbit remains in a fixed plane, but its orientation in this plane (ω) and its angular momentum (ℓ) change. The coherent resonant relaxation regime is defined by $\tau \ll 1$; imposing this condition, ω and ℓ are nearly constant, and equation (5.219b) tells us

$$\ell(\tau) \approx \ell(\tau_0) - \frac{1}{\sqrt{N}} (e \cos\omega)_0 (\tau - \tau_0), \tag{5.220}$$

that is,

$$\frac{\Delta L}{L_c} \approx -\frac{2\pi}{\sqrt{N}} (e \cos\omega)_0 \frac{\Delta t}{t_{\text{coh,M}}}, \tag{5.221}$$

similar to equation (5.199).

Equation (5.220) contains one new result: the rate of change of angular momentum is proportional to the eccentricity. This is a simple consequence of the fact that the "lever arm" of an orbit, of fixed a, is proportional to e.

Coherent resonant relaxation changes not just the eccentricity of an orbit, but also the orientation of the orbital plane—in other words, all components of $\boldsymbol{L}$. Returning to the more general equations of motion (5.218), we can resolve the changes in $\boldsymbol{L}$ into components parallel to, and perpendicular to, the original angular momentum vector. The result, after some algebra, is

$$\left(\frac{d\ell}{d\tau}\right)_{\parallel} \approx -N^{-1/2}(e\sin i\cos\omega)_0, \qquad (5.222\text{a})$$

$$\left(\frac{d\ell}{d\tau}\right)_{\perp} \approx N^{-1/2}(e\cos i)_0 \qquad (5.222\text{b})$$

in the coherent regime ($\tau \ll 1$). Changes in the orbital orientation, measured by $(d\ell/d\tau)_{\perp}$, occur on roughly the same timescale as changes in the eccentricity, measured by $(d\ell/d\tau)_{\parallel}$, and both changes scale in the same way with e. Taking averages over ω and i, one finds $\langle\Delta\ell_{\perp}^2\rangle = 2\langle\Delta\ell_{\parallel}^2\rangle$.

5.6.1.3 Three-dimensional versus two-dimensional resonant relaxation

The analysis in the previous section was based on a particular form for the torquing potential. But one feature of that analysis is generic to any treatment of resonant relaxation: the assumption of a short timescale associated with the unperturbed (Keplerian) motion. It was that assumption which justified the averaging of the motion (both test and field stars) with respect to mean anomaly, resulting in the removal of the corresponding "momentum" variable—the action I associated with the radial motion, that is, the energy, or semimajor axis—from the equations of motion of the test star.

One can imagine circumstances in which there are *two* short timescales. For instance, suppose that precession in ω is so rapid that orbits fill their annuli many times before the orbital planes have been changed by the mutual torques. In this limit, it is appropriate to carry out a *second* averaging of the equations of motion, this time over ω, effectively converting each orbit from a mass ring into a mass annulus—a circularly symmetric disk. The mutual torques between two such disks change only the direction of $\boldsymbol{L}$, not its magnitude.

Roughly speaking, a star will be in this regime when the apsidal precession time is short compared with the timescale for the $\sqrt{N}$ torques (or any other process) to change $\boldsymbol{L}$. There are two cases in which the condition on the precession rate is satisfied. Far enough from the SBH, $t_{\mathrm{coh,M}} \ll T_{\mathrm{RR,coh}}$: the mass precession time is short compared with the time for torques to change $\boldsymbol{L}$. Using equations (5.200) and (5.202), this occurs at radii where

$$\frac{M_{\bullet}}{M_{\star}(a)}P(a) \ll \frac{P(a)}{2\pi}\frac{M_{\bullet}}{m}\frac{1}{\sqrt{N}} \qquad (5.223)$$

or

$$N(<r) \gg (2\pi)^2. \qquad (5.224)$$

The other regime is near the SBH, where relativistic effects dominate the precession. Setting $t_{coh,S} \ll T_{RR,coh}$ gives[15]

$$N(< r) \ll \frac{36}{\pi^2} \left(\frac{M_\bullet}{m} \frac{r_g}{r} \right)^2 . \tag{5.225}$$

In either of these cases, precession in ω is so fast that changes in eccentricity due to the torques take place on a much longer timescale than changes in the direction of $\boldsymbol{L}$.

This can all be cast in the language of the Delaunay action-angle variables defined in section 4.2. The "momentum" variable conjugate to the mean anomaly, M, is the action $I = \sqrt{GM_\bullet a}$. Averaging the equations of motion of a test star over M, assuming a fixed torquing potential from the field stars, implies a constant I, or a; the other Delaunay variables $\{L, L_z, \Omega, \omega\}$, or equivalently $\{e, i, \Omega, \omega\}$, change with time in response to the fixed torques, as in the example worked out in the previous section (equations 5.218). The momentum variable conjugate to the argument of periastron, ω, is the angular momentum, $L = \sqrt{GM_\bullet a(1 - e^2)}$. Averaging the equations of motion of a test star a second time, over ω, implies constant $\{I, L\}$, or $\{a, e\}$. The only elements that are left to vary in response to the field-star torques in this case are the orientation variables, $\{i, \Omega\}$.

It has become commonplace to talk about these two regimes in terms of "scalar" and "vector" resonant relaxation: the former describing changes in the magnitude of $\boldsymbol{L}$ and the latter describing changes in its direction. This division is unfortunate, since there really is no regime in which the torques lead to changes only in the *magnitude* of $\boldsymbol{L}$. The more basic distinction is between evolution that causes all components of $\boldsymbol{L}$ to change, on the one hand, and evolution that is limited (due to rapid apsidal precession) to changes in the direction of $\boldsymbol{L}$ on the other hand. In this book, the former sort of evolution will be called **3d resonant relaxation** and the latter **2d resonant relaxation**.

In any case, it is standard to parametrize the changes that occur in $\boldsymbol{L}$ during the coherent regime as

$$\frac{\Delta L}{L_c} = \beta_s \sqrt{N} \frac{m}{M_\bullet} \frac{\Delta t}{P} , \tag{5.226a}$$

$$\frac{|\Delta \boldsymbol{L}|}{L_c} = \beta_v \sqrt{N} \frac{m}{M_\bullet} \frac{\Delta t}{P} , \tag{5.226b}$$

with β_s, β_v ("s" = scalar, "v" = vector) dimensionless parameters that can in principle be measured via numerical experiments. This way of characterizing the evolution is convenient numerically, since computing components of $\boldsymbol{L}$ is easier than computing the full set of Delaunay variables (say). But the division is infelicitous in other respects; for instance, β_v contains information about changes in both the magnitude and direction of $\boldsymbol{L}$ and so is not independent of β_s [131]. In any case, numerical experiments that exclude relativistic precession [223, 131] find that the values of β_s and β_v depend fairly weakly on the assumed orbital distribution $N(a)$.

---

[15]Since the rate of relativistic precession is strongly eccentricity dependent, highly eccentric orbits can be in this regime even if typical field stars with the same a are not; see section 6.4.

One study [223] found

$$\beta_s \approx 1.6e, \quad \beta_v \approx 1.8 \left(e^2 + \frac{1}{2} \right). \tag{5.227}$$

The dependence on eccentricity is qualitatively similar to what was found using the simple Hamiltonian model given above. Averaging over a "thermal" (isotropic) velocity distribution, equation (4.37), yields

$$\langle \beta_s \rangle \approx 1.1, \quad \langle \beta_v \rangle \approx 1.8. \tag{5.228}$$

As noted above, an alternative, and more complete, way to characterize the effects of coherent resonant relaxation is in terms of the full set of orbital elements that are affected by the field-star torques. Figure 5.7 shows the results of a set of numerical experiments based on this idea. In this case, the test star was taken to be S2, the bright young star near the center of the Milky Way whose orbit has been used to measure the mass of the SBH (section 4.9). S2's orbit has $a \approx 5.0$ mpc, $e \approx 0.88$, and a period of just ~ 15.8 yr (table 4.1). Figure 5.7 shows how the elements defining the eccentricity and orientation of S2's orbit change over a single period. The quantity $\Delta\theta$ plotted there measures changes in the direction of $\boldsymbol{L}$:

$$\cos(\Delta\theta) = \frac{\boldsymbol{L}_1 \cdot \boldsymbol{L}_2}{L_1 L_2} \tag{5.229}$$

with $\{\boldsymbol{L}_1, \boldsymbol{L}_2\}$ the values of $\boldsymbol{L}$ at two times separated by P. The N field stars were selected from a density profile $n(r) \propto r^{-2}$, under two assumptions about their masses: $m_\star = 10\, M_\odot$ and $50\, M_\odot$. Each of the N field-star orbits was integrated as well—in other words, no orbit averaging was carried out—and the integrator included the mutual forces between stars, as well as post-Newtonian corrections to the equations of motion. One hundred random realizations of each initial model were integrated, allowing both the mean values of the changes to S2's orbit, and their variance, to be computed.

Based on equation (5.199), we expect the variables describing the magnitude and direction of S2's angular momentum to change by average values of

$$|\Delta e| \approx K_e \sqrt{N} \frac{m}{M_\bullet}, \tag{5.230a}$$

$$\Delta\theta \approx 2\pi K_t \sqrt{N} \frac{m}{M_\bullet}, \tag{5.230b}$$

where N is understood to be the average number of stars inside $r = 9.4$ mpc, the apoapsis of S2's orbit, and $\{K_e, K_t\}$ are constants to be determined from the numerical experiments. The results are [469]

$$K_e = 1.4, \quad K_t = 1.0. \tag{5.231}$$

The results for $\Delta\omega$ deserve a bit more elaboration. One does not normally think of the argument of periapsis as a variable affected by resonant relaxation; rather, ω is the variable whose rate of change (due to mass precession, say) defines the length of the coherent regime. But equation (5.218a) for $d\omega/dt$ contains a term due to the $\sqrt{N}$ torques, representing the change in the apsidal precession rate due to field-star torques, whose amplitude is $\sqrt{N}$ smaller than the term representing

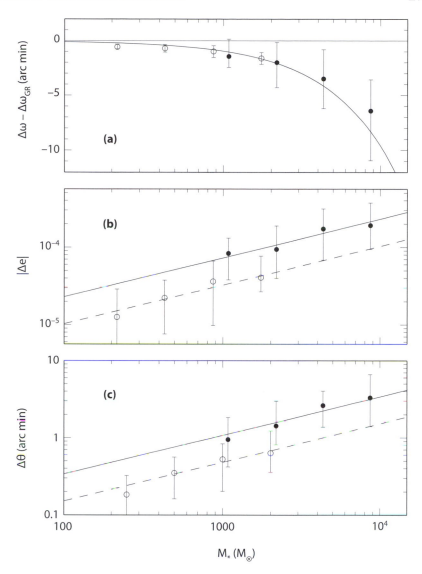

Figure 5.7 Results of numerical experiments showing changes in the orbital elements of the Galactic center star S2 over one orbital period ($\sim 16\,\mathrm{yr}$) [469]. Filled circles are from integrations assuming a field-star mass of $m_\star = 50\,M_\odot$ and open circles are for $m_\star = 10\,M_\odot$; the number of field stars was $N = \{25, 50, 100, 200\}$ for both values of $m_\star$. The abscissa is the distributed mass within S2's apoapsis at $r \approx 9.4\,\mathrm{mpc}$. In each frame, the points are median values from 100 N-body integrations, and the error bars extend from the 20th to the 80th percentile of the distribution. The curve in (a) is equation (5.233). The solid and dashed lines in (b) are equation (5.230a) with $m = 50\,M_\odot$ and $m = 10\,M_\odot$, respectively, and with $K_e = 1.4$. In (c), the solid and dashed lines are equations (5.230b) with $K_t = 1.0$.

mass precession. When considering time intervals $\ll t_{\text{coh}}$, the change in ω contains a piece due to "resonant relaxation," as well as a (generally larger) piece due to either or both of the first two precession mechanisms listed in section 5.6.1.1. In the simulations of figure 5.7, the Schwarzschild (relativistic) precession was active, and of course the mass precession due to the N field stars as well. The quantity plotted in the figure is the difference between the total change in ω, and the change due to relativity:

$$\Delta\omega - \Delta\omega_{\text{GR}} = \Delta\omega - \frac{6\pi G M_\bullet}{c^2 a (1 - e^2)}. \tag{5.232}$$

On average, one expects this quantity to have the value given by equation (4.88):

$$-2\pi G_{\text{M}}(e, \gamma)\sqrt{1 - e^2} \left[\frac{M_\star(r < a)}{M_\bullet} \right] \approx -1\rlap{.}''0 \left[\frac{M_\star(r < a)}{10^3 \, M_\odot} \right], \tag{5.233}$$

where the latter expression uses the known $\{a, e\}$ values for S2 and $M_\bullet = 4.0 \times 10^6 \, M_\odot$. The curve plotted in the top panel of figure 5.7 is equation (5.233). The figure shows that there is a significant variation from one experiment to another in the value of $\Delta\omega - \Delta\omega_{\text{GR}}$, due to random differences in the field-star torques (resonant relaxation), as well as random variations in the total enclosed mass (mass precession). If and when a shift in S2's argument of periapsis is measured, it will be necessary to account for this variation before drawing conclusions about the physical properties of the nucleus [469].

Before moving on to incoherent resonant relaxation, it is important to point out one natural source of confusion. There are always two, distinct precessional timescales that are relevant to the evolution of a test star subject to resonant relaxation: the *mean* precession time of the field stars—defined here as the coherence time; and the precession time of the test star itself. In many cases, these two times are comparable, and there is no need to distinguish between them. The distinction becomes important if the precession rate of the test star is much higher than the mean precession rate of the field stars, since it is the rate of change of the *relative* orientation of test and field stars that determines the effective coherence time. For instance, relativistic precession occurs at a rate that is proportional to $(1 - e^2)^{-1}$, and a sufficiently eccentric orbit will precess due to relativity faster than a typical field star of the same a. A test star that precesses much more rapidly than the field stars can be said to be in the "2d resonant relaxation" regime (the magnitude of L will hardly change) even if the field stars of similar a are not.

5.6.2 Incoherent resonant relaxation

The angular momentum of a test star changes linearly with time under coherent resonant relaxation, a consequence of the fact that the $\sqrt{N}$ torques driving the evolution are constant. On timescales longer than $\sim t_{\text{coh}}$, the field stars precess, and the direction of the torque changes. These changes in the direction of the torque are not really random—individual orbits precess more-or-less smoothly—but to a first approximation, we can assume that the direction of the torque is randomized after each $\Delta t \approx t_{\text{coh}}$.

In the case of 3d resonant relaxation, the accumulated change in L in the coherent regime is found by setting $\Delta t = t_{coh}$ in equation (5.199). If the coherence time is set by mass precession,

$$|\Delta L|_{coh,s} \approx \frac{\pi e}{\sqrt{N}} L_c, \tag{5.234}$$

while in the case that relativistic precession dominates,

$$|\Delta L|_{coh,s} \approx \frac{\pi}{6} \frac{a}{r_g} \frac{M_\star(a)}{M_\bullet \sqrt{N}} L_c. \tag{5.235}$$

Two-dimensional resonant relaxation, on the other hand, is not affected by in-plane precession: its coherence time is longer, $\sim t_{coh,N} \approx T_{RR,coh} \approx \sqrt{N} t_{coh,M}$. The change in L over a coherence time due to 2d resonant relaxation would therefore seem to be

$$|\Delta L|_{coh,v} \approx \pi L_c. \tag{5.236}$$

Of course, L cannot change by more than L_c! What equation (5.236) really implies is that L rotates by an angle of order unity during each coherence time.

On timescales longer than the coherence time, the angular momentum of a test star evolves approximately as a random walk. The accumulated change in L over a coherence time, $|\Delta L|_{coh}$, becomes the step size (or "mean free path" in L) for the random walk. It is because this step size is relatively large (a substantial fraction of L_c) that resonant relaxation can be more efficient over the long term than uncorrelated, or nonresonant, relaxation.

Because changes in the direction of L "saturate" already at $\sim t_{coh}$, no new timescale arises for 2d resonant relaxation in the incoherent regime. In other words, the timescale for incoherent, 2d resonant relaxation is the same as $t_{coh,N}$, equation (5.205):

$$T_{RR} \approx \frac{1}{2} \frac{M_\bullet}{m \sqrt{N}} P. \tag{5.237}$$

In the case of 3d resonant relaxation, we expect changes

$$|\Delta L| \approx |\Delta L|_{coh,s} \left(\frac{\Delta t}{t_{coh}}\right)^{1/2} \tag{5.238}$$

on timescales $\Delta t \gg t_{coh}$. We can write this as

$$\frac{|\Delta L|}{L_c} = \left(\frac{\Delta t}{T_{RR}}\right)^{1/2}, \tag{5.239a}$$

$$T_{RR} \equiv \left(\frac{L_c}{|\Delta L|_{coh}}\right)^2 t_{coh}. \tag{5.239b}$$

If t_{coh} is determined by mass precession, equations (5.202), (5.234) and (5.239) imply

$$T_{RR} \approx \left(\frac{M_\bullet}{m}\right) P. \tag{5.240}$$

In the case that relativistic precession dominates, equations (5.204), (5.235), and (5.239) give

$$T_{RR} \approx \frac{3}{\pi^2} \frac{r_g}{a} \left(\frac{M_\bullet}{m} \right)^2 \frac{P}{N}. \tag{5.241}$$

We would like to estimate the distance from an SBH at which incoherent resonant relaxation begins to "take over" from nonresonant relaxation. For this purpose, it is useful to write T_{NRR}, equation (5.61), approximately in terms of the quantities N and P. Adopting equation (3.48) for $\rho(r)$ and equation (3.63b) for $\sigma(r)$ yields

$$T_{NRR} \approx A_{NRR} \left(\frac{M_\bullet}{m} \right)^2 \frac{P}{N}, \tag{5.242a}$$

$$A_{NRR} = \frac{0.68}{(3-\gamma)(1+\gamma)^{3/2}} \frac{1}{\ln \Lambda}, \tag{5.242b}$$

where $P(a) = 2\pi \left[a^3/(GM_\bullet) \right]^{1/2}$ is the Keplerian period. (Equation 5.242 is expressed in terms of local quantities; a better quantity to compare with T_{RR} would be an orbit-averaged relaxation time within some region $r \lesssim a$.) In the case that mass precession is the source of decoherence, T_{RR} is given by equation (5.240), and $T_{RR} < T_{NRR}$ implies

$$m N(r < a) < A_{NRR} M_\bullet. \tag{5.243}$$

In the case that relativistic precession dominates, equations (5.241) and (5.242) imply

$$N(r < a) < 1.8 A_{NRR}^{1/2} \left(\frac{a}{r_g} \right)^{1/2}. \tag{5.244}$$

Note that m does not appear in this relation.

It is interesting to estimate these characteristic radii for the Milky Way nucleus. Doing so requires the adoption of a specific model for the density of stars at $r \ll r_h$. Unfortunately, little is known about the distribution of stars or stellar remnants so near to Sgr A*. One natural model—for which there is currently no observational support—would postulate a dynamically relaxed, Bahcall–Wolf cusp at $r \lesssim 0.2 r_m$ (section 5.5.2). We can normalize the cusp density by requiring it to match the density of the Milky Way nuclear star cluster (NSC) at $r = 0.2 r_m$; the density of the NSC is observed to fall off as $\rho(r) \sim r^{-2}$ for $r \gtrsim 1\,\mathrm{pc}$ [403]. Thus our "relaxed model" would have

$$\rho(r) = \begin{cases} 2 \times 10^6 \left(\frac{r}{0.5\,\mathrm{pc}} \right)^{-7/4} M_\odot\,\mathrm{pc}^{-3}, & r \le 0.5\,\mathrm{pc}, \\ 2 \times 10^6 \left(\frac{r}{0.5\,\mathrm{pc}} \right)^{-2} M_\odot\,\mathrm{pc}^{-3}, & r > 0.5\,\mathrm{pc}, \end{cases} \tag{5.245}$$

and the enclosed number of stars, assuming $m = M_\odot$, is

$$N(r) = \begin{cases} 1.6 \times 10^5 \left(\frac{r}{0.5\,\mathrm{pc}} \right)^{5/4}, & r \le 0.5\,\mathrm{pc}, \\ 1.6 \times 10^5 + 2.0 \times 10^5 \left(\frac{r}{0.5\,\mathrm{pc}} - 1 \right), & r > 0.5\,\mathrm{pc}. \end{cases} \tag{5.246}$$

In this model, the radius at which relativistic precession begins to dominate Newtonian precession for orbits of typical eccentricity is given by equation (5.206) as $r \approx 5 \times 10^{-3}$ pc ≈ 5 mpc: roughly the radius of S2's orbit (4.1). Setting $\gamma = 7/4$ and $\ln \Lambda = 15$ (equation 5.35) gives $A_{\mathrm{NRR}} \approx 0.01$; equation (5.243) then states that RR overwhelms NRR inside $r \approx 0.06$ pc $\approx 0.025 r_{\mathrm{h}}$. Referring to figure 6.5, this radius is small compared with the radius from which most (normal) stars would be scattered into the SBH, and we are probably justified in ignoring RR when computing the rate of stellar tidal disruptions [448].

As discussed in the introduction to chapter 7, number counts of late-type stars near the center of the Milky Way suggest a rather different model: a constant or slowly rising density inside ~ 0.5 pc $\approx 0.2 r_{\mathrm{h}}$. Motivated by these data, we define an "unrelaxed model" in which $n \propto r^{-1/2}$, $r \leq 0.2 r_{\mathrm{h}}$; recall that this is the shallowest power-law profile consistent with an isotropic $f(E)$. Normalizing the density at $0.2 r_{\mathrm{h}}$ in the same way as above, we find

$$
\rho(r) = \begin{cases} 2 \times 10^6 \left(\frac{r}{0.5\,\mathrm{pc}} \right)^{-1/2} M_\odot\,\mathrm{pc}^{-3}, & r \leq 0.5\,\mathrm{pc}, \\ 2 \times 10^6 \left(\frac{r}{0.5\,\mathrm{pc}} \right)^{-2}, M_\odot\,\mathrm{pc}^{-3}, & r > 0.5\,\mathrm{pc}, \end{cases} \tag{5.247}
$$

and the enclosed number is given by

$$
N(r) = \begin{cases} 8.0 \times 10^4 \left(\frac{r}{0.5\,\mathrm{pc}} \right)^{5/2}, & r \leq 0.5\,\mathrm{pc}, \\ 8.0 \times 10^4 + 2.0 \times 10^5 \left(\frac{r}{0.5\,\mathrm{pc}} - 1 \right), & r > 0.5\,\mathrm{pc}. \end{cases} \tag{5.248}
$$

In this low-density model, relativistic precession dominates mass precession already at $r \approx 0.02$ pc for orbits of typical eccentricity. We now have $A_{\mathrm{NRR}} \approx 0.016$, and $T_{\mathrm{RR}} < T_{\mathrm{NRR}}$ when $r \lesssim 0.18$ pc $\approx 0.1 r_{\mathrm{h}}$—somewhat farther out than in the "relaxed" model, but still small enough that one can reasonably ignore RR when computing the consumption rate of normal stars.

Incoherent resonant relaxation does turn out to be important when computing the rate at which compact remnants—for instance, stellar-mass BHs—are scattered into an SBH from very small distances [245]. This process is discussed in more detail in section 6.4.

5.7 ENCOUNTERS WITH A SPINNING SUPERMASSIVE BLACK HOLE

An SBH is much more massive than a star, and throughout most of this chapter, the response of the SBH to perturbations from stars has been ignored. The sole exception was the discussion of gravitational Brownian motion in section 5.4. But stars (or compact remnants, which are better able to survive tidal stresses) can have a substantial, cumulative effect on the spin angular momentum of an SBH, particularly if there is a net sense of rotation in the nucleus. Stars can affect SBH spins in at least two ways: via spin-orbit torques, the complement of the frame-dragging precession discussed in section 4.5.3; and via captures. As we will see, both mechanisms are capable of permanently changing the spin *direction* of an SBH, and capture can result in changes in the magnitude of the spin as well.

5.7.1 Spin-orbit torques

Consider a Kerr (rotating) SBH surrounded by stars or stellar remnants. Recall from chapter 4 that the spin angular momentum, $\mathcal{S}$, of the hole can be expressed as

$$\mathcal{S} = \chi \frac{GM_\bullet^2}{c},$$

where $0 \le \chi \le 1$ is the dimensionless spin parameter. The acceleration induced by this spin on a star orbiting the SBH was given, to lowest post-Newtonian order, by equation (4.214). In the orbit-averaged approximation, this acceleration was found to induce a precession of the star's orbital angular momentum $\boldsymbol{L}$ about $\mathcal{S}$ according to[16]

$$\dot{\boldsymbol{L}}_j = \boldsymbol{\nu}_j \times \boldsymbol{L}_j, \tag{5.249a}$$

$$\boldsymbol{\nu}_j = \frac{2G\mathcal{S}}{c^2 a_j^3 (1 - e_j^2)^{3/2}}. \tag{5.249b}$$

This is just equation (4.225), now with the subscript j denoting the jth star. The same torque acts back on the SBH, causing its spin to evolve (precess). The instantaneous rate of change of $\mathcal{S}$ is given, again to lowest post-Newtonian order, by [278]

$$\dot{\mathcal{S}} = \frac{2G}{c^2} \sum_{j=1}^{N} \frac{m_j}{r_j^3} \left(\boldsymbol{x}_j \times \boldsymbol{\nu}_j \right) \times \mathcal{S} \tag{5.250a}$$

$$= \frac{2G}{c^2} \sum_{j=1}^{N} \frac{\boldsymbol{L}_j \times \mathcal{S}}{r_j^3}, \tag{5.250b}$$

where m_j is the mass of the jth star whose distance from the SBH is r_j. Since the mean value of r^{-3} over the unperturbed orbit is $a^{-3}(1 - e^2)^{-3/2}$, the orbit-averaged equation for $\dot{\mathcal{S}}$ is

$$\dot{\mathcal{S}} = \boldsymbol{\nu}_\mathcal{S} \times \mathcal{S}, \tag{5.251a}$$

$$\boldsymbol{\nu}_\mathcal{S} = \frac{2G}{c^2} \sum_{j=1}^{N} \frac{\boldsymbol{L}_j}{a_j^3 (1 - e_j^2)^{3/2}}. \tag{5.251b}$$

The vector $\boldsymbol{\nu}_\mathcal{S}$ is the **spin precessional vector**.

The coupled equations (5.249) and (5.251) describe the joint evolution of the SBH spin, and the orbital angular momenta of the N stars, due to mutual spin-orbit torques. In the absence of any other mechanism that acts to change the orbital a_j and $\boldsymbol{L}_j$, these equations are complete. Of course, gravitational encounters between the stars *do* change the orbital elements, and we might expect changes in the $\boldsymbol{L}_j$ due to resonant relaxation to be particularly important. We nevertheless begin by ignoring such effects. As we will see, this "collisionless" approximation is accurate for stars sufficiently close to the SBH, where frame-dragging torques dominate torques from the mutual gravitational interactions.

---

[16]In this section, the symbol $\boldsymbol{L}$ denotes angular momentum and not angular momentum per unit mass.

Because the changes in $\boldsymbol{S}$ and $\boldsymbol{L}_j$ due to frame dragging are perpendicular to the respective vectors, the magnitudes of those vectors are conserved:

$$S \equiv |\boldsymbol{S}| = \text{constant}, \tag{5.252a}$$

$$L_j \equiv |\boldsymbol{L}_j| = \text{constant}, \quad j = 1, \ldots, N. \tag{5.252b}$$

We might also expect conservation of the total (spin plus orbital) angular momentum, defined as

$$\boldsymbol{J}_{\text{tot}} \equiv \boldsymbol{S} + \sum_j \boldsymbol{L}_j = \boldsymbol{S} + \boldsymbol{L}_{\text{tot}}. \tag{5.253}$$

This turns out to be correct:

$$\dot{\boldsymbol{J}}_{\text{tot}} = \dot{\boldsymbol{S}} + \sum_j \dot{\boldsymbol{L}}_j$$

$$= \frac{GM_\bullet^2}{c} (\boldsymbol{v}_S \times \boldsymbol{\chi}) + \sum_j (\boldsymbol{v}_j \times \boldsymbol{L}_j)$$

$$= \frac{2G^2 M_\bullet^2}{c^3} \sum_j \frac{\boldsymbol{L}_j \times \boldsymbol{\chi}}{a_j^3 (1 - e_j^2)^{3/2}} + \frac{2G^2 M_\bullet^2}{c^3} \sum_j \frac{\boldsymbol{\chi} \times \boldsymbol{L}_j}{a_j^3 (1 - e_j^2)^{3/2}}$$

$$= 0.$$

Note that $\boldsymbol{L}_{\text{tot}}$ is *not* conserved, either in magnitude or direction, nor is the spin precessional vector $\boldsymbol{v}_S$. As the SBH precesses, both the magnitude and direction of $\boldsymbol{L}_{\text{tot}}$ may change in order to keep $\boldsymbol{J}_{\text{tot}}$ constant.

As an instructive example, consider the case that all the stars have the same a and e; for instance, the orbits could all lie in a circular ring. There is no differential precession in this case, and

$$\dot{\boldsymbol{S}} = \boldsymbol{v}_0 \times \boldsymbol{S}, \quad \dot{\boldsymbol{L}}_{\text{tot}} = \boldsymbol{v}_0 \times \boldsymbol{L}_{\text{tot}}, \tag{5.254}$$

where

$$\boldsymbol{v}_0 = \frac{J_{\text{tot}}}{S} \boldsymbol{v}_{\text{LT}}, \tag{5.255a}$$

$$\boldsymbol{v}_{\text{LT}} = \frac{2G^2 M_\bullet^2 \chi}{c^3 a^3 (1 - e^2)^{3/2}}$$

$$\approx \frac{(7.0 \times 10^5)^{-1}}{(1 - e^2)^{3/2}} \chi \left(\frac{M_\bullet}{10^6 \, M_\odot} \right)^2 \left(\frac{a}{1 \, \text{mpc}} \right)^{-3} \text{yr}^{-1}. \tag{5.255b}$$

The frequency ν_{LT} is the same quantity that was called ν_J in section 4.5.3: the Lense–Thirring precession frequency for the stars in the ring. In this idealized case, $\boldsymbol{L}_{\text{tot}}$ *is* conserved, and both $\boldsymbol{S}$ and $\boldsymbol{L}_{\text{tot}}$ precess with frequency ν_0 about the fixed vector $\boldsymbol{J}_{\text{tot}}$. The controlling parameter is $\Theta \equiv L_{\text{tot}}/S$. If $\Theta \ll 1$, $\dot{\boldsymbol{S}} \approx 0$ and $\boldsymbol{L}_{\text{tot}}$ precesses about the nearly fixed SBH spin vector at the Lense–Thirring rate. If $\Theta \gg 1$, $\dot{\boldsymbol{L}}_{\text{tot}} \approx 0$ and $\boldsymbol{S}$ precesses about the nearly fixed angular-momentum vector of the stars with frequency $\Theta \times \nu_{\text{LT}} \gg \nu_{\text{LT}}$.

Is it reasonable that the stars in the nucleus should contain more angular momentum than the SBH? The answer, of course, is yes, at least if stars sufficiently far from the center are included, and if there is a nontrivial degree of net rotation in the stellar motions. Consider, for instance, the "clockwise disk" at the Galactic center. Its parameters are [29, 397, 415]

$$5000 \, M_\odot \lesssim M_{\mathrm{CWD}} \lesssim 15,000 \, M_\odot,$$

$$r_{\mathrm{inner}} \approx 0.05 \, \mathrm{pc}, \quad r_{\mathrm{outer}} \approx 0.5 \, \mathrm{pc},$$

$$\langle e \rangle \approx 0.2.$$

The total angular momentum of the disk is roughly

$$L_{\mathrm{tot}} \approx M_{\mathrm{CWD}} \sqrt{G M_\bullet r_{\mathrm{CWD}}},$$

and so

$$\Theta \approx \frac{1}{\chi} \frac{M_{\mathrm{CWD}}}{M_\bullet} \sqrt{\frac{r_{\mathrm{CWD}}}{r_{\mathrm{g}}}} \tag{5.256a}$$

$$\approx \frac{2}{\chi} \left(\frac{M_{\mathrm{CWD}}}{10^4 \, M_\odot} \right) \left(\frac{r_{\mathrm{CWD}}}{0.1 \, \mathrm{pc}} \right)^{1/2}. \tag{5.256b}$$

Even if χ is as large as one, the clockwise disk still contains roughly as much angular momentum as the SBH. Evidently, this structure torques the SBH about as much as it is torqued by it! However, the mutual precession time is long:

$$\frac{\pi}{\nu_{\mathrm{LT}}} \approx 8 \times 10^{10} \, \chi^{-1} \left(\frac{R_{\mathrm{CWD}}}{0.1 \, \mathrm{pc}} \right)^3 \, \mathrm{yr}, \tag{5.257}$$

much longer than the $\sim 10^7$ yr age of the disk inferred from the properties of its stars. Nevertheless, this example demonstrates that identified structures near the Galactic center can easily contain a net orbital angular momentum that exceeds $\mathcal{S}$, and the same may well be true in other nuclei. But if timescales associated with spin precession are to be interestingly short, then (at least in a galaxy like the Milky Way) there must be a significant amount of rotation in stars that are somewhat closer to the SBH than 0.1 pc.

Suppose that the nucleus is approximately spherical. A reasonable guess for the distribution of orbital elements near the SBH is equation (4.36):

$$N(a, e) \, da \, de = N_0 \, a^{2-\gamma} \, da \, e \, de, \tag{5.258}$$

which corresponds to a configuration-space density $\rho(r) = \rho_0 (r/r_0)^{-\gamma}$, with

$$m_\star N_0 = \frac{8 \pi^{3/2}}{2^\gamma} \frac{\Gamma(\gamma + 1)}{\Gamma(\gamma - 1/2)} \rho_0 r_0^\gamma, \quad \gamma > 1/2, \tag{5.259}$$

and a "thermal" distribution of eccentricities. Rotation of such a cluster about an axis—the z-axis, say—can be induced by identifying those orbits with $L_{j,z} < 0$ and "flipping" some fraction, F, of them, that is, reversing their motion, and therefore changing the sign of $\mathbf{L}_j$. For such a model, the spin precessional vector is given by

$$v_S = \frac{2G}{c^2}(Fe_L) \sum_j \frac{m_j \left[GM_{\bullet} a_j (1 - e_j^2) \right]^{1/2}}{a_j^3 (1 - e_j^2)^{3/2}} \qquad (5.260a)$$

$$\to \frac{2G^{3/2} M_{\bullet}^{1/2}}{c^2}(Fe_L) N_0 m_{\star} \int \int \frac{da\, e\, de}{a^{1/2+\gamma}(1 - e^2)}, \qquad (5.260b)$$

where e_L is a unit vector in the direction of L_{tot}.

Formally, equation (5.260) states that stars with sufficiently small a or large e can cause S to precess at an arbitrarily high rate. But in reality, stars nearest to the SBH will precess so quickly, compared with the rate of precession of S, that their time-averaged torque on the SBH is effectively zero. Furthermore, there will be a region around the SBH in which the *instantaneous* torque from the stars will drop to zero, as differential precession distributes the L_j uniformly about S. Roughly speaking, this will be the region containing a total $|L|$ equal to S, since stars in this region precess faster than the SBH is precessed by the stars.

This argument suggests that an important quantity is a_L, defined as the value of a such that

$$L_{\text{tot}}(a < a_L) = S = \chi \frac{GM_{\bullet}^2}{c}. \qquad (5.261)$$

Figure 5.8 shows how a_L varies with nuclear properties, assuming the orbital distribution of equation (5.258), and two different parametrizations of the density: in terms of the stellar mass $M_{0.1}$ within 0.1 pc; and in terms of r_m, the gravitational influence radius, assuming that r_m scales with $M_{\bullet}$ as in equation (2.16). The former parametrization is most suitable for low-mass galaxies containing (mostly unresolved) NSCs, the latter for massive galaxies with cores. In massive galaxies, and for $\chi/F \approx 1$,

$$10^{-2} r_m \lesssim a_L \lesssim 10^{-1} r_m.$$

Figure 5.9 shows the results of integrating the coupled equations (5.249) and (5.251). The model parameters are $M_{\bullet} = 10^6 M_{\odot}$, $\chi = 1$, $\gamma = 1$, $F = 1/2$, $M_{0.1} \approx 6 \times 10^4 M_{\odot}$ and $a_L \approx 15$ mpc. Because the total stellar angular momentum in this model greatly exceeds S, one might naturally expect the evolution to consist of nearly uniform precession of S about L_{tot}. Such evolution can occur, as shown in the lower set of figures. But changing the initial conditions just slightly— decreasing the initial angle between S and L_{tot} from $70°$ to $40°$—produces a qualitatively different sort of evolution, in which S, L_{tot} and v_S reach nearly complete alignment after less than one precessional period in S. Evolution of the second sort, or "damped precession," is not excluded by the conservation laws (5.253), (5.252) for any initial conditions, but numerical experiments show that it is usually associated with small initial angles between S and L_{tot} [379].

In the examples of figure 5.9, spin precessional times are long compared with the times over which stellar angular momenta might be expected to change as a result of star–star interactions, via the mechanism of resonant relaxation discussed in the

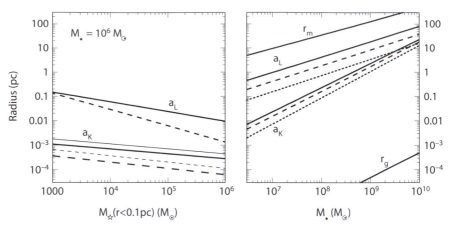

Figure 5.8 Radii associated with spin-orbit torques in galactic nuclei [379]. a_L, equation
(5.261), is the radius containing a total angular momentum in stars equal to S,
computed assuming $F = 1/2$ and $\chi = 1$ (maximum rotation of SBH and stellar
cluster). a_K, equation (5.263), is the radius of rotational influence of the SBH,
assuming $\chi = 1$. The left panel assumes $M_\bullet = 10^6 M_\odot$ and the stellar density
is parametrized in terms of $M_{0.1}$, the mass within 0.1 pc, and γ, the power-law
index; solid lines are for $\gamma = 1$ and dashed lines for $\gamma = 2$. In the case of a_K, two
values are assumed for the stellar mass: $m_\star = 1 M_\odot$ (thin lines) and $m_\star = 10 M_\odot$
(thick lines). The right panel, for massive galaxies, assumes the relation (2.16)
between $M_\bullet$ and the influence radius r_m; solid, dashed and dotted lines are for
$\gamma = 5/8$, 1, and 3/2, respectively. The curves for a_K in the right panel assume
$m_\star = 1 M_\odot$.

previous section. Recall that the shortest such timescale was associated with (coherent or incoherent) "2d resonant relaxation" (2dRR): the time for mutual torques
between stars to randomize orbital planes, that is, the directions of the L_j. Indeed,
in section 4.9 it was pointed out that this time becomes shorter than the Lense–
Thirring time for orbits closer than $\sim 10^{-3}$ pc from the Milky Way SBH, and figure 4.25 showed a numerical integration of orbits inside the region where changes
in the L_j due to gravitational encounters overwhelm changes due to frame dragging.

  If we adopt equation (5.200) as an estimate of the time for 2dRR to randomize
orbital planes, the condition that this time be longer than the Lense–Thirring time,
equation (5.257), is

$$\left(1 - e^2\right)^3 \left(\frac{a}{r_g}\right)^3 \lesssim \frac{16\chi^2}{N(a)} \left(\frac{M_\bullet}{m_\star}\right)^2, \tag{5.262}$$

where $N(a)$ is the number of stars with semimajor axes less than a. Approximating
$N(a)$ as

$$N(a) = \frac{2M_\bullet}{m_\star} \left(\frac{a}{r_m}\right)^{3-\gamma},$$

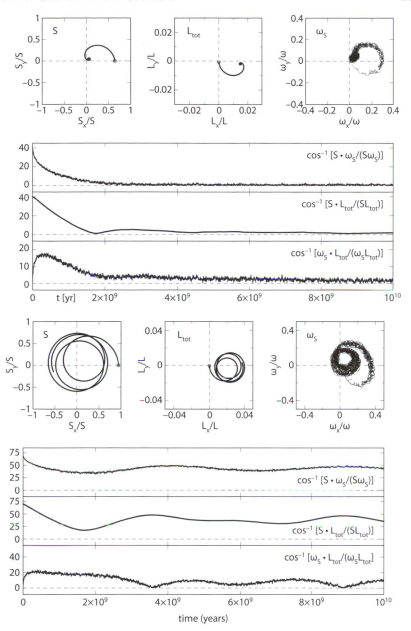

Figure 5.9 This figure illustrates the joint evolution of $\boldsymbol{S}$, $\boldsymbol{L}_{\rm tot}$, and the spin precessional vector (written $\boldsymbol{\omega}_S$) due to mutual, spin-orbit torques, equations (5.249) and (5.251) [379]. Stars were distributed initially according to equation (5.258), a sphere in configuration space, with rotation induced by requiring all stars to orbit in the same sense around the z-axis. The other model parameters are given in the text. The only difference between top and bottom is the initial angle between $\boldsymbol{S}$ and $\boldsymbol{L}_{\rm tot}$: $40°$ at the top and $70°$ at the bottom. In the first case, all vectors align in roughly one precessional time, while in the second, the SBH continues to precess.

we can write the condition (5.262) as

$$\left(1 - e^2\right)^3 \left(\frac{a}{a_{\mathrm{K}}}\right)^{6-\gamma} \lesssim 1, \tag{5.263a}$$

$$a_{\mathrm{K}} = r_{\mathrm{g}} \left(8\chi^2 \frac{M_\bullet}{m_\star}\right)^{1/(6-\gamma)} \left(\frac{r_{\mathrm{m}}}{r_{\mathrm{g}}}\right)^{(3-\gamma)/(6-\gamma)}. \tag{5.263b}$$

The radius a_{K} is the **rotational influence radius** of the SBH. Just as r_{h} or r_{m} define the sphere inside of which the gravitational *force* from the SBH dominates the force from the stars, so a_{K} defines the size of the region inside of which the *torque* exerted by the (spinning) SBH dominates the collective torque from the other stars.

To get an idea of the magnitude of a_{K}, we can set $\gamma = 1$ in equation (5.263), and adopt the empirical relation (2.16) between $M_\bullet$ and r_{m}. The result—appropriate for massive galaxies—is

$$a_{\mathrm{K}} \approx 0.16 \, \chi^{2/5} \left(\frac{M_\bullet}{10^8 \, M_\odot}\right)^{1.0} \left(\frac{m_\star}{M_\odot}\right)^{-1/5} \mathrm{pc} \quad (\gamma = 1). \tag{5.264}$$

The scaling turns out to be nearly linear with $M_\bullet$, allowing us to say that for a rapidly spinning SBH, the radius of rotational influence extends $\sim 10^4$ times farther than r_{g} (figure 5.8). This is yet another example of how relativistic effects in galactic nuclei can be important far from the SBH event horizon.

Angular momenta of stars satisfying the condition (5.263) evolve "collisionlessly" in response to frame dragging, unaffected by perturbations from other stars. Since $a_{\mathrm{K}} < a_{\mathrm{L}}$, differential precession will allow these stars to distribute their angular momentum vectors uniformly about $\mathcal{S}$ in a time shorter than the precession time for $\mathcal{S}$. Orbits of stars beyond a_{K} evolve essentially independently of $\mathcal{S}$, in response to mutual gravitational perturbations. But although the L_j of stars in this region are randomized by the mutual torques, gravitational encounters, by themselves, leave L_{tot} unchanged for these stars. Now, the torque that the stars in this outer region exert on the SBH is determined by $v_{\mathcal{S}}$, not by L_{tot}; but conservation of L_{tot} implies that the spin precessional vector will fluctuate, stochastically, about some mean vector that is essentially constant over time and that points in roughly the same direction as L_{tot}. Furthermore, since these fluctuations occur with a characteristic time that is short compared with $v_{\mathcal{S}}^{-1}$, the SBH takes little notice of them, precessing smoothly about the mean $v_{\mathcal{S}}$. Detailed modeling [379] suggests that typical spin precessional periods are $\sim 10^7$–10^8 yr for low-mass SBHs in dense nuclei, $\sim 10^8$–10^{10} yr for SBHs with masses $\sim 10^8 \, M_\odot$, and $\sim 10^{10}$–10^{11} yr for the most massive SBHs.

As discussed briefly in chapter 2, evidence for SBH precession does exist, in the form of changes over time in the directions of radio jets or lobes in active galaxies. According to current models, production of the jets requires the presence of a gaseous accretion disk, and models for jet precession have most often invoked a misaligned accretion disk as the source of the torque on the spinning SBH [23]. This is reasonable, since an accretion disk, if present, would probably dominate the torque from the stars near the SBH.

Structures like the "clockwise disk" in the Milky Way are young, and their presence implies that infall of gas, followed by star formation, probably occurs episodically. If this is the case, the direction of the spin-orbit torques acting on the SBH will change in some near-random fashion after each star formation event. The consequences of such evolution have yet to be explored.

5.7.2 Capture

The spin-orbit torques described above leave the magnitude of the hole's spin unchanged. But capture of a star or stellar remnant will change both the direction and the magnitude of $\mathcal{S}$, since the orbital angular momentum and mass-energy of the star are added to those of the SBH. The change in χ after a single capture is computed by evaluating the star's energy and angular momentum on the capture orbit and adding them to the energy and angular momentum of the SBH; the very slight losses due to gravitational radiation during the final plunge can safely be ignored.

In section 4.6, we found that L_c, the angular momentum of the innermost stable circular orbit (ISCO), varies from $L_c/m_\star = \sqrt{12}r_g c$ for a test mass around a non-spinning hole, to $L_c/m_\star = 1(9)r_g c$ for direct (retrograde) orbits in the equatorial plane of a maximally spinning hole. The much larger value of L_c in the case of retrograde capture implies that a rapidly rotating hole will typically spin down if capture occurs from random directions [124, 197, 572].

The parameters defining the ISCO depend on an orbit's inclination with respect to the SBH spin direction. The inclination dependence was derived in section 4.6 in the limit of large orbital eccentricities. Here we restrict ourselves to the simpler case of capture from circular orbits [247]. If one repeats the derivation leading to equation (4.255), now allowing for arbitrary inclination, the numerical results are found to be reasonably well approximated by

$$\zeta(\mu) \approx |\zeta_-| + \frac{1}{2}(\mu + 1)(\zeta_+ - |\zeta_-|), \tag{5.265}$$

where ζ can stand for r_c, $E_{\infty,c}$, or L_c, the subscripts $+$ and $-$ refer to direct and retrograde orbits in the equatorial plane, and

$$\mu \equiv \cos i = \frac{L_z}{L_z^2 + L_\perp^2} \tag{5.266}$$

with $L_z = \mu L$, $L_\perp = \sqrt{1 - \mu^2}L$. If the mass and spin of the SBH are M and $\mathcal{S} = (GM^2/c)\chi \boldsymbol{e}_z$ before capture, their values after capture are

$$M' = M\left[1 + qcE_{\infty,c}(\chi, \mu)\right], \tag{5.267a}$$

$$S'_z = \frac{GM^2}{c}\chi + qL_{z,c}(\chi, \mu), \tag{5.267b}$$

$$S'_\perp = qL_{\perp,c}(\chi, \mu), \tag{5.267c}$$

where $q \equiv m/M$. The new value of the dimensionless spin is

$$\chi' = \frac{c\sqrt{S'^2_z + S'^2_\perp}}{GM'^2} \tag{5.268}$$

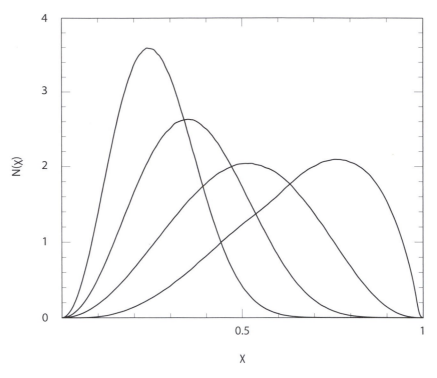

Figure 5.10 Steady-state spin distributions produced by successive capture from random directions at fixed mass ratio q, for $q = (1/32, 1/16, 1/8, 1/4)$. Curves were generated using Monte Carlo experiments based on the test-particle approximation, $q \ll 1$.

and the new spin vector is inclined by an angle $\delta\theta$ with respect to the original spin, where

$$\cos \delta\theta = \frac{\mathcal{S}'_z}{\sqrt{\mathcal{S}'^2_z + \mathcal{S}'^2_\perp}}. \tag{5.269}$$

We are interested here in the limiting forms of these expressions for small q. (Indeed, since our treatment is based on the test-particle equations of motion, it is only valid in this limit.) The change in the dimensionless spin works out to be [247]

$$\delta\chi \approx -2q\chi + \frac{L_{c,z}}{r_g Mc}, \quad (q \ll 1). \tag{5.270}$$

The first term in equation (5.270) describes conservation of spin angular momentum of the SBH as its mass grows, $\chi \propto M^{-2}$, while the second term accounts for the additional angular momentum brought in by the smaller body. The change in orientation is, to lowest order in q,

$$\chi \delta\theta = \frac{q}{r_g Mc} L(\chi, \mu)\sqrt{1 - \mu^2}, \quad q \ll 1. \tag{5.271}$$

Assuming successive mergers from random directions with fixed mass ratio q (equivalent to assuming that $m_\star$ grows proportionately to $M_\bullet$—mathematically simple but not very realistic) leads to a steady-state spin distribution $N(\chi)$ that is uniquely determined by q. For small q, this distribution can be derived from a Fokker–Planck equation [247]:

$$N(\chi)d\chi \approx N_0 \chi^2 e^{-3\chi^2/2\chi_{\text{rms}}^2}d\chi, \quad \chi_{\text{rms}} \approx 1.58\sqrt{q}. \tag{5.272}$$

Equation (5.272) reflects the competition between captures that increase spin and captures that decrease spin. For small χ, equation (5.270) implies that $\delta\chi > 0$, while for $\chi \gtrsim \sqrt{q}$ the typical effect of capture from random directions is to decrease spin; the SBH spin evolves to a value such that spinup and spindown occur with equal frequency on average. This situation is similar to the Brownian motion of a massive particle, in the sense that the "frictional" force (due in this case to retrograde captures) increases with the amplitude of the velocity (i.e., the spin).

Figure 5.10 shows the steady state $N(\chi)$ for various values of q. The Gaussian form of equation (5.272) is seen to be accurate for $q \lesssim 0.1$.

The orientation evolves as a random walk; there is no "resistive" term. The typical change in spin direction, after the SBH has grown by a mass ΔM, is [247]

$$\langle\delta\theta\rangle \approx 2.7\sqrt{\frac{\Delta M}{M}\frac{q}{\chi^2}}. \tag{5.273}$$

Chapter Six

Loss-Cone Dynamics

A supermassive black hole (SBH) at the center of a galaxy acts like a sink, removing stars that come sufficiently close to it. This removal can occur in one of two ways, depending on the mass of the SBH and on the physical properties of the passing star. At one extreme, the "star" can itself be a gravitationally compact object: a stellar-mass black hole or a neutron star. For such objects, tidal stresses from the SBH are unimportant, and removal occurs only when the object finds itself on an orbit that takes it inside the SBH event horizon. The properties of capture orbits were derived in section 4.6. In the case of circular orbits around nonspinning (Schwarzschild) holes, the innermost stable radius is $6r_g$ where $r_g \equiv GM_\bullet/c^2$ is the gravitational radius of the SBH. This changes to $1(9)r_g$ in the case of prograde (retrograde) circular orbits in the equatorial plane of a maximally spinning SBH. Circular orbits are not terribly likely, however, unless the mass of the infalling object is large enough that some process like dynamical friction, or gravitational-wave energy loss, can circularize the orbit prior to capture. In the case of stellar-mass objects, capture is more likely to be preceded by scattering onto an eccentric orbit. For such orbits, it was shown in section 4.6 that the critical angular momentum for capture by a nonrotating SBH is $\sim 4GM_\bullet/c$; the periapsis of a Newtonian orbit with this value of L is $\sim 8r_g$. This critical radius changes to $\sim 2(12)r_g$ for prograde (retrograde) orbits around maximally rotating SBHs.

Ordinary stars can also be swallowed whole, but only if they manage to resist being pulled apart by tidal stresses from the SBH. The strength of a tidal encounter can be expressed in terms of the parameter η, the square root of the ratio between the surface gravity of the star and the tidal acceleration due to the SBH near periapsis passage, $r = r_p$:

$$\eta \equiv \left(\frac{r_p^3}{GM_\bullet R_\star} \frac{Gm_\star}{R_\star^2} \right)^{1/2}, \tag{6.1}$$

where $m_\star$ and $R_\star$ are the mass and radius of the star. The right-hand side of equation (6.1) is also the ratio between the duration of periapsis passage and the hydrodynamic timescale of the star [437]. It is customary to define the **tidal disruption radius**, r_t, as the value of r_p that satisfies this expression, or

$$r_t = \left(\eta^2 \frac{M_\bullet}{m_\star} \right)^{1/3} R_\star \tag{6.2a}$$

$$\approx 1.1 \times 10^{-5} \eta^{2/3} \left(\frac{M_\bullet}{10^8 \, M_\odot} \right)^{1/3} \left(\frac{m_\star}{M_\odot} \right)^{-1/3} \left(\frac{R_\star}{R_\odot} \right) \, \text{pc}. \tag{6.2b}$$

The quantity η then becomes a "form factor," of order unity, that can be calculated given the internal properties of the star.

Main-sequence stars can be modeled reasonably well as stellar-dynamical **polytropes**, with gaseous equations of state $P = K\rho^\gamma = K\rho^{(n+1)/n}$; n is the "polytropic index" [77]. For stars like the Sun, $n \approx 3$ ($\gamma \approx 4/3$), and one finds $\eta \approx 0.844$. The following are values of η corresponding to other polytropic indices [120].

n:	3	2	1.5	1	0
η:	0.844	1.482	1.839	2.223	3.074

Rewriting equation (6.2b) as

$$\Theta \equiv \frac{r_t}{r_g} = \eta^{2/3} \left(\frac{M_\bullet}{m_\star} \right)^{1/3} \frac{R_\star}{r_g}$$

$$\approx 2.2 \, \eta^{2/3} \left(\frac{M_\bullet}{10^8 \, M_\odot} \right)^{-2/3} \left(\frac{m_\star}{M_\odot} \right)^{-1/3} \frac{R_\star}{R_\odot}, \qquad (6.3)$$

we see that tidal disruption occurs outside of the SBH's event horizon for solar-type stars when $M_\bullet \lesssim 10^8 \, M_\odot$; but when $M_\bullet \gtrsim 10^8 \, M_\odot$, stars like the Sun can be captured whole. Figure 6.1, which is based on detailed models of stellar interiors, shows the maximum value of $M_\bullet$ for which tidal disruption of main-sequence stars can occur. The most massive SBHs, $M_\bullet \approx 10^9 \, M_\odot$, can only disrupt main-sequence stars with $m_\star \gtrsim 10^2 \, M_\odot$. Such massive stars have very short lifetimes on the main sequence, and in the absence of ongoing star formation, the only stars with comparable radii that would be present are red giants or asymptotic-giant-branch stars. The relatively short lifetimes of these giant phases, plus the fact that "disruption" of a red giant may leave its structure nearly unchanged, complicates the calculation of tidal event rates in giant galaxies, as discussed in more detail in section 6.1.4.

SBHs of sufficiently *low* mass can disrupt objects even smaller than the Sun. For instance, degenerate dwarfs have radii [396]

$$\frac{R}{R_\odot} \approx 1.1 \times 10^{-2} \frac{\left[1 - (m_{\rm WD}/M_{\rm Ch})^{4/3} \right]^{1/2}}{(m_{\rm WD}/M_{\rm Ch})^{1/3}}, \qquad M_{\rm Ch} \approx 1.4 \, M_\odot, \qquad (6.4)$$

or $\sim 10^{-2} \, R_\odot$ for $m_{\rm WD} \approx 1 \, M_\odot$. White dwarfs are well approximated as polytropes, with $n \approx 3/2$ ($\eta \approx 1.8$) at low masses, increasing to $n \approx 3$ ($\eta \approx 0.8$) for $m \lesssim M_{\rm Ch}$ [492]. Thus,

$$\Theta_{\rm WD} \approx 0.5 \left(\frac{M_\bullet}{10^6 \, M_\odot} \right)^{-2/3} \left(\frac{m_{\rm WD}}{M_{\rm Ch}} \right)^{-1/3} \left(\frac{R_{\rm WD}}{10^{-2} \, R_\odot} \right). \qquad (6.5)$$

In the Milky Way, $M_\bullet \approx 4 \times 10^6 \, M_\odot$, and equations (6.4)–(6.5) imply that tidal disruption of a white dwarf is only possible for $m_{\rm WD} \lesssim 0.2 \, M_\odot$. In fact, a more accurate calculation [579] gives for the maximum radius of a carbon white dwarf $R_{\rm WD} \approx 3.9 \times 10^{-2} \, R_\odot$ corresponding to a mass of $\sim 2.2 \times 10^{-3} \, M_\odot$. Even for such an extreme object, equation (6.5) implies $\Theta \approx 0.6$, and for this reason, tidal disruption of degenerate dwarfs is usually only discussed in the context of (hypothetical) intermediate-mass black holes with $M_\bullet \lesssim 3 \times 10^5 \, M_\odot$. However, it has been argued that tidal stresses can significantly change the internal structure of a white dwarf even if it is not fully disrupted [73].

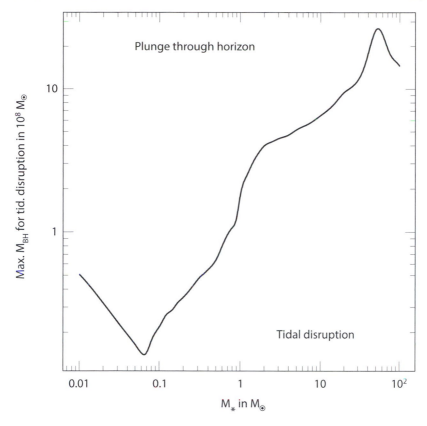

Figure 6.1 Maximum mass of the SBH for which tidal disruption of stars on the main sequence can occur, as a function of main-sequence mass [171].

Tidal disruption of stars is important as a test of the SBH paradigm. Disrupted
stars are expected to produce X- and UV radiation with luminosities of
$\sim 10^{44}$ erg s^{-1}, potentially outshining their host galaxies [451]. Debris from a disrupted star is launched onto orbits which span an energy range $\delta E \approx GM_\bullet R_\star / r_t^2$
[450]. This energy range is large compared with the energy of the (highly eccentric)
initial orbit, and so roughly half the liberated material will be unbound while the
other half will begin to fall back onto the SBH. For main-sequence stars, the fall-
back rate at late times should decline as $\sim t^{-5/3}$ [429]. A handful of X-ray flaring
events have been observed that have the expected signatures [291], and the number
of detections is at least roughly consistent with theoretical estimates of the event
rate [123]. Neutron stars or stellar-mass black holes would not be disrupted, but
these objects can emit gravitational waves at potentially observable amplitudes and
frequencies before spiraling in, and may dominate the event rate for low-frequency
(space-based) gravitational-wave interferometers [496].

The rate of either sort of event is determined by the rate at which stars or
compact remnants pass within some critical distance from the SBH. We define the

loss-cone radius r_{lc} to be the larger of the tidal disruption radius, r_t, or the radius of capture, r_c, for stars of a given type. An orbit that just grazes the sphere at $r = r_{\mathrm{lc}}$ has angular momentum

$$L_{\mathrm{lc}}^2(E) = 2r_{\mathrm{lc}}^2 [E - \Phi(r_{\mathrm{lc}})] \approx 2GM_\bullet r_{\mathrm{lc}}; \tag{6.6}$$

the latter expression assumes $|E| \ll GM_\bullet/r_{\mathrm{lc}}$, that is, that the star is on an orbit with semimajor axis much greater than r_{lc}. Within the SBH influence sphere, orbits are nearly Keplerian, and equations (6.3) and (6.6) imply a limiting eccentricity

$$1 - e_{\mathrm{lc}}^2 \approx \frac{L_{\mathrm{lc}}^2}{GM_\bullet a} \approx 2\Theta \left(\frac{a}{r_{\mathrm{g}}}\right)^{-1}, \quad a \ll r_{\mathrm{h}} \tag{6.7}$$

with Θ now defined in terms of r_{lc} rather than r_t. Orbits with $L \leq L_{\mathrm{lc}}$ are called **loss-cone orbits**, and the ensemble of such orbits is sometimes called simply the "loss cone."[1]

The loss cone can be visualized as the set of velocity vectors, at some distance r from the SBH, that are associated with orbits that pass within r_{lc}. To satisfy this condition, a star's velocity vector must lie within a cone of half-angle θ_{lc} that is given approximately by

$$\theta_{\mathrm{lc}} \approx \begin{cases} (r_{\mathrm{lc}}/r)^{1/2}, & r \lesssim r_{\mathrm{h}}, \\ (r_{\mathrm{lc}} r_{\mathrm{h}}/r^2)^{1/2}, & r \gtrsim r_{\mathrm{h}} \end{cases} \tag{6.8}$$

(figure 6.2a). These relations follow from equation (6.6), $L_{\mathrm{lc}}^2 \approx 2GM_\bullet r_{\mathrm{lc}}$, the first after setting $v(r) \approx \sqrt{GM_\bullet/r}$ and the second after setting $v \approx \sigma$.

In a spherical galaxy, the number of stars with angular momenta small enough to satisfy equation (6.6) would ordinarily be small; furthermore, these stars would be removed at the first periapsis passage, that is, after a single orbital period. Continued supply of stars to the SBH requires some mechanism for **loss-cone repopulation**: new stars need to be transferred onto loss-cone orbits, and the rate of supply of stars to the SBH will be determined by the efficiency of the resupply process.

An upper limit to the consumption rate in a spherical galaxy comes from assuming that stars are *instantaneously* replaced after being consumed by the SBH. In this **full-loss-cone** model, stars at energy E are consumed at a constant rate (stars per unit time) of

$$F^{\mathrm{flc}}(E) \, dE \approx P(E)^{-1} N_{\mathrm{lc}}(E) \, dE, \tag{6.9}$$

where $P(E)$ is the period of a nearly radial orbit of energy E and $N_{\mathrm{lc}}(E) \, dE$ is the number of stars at energies E to $E + dE$ on orbits with $L \leq L_{\mathrm{lc}}$. In a spherical galaxy, we know from equation (5.159) that

$$N(E, L)dE \, dL = 8\pi^2 L f(E, L) P(E, L) dE \, dL,$$

---

[1]This term derives from plasma physics: in a magnetic-mirror machine, particles are trapped if their velocity vectors lie outside a cone in velocity space whose opening angle is determined by the properties of the magnetic field [301].

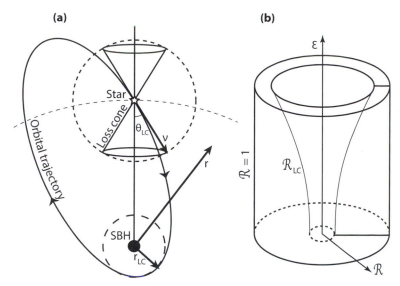

Figure 6.2 Two representations of the loss cone. (a) Orbits with velocity vectors that fall within the cone $\theta \leq \theta_{\mathrm{lc}}$ will pass within the capture/disruption sphere at $r = r_{\mathrm{lc}}$; the angle θ_{lc} is given approximately by equation (6.8). (b) In energy–angular-momentum space, the loss cone consists of orbits with $L \leq L_{\mathrm{lc}}$ (equation 6.6). In this figure, $\mathcal{E} \equiv -E$ and $\mathcal{R} \equiv L^2/L_c(E)^2$; $\mathcal{R} = 0$ corresponds to radial orbits and $\mathcal{R} = 1$ to circular orbits. The representation of the loss cone as a cylinder is motivated by the fact that the differential equation describing SBH feeding in a spherical galaxy, equation (6.29), has the same form mathematically as the equation describing the flow of heat in an infinite cylinder.

and so

$$N_{\mathrm{lc}}(E)\, dE \approx 4\pi^2 L_{\mathrm{lc}}^2(E) P(E) f(E)\, dE, \tag{6.10a}$$

$$F^{\mathrm{flc}}(E) \approx 4\pi^2 L_{\mathrm{lc}}^2(E) f(E) \approx 8\pi^2 GM_\bullet r_{\mathrm{lc}} f(E), \tag{6.10b}$$

where isotropy in velocity space has been assumed. Using equation (3.54), this can also be written as

$$F^{\mathrm{flc}}(E) \approx \frac{N(E)}{P(E)} \frac{L_{\mathrm{lc}}^2(E)}{L_c^2(E)}, \tag{6.11}$$

showing that a fraction $\sim L_{\mathrm{lc}}^2/L_c^2$ of the stars at E get into the SBH each radial period.

Within the gravitational influence sphere, if the density follows a power law in radius, $n \propto r^{-\gamma}$, equation (3.49) tells us that $f \propto |E|^{\gamma-3/2}$. For $\gamma \gtrsim 3/2$, f increases inward (i.e., toward larger $|E|$), while in galaxies with flat nuclear density profiles, f typically peaks at energies near $\Phi(r_{\mathrm{h}})$ then falls outward, since stellar densities generally fall more rapidly than $r^{-3/2}$ beyond $\sim r_{\mathrm{h}}$. In either case, most of the flux comes from stars inside the influence sphere. Inserting $f(E)$ from equation (3.49) into (6.10) then yields

$$F^{\text{flc}}(E) \approx (3 - \gamma)\sqrt{\frac{2}{\pi}}\frac{\Gamma(\gamma+1)}{\Gamma(\gamma-\frac{1}{2})}\frac{M_\bullet}{m_\star}\frac{r_{\text{lc}}}{\sqrt{GM_\bullet r_{\text{m}}^3}}\left(\frac{|E|}{\phi_0}\right)^{\gamma-3/2}, \qquad (6.12)$$

$$\frac{1}{2} < \gamma < 3,$$

for $|E| \gtrsim \phi_0$; recall that r_{m} is the radius containing a stellar mass equal to $2M_\bullet$, and $\phi_0 \equiv GM_\bullet/r_{\text{m}} \approx \sigma^2$.

Integrating this expression with respect to energy gives the total consumption rate (stars lost per unit time) in the full-loss-cone model:

$$\dot{N}^{\text{flc}} \approx \int_{-\infty}^{\phi_0} F^{\text{flc}}(E)dE. \qquad (6.13)$$

For $\gamma > 1/2$, the integral diverges at large $|E|$ due to the rapid rate of consumption of the most bound stars. On the other hand, $\gamma = 1/2$ is the smallest value consistent with an isotropic $f(E)$ around an SBH! This means that the consumption rate corresponding to a "full loss cone" is a rather poorly defined quantity—perhaps helping to explain why different authors give rather different estimates for this quantity.

The divergence in the total consumption rate is a result of the very short orbital periods near the SBH. In reality, it is unlikely that those very bound orbits would be refilled as efficiently as orbits near, say, $r = r_{\text{h}}$. We can define an alternative, and more conservative, full-loss-cone consumption rate as the number of stars on loss-cone orbits within the SBH influence radius, divided by the orbital period *at that radius*. The first quantity is

$$\int_{-\infty}^{\phi_0} 4\pi^2 L_{\text{lc}}^2 P f dE \approx \pi^{1/2}\frac{(3-\gamma)}{(2-\gamma)}\frac{\Gamma(\gamma+1)}{\Gamma(\gamma-1/2)}\frac{M_\bullet}{m_\star}\frac{r_{\text{lc}}}{r_{\text{m}}}, \quad \frac{1}{2} < \gamma < 2. \qquad (6.14)$$

Assuming that these stars are lost in a time equal to the Keplerian orbital period when $|E| = \phi_0$, or

$$\frac{\pi}{\sqrt{2}}\frac{GM_\bullet}{\phi_0^{3/2}} = \frac{\pi}{\sqrt{2}}\left(\frac{r_{\text{m}}^3}{GM_\bullet}\right)^{1/2},$$

gives a (mass) consumption rate of

$$m_\star\dot{N} \approx \sqrt{\frac{2}{\pi}}\frac{(3-\gamma)}{(2-\gamma)}\frac{\Gamma(\gamma+1)}{\Gamma(\gamma-1/2)}\frac{r_{\text{lc}}}{r_{\text{m}}}\left(\frac{GM_\bullet}{r_{\text{m}}^3}\right)^{1/2}M_\bullet, \quad \frac{1}{2} < \gamma < 2. \qquad (6.15)$$

For $M_\bullet \gtrsim 10^8 M_\odot$ and solar-type stars, we can equate r_{lc} with the SBH capture radius $\sim 8GM_\bullet/c^2$. If we are willing to ignore the distinction between r_{m} and $r_{\text{h}} \equiv GM_\bullet/\sigma^2$, the dimensional quantities in equation (6.15) then become

$$\frac{r_{\text{lc}}}{r_{\text{m}}}\left(\frac{GM_\bullet}{r_{\text{m}}^3}\right)^{1/2}M_\bullet \approx 8\frac{\sigma^5}{Gc^2}$$

$$\approx \frac{7\times 10^7}{10^{10}}\left(\frac{\sigma}{200\,\text{km s}^{-1}}\right)^5 M_\odot\,\text{yr}^{-1}. \qquad (6.16)$$

While it took a fair bit of hand waving to get to this point, the result is remarkable: a growth rate that reproduces the $M_\bullet$–σ relation after $10\,\mathrm{Gyr}$, both in slope and normalization [581]!

Models that invoke a full loss cone have an obvious weakness: they require some mechanism for repopulating the depleted orbits on a timescale that is comparable with orbital periods. At least in spherically symmetric or axisymmetric nuclei, it is hard to think of such a mechanism. Efficient loss-cone repopulation is more plausible during periods of rapid potential change, for example, during the galaxy mergers that are thought to accompany quasar activity. Some early discussions of quasar fueling [236, 410] invoked a full loss cone; for instance, in the "black tide" model [574], gas from tidally disrupted stars radiates as it spirals into the SBH, and the SBH shines as a quasar until its mass reaches $\sim 10^8\,M_\odot$ at which point stars are swallowed whole and the quasar fades.

Under more steady-state conditions, gravitational encounters can always be counted on to repopulate loss-cone orbits, although at a rate that may be very low. Gravitational encounters also drive the distribution of orbital energies toward the Bahcall–Wolf form near the SBH: $f \sim |E|^{1/4}$, $n \sim r^{-7/4}$. But as discussed in chapter 5, the Bahcall–Wolf solution is characterized by a flux of stars into the SBH that is essentially zero, in the sense that $|F| \ll N(< r_\mathrm{h})/T_r(r_\mathrm{h})$—indeed we derived that solution by setting the flux in energy space *precisely* to zero. To the extent that stars find their way into the SBH via diffusion in energy, the feeding rate implied by these models would appear to be many orders of magnitude smaller than the full-loss-cone rate.

This discouraging conclusion led almost immediately [168, 323, 573] to the realization that the most efficient way to scatter stars into an SBH is via changes in angular momentum, not energy. Changes in L were ignored in the original derivation of the Bahcall–Wolf solution, which was based on the isotropic Fokker–Planck equation. Generalizing that solution to the anisotropic case, $f = f(E, L)$, turns out to leave the energy dependence of the solution, and hence the radial density profile, nearly unchanged [91]. But allowing stars to diffuse in angular momentum implies a much greater flux of stars into the loss cone: roughly speaking, *all* stars within a sphere of radius r will diffuse into the SBH in one relaxation time at r. The total consumption rate is therefore $\sim (M_\bullet/m_\star)/T_r(r_\mathrm{h})$—smaller (typically) than the full-loss-cone rate but much larger than the rate implied by energy diffusion alone.

Figure 6.2b illustrates the loss cone in a spherical galaxy in terms of the energy–angular momentum variables defined in chapter 5:

$$\mathcal{E} \equiv -E = -v^2/2 + \psi(r), \quad \mathcal{R} \equiv L^2/L_c^2(E)$$

with $\psi(r) = -\Phi(r)$; near the SBH, $L_c^2 \approx (GM_\bullet)^2/(2\mathcal{E})$. To a first approximation, the phase-space density must vanish for orbits that approach closer to the SBH than r_lc. The energy of a circular orbit at $r = r_\mathrm{lc}$ is $\mathcal{E}_\mathrm{lc} = GM_\bullet/(2r_\mathrm{lc})$; this defines the maximum $\mathcal{E}$ for which f can be nonzero. For $\mathcal{E} < \mathcal{E}_\mathrm{lc}$ the lower limit on $\mathcal{R}$ is given by equation (6.6):

$$\mathcal{R}_\mathrm{lc}(\mathcal{E}) = \frac{2r_\mathrm{lc}^2}{L_c^2(\mathcal{E})} \left[\psi(r_\mathrm{lc}) - \mathcal{E}\right] \approx 2\frac{\mathcal{E}}{\mathcal{E}_\mathrm{lc}} \left(1 - \frac{1}{2}\frac{\mathcal{E}}{\mathcal{E}_\mathrm{lc}}\right), \tag{6.17}$$

where the last expression is valid for large $\mathcal{E}$, that is, near the SBH. We can also express $\mathcal{R}_{lc}$ approximately in terms of the angle θ_{lc} defined in figure 6.2: $\mathcal{R}_{lc} \approx \theta_{lc}^2$, or

$$\mathcal{R}_{lc} \approx \begin{cases} r_{lc}/r, & r \lesssim r_h, \\ r_{lc}r_h/r^2, & r \gtrsim r_h. \end{cases} \tag{6.18}$$

Once inside the loss cone, stars are lost in a time $\sim P(\mathcal{E})$. Near the SBH, orbital periods are very short, and stars hardly penetrate beyond the loss-cone boundary before they are consumed or destroyed. At these energies, the phase-space density vanishes throughout the loss cone except for a very small region near the boundary. For small $\mathcal{E}$, on the other hand, P is large and $\mathcal{R}_{lc}$ is small; at these energies it is possible for a star to diffuse across the loss cone by gravitational encounters during a *single* orbital period. Consider an orbit inside the loss cone. At $r = r_{lc}$ on such an orbit, there are no stars moving in an outward direction; but if one were to follow the orbit outward, stars from neighboring orbits would be scattered onto it. This argument suggests that even orbits inside the loss cone will be populated, and to an increasing degree as $\mathcal{E} \to 0$.

Let $T_r(\mathcal{E})$ be the (orbit-averaged) relaxation time for orbits of energy $\mathcal{E}$. The typical change in L^2 in a time T_r is $\sim L_c^2$. Assuming that L evolves as a random walk, then in a single orbital period, the rms change in L is roughly

$$\delta L \approx (P/T_r)^{1/2} L_c. \tag{6.19}$$

If $\delta L \ll L_{lc}$, a star on the edge of the loss cone will execute many orbits before suddenly disappearing. This is the diffusive, or **empty-loss-cone** regime. The consumption rate of stars in this regime is set by the diffusion coefficients that appear in the orbit-averaged Fokker–Planck equation and by the gradients in f with respect to L at $L \approx L_{lc}$. On the other hand, if $\delta L \gg L_{lc}$, orbits are repopulated so efficiently by gravitational encounters that $f > 0$ even for orbits with $L \ll L_{lc}(\mathcal{E})$; the feeding rate approaches the full-loss-cone rate defined above, independent of the relaxation time. This is called the pinhole,[2] or **full-loss-cone** regime. The energy separating the inner, diffusive regime from the outer, pinhole regime is $\mathcal{E}_{crit}$, defined as the energy for which

$$\delta L \approx (P/T_r)^{1/2} L_c = L_{lc}. \tag{6.20}$$

In steady-state solutions like Bahcall and Wolf's, one typically finds that the total consumption rate is dominated by stars with energies greater than $\mathcal{E}_{crit}$; in other words, most captures take place in the diffusive regime.

As emphasized throughout this book, nuclear relaxation times are long, and there is no a priori reason to suppose that the distribution of orbital energies near the SBH is close to the Bahcall–Wolf form. But even in an unrelaxed nucleus, it can still make sense to talk about steady-state distributions with respect to orbital angular momenta near the loss cone. The time for L to change by an amount $\sim L_{lc}$ due to encounters is

$$\Delta t_{lc} \approx \left(\frac{L_{lc}}{L_c}\right)^2 T_r \ll T_r. \tag{6.21}$$

---

[2]So called because far from the SBH, the capture sphere looks like a point.

Even if T_r is longer than the age of the universe, Δt_{lc} need not be. In this chapter, a **steady-state loss cone** will be defined as one for which the distribution of angular momenta near the loss-cone boundary is in a stationary state at each $\mathcal{E}$, even if the distribution of orbital *energies* is still evolving. Time-dependent loss cones are defined similarly as those for which f is still evolving with respect to L. Evolution of nuclei on timescales of $\sim T_r$—that is, evolution in the energy distribution—will be dealt with separately in chapter 7.

Loss-cone theory is less well developed in the context of axisymmetric or triaxial nuclei. In a triaxial nucleus, a fraction of order unity of the stars can follow centrophilic orbits like the pyramids (chapter 3). A star on such an orbit will eventually come very close to the SBH, even in the absence of gravitational encounters. The "loss cone" in such a nucleus could reasonably be said to include all such stars; however, the time required for a star on a centrophilic orbit to pass within r_{lc} can be much longer than the period, P, of radial oscillations. Nevertheless, consumption rates in triaxial nuclei can be high—of the same order as full-loss-cone rates in spherical galaxies—and can remain high for relatively long times, even in the absence of collisional loss-cone repopulation [371].

Stars like the Sun are disrupted outside the SBH's event horizon for $M_{\bullet} \lesssim 10^8 \, M_{\odot}$, but compact remnants like neutron stars and stellar-mass black holes can survive much farther in. For such objects, capture can occur in one of two ways: directly, by scattering onto a loss-cone orbit, or indirectly, by scattering onto an orbit for which changes in energy and angular momentum due to gravitational-wave emission occur more quickly than changes due to gravitational encounters. The first sort of event is called a "plunge," the second an "EMRI"—an **extreme-mass-ratio inspiral**. This terminology originated within the community of physicists hoping to detect gravitational waves. Plunges are of less interest to such researchers since the gravitational waves are only emitted once, in a short burst near periapsis. In the case of EMRIs, inspiral is driven by the gravitational waves themselves, and the signal can be built up over time, making it much easier to detect and model the source. The gravitational interactions that result in EMRIs differ from those that produce tidal disruption events in that they take place much closer to the SBH. This means that resonant relaxation needs to be considered in addition to nonresonant relaxation, and also that relativistic precession of orbits cannot be ignored.

6.1 SPHERICAL SYMMETRY

6.1.1 Basic relations

In a spherical galaxy, the orbit-averaged Fokker–Planck equation expressed in $(\mathcal{E}, \mathcal{R})$ variables is (section 5.5.1)

$$\frac{\partial N}{\partial t} = -\frac{\partial}{\partial \mathcal{E}} \left(N \langle \Delta \mathcal{E} \rangle_t \right) + \frac{1}{2} \frac{\partial^2}{\partial \mathcal{E}^2} \left(N \langle (\Delta \mathcal{E})^2 \rangle_t \right) - \frac{\partial}{\partial \mathcal{R}} \left(N \langle \Delta \mathcal{R} \rangle_t \right)$$
$$+ \frac{1}{2} \frac{\partial^2}{\partial \mathcal{R}^2} \left[N \langle (\Delta \mathcal{R})^2 \rangle_t \right] + \frac{\partial^2}{\partial \mathcal{E} \partial \mathcal{R}} \left[N \langle (\Delta \mathcal{E}) \Delta \mathcal{R} \rangle_t \right], \qquad (6.22)$$

where N, the number density in $(\mathcal{E}, \mathcal{R})$ space, is related to the distribution function f via equation (5.166):

$$N(\mathcal{E}, \mathcal{R}, t)d\mathcal{E}\, d\mathcal{R} = 4\pi^2 P(\mathcal{E}, \mathcal{R})L_c^2(\mathcal{E})\, f(\mathcal{E}, \mathcal{R}, t)d\mathcal{E}\, d\mathcal{R}. \qquad (6.23)$$

The subscripts t on the diffusion coefficients denote orbit averages, for example,

$$\langle \Delta \mathcal{E} \rangle_t \equiv \frac{2}{P} \int_{r_-}^{r_+} \frac{dr}{v_r} \langle \Delta \mathcal{E} \rangle$$

(equation 5.155).

As discussed in the introduction to this chapter, the timescale over which N attains a steady state with regard to $\mathcal{R}$, near the loss cone, is expected to be much shorter than the time required for N to reach a steady state with respect to $\mathcal{E}$. At any given $\mathcal{E}$, therefore, we can seek solutions to

$$\frac{\partial N}{\partial t} = -\frac{\partial}{\partial \mathcal{R}} \left(N \langle \Delta \mathcal{R} \rangle_t \right) + \frac{1}{2} \frac{\partial^2}{\partial \mathcal{R}^2} \left[N \langle (\Delta \mathcal{R})^2 \rangle_t \right] \qquad (6.24)$$

while ignoring changes in energy. Using equations (5.125) and (5.167), the local diffusion coefficients $\langle \Delta \mathcal{R} \rangle$, $\langle (\Delta \mathcal{R})^2 \rangle$ that appear in this equation can be expressed in terms of the velocity-space diffusion coefficients defined in equation (5.94). Furthermore, except at energies close to $\psi(r_{\text{lc}})$, the angular momentum of a loss-cone orbit is small compared with $L_c(\mathcal{E})$; that is, $\mathcal{R} \ll 1$. Retaining only the leading terms in $\mathcal{R}$, we find for the local diffusion coefficients,

$$\langle \Delta \mathcal{R} \rangle = \frac{r^2}{L_c(\mathcal{E})^2} \langle \Delta v_\perp^2 \rangle + O(\mathcal{R}), \qquad (6.25a)$$

$$\langle (\Delta \mathcal{R})^2 \rangle = \frac{2r^2}{L_c(\mathcal{E})^2} \mathcal{R} \langle \Delta v_\perp^2 \rangle + O(\mathcal{R}^2). \qquad (6.25b)$$

To lowest order in $\mathcal{R}$,

$$\langle \Delta \mathcal{R} \rangle = \frac{1}{2} \frac{\partial}{\partial \mathcal{R}} \langle (\Delta \mathcal{R})^2 \rangle, \qquad (6.26)$$

a relation that is also valid if the local diffusion coefficients are replaced by their orbit-averaged counterparts. Using (6.26), equation (6.24) can be written

$$\frac{\partial N}{\partial t} \approx \frac{1}{2} \frac{\partial}{\partial \mathcal{R}} \left[\langle (\Delta \mathcal{R})^2 \rangle_t \frac{\partial N}{\partial \mathcal{R}} \right], \quad \mathcal{R} \ll 1. \qquad (6.27)$$

It is convenient to define

$$\mathcal{D}(\mathcal{E}) \equiv \lim_{\mathcal{R} \to 0} \frac{\langle (\Delta \mathcal{R})^2 \rangle_t}{2\mathcal{R}}. \qquad (6.28)$$

In terms of $\mathcal{D}$, equation (6.27) becomes

$$\frac{\partial N}{\partial t} \approx \mathcal{D} \frac{\partial}{\partial \mathcal{R}} \left(\mathcal{R} \frac{\partial N}{\partial \mathcal{R}} \right), \quad \mathcal{R} \ll 1, \qquad (6.29)$$

where the $\mathcal{E}$ dependence is understood. Equation (6.29), after a trivial change of variables, has the same mathematical form as the equation governing transfer of heat in a cylindrical rod. Because of this, some authors prefer the term "loss cylinder" to "loss cone," and that is why figure 6.2b adopts a cylindrical geometry.

Using the relations derived in chapter 5, and assuming an isotropic field-star distribution, we can write

$$\langle (\Delta v_\perp)^2 \rangle = \frac{32\pi^2 G^2 m^2 \ln \Lambda}{3} v \left[3 I_2(v) - I_4(v) + 2 J_1(v) \right], \quad (6.30a)$$

$$I_n(v) = \int_0^v \left(\frac{v_f}{v} \right)^n f(v_f) \, dv_f, \quad (6.30b)$$

$$J_n(v) = \int_v^\infty \left(\frac{v_f}{v} \right)^n f(v_f) \, dv_f, \quad (6.30c)$$

and

$$\mathcal{D}(\mathcal{E}) = \frac{2}{L_c^2(\mathcal{E}) P(\mathcal{E})} \int_0^{\psi^{-1}(\mathcal{E})} \frac{r^2 dr}{v_r} \langle (\Delta v_\perp)^2 \rangle. \quad (6.31)$$

If there is a distribution of field-star masses, $m^2 f(v_f)$ in equations (6.30) is replaced by $\mu(v_f) \equiv \int dm\, m^2 f(v_f, m)$.

We wish to find steady-state solutions to (6.29). Setting $N \propto \ln \mathcal{R}$ implies $\partial N / \partial t = 0$. As a boundary condition, it makes sense to require that N fall to zero at some small angular momentum $\mathcal{R}_0(\mathcal{E})$; we expect that $\mathcal{R}_0 \approx \mathcal{R}_{lc} \equiv L_{lc}^2 / L_c^2$, at least in the "empty-loss-cone" regime. The steady-state solution then becomes

$$N(\mathcal{R}; \mathcal{E}) = \frac{\ln(\mathcal{R}/\mathcal{R}_{lc})}{\ln(1/\mathcal{R}_{lc}) + \mathcal{R}_{lc} - 1} \bar{N}(\mathcal{E}), \quad (6.32)$$

where

$$\bar{N}(\mathcal{E}) = \int_{\mathcal{R}_{lc}}^1 N(\mathcal{E}, \mathcal{R}) d\mathcal{R} \quad (6.33)$$

is a number-weighted average of N over angular momentum. $\bar{N}(\mathcal{E})$, and the corresponding phase-space density

$$\bar{f}(\mathcal{E}) \approx \frac{\bar{N}(\mathcal{E})}{4\pi^2 L_c^2(\mathcal{E}) P(\mathcal{E})}, \quad (6.34)$$

are approximately the N and f that would be inferred for an observed galaxy if it were modeled assuming an isotropic velocity distribution.

Let $F(\mathcal{E})d\mathcal{E}$ be the flux of stars (number per unit time) in energy interval $d\mathcal{E}$ centered on $\mathcal{E}$, into the loss cone.[3] In general,

$$F(\mathcal{E})d\mathcal{E} = -\frac{d}{dt} \left[\int_{\mathcal{R}_{lc}}^1 N(\mathcal{E}, \mathcal{R}) d\mathcal{R} \right] d\mathcal{E}. \quad (6.35)$$

Substituting equation (6.29) into equation (6.35), and requiring $\partial N / \partial \mathcal{R} = 0$ at $\mathcal{R} = 1$, we find

$$F(\mathcal{E}) = \mathcal{D}(\mathcal{E}) \mathcal{R}_{lc} \left(\frac{\partial N}{\partial \mathcal{R}} \right)_{\mathcal{R}_{lc}} \approx \frac{\bar{N}(\mathcal{E}) \mathcal{D}(\mathcal{E})}{\ln(1/\mathcal{R}_{lc})}, \quad (6.36)$$

---

[3] Not to be confused with $\mathcal{F}_\mathcal{E}$ or $\mathcal{F}_\mathcal{R}$, the functions that appear in the flux-conservation form of the Fokker–Planck equation.

where the latter expression assumes $L_{lc} \ll L_c$. Now it is clear from its definition that $\mathcal{D}$ is the inverse of an orbit-averaged relaxation time: the time over which $\mathcal{R}$ changes by of order unity, that is, L changes by order L_c. Equation (6.36) states therefore that a fraction $\sim 1/|\ln(\mathcal{R}_{lc})|$ of stars at energies $\mathcal{E}$ to $\mathcal{E} + d\mathcal{E}$ are scattered into the loss cone each relaxation time.

Before proceeding, we must return to a point made in the introduction. If the change in L over one orbital period, δL, is comparable to L_{lc}, the diffusive, orbit-averaged approximation on which equation (6.36) is based breaks down. Instead, a star will enter and exit the loss cone multiple times over a single orbital period, and the loss rate should approach the full-loss-cone rate defined above. We can parametrize the goodness of the diffusive approximation in terms of $q(\mathcal{E}) \equiv (\delta L/L_{lc})^2$. Using equation (6.19), and identifying T_r in that expression with the orbit-averaged time $\mathcal{D}^{-1}$, we are led to the definition

$$q(\mathcal{E}) \equiv \frac{P(\mathcal{E})\mathcal{D}(\mathcal{E})}{\mathcal{R}_{lc}(\mathcal{E})}, \tag{6.37}$$

where $P(\mathcal{E}) \equiv P(\mathcal{E}, \mathcal{R})_{\mathcal{R} \to 0}$. Evidently, $q \ll 1$ is the diffusive (empty-loss-cone) regime and $q \gg 1$ is the pinhole (full-loss-cone) regime. In terms of q, for the flux in the diffusive regime, equation (6.36) gives

$$F(\mathcal{E}) \approx \frac{q}{\ln(1/\mathcal{R}_{lc})} \frac{\bar{N}\mathcal{R}_{lc}}{P}, \quad q \ll 1. \tag{6.38}$$

Since $N_{lc}(\mathcal{E}) \approx N(\mathcal{E})\mathcal{R}_{lc}(\mathcal{E})$, this capture rate is smaller by a factor $\sim q/\ln(1/\mathcal{R}_{lc})$ than the full-loss-cone capture rate defined in equation (6.10).

As we will see in section 6.1.3, in physically reasonable models for the distribution of stars in galactic nuclei, $q \ll 1$ at very bound energies (i.e., near the SBH), while $q \gg 1$ far from the SBH. Furthermore, the energy at which $q = 1$ is roughly the energy at which $F(\mathcal{E})$ peaks. This means that feeding of stars to the SBH occurs, in roughly equal numbers, from stars in the diffusive and pinhole regimes. We therefore must develop an understanding of how stars get into the loss cone when the orbit-averaged approximation breaks down.

6.1.2 The Cohn–Kulsrud boundary layer

Cohn and Kulsrud [91] derived a steady-state expression for f near the loss cone in a spherical galaxy that is valid whether or not $q \ll 1$. Their solution was numerical. Here we derive the **Cohn–Kulsrud boundary layer** solution in a more transparent way, making use of the fact that the governing equation is mathematically equivalent to the equation describing transfer of heat in a cylinder.[4] The same approach will be useful later in this chapter when the time-dependent loss-cone problem is addressed.

Our starting point is the local (r-dependent) Fokker–Planck equation (5.121). We express the collision term in that equation in terms of velocity-space variables $\mathcal{E}$ and $\mathcal{R}$ using the volume element $a^{1/2} = L_c^2/(r^2 v_r)$ (equation 5.165). As justified

---

[4]Much of the material in this section is based on unpublished work by M. Milosavljevic, who kindly gave permission for its use here.

above, we ignore changes in energy so that $\mathcal{E}$ may be treated as a parameter rather than a variable. Since $v_r(\mathcal{E}, \mathcal{R}) \approx v_r(\mathcal{E})$ when $\mathcal{R} \ll 1$, v_r may be taken outside the derivative with respect to $\mathcal{R}$, yielding

$$\left(\frac{\partial f}{\partial t}\right)_c = -\frac{\partial}{\partial \mathcal{R}} (f\langle\Delta\mathcal{R}\rangle) + \frac{1}{2}\frac{\partial^2}{\partial \mathcal{R}^2} \left(f\langle(\Delta\mathcal{R})^2\rangle\right) \tag{6.39a}$$

$$\approx \frac{1}{2}\frac{\partial}{\partial \mathcal{R}} \left[\langle(\Delta\mathcal{R})^2\rangle\frac{\partial f}{\partial \mathcal{R}}\right]; \tag{6.39b}$$

the second expression used equation (6.26). In a steady state, this must equal the left-hand side of the Boltzmann equation with $\partial f/\partial t$ set to zero:

$$\sum_i v_i \frac{\partial f}{\partial x_i} + \sum_i a_i \frac{\partial f}{\partial v_i} = \left(\frac{\partial f}{\partial t}\right)_c. \tag{6.40}$$

Expressed in terms of $(\mathcal{E}, \mathcal{R})$ as velocity-space variables, the left-hand side becomes simply $v_r \partial f/\partial r$, and so

$$\frac{\partial f}{\partial r} = v_r^{-1}\frac{\partial}{\partial \mathcal{R}} \left[\frac{\langle(\Delta\mathcal{R})^2\rangle}{2\mathcal{R}}\mathcal{R}\frac{\partial f}{\partial \mathcal{R}}\right]. \tag{6.41}$$

Equation (6.41) describes how the phase-space density varies *along an orbit*, as stars are scattered onto it and off of it.

It is useful to introduce the timelike variable τ:

$$\tau \equiv \left[\int_{r_-}^r \frac{dr}{v_r}\frac{\langle(\Delta\mathcal{R})^2\rangle}{2\mathcal{R}}\right] \Big/ \left[\oint \frac{dr}{v_r}\frac{\langle(\Delta\mathcal{R})^2\rangle}{2\mathcal{R}}\right]$$

$$= (P\mathcal{D})^{-1} \int_{r_-}^r \frac{dr}{v_r}\frac{\langle(\Delta\mathcal{R})^2\rangle}{2\mathcal{R}}. \tag{6.42}$$

The integrals are along the orbit, and the "orbital integral" of a function $F(r, v_r)$ is defined as

$$\oint dr\, F(r, v_r) = \int_{r_-}^{r_+} F(r, v_r)dr + \int_{r_+}^{r_-} F(r, -v_r)dr. \tag{6.43}$$

As τ varies from 0 to 1, r increases from r_- to r_+ and back to r_- again. In terms of τ, equation (6.41) becomes

$$\frac{\partial f}{\partial \tau} = P(\mathcal{E})\mathcal{D}(\mathcal{E})\frac{\partial}{\partial \mathcal{R}} \left(\mathcal{R}\frac{\partial f}{\partial \mathcal{R}}\right), \tag{6.44}$$

where use has been made of the fact that for small $\mathcal{R}$, $\langle(\Delta\mathcal{R})^2\rangle/2\mathcal{R}$ is independent of $\mathcal{R}$ and can be commuted outside of $\partial/\partial \mathcal{R}$. As a final simplification, we define a new angular momentum variable:

$$y \equiv \frac{\mathcal{R}}{P(\mathcal{E})\mathcal{D}(\mathcal{E})}, \tag{6.45}$$

in terms of which equation (6.44) becomes

$$\frac{\partial f}{\partial \tau} = \frac{\partial}{\partial y} \left(y\frac{\partial f}{\partial y}\right). \tag{6.46}$$

The dimensionless angular momentum corresponding to the geometric boundary of the loss cone, at radius r_{lc}, is $\mathcal{R} = \mathcal{R}_{lc}$; in terms of y,

$$y_{lc} = \frac{\mathcal{R}_{lc}}{P(\mathcal{E})\mathcal{D}(\mathcal{E})}. \tag{6.47}$$

Now, for $y > y_{lc}$, the phase-space density $f(\tau, y)$ is a periodic function of τ:

$$f(0, y) = f(1, y), \quad y > y_{lc}. \tag{6.48}$$

Inside the loss cone, we know that no stars exit the zone of destruction around the SBH:

$$f(0, y) = 0, \quad y < y_{lc}, \tag{6.49}$$

while the boundary condition at the other end of the orbit is

$$f(1, y) \geq 0. \tag{6.50}$$

The rate at which stars enter a sphere of radius r, per unit of $\mathcal{E}$ and $\mathcal{R}$, is

$$\frac{1}{2} \times 4\pi r^2 \times \frac{2\pi L_c^2}{v_r r^2} f(1, \mathcal{R}) \times v_r. \tag{6.51}$$

Integrating this expression over $\mathcal{R}$ gives the flux per unit of energy into the SBH:

$$F(\mathcal{E})d\mathcal{E} = 4\pi^2 P(\mathcal{E})\mathcal{D}(\mathcal{E})L_c^2(\mathcal{E}) \left[\int_0^{y_{lc}} f(1, y)dy \right] d\mathcal{E}. \tag{6.52}$$

To calculate the flux, we need to solve for $f(\tau, y)$ subject to the boundary conditions (6.48) and (6.49), with an additional boundary condition stipulating the smoothness of f at $y = 0$:

$$\frac{\partial f}{\partial L} \propto \frac{\partial f}{\partial \sqrt{y}} = 0 \quad (y = 0). \tag{6.53}$$

Equation (6.46) is solved separately inside and outside the geometric loss-cone boundary and the two solutions are matched at $y = y_{lc}$. Outside the boundary, one can assume that Jeans's theorem applies and $f(\tau, y)$ is independent of τ. Inside the loss cone, the solution can be obtained using the method of separation of variables. The problem is mathematically equivalent to the transfer of heat in a solid cylinder of radius $y_{lc}^{1/2}$ that is initially at zero temperature, the surface of which is kept at the constant temperature $f(y_{lc})$. The solution is [411]

$$f(\tau, y) = f(y_{lc}) \left[1 - \frac{2}{\sqrt{y_{lc}}} \sum_{m=1}^{\infty} \frac{e^{-\beta_m^2 \tau/4}}{\beta_m} \frac{J_0(\beta_m \sqrt{y})}{J_1(\beta_m \sqrt{y_{lc}})} \right]. \tag{6.54}$$

Here J_0 and J_1 are Bessel functions of the first kind, and β_m yield consecutive zeros of the equation

$$J_0(\beta \sqrt{y_{lc}}) = 0. \tag{6.55}$$

Now,

$$\int_0^{y_{lc}} f(\tau, y)dy = f(y_{lc})\left(y_{lc} - 4\sum_{m=1}^{\infty} \frac{e^{-\beta_m^2/4}}{\beta_m^2}\right)$$

$$= f(y_{lc})y_{lc}\left(1 - 4\sum_{m=1}^{\infty} \frac{e^{-\alpha_m^2/4 y_{lc}}}{\alpha_m^2}\right), \quad (6.56)$$

where the α_m are the consecutive zeros of the Bessel function $J_0(\alpha)$. Thus, the average flux traveling across y_{lc} is

$$F(\mathcal{E}) = 4\pi^2 P(\mathcal{E})\mathcal{D}(\mathcal{E})L_c^2(\mathcal{E})\int_0^{y_{lc}} f(1, y)dy$$

$$= 4\pi^2 L_c^2 \mathcal{R}_{lc} f(\mathcal{R}_{lc})\left(1 - 4\sum_{m=1}^{\infty} \frac{e^{-\alpha_m^2 q/4}}{\alpha_m^2}\right), \quad (6.57)$$

where the $\mathcal{E}$ dependence is understood. Here, the quantity q,

$$q(\mathcal{E}) \equiv \frac{P(\mathcal{E})\mathcal{D}(\mathcal{E})}{\mathcal{R}_{lc}(\mathcal{E})},$$

is defined precisely as in equation (6.37): it is the orbital period divided by the time a star takes to diffuse across the loss cone. We can rewrite equation (6.57) as

$$F(\mathcal{E}) = 4\pi^2 L_c^2 \mathcal{R}_{lc} f(\mathcal{R}_{lc})\xi(q), \quad (6.58a)$$

$$\xi(q) \equiv 1 - 4\sum_{m=1}^{\infty} \frac{e^{-\alpha_m^2 q/4}}{\alpha_m^2}. \quad (6.58b)$$

For small q, $\xi \approx (2/\sqrt{\pi})\sqrt{q} \approx 1.13\sqrt{q}$; a good approximation for arbitrary q is $\xi \approx (q^2 + q^4)^{1/4}$.

The value of f at $\mathcal{R}_{lc}$ can be determined from the flux matching condition (6.53). Setting $\partial f/\partial \tau = 0$ in equation (6.44), we find

$$f(\mathcal{R}) = f(\mathcal{R}_{lc}) + \frac{f(1) - f(\mathcal{R}_{lc})}{\ln(1/\mathcal{R}_{lc})}\ln(\mathcal{R}/\mathcal{R}_{lc}). \quad (6.59)$$

This implies a flux just outside of the loss cone of

$$F(\mathcal{E}) = 4\pi^2 L_c^2 PD\mathcal{R}_{lc}\left.\frac{\partial f}{\partial \mathcal{R}}\right|_{\mathcal{R}_{lc}}$$

$$= 4\pi^2 L_c^2 PD\frac{f(1) - f(\mathcal{R}_{lc})}{\ln(1/\mathcal{R}_{lc})}. \quad (6.60)$$

Equating the external flux (6.60) with the internal flux (6.58) and solving for $f(\mathcal{R}_{lc})$,

$$f(\mathcal{R}_{lc}) = \frac{f(1)}{1 + q^{-1}\xi(q)\ln(1/\mathcal{R}_{lc})}. \quad (6.61)$$

This can be substituted into equation (6.58) to yield

$$F(\mathcal{E}) = 4\pi^2 L_c^2 \mathcal{R}_{lc}\frac{\overline{f}(\mathcal{E})}{\xi(q)^{-1} + q^{-1}\ln(1/\mathcal{R}_{lc})}, \quad (6.62)$$

where we have approximately identified

$$f(1) \approx \overline{f}(\mathcal{E}) \approx \int_0^1 f(\mathcal{E}, \mathcal{R}) d\mathcal{R}. \tag{6.63}$$

It is customary to represent the flux in a form similar to that of equation (6.36):

$$F(\mathcal{E}) = \frac{\overline{N}(\mathcal{E})\mathcal{D}(\mathcal{E})}{\ln(1/\mathcal{R}_0)} = 4\pi^2 q L_c^2 \mathcal{R}_{lc} \frac{\overline{f}(\mathcal{E})}{\ln(1/\mathcal{R}_0)}, \tag{6.64}$$

where $\mathcal{R}_0$ is the $f = 0$ intercept of the external solution (6.59) extrapolated inside the loss cone. Comparison of equations (6.62) and (6.64) reveals the relation between $\mathcal{R}_0$ and $\mathcal{R}_{lc}$:

$$\mathcal{R}_0(q) = \mathcal{R}_{lc} e^{-q/\xi(q)}. \tag{6.65}$$

Since $\xi(q) \approx 1$ except when $q \ll 1$, the dependence of $\mathcal{R}_0$ on $\mathcal{R}_{lc}$ is exponential to a good approximation. Based on their numerical solution, Cohn and Kulsrud [91] suggested the form

$$\mathcal{R}_0(q) = \mathcal{R}_{lc} \times \begin{cases} e^{-q}, & q \gg 1, \\ e^{-0.186q - 0.824\sqrt{q}}, & q \ll 1. \end{cases} \tag{6.66}$$

Figure 6.3 shows that Cohn and Kulsrud's approximation is good at both large and small q; it is poorest at $q \approx 1$ where it overestimates $\mathcal{R}_0/\mathcal{R}_{lc}$ by $\sim 20\%$.

In terms of $\mathcal{R}_0$, the $\mathcal{R}$ dependence of f is given by (equations 6.59, 6.61)

$$f(\mathcal{R}) = \overline{f}(\mathcal{E}) \frac{\ln(\mathcal{R}/\mathcal{R}_0)}{\ln(1/\mathcal{R}_0) - 1 + \mathcal{R}_0} \approx \overline{f}(\mathcal{E}) \frac{\ln(\mathcal{R}/\mathcal{R}_0)}{\ln(1/\mathcal{R}_0)}. \tag{6.67}$$

Near the SBH, $q \ll 1$ and $\mathcal{R}_0 \approx \mathcal{R}_{lc}$: relaxation effects are small and the phase-space density falls to zero just inside the loss-cone boundary. Far from the SBH, relaxation dominates, and f only falls to zero for orbits with $L \ll L_{lc}$ (i.e., $\mathcal{R}_0 \ll \mathcal{R}_{lc}$); at these energies, the loss cone is essentially full (figure 6.4).

A compact way to express the flux is in terms of the quantity F^{max}, where

$$F^{max}(\mathcal{E}) \equiv N(\mathcal{E})\mathcal{D}(\mathcal{E}). \tag{6.68}$$

F^{max} is roughly equal to the maximum flux that can be driven, by gravitational encounters, through a surface of constant $\mathcal{E}$ [323]. In terms of F^{max},

$$F(\mathcal{E}) \approx \frac{F^{max}(\mathcal{E})}{\ln(1/\mathcal{R}_0)}. \tag{6.69}$$

In the limits $q \ll 1$ and $q \gg 1$ this becomes

$$F \approx F^{max} \times \begin{cases} |\ln \mathcal{R}_{lc}|^{-1}, & q \ll -\ln |\mathcal{R}_{lc}|, \\ q^{-1}, & q \gg -\ln \mathcal{R}_{lc}. \end{cases} \tag{6.70}$$

Alternatively, we can express F in terms of the full-loss-cone flux defined in equation (6.10):

$$F(\mathcal{E}) \approx q(\mathcal{E}) \frac{F^{flc}(\mathcal{E})}{\ln(1/\mathcal{R}_0)} \tag{6.71}$$

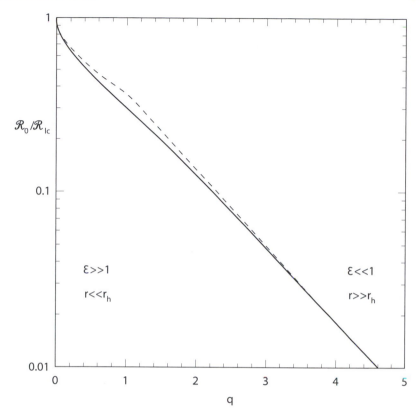

Figure 6.3 $\mathcal{R}_0 \equiv L_0^2/L_c^2(\mathcal{E})$ is the dimensionless angular momentum at which the phase-space density falls to zero, for orbits that pass inside the loss cone of a spherical galaxy. It is plotted here as a fraction of $\mathcal{R}_{lc}$, the angular momentum of an orbit that just grazes the loss cone. The parameter $q = q(\mathcal{E})$ (equation 6.58) measures the degree to which the diffusive approximation holds: $q \ll 1$ is the diffusive (empty-loss-cone) limit, $q \gg 1$ is the pinhole (full-loss-cone) limit. The solid line is the exact relation, equation (6.65), and the dashed line is the approximation of equation (6.66).

which, in the large-q and small-q limits, becomes

$$F \approx F^{\text{flc}} \times \begin{cases} q|\ln \mathcal{R}_{lc}|^{-1}, & q \ll -\ln \mathcal{R}_{lc}, \\ 1, & q \gg -\ln \mathcal{R}_{lc}. \end{cases} \tag{6.72}$$

The transition from empty- to full-loss-cone regimes can be said to occur at the energy $\mathcal{E}_{\text{crit}}$ where $q(\mathcal{E}_{\text{crit}}) = |\ln \mathcal{R}_{lc}(\mathcal{E})|$; in other words, where the flux from the two regimes is equal.[5] The radius at which $\psi(r) = \mathcal{E}_{\text{crit}}$ is called the **critical radius**, r_{crit}. An approximate expression for r_{crit} is given in section 6.1.4.

[5] Some authors prefer to define $q(\mathcal{E}_{\text{crit}}) = 1$.

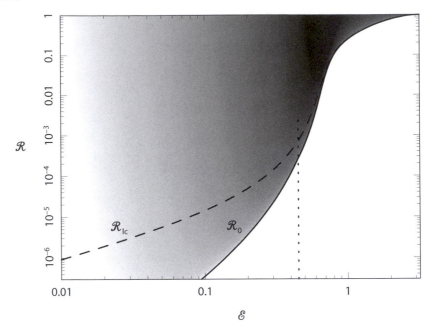

$\mathcal{R}$

$\mathcal{E}$

Figure 6.4 Illustrating the Cohn–Kulsrud steady-state solution for $f(\mathcal{R})$. The gray scale is
proportional to the logarithm of f. $\mathcal{R}_{lc}$ is the dimensionless angular momentum
of an orbit that grazes the loss sphere. The vertical dotted line marks the energy
at which $q = 1$; to the left, $q \gg 1$ and $\mathcal{R}_{lc} \gg \mathcal{R}_0$ (full loss cone); to the right,
$q \ll 1$ and $\mathcal{R}_{lc} \ll \mathcal{R}_0$ (empty loss cone).

6.1.3 Tidal disruption rates in isothermal nuclei

It is instructive to apply the equations derived so far to a concrete nuclear model.
A density profile of the form

$$\rho(r) = m_\star n(r) = \frac{\sigma^2}{2\pi G r^2} \tag{6.73}$$

is a fair description of the distribution of (old) stars near the center of the Milky
Way, at distances $0.5\,\mathrm{pc} \lesssim r \lesssim 10\,\mathrm{pc}$ from Sgr A*—the "nuclear star cluster"
[403]. The density law (6.73) is sometimes called the "singular isothermal sphere,"
since in the absence of a central point mass, the velocity dispersion computed from
the spherical Jeans equation is constant and equal to the parameter σ. Setting $\sigma =
90\,\mathrm{km\,s^{-1}}$—roughly the value measured in the Milky Way just beyond the SBH
influence radius [184]—in equation (6.73) yields a mass density

$$\rho(r) \approx 5 \times 10^5 \left(\frac{r}{1\,\mathrm{pc}}\right)^{-2} M_\odot\,\mathrm{pc}^{-3} \tag{6.74}$$

consistent with the density derived from detailed modeling of the observed number
counts and velocities in the inner few parsecs [481, 403].

As discussed in chapter 2, compact nuclear star clusters (NSCs) like the one observed in the Milky Way appear to be common components of low-luminosity stellar spheroids. But in galaxies beyond the Local Group, NSCs are too small to be spatially resolved, which seriously limits how well we can compute SBH feeding rates in these systems. A reasonable starting point is to assume that equation (6.73) describes the central density in all galaxies containing compact nuclear clusters. If we are bold enough to assume that σ is these galaxies is related to $M_\bullet$ via the $M_\bullet$–σ relation, then all the relevant properties of the nucleus are determined by a single parameter. Of course, we should always keep in mind that the Milky Way SBH is the smallest with a well-determined mass.

The gravitational potential corresponding to the density (6.73) is

$$\psi(r) = \frac{GM_\bullet}{r} - 2\sigma^2 \ln\left(\frac{r}{r_h}\right); \qquad (6.75)$$

the arbitrary additive constant has been chosen so that the stellar potential is equal to zero at $r = r_h$, which we define in the usual way as $GM_\bullet/\sigma^2$—in this case, σ is the parameter appearing in equation (6.73), and not the more vaguely defined quantity that normally appears in the definition of r_h (section 2.2).

The isotropic distribution function (number of stars per unit phase-space volume) that reproduces the density (6.73) in the potential (6.75) is easily found from Eddington's formula (3.47):

$$f(\mathcal{E}) = \frac{1}{\sqrt{8}\pi^2 m_\star} \frac{d}{d\mathcal{E}} \int_{-\infty}^{\mathcal{E}} \frac{d\rho}{d\psi} \frac{d\psi}{\sqrt{\mathcal{E} - \psi}}$$

$$= \frac{1}{r_h^3 \sigma^3} \left(\frac{M_\bullet}{m_\star}\right) g(\mathcal{E}^*), \qquad (6.76a)$$

$$g(\mathcal{E}^*) = \frac{\sqrt{2}}{4\pi^3} \int_{-\infty}^{\mathcal{E}^*} \frac{\mathcal{L}^2(u)\,[2 + \mathcal{L}(u)]}{[1 + \mathcal{L}(u)]^3} \frac{d\psi^*}{\sqrt{\mathcal{E}^* - \psi^*}}, \qquad (6.76b)$$

$$u(\psi^*) \equiv \frac{1}{2} e^{\psi^*/2}.$$

$\mathcal{L}(u)$ is the Lambert function (also called the W function) defined implicitly via $u = \mathcal{L}e^{\mathcal{L}}$. The superscripts "$*$" denote dimensionless quantities, defined by setting $M_\bullet, \sigma$ and G to 1, for example, $\mathcal{E}^* = \mathcal{E}/\sigma^2$.

Given f and ψ, q can be computed from equations (6.30)–(6.31) and (6.58):

$$q(\mathcal{E}^*) = \frac{32\pi^2}{3\sqrt{2}} \ln\Lambda \left(\frac{m_\star}{M_\bullet}\right) \frac{h(\mathcal{E}^*)}{\psi^*(r_{lc}) - \mathcal{E}^*} \left(\frac{r_{lc}}{r_h}\right)^{-2}, \qquad (6.77)$$

where

$$h(\mathcal{E}^*) \equiv 2h_0(\mathcal{E}^*) + 3h_{1/2}(\mathcal{E}^*) - h_{3/2}(\mathcal{E}^*), \qquad (6.78)$$

$$h_0(\mathcal{E}^*) = \left[\int_0^{r^*(\mathcal{E}^*)} \frac{dr^* r^{*2}}{\sqrt{\psi^*(r^*) - \mathcal{E}^*}}\right] \left[\int_{-\infty}^{\mathcal{E}^*} g(\mathcal{E}^{*\prime}) d\mathcal{E}^{*\prime}\right],$$

$$h_{n/2}(\mathcal{E}^*) = \int_0^{r^*(\mathcal{E}^*)} \frac{dr^* r^{*2}}{\sqrt{\psi^*(r^*) - \mathcal{E}^*}} \int_{\mathcal{E}^*}^{\psi^*(r^*)} \left[\frac{\psi^*(r^*) - \mathcal{E}^{*\prime}}{\psi^*(r^*) - \mathcal{E}^*}\right]^{n/2} g(\mathcal{E}^{*\prime}) d\mathcal{E}^{*\prime}.$$

The flux then follows, from equation (6.64):

$$F(\mathcal{E}^*) = \frac{256\pi^4}{3\sqrt{2}} \frac{\ln \Lambda}{\sigma r_{\rm h}} \frac{1}{\ln R_0^{-1}} g(\mathcal{E}^*) h(\mathcal{E}^*). \tag{6.79}$$

The quantity $\mathcal{R}_0(\mathcal{E})$ is given by equation (6.66), with

$$\mathcal{R}_{\rm lc}(\mathcal{E}^*) = 2\left(\frac{r_{\rm lc}}{r_{\rm h}}\right)^2 \frac{\psi^*(r_{\rm lc}^*) - \mathcal{E}^*}{\left(2 + r_c^{*-1}\right) r_c^{*2}} \approx \frac{r_{\rm lc}}{r_{\rm h}} \frac{\mathcal{E}}{\sigma^2}, \tag{6.80a}$$

$$r_c^* = \frac{1}{4\mathcal{L}\left[e^{(1+\mathcal{E}^*)/2}\right]}; \tag{6.80b}$$

$r_c(\mathcal{E})$ is the radius of a circular orbit of energy $\mathcal{E}$ and the second expression for $\mathcal{R}_{\rm lc}$ is valid far from the loss sphere.

Here we note a surprising result: the stellar mass $m_\star$ does not appear explicitly in the expression for the flux. The reason is that, for a given σ and $M_\bullet$, the number of stars scales as $m_\star^{-1}$ while the scattering rate scales as $m_\star \rho \propto m_\star \sigma^2 \propto m_\star$. There is a dependence of F on $m_\star$, but it is indirect, via q and $r_{\rm lc}$, both of which appear in the expressions for $\mathcal{R}_0$. In fact, the dimensionless flux depends on just two parameters: $M_\bullet/m_\star$ and $r_{\rm lc}/r_{\rm h}$. The latter is given by equation (6.2b):

$$\frac{r_t}{r_{\rm h}} \approx 4.7 \times 10^{-2} \left(\frac{\eta}{0.844}\right)^{2/3} \tag{6.81a}$$

$$\times \left(\frac{M_\bullet}{m_\star}\right)^{-2/3} \left(\frac{\sigma}{100\,{\rm km\,s^{-1}}}\right)^2 \left(\frac{m_\star}{M_\odot}\right)^{-1} \left(\frac{R_\star}{R_\odot}\right)$$

$$\approx 1.5 \times 10^{-6} \left(\frac{\eta}{0.844}\right)^{2/3} \tag{6.81b}$$

$$\left(\frac{\sigma}{100\,{\rm km\,s^{-1}}}\right)^{-1.24} \times \left(\frac{m_\star}{M_\odot}\right)^{-1/3} \left(\frac{R_\star}{R_\odot}\right),$$

where (6.81b) has used the $M_\bullet$–σ relation, equation (2.33), to express σ in terms of $M_\bullet$.

Figure 6.5 plots $F(\mathcal{E})$ and $q(\mathcal{E})$ for various values of $M_\bullet$, assuming $m_\star = M_\odot$ and $R_\star = R_\odot$, and equating $r_{\rm lc}$ with r_t. The flux exhibits a mild maximum at $\mathcal{E}^* \approx 1$ (i.e., $\mathcal{E} \approx \sigma^2$) and falls off slowly toward large (more bound) energies: in other words, most of the disruptions occur from orbits within the gravitational influence sphere, regardless of the value of $M_\bullet$. The plot of $q(\mathcal{E})$ shows that, for $M_\bullet \approx 10^8\,M_\odot$, the entire influence sphere lies within the empty-loss-cone regime; that is, $q < 1$ for $\mathcal{E} \gtrsim \sigma^2$. As $M_\bullet$ is reduced, more and more of the loss cone is full.

Figure 6.6 shows the total consumption rate $\dot{N} = \int F(\mathcal{E})d\mathcal{E}$ as a function of $M_\bullet$ under two assumptions about σ: $\sigma = 100\,{\rm km\,s^{-1}}$, or σ is related to $M_\bullet$ via the $M_\bullet$–σ relation. For fixed σ, figure 6.6 shows that $\dot{N} \sim M_\bullet^{-1}$, while allowing σ to vary with $M_\bullet$ implies a weaker (but still inverse) dependence of $\dot{N}$ on $M_\bullet$.

The dependence of the consumption rate on $M_\bullet$ and σ appears from figure 6.6 to be fairly simple, and it is useful to derive analytic approximations. Figure 6.5 shows that over a wide range of $M_\bullet$ values, most of the flux comes from stars within the

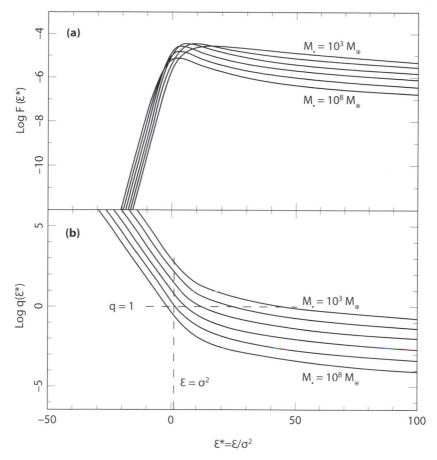

Figure 6.5 (a) Energy-dependent flux of stars into the loss cone of an SBH embedded in an "isothermal" nucleus, with density given by equation (6.73). Solar-type stars were assumed, and the $M_\bullet$–σ relation was used to relate σ to $M_\bullet$. (b) The dimensionless function $q(\mathcal{E})$, equation (6.77), that describes the degree to which the loss cone is filled by gravitational scattering. For small $M_\bullet$, most of the stars inside the SBH influence sphere are in the full-loss-cone regime. (Adapted from [555].)

gravitational influence sphere, $\mathcal{E} \gtrsim \sigma^2$. In this energy interval, $\psi(r) \approx GM_\bullet/r$ and

$$g(\mathcal{E}^*) \approx \frac{1}{\sqrt{2}\pi^3}\mathcal{E}^{*1/2}, \qquad h(\mathcal{E}^*) \approx \frac{5\sqrt{2}}{24\pi^2}\mathcal{E}^{*-2}; \tag{6.82}$$

the latter expression makes use of the fact that $h \approx h_0$, that is, most of the flux comes from scattering by stars with energies less than that of the test star. It follows that

$$q(\mathcal{E}^*) \approx \frac{20}{9}\ln\Lambda\left(\frac{m_\star}{M_\bullet}\right)\left(\frac{r_h}{r_t}\right)\mathcal{E}^{*-2} \tag{6.83}$$

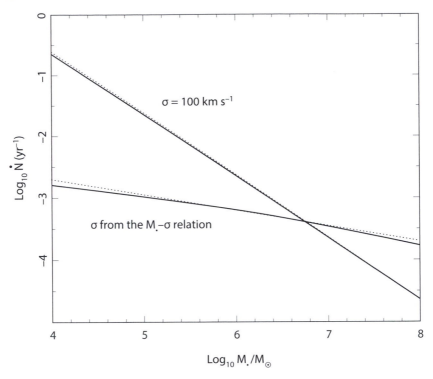

Figure 6.6 Tidal disruption rates as a function of $M_\bullet$ in "isothermal" nuclei, for two assumptions about the relation between the velocity dispersion σ and the SBH mass $M_\bullet$. Dotted lines are equation (6.86). All curves assume $\{m_\star, R_\star\} = \{M_\odot, R_\odot\}$. (Adapted from [555].)

and $R_{\rm lc} \approx 4(r_t/r_{\rm h})\mathcal{E}^*$. The dimensionless flux is then

$$F^*(\mathcal{E}^*) \approx \frac{160 \ln \Lambda}{9\sqrt{2\pi}} \mathcal{E}^{*1/2} \left[A + \mathcal{E}^{*2} \ln \left(\frac{B}{\mathcal{E}^*} \right) \right]^{-1}, \tag{6.84}$$

$$A = \frac{20}{9} \ln \Lambda \left(\frac{m_\star}{M_\bullet} \right) \left(\frac{r_{\rm h}}{r_t} \right), \quad B = \frac{r_{\rm h}}{4r_t}.$$

Ignoring the weak energy dependence of the logarithmic term and taking $r_{\rm h}/r_t$ from equation (6.81), we find, for $\{m_\star, R_\star\} = \{M_\odot, R_\odot\}$,

$$\dot{N} = \int F(E)dE \propto A^{-1/4} \propto \sigma^{7/2} M_\bullet^{-11/12}. \tag{6.85}$$

It turns out that the following, slightly different scaling,

$$\dot{N} \approx 4.3 \times 10^{-4} \left(\frac{\sigma}{90\,{\rm km\,s^{-1}}} \right)^{7/2} \left(\frac{M_\bullet}{4 \times 10^6\,M_\odot} \right)^{-1} {\rm yr^{-1}}, \tag{6.86}$$

provides a better fit to the exact feeding rates plotted in figure 6.6. Equation (6.86), combined with the $M_\bullet$–σ relation, implies $\dot{N} \sim M_\bullet^{-0.25}$. The dependence of the loss rate on $\{m_\star, R_\star\}$ can be computed using equation (6.84); unfortunately this dependence is not simply expressible.

Tidal disruption rates as high as 10^{-4} yr^{-1} in nuclei with $M_\bullet = 10^6\,M_\odot$ imply a liberated mass comparable to $M_\bullet$ after 10 Gyr. This is not necessarily a problem, since only a fraction of the gas removed from stars is expected to find its way into the hole [451]. Nevertheless, the high values of $\dot{N}$ predicted for low-luminosity galaxies suggest that matter tidally liberated from stars might contribute substantially to SBH growth in these galaxies. If SBHs are common in dwarf galaxies and in the bulges of late-type spiral galaxies (both very uncertain hypotheses), these systems would dominate the total tidal flaring rate, due both to their large numbers and to their high individual event rates [555].

6.1.4 Consumption rates in giant galaxies

In galaxies with $M_\bullet \gtrsim 10^8\,M_\odot$, figure 6.1 shows that stars like the Sun are "swallowed whole," while giant stars—red giants, asymptotic giant branch stars, or (in the case of ongoing star formation) massive main-sequence stars—can be tidally disrupted. Estimating feeding rates in these galaxies is both easier and harder than in the case of galaxies with nuclear star clusters. On the one hand, the nuclear structure of bright galaxies is better constrained: core radii can exceed 10^2 pc, sometimes large enough to be resolved beyond the Local Group. As discussed in chapter 2, the density of starlight in giant galaxies follows weakly rising power laws, $j \sim r^{-\gamma}$, $\gamma \lesssim 1$, inside cores whose linear sizes are comparable with r_h or r_m.

On the negative side, central densities in these galaxies are so low, and relaxation times so long, that it is very unlikely that $f(E, L)$ has reached a steady state, even with regard to angular momenta (and this is certainly the case with regard to energies). One could choose to ignore this complication and apply the Cohn–Kulsrud boundary layer solution anyway [334]. But doing so implies a greater degree of certainty about the phase-space structure of these galaxies than is probably justified. Instead, we will be satisfied here to get a rough idea of how consumption rates vary with nuclear properties in giant galaxies, using a simplified approach that is based loosely on the more exact relations derived earlier in this chapter [168, 509]. The presentation will be simple enough that the reader should have no difficulty repeating the calculations for different galaxy models, different assumptions about the distribution of stellar masses, etc. Loss-cone feeding in a galaxy in which $f(E, L)$ is evolving due to encounters is discussed in the next section.

Throughout this section, the periapsis of a capture orbit will be assumed to lie at $r_p = 8r_g$. Recall from chapter 4 that this is the minimum (Keplerian) periapsis of an eccentric orbit around a nonspinning hole. The more general capture condition, which depends on the hole's spin and orientation of the orbit with respect to the hole's equatorial plane, is also given in that chapter. However, most of the results presented in this section depend only logarithmically on r_c.

We begin by noting that core hydrogen burning on the main sequence continues for a time [227]

$$T_{\rm MS} \approx \frac{10^{10}(m/M_\odot)}{L/L_\odot}\,{\rm yr} \approx 10^{10}\left(\frac{m_\star}{M_\odot}\right)^{-2.5}{\rm yr}. \qquad (6.87)$$

Assuming that star formation ceased many Gigayears ago, the stars that are available to be captured or tidally disrupted will consist of two types: main-sequence stars with masses $m_\star < m_{\mathrm{to}} \approx 1\,M_\odot$, the main-sequence turnoff mass; and evolved giant stars, with masses a little above the turnoff mass. Figure 6.1 shows that the minimum mass of a star on the main sequence that is susceptible to tidal disruption (as opposed to capture) is given approximately by

$$\frac{m_\star}{M_\odot} \approx \frac{M_\bullet}{10^8\,M_\odot}, \qquad 0.08 \lesssim m_\star/M_\odot \lesssim 1. \tag{6.88}$$

Thus, for $M_\bullet \lesssim 10^8\,M_\odot$, stars uppermost on the main sequence will be disrupted, while for $M_\bullet \gtrsim 10^8\,M_\odot$, stars on the main sequence can only be captured.

6.1.4.1 Single stellar mass

In a steady-state nucleus, equation (6.69) states

$$F(\mathcal{E}) = \frac{F^{\max}(\mathcal{E})}{\ln R_0^{-1}} \approx \frac{N(\mathcal{E})}{T_r(\mathcal{E})}\frac{1}{|\ln R_0|}. \tag{6.89}$$

In the final expression, $\mathcal{D}(\mathcal{E})^{-1}$ has been identified with $T_r(\mathcal{E})$, an orbit-averaged relaxation time for stars of energy $\mathcal{E}$. We can divide the total consumption rate into two pieces, corresponding to stars in the empty- and full-loss-cone regimes. Referring to equations (6.11) and (6.36),

$$\dot{N} \approx \dot{N}_{\mathrm{empty}} + \dot{N}_{\mathrm{full}}$$

$$\approx \int_{\mathcal{E}_{\mathrm{crit}}}^{\infty} \frac{N(\mathcal{E})}{T_r(\mathcal{E})}\frac{1}{|\ln R_{\mathrm{lc}}|}d\mathcal{E} + \int_0^{\mathcal{E}_{\mathrm{crit}}} \frac{N(\mathcal{E})}{P(\mathcal{E})}\mathcal{R}_{\mathrm{lc}}(\mathcal{E})d\mathcal{E}, \tag{6.90}$$

where $\mathcal{E}_{\mathrm{crit}}$ is the energy at which $P/(T_r \mathcal{R}_{\mathrm{lc}}) \approx q(\mathcal{E}) = |\ln \mathcal{R}_{\mathrm{lc}}(\mathcal{E})|$; note that, at this energy, the two integrands in equation (6.90) are equal. To further simplify the computations, let us replace the energy by the radius as integration variable:

$$\dot{N} \approx \frac{4\pi}{|\ln \mathcal{R}_{\mathrm{lc}}|} \int_0^{r_{\mathrm{crit}}} \frac{n(r)}{T_r(r)}r^2 dr + 4\pi \int_{r_{\mathrm{crit}}}^{\infty} \frac{n(r)}{P(r)}\mathcal{R}_{\mathrm{lc}}(r)r^2 dr. \tag{6.91}$$

The approximate dependence of $\mathcal{R}_{\mathrm{lc}}$ on r was given in equation (6.18). Equation (6.91) is expressed purely in terms of configuration-space quantities, which is reasonable given our inevitable uncertainty about the detailed phase-space structure of these unrelaxed systems.

   In order to apply equation (6.91), we need to know how r_{crit} depends on galaxy properties. It will turn out that—in these giant galaxies—r_{crit} is large compared with r_{h} or r_{m}. We can therefore define r_{crit} as the solution to

$$|\ln \mathcal{R}_{\mathrm{lc}}(r)|\,T_r(r) = \frac{r^2}{r_{\mathrm{lc}}r_{\mathrm{h}}} P(r). \tag{6.92}$$

   All that remains is to specify the galaxy model. A simple, but not too unrealistic, model for the distribution of stellar mass in a giant galaxy is

$$\rho(r) = \begin{cases} \rho_c, & r \le R_c, \\ \rho_c\,(r/R_c)^{-\gamma}, & r > R_c, \end{cases} \tag{6.93}$$

where R_c is the core radius (figure 2.2), and $\gamma \gtrsim 2$.[6] It is also not too bad an approximation to identify the core radius with $r_{\rm m}$, the radius containing a stellar mass equal to twice $M_\bullet$ (section 8.2.2). Then

$$\rho_c = \frac{3}{2\pi} \frac{M_\bullet}{r_{\rm m}^3} \tag{6.94}$$

and we can write the distribution of enclosed mass (including the mass of the SBH) as

$$M(<r) = \begin{cases} M_\bullet \left[1 + 2(r/r_{\rm m})^3\right], & r \le r_{\rm m}, \\ 3M_\bullet \left[1 - \gamma + 2(r/r_{\rm m})^{3-\gamma}\right]/(3-\gamma), & r > r_{\rm m}. \end{cases} \tag{6.95}$$

The orbital period at r can be approximated as the period of a circular orbit:

$$P(r) = \frac{2\pi r^{3/2}}{\sqrt{GM_\bullet}} \begin{cases} \left(1 + 2x^3\right)^{-1/2}, & r \le r_{\rm m}, \\ \left[\frac{3-\gamma}{3(1-\gamma+2x^{3-\gamma})}\right]^{-1/2}, & r > r_{\rm m}, \end{cases} \tag{6.96}$$

where $x \equiv r/r_{\rm m}$. (The reader can easily derive the expressions for $M(r)$ and $P(r)$ in the special case $\gamma = 3$.)

To compute the dependence of the relaxation time T_r on r, we need to know how the velocity dispersion σ varies with radius. Beyond $r_{\rm m}$, it is not a bad approximation to set $\sigma(r)$ to a constant, σ_c, the same quantity that appears in the $M_\bullet$–σ relation. This statement is based both on observed velocity dispersion profiles, and also on the fact that for $\gamma = 2$ the Jeans equation gives $\sigma(r) = $ constant. We can therefore write

$$\sigma^2(r) \approx \frac{GM_\bullet}{r} + \sigma_c^2 \approx \sigma_c^2 \left(1 + \frac{r_{\rm h}}{r}\right). \tag{6.97}$$

Using equation (3.2), the relaxation time at r is then

$$T_r = \frac{0.34\sigma^3}{G^2 \rho m_\star \ln \Lambda} \tag{6.98a}$$

$$\approx \left(1 + \frac{r_{\rm h}}{r}\right)^{3/2} \frac{\sigma_c^3}{3G^2 m_\star \rho_c \ln \Lambda} \times \begin{cases} 1, & r \le r_{\rm m}, \\ x^\gamma, & r > r_{\rm m}. \end{cases} \tag{6.98b}$$

Finally, we invoke the phenomenological correlations that have been presented elsewhere in this book to relate $r_{\rm m}$ and σ_c to $M_\bullet$: the $M_\bullet$–σ relation (equation 2.33), and the empirical relation between $r_{\rm m}$ and $M_\bullet$ (equation 2.16):

$$\sigma_c \approx \sigma_0 M_8^{1/\alpha}, \qquad \sigma_0 \approx 200\,{\rm km\,s}^{-1}, \qquad \alpha \approx 4.86,$$
$$r_{\rm m} \approx r_0 M_8^{1/\beta}, \qquad r_0 \approx 35\,{\rm pc}, \qquad \beta \approx 1.79, \tag{6.99}$$

where $M_8 \equiv M_\bullet/10^8\,M_\odot$.

Figure 6.7 plots the characteristic radii as functions of $M_\bullet$, assuming $r_{\rm lc} = 8r_{\rm g}$ and $m_\star = M_\odot$. The critical radius separating full- and empty-loss-cone regimes

---

[6]In reality, observed brightness profiles often imply a rising density inside R_c, but the radial dependence is weak [95]. Note that a constant density inside $r_{\rm h}$ is not strictly consistent with an isotropic f.

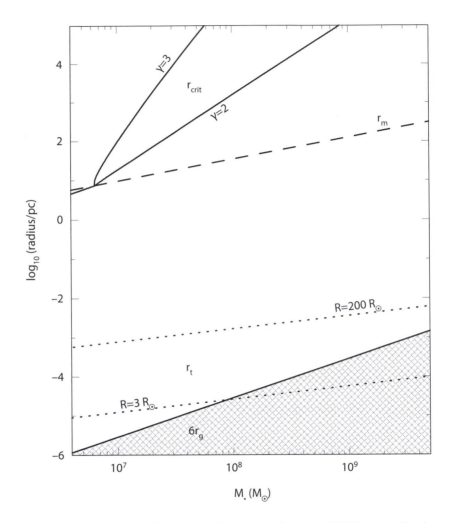

Figure 6.7 Characteristic radii in giant galaxies as a function of SBH mass. The lower
hatched region lies inside $6r_g = 6GM_\bullet/c^2$, the radius of the innermost stable
circular orbit around a nonspinning hole; the minimum periapsis of a highly eccentric orbit around a nonspinning SBH is $8r_g$. The lines labeled r_t show the
tidal disruption radii of a 2 $M_\odot$ red giant star at the start and end of the red giant
phase (equation 6.116). Giant stars like these are the only ones amenable to tidal
disruption when $M_\bullet > 10^8 M_\odot$. The line labeled r_m is the gravitational influence
radius, i.e., the radius containing a mass in stars equal to $2M_\bullet$ (equation 2.16).
The curves near the top show r_{crit}, the radius separating the empty- and full-
loss-cone regimes, assuming the density model of equation (6.93), which has a
core of radius r_m and $\rho \propto r^{-\gamma}$ outside the core, and assuming $m_\star = 1 M_\odot$. For
$M_\bullet > 10^7 M_\odot$, the critical radius lies outside the influence radius, and most of
the flux into the loss cone originates in the empty-loss-cone regime. Critical radii
corresponding to tidal disruption of red giant stars are even greater.

turns out to be roughly equal to the core radius, or to r_m, when $M_\bullet \approx 10^7 \, M_\odot$, but the ratio r_{crit}/r_m increases with increasing $M_\bullet$. Approximate expressions are

$$\left(\frac{r_{crit}}{r_m}\right)^{3\gamma/2-5} \approx \frac{9\sqrt{6}}{\sqrt{3-\gamma}} \frac{G\tilde{m}}{\sqrt{GM_\bullet}} \frac{\ln\Lambda}{|\ln\mathcal{R}_{lc}|} \frac{r_m^{1/2}}{r_{lc}\sigma_c} \tag{6.100a}$$

$$\approx 0.04 M_8^{-1.43} \left(\frac{r_g}{r_{lc}}\right), \tag{6.100b}$$

where the second relation uses the empirical correlations (6.99), and sets the ratio of the logarithmic terms to unity. A robust conclusion is that most of the flux into the loss cone of a (spherical) giant galaxy will originate from the diffusive, or empty-loss-cone, regime.

The implied loss rate, equation (6.91), is shown in figure 6.8 for two values of γ. The figure confirms our expectation that most captures take place from the empty-loss-cone region when $M_\bullet \gtrsim 10^7 \, M_\odot$. The total loss rate declines gradually with increasing $M_\bullet$, with values in the approximate range $10^{-6} \, \mathrm{yr}^{-1} \lesssim \dot{N} \lesssim 10^{-5} \, \mathrm{yr}^{-1}$.

It is interesting to compute the radius, r_{peak}, from which most stars are scattered into the SBH. In the case of isothermal nuclei, we saw in the previous section (figure 6.5) that this radius is roughly equal to the influence radius. It turns out that for the galaxy models being considered here, the integrand of the first integral in equation (6.91), $r^2 n/T_r$, always has its maximum at $r_{peak} \approx r_m$. Since $r_{crit} > r_m$, the loss cone is empty at this radius. We can make use of this result to derive an even simpler, approximate expression for the total loss rate. Ignoring the second (full-loss-cone) term in equation (6.91), we can write

$$\dot{N} \approx \frac{1}{|\ln\mathcal{R}_{lc}(r_{peak})|} \frac{N(r < r_{peak})}{T_r(r_{peak})} \tag{6.101a}$$

$$\approx \frac{2M_\bullet/m}{|\ln(r_m/r_{lc})|} \frac{1}{T_r(r_m)}. \tag{6.101b}$$

For the galaxy models used in figure 6.8, this works out to be

$$\dot{N} \approx 1.6 \times 10^{-6} M_8^{-0.29} \, \mathrm{yr}^{-1}. \tag{6.102}$$

This result is plotted as the dot-dashed line in figure 6.8. Not surprisingly, it falls a little below the more exact calculation, but the dependence on $M_\bullet$ is virtually the same.

It is worth emphasizing again that in these giant galaxies, most stars scattered into the loss cone are directly captured, not tidally disrupted. In a subsequent section, the possibility of tidal disruption of red giant stars is considered in detail.

6.1.4.2 Distribution of stellar masses

The equations derived so far are appropriate when all stars are of a single type. We would like to generalize these expressions to allow for a distribution of stellar masses and radii; this will be particularly important when we consider giant stars in the next section. For the moment, however, we continue limiting our consideration to main-sequence stars.

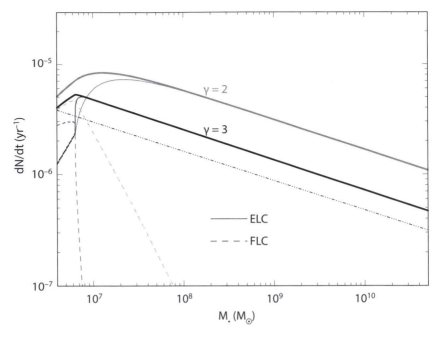

Figure 6.8 Consumption rates in spherical galaxies as a function of SBH mass, computed
using equation (6.91), and the same galaxy model as in figure 6.7. A single stellar
mass ($m_\star = 1\,M_\odot$) was assumed and the loss-cone radius was fixed at $r_{\mathrm{lc}} = 8r_{\mathrm{g}}$;
no distinction is made here between tidal disruptions and direct captures. The
thick curves show the total consumption rates; the contributions from the empty-
and full-loss-cone regimes are plotted independently as the thin and the dashed
curves. In giant galaxies, almost all of the flux into the SBH comes from the
empty-loss-cone region. The dot-dashed line is the analytical estimate of equation (6.102); this estimate is independent of γ, the parameter that defines the
falloff in density outside the core.

Throughout this section, we adopt the simplified expression (6.101) for $\dot{N}$ and
assume that $r_{\mathrm{peak}} = r_{\mathrm{m}}$.

On the main sequence, stars obey a mass–radius relation of the form [288]

$$R(m) \approx \left(\frac{m}{M_\odot}\right)^{0.8} R_\odot. \tag{6.103}$$

The number of stars per unit of mass on the main sequence is identical to the initial
mass function (IMF) defined in chapter 7; in the mass range $0.08 \lesssim m_\star/M_\odot \lesssim 0.5$,
the IMF is believed to have the form

$$n(m)\,dm \propto \left(\frac{m}{M_\odot}\right)^{-\alpha} dm, \quad \alpha \approx 1.3 \tag{6.104}$$

(equation 7.21). We assume that this mass function holds in the mass range $m_0 \leq
m \leq m_{\mathrm{to}}$, where $m_0 \approx 0.08\,M_\odot$ and m_{to} is the main-sequence turnoff mass given

by equation (6.87):

$$m_{to} \approx \left(\frac{T}{10^{10} \text{ yr}} \right)^{-0.4} M_\odot.$$

Let $N(m)dm$ be the number of stars, at radii $r \leq r_m$, in the mass range m to $m + dm$. By definition, the total stellar mass in this region is $2M_\bullet$. We associate this mass, somewhat arbitrarily, with stars whose initial masses were in the range $m_0 \leq m \leq 1 M_\odot$. This assumption allows us to normalize $N(m)$, yielding

$$N(m)dm \approx \begin{cases} (2 - \alpha) \left(\frac{2M_\bullet}{M_\odot} \right) \left(\frac{m}{M_\odot} \right)^{-\alpha} \frac{dm}{M_\odot}, & m \leq m_{to}, \\ 0, & m > m_{to}. \end{cases} \quad (6.105)$$

The rate of loss of stars, per unit of m, is then given by replacing $N(r < r_{peak})$ in equation (6.101a) by equation (6.105). One additional change is required. The stellar mass, $m_\star$, that appears in the expression for T_r, equation (6.98), must be replaced by $\tilde{m}$, as defined in equation (5.62b). It is argued in chapter 7 that $\tilde{m} \approx 0.5 M_\odot$ for an old stellar population; note that this is one half the mass that was assigned to $m_\star$ in the previous sections.

Finally, we should distinguish between stars that are captured, and stars that are tidally disrupted. The minimum stellar mass that is amenable to tidal disruption, $m_{min}(M_\bullet)$, is given by equation (6.88). For a given $M_\bullet$ and T, this minimum mass may be greater or less than the turnoff mass m_{to}; if the former, there can be no tidal disruptions of main-sequence stars. This will be the case if

$$\left(\frac{M_\bullet}{10^8 M_\odot} \right)^{2-\alpha} \left(\frac{T}{10^{10} \text{ yr}} \right)^{0.4(2-\alpha)} \gtrsim 1. \quad (6.106)$$

When this condition is satisfied, the rate of tidal disruptions is zero, and the total capture rate is given by integrating the expression for $\dot{N}(m)$ until m_{to}.

Rather than quote results for $\dot{N}$—which can be difficult to interpret when there is a mass spectrum—we choose instead to compute $\dot{M}$, the rate of loss of mass:

$$\dot{M} = \int \dot{N}(m) \, m \, dm. \quad (6.107)$$

Given our assumptions, this becomes

$$\dot{M} \approx \frac{1}{|\ln(r_m/r_{lc})|} \frac{2M_\bullet}{T_r(r_m)} \left[(2 - \alpha) \int x^{1-\alpha} dx \right], \quad x \equiv m/M_\odot, \quad (6.108)$$

$$\approx 5.3 \times 10^{-7} \left(\frac{M_\bullet}{10^8 M_\odot} \right)^{-0.29} \left[(2 - \alpha) \int x^{1-\alpha} dx \right] M_\odot. \quad (6.109)$$

The latter expression uses the empirical relations noted in the previous section to express $T_r(r_m)$ in terms of $M_\bullet$; in addition, $\tilde{m}$ was set to $0.5 M_\odot$. The limits on the integral depend on the age of the galaxy, and are different depending on whether capture or tidal disruption is being considered.

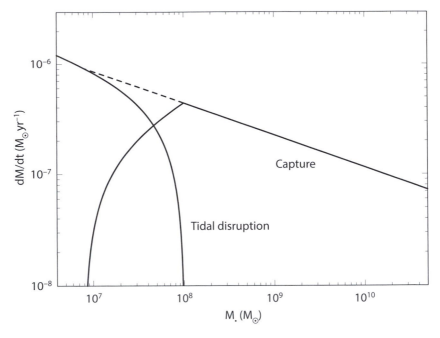

Figure 6.9 Rates of mass consumption in spherical galaxies as a function of SBH mass, assuming a distribution of stellar masses, equation (6.104), below a turnoff mass of $1\,M_\odot$ and above a mass of $0.08\,M_\odot$. The "scattering mass," $\tilde{m}$, has been set to $0.5\,M_\odot$. The contribution from red giant stars is not included here. Above $\sim 10^8\,M_\odot$, there are no tidal disruptions, since main-sequence stars are captured whole.

In the case of captures, we find

$$\dot{M}_{\mathrm{capture}} \approx 5.3 \times 10^{-7} M_8^{-0.29}\, M_\odot \tag{6.110}$$

$$\times \begin{cases} \left[\left(\dfrac{m_{\min}}{M_\odot} \right)^{2-\alpha} - \left(\dfrac{m_0}{M_\odot} \right)^{2-\alpha} \right], & m_{\min} \le m_{\mathrm{to}}, \\[2ex] \left[\left(\dfrac{m_{\mathrm{to}}}{M_\odot} \right)^{2-\alpha} - \left(\dfrac{m_0}{M_\odot} \right)^{2-\alpha} \right], & m_{\min} > m_{\mathrm{to}}, \end{cases} \tag{6.111}$$

where $M_8 \equiv M_\bullet / (10^8\, M_\odot)$.

The rate of tidal disruption of main-sequence stars is given by a similar set of expressions:

$$\dot{M}_{\mathrm{tidal}} \approx 5.3 \times 10^{-7} M_8^{-0.29}\, M_\odot \tag{6.112}$$

$$\times \begin{cases} \left[\left(\dfrac{m_{\mathrm{to}}}{M_\odot} \right)^{2-\alpha} - \left(\dfrac{m_{\min}}{M_\odot} \right)^{2-\alpha} \right], & m_{\min} \le m_{\mathrm{to}}, \\[2ex] 0, & m_{\min} > m_{\mathrm{to}}. \end{cases} \tag{6.113}$$

Figure 6.9 plots these functions assuming $T = 10\,\mathrm{Gyr}$, $\alpha = 1.3$, $m_{\mathrm{to}} = 1\,M_\odot$.

6.1.4.3 Giant stars

The only stars large enough to be tidally disrupted by SBHs with $M_{\bullet} \gtrsim 10^8 M_{\odot}$ are giant stars: either upper-main-sequence stars, or stars that are ascending the red giant or asymptotic giant branches. The times spent by stars in these giant phases are relatively short. Nevertheless, such stars deserve special attention, since they are the *only* stars that might be observed as tidal flaring events in giant galaxies [509].

For a $10 M_{\odot}$ star, which is the lowest mass on the main sequence that can be disrupted by a $10^9 M_{\odot}$ SBH (figure 6.1), $T_{MS} \approx 3 \times 10^7$ yr. Unless there is ongoing star formation in the nucleus, the contributions of massive stars on the main sequence to tidal event rates can generally be ignored.

Of more interest are the post-main-sequence giant phases reached by stars of lower initial mass, $0.5 M_{\odot} \lesssim m \lesssim 10 M_{\odot}$. In these stars, exhaustion of hydrogen in the core is followed by burning of hydrogen to helium in a thin shell; the mass of the helium core gradually increases, and the star's radius and luminosity shoot up as the star climbs the red giant branch.[7] In the evolutionary models, it turns out that the radius and luminosity during the giant phase are determined almost uniquely by the mass, m_c, of the (helium) core. Approximate relations are

$$\frac{L}{L_{\odot}} \approx \frac{10^{5.3}\mu^6}{1 + 10^{0.4}\mu^4 + 10^{0.5}\mu^5}, \tag{6.114a}$$

$$\frac{R}{R_{\odot}} \approx \frac{3.7 \times 10^3 \mu^4}{1 + \mu^3 + 1.75\mu^4}, \tag{6.114b}$$

where $\mu \equiv m_c/M_{\odot}$ [263]. The dominant energy source in these evolutionary phases is the p–p chain, and so the rate of energy production is tied to the rate of increase of core mass by

$$L \approx 7 \times 10^{-3}c^2 \times \frac{d\mu}{dt} M_{\odot}. \tag{6.115}$$

Equations (6.114)–(6.115) can be solved for the dependence of L and R on time. Figure 6.10 shows the results, assuming an initial core mass of $0.17 M_{\odot}$. In the late stages, mass loss becomes important, and the maximum radius is believed to be about $200 R_{\odot}$ [61]. The total length of the red giant phase is about 7×10^8 yr and the time-averaged radius during the giant phase is about $12 R_{\odot}$.

As discussed above, the degenerate core would be expected to survive tidal stresses from all but the smallest SBHs. As for the envelope, it turns out that much of it can be fit locally to a polytropic model with index $n \equiv d \ln P/d \ln T \approx 3/2$ [130]. Setting $\eta = 1.8$ in equation (6.3), disruption of the envelope would be expected to occur at a distance r_t from the SBH such

---

[7] The causes of this expansion are still debated [129].

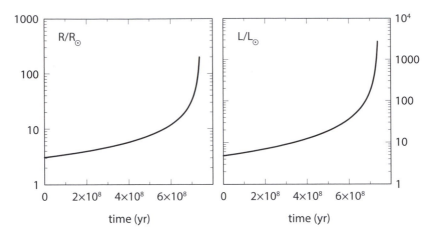

Figure 6.10 Radius (left) and luminosity (right) of stars as they climb the red giant branch. The initial core mass was set to $0.17\,M_\odot$ and the final core mass is $0.8\,M_\odot$.

that

$$\Theta_{\text{RG}} \equiv \frac{r_t}{r_g} \approx 5 \left(\frac{M_\bullet}{10^9\,M_\odot} \right)^{-2/3} \left(\frac{m}{M_\odot} \right)^{-1/3} \left(\frac{R_{\text{RG}}}{10\,R_\odot} \right) \qquad (6.116)$$

$$\approx \begin{cases} 1.5 \left(\dfrac{M_\bullet}{10^9\,M_\odot} \right)^{-2/3} \left(\dfrac{m}{M_\odot} \right)^{-1/3}, & R_{\text{RG}} = 3\,R_\odot, \\[2ex] 95 \left(\dfrac{M_\bullet}{10^9\,M_\odot} \right)^{-2/3} \left(\dfrac{m}{M_\odot} \right)^{-1/3}, & R_{\text{RG}} = 200\,R_\odot. \end{cases}$$

However, even if the star passes within r_t, some fraction of the envelope can remain bound to the core (figure 6.11). A red giant that has been partially disrupted in this way tends to return, on a thermal evolution timescale, to the radius–core-mass and luminosity–core-mass relations given above, and this is progressively more true the farther up the red giant branch the star is at the time of disruption [97]. A likely consequence is episodic tidal flaring, as the star returns again and again to orbital periapsis [7].

Equation (6.100) implies that r_{crit}, the critical radius separating the empty- and full-loss-cone regimes, would be even greater for red giants than for stars like the Sun, and so it is clear that the full-loss-cone contribution to the capture rate is ignorable. In a moment, we will argue that even the *empty*-loss-cone contribution is small in comparison to an additional term that we have neglected up till now. But before making that argument, we point out that the steep dependence of R_{RG} on time implies that there are really two values for r_{crit}, $r_{\text{crit,min}}$, and $r_{\text{crit,max}}$, defined in terms of the minimum and maximum radii on the red giant branch, $R_{\text{RG}} = (3\,R_\odot, 200\,R_\odot)$. Below $\sim r_{\text{crit,min}}$, the loss cone is empty for all red giants, and above $\sim r_{\text{crit,max}}$, the red giant loss cone is always full. Between these two radii, the loss cone as seen by a single red giant is full until such time as the star reaches a size for which r_{crit} is greater than its distance from the center.

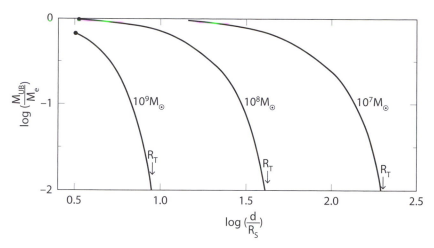

Figure 6.11 Tidal disruption of red giant stars by SBHs [452]. The initial stellar model con-
        tained a degenerate core of mass $0.33\,M_\odot$ and an extended envelope of mass
        $M_e = 0.97\,M_\odot$. In these calculations, the star was assumed to fall inward on an
        essentially radial orbit. The curves, which are labeled by $M_\bullet$, show the fraction
        of the mass of the stellar envelope that is unbound as a function of distance, d,
        from the SBH, in units of $R_S \equiv 2GM_\bullet/c^2$. The radius labeled R_T is an estimate
        of the radius of tidal disruption based on an equation similar to (6.2b). When
        $M_\bullet \approx 10^9\,M_\odot$, only $\sim$70% of the envelope is removed before the core of the
        star is swallowed.

However, there is a second way that red giants can enter the loss cone [509]. Stars
on orbits that pass within the tidal disruption sphere corresponding to $R_\star \approx 200\,R_\odot$,
the maximum radius of a red giant, can "grow onto the loss cone" after they begin
ascending the red giant branch, even in the absence of gravitational scattering. At
some early time, the fraction of stars on such orbits is given by

$$\frac{N_{\mathrm{lc}}(\mathcal{E})}{N(\mathcal{E})} \approx \frac{L_{\mathrm{lc}}^2(\mathcal{E})}{L_c^2(\mathcal{E})} \approx \frac{2GM_\bullet r_t}{r^2\sigma^2} \approx \theta_{\mathrm{lc}}^2 \qquad (6.117)$$

(equations 6.8, 6.6, and 6.10), which is a factor $\sim$300 times larger than the equivalent expression for stars like the Sun. It turns out [509] that "growth onto the loss
cone" dominates diffusion onto the loss cone for red giants, and henceforth we will
only consider the former process.

The rate at which red giants grow onto the loss cone is determined by the rate at
which their progenitor, main-sequence stars leave the main sequence. Call that rate
$\dot{N}_{\mathrm{to}}$, and define $d\dot{N}_{\mathrm{to}}/dM$ as the red giant production rate per unit of total stellar
mass. The rate of tidal disruption events is then given by an integral over the galaxy:

$$\dot{N}_{\mathrm{RG}} \approx \int dM \theta_{\mathrm{lc}}^2 \frac{d\dot{N}_{\mathrm{to}}}{dM} \qquad (6.118a)$$

$$\approx 4\pi \int r^2 \rho(r) \frac{r_h r_t}{r^2} \frac{d\dot{N}_{\mathrm{to}}}{dM} dr \qquad (6.118b)$$

$$\approx 4\pi\, r_h r_t \frac{d\dot{N}_{\mathrm{to}}}{dM} \int \rho(r)\, dr. \qquad (6.118c)$$

For the galaxy models considered here, the integral can be approximated by $\rho_c r_{\rm m}$. Using equations (6.94) and (6.116), this becomes

$$\dot{N}_{\rm RG} \approx 6\,\Theta_{\rm RG}\frac{G^2 M_\bullet^3}{c^2\sigma^2 r_c^2}\frac{d\dot{N}_{\rm to}}{dM}. \tag{6.119}$$

To compute the red giant production rate, we combine an initial mass function appropriate to stars with $m_\star \gtrsim 0.5\,M_\odot$, $n(m)dm \propto m^{-\delta}dm$, $\delta = 2.3$ (section 7.1.2.1), with equation (6.87) for the main-sequence lifetime, yielding

$$\frac{d\dot{N}_{\rm to}}{dM} \approx \frac{1}{50}\left(\frac{m_{\rm to}}{M_\odot}\right)^{3.5-\delta}\,{\rm Gyr}^{-1}M_\odot^{-1}. \tag{6.120}$$

Substituting this expression into equation (6.119) then gives

$$\dot{N}_{\rm RG} \approx \frac{\Theta_{\rm RG}}{8.3}\left(\frac{GM_\bullet}{c\sigma_c r_c}\right)^2\left(\frac{M_\bullet}{M_\odot}\right)\left(\frac{m_{\rm to}}{M_\odot}\right)^{3.5-\delta}\,{\rm Gyr}^{-1} \tag{6.121a}$$

$$\approx \frac{\Theta_{\rm RG}}{2.0}M_8^{1.47}\left(\frac{m_{\rm to}}{M_\odot}\right)^{1.2}\,{\rm Gyr}^{-1} \tag{6.121b}$$

$$\approx 2\times10^{-7}M_8^{0.8}\left(\frac{m_{\rm to}}{M_\odot}\right)^{1.2}\,{\rm yr}^{-1}. \tag{6.121c}$$

The second of these three expressions assumes the empirical relations given above between $\{\sigma_c, r_{\rm m}\}$ and $M_\bullet$, and the third adopts the r_t value corresponding to $R_{\rm RG} = 200\,R_\odot$ in equation (6.116). While very approximate, this calculation suggests that tidal disruption of giant stars would occur at a rate as high as a few per Megayear in the galaxies with the most massive SBHs.

6.1.5 Time-dependent loss-cone dynamics

Loss-cone theory was originally directed toward understanding the observable consequences of massive black holes at the centers of globular clusters [168, 323]. Globular clusters are many relaxation times old, and this assumption was built into the theory, by requiring the stellar phase-space density near the hole to have reached an approximate steady state under the influence of gravitational encounters. In a collisionally relaxed cluster around a massive black hole, the dependence of f on E is given by the Bahcall–Wolf "zero-flux" solution, $f \sim |E|^{1/4}$, and the dependence of f on L near the loss cone is described by the Cohn–Kulsrud boundary-layer solution. The rate of supply of stars to the hole is fixed by these assumptions [91].

Relaxation times in the nuclei of giant galaxies often exceed 10 Gyr (figure 3.1). One consequence is that the stellar density profile near the SBH need not have the Bahcall–Wolf form. But the fact that galactic nuclei are not collisionally relaxed also has implications for the more detailed form of the phase-space density near the loss-cone boundary, and hence for the SBH feeding rate. In a spherical galaxy, the characteristic time for gravitational encounters to set up a steady-state distribution in angular momentum for orbits with $L \lesssim L_0$ is

$$t_L \approx \frac{L_0^2}{L_c^2}T_r. \tag{6.122}$$

If L_0 is equal to L_{lc}, the angular momentum of tidal disruption or capture, then $t_L \ll T_r$. But giant galaxies almost universally exhibit cores, with sizes 10^1–10^2 pc, and one widely discussed model attributes cores to the ejection of stars by a binary SBH during a galaxy merger. As discussed in chapter 8, the binary carves out a hole in phase space corresponding to orbits with periapsis distances less than the "hard binary" separation $a_h \approx \nu r_h$, where $\nu \equiv M_1 M_2/(M_1 + M_2)^2$ is the reduced mass ratio of the binary. Replacing L_0^2 by $2GM_\bullet a_h$ in equation (6.122), and writing $L_c^2 \approx GM_\bullet r_h$, appropriate for stars at a distance $\sim r_h$ from the SBH, yields

$$\frac{t_L}{T_r(r_h)} \approx \frac{a_h}{r_h} \approx \frac{M_2}{M_1}. \tag{6.123}$$

Since $T_r(r_h)$ can be much greater than 10^{10} yr in bright elliptical galaxies, even binary SBHs with $M_2 \ll M_1$ can open up phase-space gaps that would not be refilled in a galaxy's lifetime [380].

Our starting point for describing time-dependent loss cones is equation (6.29):

$$\frac{\partial N}{\partial t} \approx \mathcal{D} \frac{\partial}{\partial \mathcal{R}} \left(\mathcal{R} \frac{\partial N}{\partial \mathcal{R}} \right).$$

Ignoring evolution in $\mathcal{E}$ is justified if $t_L \ll T_r$. Equation (6.29) assumes diffusive evolution—a very good approximation here if the physical size of the evacuated region is much larger than, say, r_t. Changing variables to $\ell^2 = \mathcal{R} \approx 1 - e^2$,

$$\frac{\partial N}{\partial t} = \frac{\mu}{\ell} \frac{\partial}{\partial \ell} \left(\ell \frac{\partial N}{\partial \ell} \right), \tag{6.124}$$

where $\mu(\mathcal{E}) \equiv \mathcal{D}(\mathcal{E})/4$.

Equation (6.124) is the heat conduction equation in cylindrical coordinates, with radial variable ℓ and diffusivity μ [411]. To find its solution, we first need to specify boundary conditions in ℓ at every $\mathcal{E}$. A reasonable boundary condition at $\ell = 1$ is

$$\left. \frac{\partial N}{\partial \ell} \right|_{\ell=1} = 0. \tag{6.125}$$

For small ℓ, a perfectly absorbing, Dirichlet boundary condition is appropriate:

$$N(\mathcal{E}, \ell) = 0, \quad \ell \leq \ell_{lc}(\mathcal{E}) = \mathcal{R}_{lc}(\mathcal{E})^{1/2}, \tag{6.126}$$

where $\mathcal{R}_{lc} \ll 1$ is the angular momentum associated with the tidal disruption or capture sphere, of radius r_{lc}.

Given an initial $N(\mathcal{E}, \ell, 0)$ and the boundary conditions (6.125)–(6.126), the solution to equation (6.124) can be obtained by means of Fourier–Bessel synthesis:

$$N(\mathcal{E}, \ell, t) = \frac{\pi^2}{2} \sum_{m=1}^{\infty} \frac{[\beta_m J_0(\beta_m \ell_{lc})]^2}{[J_0(\beta_m \ell_{lc})]^2 - [J_1(\beta_m)]^2}$$

$$\times A(\beta_m; \ell) e^{-\mu \beta_m^2 t} \int_{\ell_{lc}}^1 \ell' A(\beta_m; \ell') N(\mathcal{E}, \ell', 0) d\ell', \tag{6.127}$$

where J_n and Y_n ($n = 0, 1$) are Bessel functions of the first and second kind, $A(x; y)$ is a combination of the Bessel functions defined via

$$A(x; y) \equiv J_0(xy) Y_1(x) - J_1(x) Y_0(xy),$$

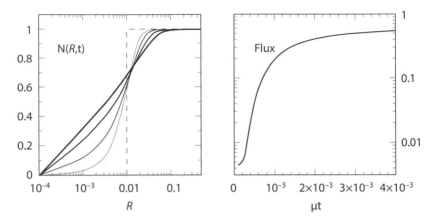

Figure 6.12 The left panel shows the evolution of $N(\mathcal{E}, \mathcal{R}, t)$ at one $\mathcal{E}$, computed using equation (6.127). The right panel shows the flux (per unit of $\mathcal{E}$) into the loss cone at this $\mathcal{E}$, computed using equation (6.128). The initial $N(\mathcal{R})$ is shown at left as the dashed line: $N(\mathcal{E}, \mathcal{R}, 0) = 0$ for $\mathcal{R} \leq 0.01$. The angular momentum of the loss cone was fixed at $\mathcal{R}_{lc} = 10^{-4}$. In the left panel, times shown are $\mu t = (0, 0.05, 0.1, 0.2, 0.4) \times 10^{-2}$; line width increases with time. The steady-state solution is nearly reached in a time of $\sim 10^{-2} T_r$, consistent with the estimate of equation (6.122).

while the β_m are consecutive solutions of the equation

$$A(\beta_m; \ell_{lc}) = 0.$$

The flux into the loss cone at energy $\mathcal{E}$ is

$$F(\mathcal{E}, t)d\mathcal{E} = -\frac{d}{dt}\left[2\int_{\ell_{lc}}^{1} N(\mathcal{E}, \ell, t)\ell\, d\ell\right]d\mathcal{E} \tag{6.128}$$

$$= 4\mu \frac{dN}{d\ln\mathcal{R}}\bigg|_{\mathcal{R}_{lc}}$$

$$= -\pi^2 \sum_{m=1}^{\infty} \frac{\mu\beta_m^3 \ell_{lc}\, [J_0(\beta_m\ell_{lc})]^2}{[J_0(\beta_m\ell_{lc})]^2 - [J_1(\beta_m)]^2}$$

$$\times B(\beta_m; \ell_{lc})e^{-\mu\beta_m^2 t}\left[\int_{\ell_{lc}}^{1} \ell' A(\beta_m; \ell')N(\mathcal{E}, \ell', 0)d\ell'\right]d\mathcal{E},$$

where $B(x; y)$ is another combination of the Bessel functions:

$$B(x; y) \equiv J_1(xy)Y_1(x) - J_1(x)Y_1(xy).$$

Figure 6.12 illustrates the evolution described by these equations. The initially steep phase-space gradients decay on the expected timescale of $\sim \mathcal{R}_0 T_r \sim 10^{-2}\mu^{-1}$. At the final time, $N(\mathcal{R})$ has nearly attained the exponential form expected for the steady-state solution outside of an empty loss cone, equation (6.32).

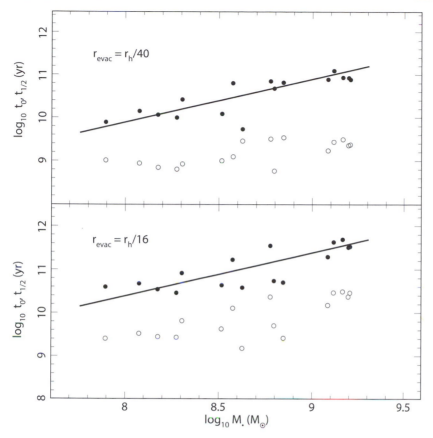

Figure 6.13 Two characteristic times associated with loss-cone refilling in a sample of elliptical galaxies, assuming spherical symmetry, and that initially no stars were present with periapsides inside r_{evac}. t_0 (open circles) is the elapsed time before the first star is scattered into the loss cone, and $t_{1/2}$ (filled circles) is the time for the loss-cone flux to reach $1/2$ of its steady-state value. Solid lines are the approximate fitting function for $t_{1/2}$, equation (6.129). (Adapted from [380].)

Similar calculations can be used to estimate whether loss-cone feeding rates in observed galaxies are likely to be close to their steady-state values [380]. The initial normalization of $N(\mathcal{R})$ at each $\mathcal{E}$ is fixed by the requirement that the *final*, $\mathcal{R}$-averaged f be equal to the f inferred from the galaxy's luminosity profile, assuming velocity isotropy. Figure 6.13 shows the results for a sample of elliptical galaxies, and for two values of r_{evac}, assuming that stars are initially absent from orbits with periapsides inside r_{evac}. The values of r_{evac} are roughly what would be expected if the current SBH were preceded by a binary with mass ratio of 0.1 or 1. Two characteristic times are plotted: the elapsed time before a single star would be scattered into the loss cone; and the time before the flux (integrated over energies) reaches one half of its steady-state value. The latter time is found to be given roughly by

$$\frac{t_{1/2}}{10^{11} \text{ yr}} \approx 4 \frac{r_{\text{evac}}}{r_{\text{h}}} \frac{M_\bullet}{10^8 \, M_\odot}. \tag{6.129}$$

Evidently, it would be dangerous to assume steady-state feeding rates in galaxies with SBHs more massive than $\sim 10^8 \, M_\odot$.

6.2 NONSPHERICAL NUCLEI

So far in our discussion of loss-cone dynamics, we have ignored one population of stars that might in principle contribute to the feeding of an SBH. If the orbits in a spherical galaxy were populated at some early time without regard to the presence of a central "sink," a certain number of stars would find themselves on loss-cone orbits. Stars on such orbits will pass inside r_t or r_c simply as a consequence of their unperturbed motion. This process is sometimes called **orbit draining**; in a spherical galaxy, the rate of passage of stars into the SBH due to orbit draining is of course just equal to the full-loss-cone rate defined above.

Orbit draining is usually ignored in the context of spherical galaxies, for two (fairly obvious) reasons. First, the total number of stars that were initially on loss-cone orbits is likely to have been small: a fraction $\sim L_{\text{lc}}^2/L_c^2(E)$ of all the stars at any E, and far fewer than $M_\bullet/m_\star$ in total. Second, these stars would have been consumed very soon after the SBH was in place. As argued in the previous section, if the SBH was preceded by a massive binary, even orbits well outside the SBH's loss cone might have been evacuated before the single SBH formed.

These arguments need to be modified in the case of nonspherical nuclei. Torques from a flattened potential cause orbital angular momenta to change, even in the absence of gravitational encounters. This means that—compared with spherical nuclei—a potentially much larger fraction of stars can be on orbits that will eventually bring them inside the sphere of destruction. The timescale over which stars on such orbits pass within r_{lc} is typically long compared with *radial* orbital periods, but it may still be much shorter than the timescale for gravitational encounters to act.[8]

Consider first axisymmetric nuclei. Orbits conserve the energy E and the component L_z of the angular momentum parallel to the symmetry axis. In the absence of encounters, a minimum condition for a star to find its way into the SBH is $\ell_z \equiv L_z/L_c(E) < \ell_{\text{lc}}$. But as discussed in chapter 4, small-ℓ_z orbits near the SBH in axisymmetric nuclei need not conserve total angular momentum, even approximately; they are often "saucers," orbits whose inclination and angular momentum oscillate in such a way that $\ell_z = \ell \cos i$ is conserved. The maximum, instantaneous angular momentum of a saucer orbit is $\ell \approx \epsilon^{1/2}$ (equation 4.110), where $\epsilon \approx (1-q)/2$ is a measure of the short-to-long axis ratio q of the stellar density. A substantial fraction of stars with $\ell_z < \ell_{\text{lc}}$, and with *instantaneous* angular momenta less than $\epsilon^{1/2} \gg \ell_{\text{lc}}$, will pass eventually within r_{lc}. If the population of low-ℓ

---

[8]In principle, one could call all such orbits "loss-cone orbits." In this text, that term is reserved for orbits that will pass within r_{lc} in at most one radial period.

orbits is not too different from the population in an isotropic, spherical galaxy with the same radial mass distribution, the fraction of stars at any E that are destined to pass within r_{lc} is

$$\sim \int_0^{\ell_{lc}} d\ell_z \int_0^{\sqrt{\epsilon}} d\ell \approx \sqrt{\epsilon}\ell_{lc} \tag{6.130}$$

(equation 3.114), compared with the much smaller fraction $\sim \ell_{lc}^2$ in a spherical galaxy. The timescale over which these orbits will be drained is the longer of the radial period and the period associated with circulation of saucer orbits about the fixed-point orbit that generates the saucers. As shown in chapter 4, the latter time is roughly $\sim \epsilon^{-1/2}$ times the "mass precession time" $t_M \approx P M_\bullet/M_\star$. Near the influence radius, $M_\star \approx M_\bullet$, and in a nucleus of moderate flattening, $\epsilon^{-1/2} t_M$ will be of order or somewhat longer than $P(r_h)$. While longer than the time, P, required for loss-cone draining in spherical galaxies, this time is still short enough that the saucer orbits within $\sim r_h$ would probably be drained soon after the SBH was in place.

The situation is likely to be rather different in nuclei that are fully triaxial. Saucer orbits still exist in triaxial potentials, but much of the phase space can be occupied by an additional family of orbits: the pyramids, whose fixed-point orbit is the short axis of the triaxial figure. The angular momentum of a pyramid orbit reaches zero at the corners of the pyramid, and so *every* star on a pyramid orbit will eventually pass within r_{lc}—unlike saucer orbits, which are limited by conservation of ℓ_z to a minimum radius of periapsis. As discussed in chapters 3 and 4, a fraction of order ϵ of stars in a triaxial nucleus may be on pyramid orbits—much larger than the fractions $\sim \ell_{lc}^2$ or $\sim \sqrt{\epsilon}\ell_{lc}$ of stars in spherical or axisymmetric nuclei that pass within r_{lc}. But the time required for a star on a pyramid orbit to reach a given $\ell = \ell_{lc} \ll 1$ can be much longer than for saucers, since the angular momentum of a pyramid orbit oscillates with two independent frequencies, and the time to reach $\ell < \ell_{lc}$ is roughly ℓ_0/ℓ_{lc} times the oscillation period in either direction, where ℓ_0 is the orbit's typical eccentricity. This time is long enough that some stars on pyramid orbits, even within r_h, can be expected to survive for times comparable to galaxy lifetimes.

We can summarize these arguments as follows. In going from spherical to axisymmetric to triaxial geometries, the fraction of orbits that pass within r_{lc} increases from negligible (spherical) to $O(1)$ (triaxial), while the time for these orbits to drain increases from $\sim P$ (spherical) to $\gg P$ (triaxial). It is probably reasonable to assume that the feeding of SBHs in spherical galaxies is dominated by gravitational scattering, as discussed in the previous sections of this chapter, while in triaxial galaxies, gravitational encounters are of secondary importance compared with the draining of centrophilic orbits like the pyramids.

Precisely axisymmetric nuclei are problematic. It has been argued [334] that feeding rates in axisymmetric galaxies can be dominated by orbital draining, even at very late times ($\gtrsim 10\,\text{Gyr}$) after formation of the SBH, implying capture rates that are essentially the same as in fully triaxial galaxies. This argument assumes that there exists in axisymmetric potentials a substantial population of centrophilic—and, presumably, stochastic—orbits at $r \gg r_h$ and that these orbits are not drained at some early time. Whether or not this is correct, gravitational scattering of stars into the loss cone in axisymmetric nuclei will be affected by the presence of the saucers, even after draining is complete, implying modestly larger capture rates than in equivalent spherical nuclei.

It is probably fair to say that no very satisfactory analysis of the loss-cone problem has yet been carried out in nonspherical geometries. For instance, there exists no treatment of the steady-state loss-cone boundary condition in nonspherical nuclei comparable to Cohn and Kulsrud's analysis in spherical galaxies. Regrettably, the treatment of nonspherical loss cones that is presented in this section is no better than what is currently available in the literature. Indeed the problem will be simplified even further. Feeding rates in axisymmetric galaxies will be assumed to be determined by gravitational scattering alone; orbit draining will be ignored. In the triaxial geometry, *only* orbit draining will be considered; gravitational encounters will be ignored. It is hoped that these idealized treatments will motivate more work on this important problem in the future.

6.2.1 Axisymmetric nuclei

The character of the motion near an SBH in an axisymmetric nucleus was discussed in section 4.4.2; the main results are summarized here. Orbits fall into one of two classes, the tubes and the saucers. Except near the separatrices separating the two families, tube orbits behave in a manner similar to the annular orbits in spherical potentials: the amplitude of the total angular momentum, L, is nearly fixed, and there is a minimum distance of closest approach to the SBH that is related to this nearly constant L by an equation similar to (6.6). Saucer orbits, on the other hand, can exhibit large angular momentum variations. Conservation of L_z, the component of angular momentum parallel to the symmetry axis, implies

$$L \cos i = \text{constant}, \tag{6.131}$$

where i is the inclination with respect to the symmetry plane; saucer orbits undergo large oscillations in i and accordingly in L, allowing them to approach much more closely to the SBH than would be implied by their average, or minimum, eccentricity.

Saucer orbits exist for

$$\ell_z \equiv \frac{L_z}{L_c(E)} < \ell_{\text{sep}} \approx \sqrt{\epsilon}, \tag{6.132}$$

where ϵ is related to the isodensity axis ratios q by $q \approx 1 - 2\epsilon$. When L_z satisfies this condition, there exists a one-parameter set of saucer orbits at each E and L_z defined by the third integral, or equivalently by the maximum and minimum values of ℓ, $\{\ell_+, \ell_-\}$, reached during a single period of oscillation in ℓ or i. On the separatrix, $\ell_+ = \ell_{\text{sep}}$ and $\ell_- = \ell_z$; away from the separatrix the variations in ℓ are smaller. The period of oscillation in ℓ is very long near the separatrix but drops rapidly away from it, with a typical value of $\sim \epsilon^{-1/2}$ times the "mass precession time" $P M_\bullet / M_\star (r < a)$: much longer than orbital periods, but probably shorter than the timescale for gravitational encounters to change L or L_z.

In the spherical geometry, an orbit must satisfy $L < L_{lc}$ if the star is to go into the SBH. In the axisymmetric geometry, orbits satisfying the weaker condition $L_z < L_{lc}$ can be captured, as long as $L \lesssim L_{sep}$; but since L_{sep} is typically much greater than L_{lc}, the number of stars available for capture can be much larger than in the spherical case. Because the condition for capture is less stringent, the rate of diffusion onto loss-cone orbits will be higher.

We can estimate how much higher by the following argument [334]. Consider diffusion only in L_z. Equations (5.196) give expressions for the diffusion coefficients assuming an isotropic field-star distribution. In the limit of small L_z, these become

$$\langle \Delta L_z \rangle = \frac{L_z}{v} \langle \Delta v_\parallel \rangle \approx 0, \tag{6.133a}$$

$$\langle (\Delta L_z)^2 \rangle = \frac{L_z^2}{v^2} \langle (\Delta v_\parallel)^2 \rangle + \frac{1}{2} \frac{(\varpi^2 v^2 - L_z^2)}{v^2} \langle (\Delta v_\perp)^2 \rangle$$

$$\approx \frac{1}{2} \varpi^2 \langle (\Delta v_\perp)^2 \rangle. \tag{6.133b}$$

Recall that in the spherical geometry, diffusion in L at low L is determined by the quantity

$$D(E) \equiv \lim_{\mathcal{R} \to 0} \frac{\langle (\Delta \mathcal{R})^2 \rangle}{2\mathcal{R}}, \tag{6.134}$$

the orbit-average of which appears in the spherical diffusion equation (6.29). Comparing equations (6.25b) and (6.133b), we can write for the (local) L_z diffusion coefficient

$$\frac{\langle (\Delta L_z)^2 \rangle}{L_c^2} = \frac{D}{2} \frac{\varpi^2}{r^2} = \frac{D}{2} \sin^2 \theta, \tag{6.135}$$

where θ is the (instantaneous) colatitude. We desire an orbit-averaged expression for this coefficient. For a single, eccentric orbit, it is easy to show that θ is nearly independent of the mean anomaly, and so the averaging would be carried out with respect to ω and i. Since we are seeking an estimate of the typical diffusion rate for saucer orbits of specified E and L_z, an additional averaging is required with respect to the third integral. In the absence of detailed knowledge about the distribution over that integral, we simply assume that θ is a uniformly populated variable over its allowed range. Making use of the fact that θ varies nearly from 0 to π for saucers with low L_z (figure 4.4), we can write

$$\langle (\Delta L_z)^2 \rangle_t \approx \frac{\mathcal{D} L_c^2}{4}. \tag{6.136}$$

The orbit-averaged diffusion equation for $N(E, L_z)$ is

$$\frac{\partial N}{\partial t} = \frac{1}{2} \frac{\partial^2}{\partial L_z^2} \left(N \langle (\Delta L_z)^2 \rangle_t \right) \tag{6.137}$$

(cf. equation 5.195). Again acknowledging that we do not know the distribution over the third integral, we can find the approximate relation between $N(E, L_z)$ and

$f(E, L_z)$ by integrating equation (3.114) with respect to angular momentum from 0 to L_{sep}:

$$N(E, L_z) \, dE \, dL_z \approx 4\pi^2 P L_{\text{sep}} f(E, L_z) dE \, dL_z \qquad (6.138)$$

with P the period of a nearly radial orbit. Then a steady state with respect to L_z near the loss cone implies $f(E, L_z) = a(E) + b(E)|L_z|$: a linear dependence, rather than the logarithmic-dependence characteristic of the (E, L) loss cone. In terms of the "isotropized" f, $\overline{f} \equiv L_c^{-1} \int f \, dL_z$,

$$f(E, L_z) = \overline{f}(E) \left[\frac{1 + (b/a)L_z}{1 + (b/2a)L_c} \right]. \qquad (6.139)$$

In the spherical problem, we were able to determine the second integration constant by requiring $f = 0$ at $L = 0$. In the axisymmetric geometry, $L_z = 0$ corresponds to orbits with a range of values for the third integral, only some of which penetrate to $r = r_{\text{lc}}$. Hence $f(L_z = 0) \neq 0$ and we must use another argument to determine the quantity $b(E)/a(E)$ in equation (6.139).

The flux of stars into the loss cone is

$$F(E)dE = -\frac{d}{dt} \left[\int_{L_{\text{lc}}}^{L_{\text{sep}}} dL_z N(E, L_z) \right] dE \qquad (6.140a)$$

$$\approx \frac{1}{2} \langle (\Delta L_z)^2 \rangle \left[\left(\frac{\partial N}{\partial L_z} \right)_{L_{\text{lc}}} \right] dE \qquad (6.140b)$$

$$\approx \frac{\pi^2}{2} \mathcal{D} P L_c^2 L_{\text{sep}} \frac{\partial f}{\partial L_z} dE, \qquad (6.140c)$$

where we are assuming $\partial f/\partial L_z$ is constant within the loss region. Now, at sufficiently large E, the diffusive approximation breaks down, and the loss-cone flux should approach the full-loss-cone value given by equation (6.10):

$$F^{\text{flc}}(E) \approx 4\pi^2 L_{\text{lc}}^2(E) f(E). \qquad (6.141)$$

At these large energies, $b(E) \to 0$ and $f \to a(E)$. Setting $f = a$ in equation (6.141) and identifying F^{flc} with the flux in equation (6.140a) then gives

$$a(E) \approx \frac{\mathcal{P}\mathcal{D}L_c^2}{8L_{\text{lc}}^2} L_{\text{sep}} b(E) \qquad (6.142a)$$

$$\approx q_z(E) L_{\text{sep}}(E) b(E), \qquad (6.142b)$$

where

$$q_z(E) \equiv \frac{P \langle (\Delta L_z)^2 \rangle_t}{2 L_{\text{lc}}^2} = \frac{\mathcal{P}\mathcal{D}L_c^2}{8L_{\text{lc}}^2} = \frac{q}{8} \qquad (6.143)$$

plays the role of $q(E)$ in the spherical geometry. Finally, replacing a/b in equation (6.139) by $q_z L_{\text{sep}}$ yields

$$f(E, L_z) = \overline{f}(E) \left[\frac{1 + |L_z|/(q_z L_{\text{sep}})}{1 + L_c/(2q_z L_{\text{sep}})} \right] \qquad (6.144)$$

and substituting the result into equation (6.140c) gives for the diffusive loss-cone flux,

$$F(E)\,dE \approx 4\pi^2 L_{lc}^2(E)\frac{\overline{f}(E)\,dE}{1 + L_c/(2q_z L_{sep})}. \tag{6.145}$$

The transition from "empty" to "full" loss cones takes place at $q_z \approx 0.5 L_c/L_{sep}$ (i.e., $q \approx 4L_c/L_{sep}$), rather than at $q \approx \ln(1/\mathcal{R}_{lc})$ as in the spherical case.

Referring to equation (6.70), we can write the diffusive flux in the spherical and axisymmetric geometries in terms of the quantity F^{max} defined in equation (6.68):

$$F \approx F^{max} \times \begin{cases} \left[2\ln(L_c/L_{lc})\right]^{-1}, & \text{spherical,} \\ (4L_c/L_{sep})^{-1}, & \text{axisymmetric.} \end{cases} \tag{6.146}$$

Thus, the fraction of stars of energy E that are lost in one relaxation time is $\sim 1/\ln(L_c/L_{lc})$ in the spherical geometry and $\sim L_{sep}/L_c$ in the axisymmetric geometry. The different functional dependencies reflect the fact that diffusion is two-dimensional in the spherical case and effectively one-dimensional in the axisymmetric case [323].

Even in the spherical geometry, the logarithmic factor in equation (6.146a) is of order unity if evaluated at the influence radius, and it is not clear from this comparison that feeding rates in the axisymmetric geometry should be much greater than in the spherical case. In fact, computations based essentially on the formalism just presented [334] suggest that accounting for axisymmetry increases the feeding rate by less than a factor of two in most galaxies. Such an enhancement is not insignificant, but is probably small compared with the combined effects of the various systematic uncertainties; for instance, about the degree to which the steady-state assumption is valid (section 6.1.5). For most purposes, the effect of axisymmetric distortions on diffusive feeding rates can probably be ignored.

6.2.2 Triaxial nuclei

In axisymmetric nuclei, the number of stars on saucer orbits is expected to be small compared with the number on tube orbits. In triaxial nuclei, on the other hand, centrophilic orbits like the pyramids can dominate the orbital population of self-consistent models (figure 3.18). In particular, the mass of stars on pyramid orbits can easily exceed $M_\bullet$, and can greatly exceed the mass on loss-cone orbits in the spherical or axisymmetric geometries. A reasonable estimate of the feeding rate in triaxial nuclei can be obtained by simply ignoring collisional loss-cone refilling and counting the rate at which stars on centrophilic orbits pass within a distance r_{lc} from the SBH.

Giant galaxies have very long central relaxation times (figure 3.1). These are also the galaxies in which the evidence for triaxiality tends to be strongest. For both these reasons, it is very likely that SBH feeding in giant galaxies is dominated by the draining of centrophilic orbits. Furthermore, since the SBHs in these galaxies are likely to have masses $\gtrsim 10^8\,M_\odot$, tidal disruption will only occur for giant stars; stars like the Sun will be "swallowed whole."

Referring to section 4.4.3, we recall the following property of the pyramid orbits: as long as the frequencies of oscillation in the two directions $\{x, y\}$ about the
short axis are incommensurate, the vector (e_x, e_y), which points along the major
axis of the osculating ellipse, equation (4.145), densely fills the whole available
area, which has the form of a distorted rectangle. The corner points correspond
to zero angular momentum; near the corners, the periapsis distance is small. The
"drainage area" is therefore similar to the four holes in the corners of a billiard
table.

Unless otherwise noted, in this section we adopt the simple harmonic oscillator
(SHO) approximation to the $\{e_x, e_y\}$ motion, that is, we use the simplified Hamiltonian (4.146) and its solutions (4.151); these orbits have $e_x^2 + e_y^2 \ll 1$ and they
form a rectangle in the e_x–e_y plane, with sides $\{2e_x, 2e_y\}$. As long as the motion is
integrable, the results for arbitrary pyramids with $\{e_x, e_y\} \lesssim 1$ will be qualitatively
similar.

Figure 6.14 shows a two-torus describing oscillations in $\{e_x, e_y\}$ for a pyramid
orbit. In the SHO approximation, solutions are given by equation (4.151). If the
two frequencies $\{v_x^{(0)}, v_y^{(0)}\}$ are incommensurate, the motion will fill the torus; in
this case, we are free to shift the time coordinate so as to make both phase angles
$\{\phi_1, \phi_2\}$ zero, yielding

$$\ell^2(\tau) = \ell_{x0}^2 \sin^2(v_x^{(0)}\tau) + \ell_{y0}^2 \sin^2(v_y^{(0)}\tau) \tag{6.147a}$$

$$= \ell_{x0}^2 \sin^2 \theta_1 + \ell_{y0}^2 \sin^2 \theta_2. \tag{6.147b}$$

In the case of commensurability (i.e., $m_1 v_x^{(0)} + m_2 v_y^{(0)} = 0$ with $\{m_1, m_2\}$ integers), the trajectory will avoid certain regions of the torus and such a shift may
not be possible; we ignore that possibility here. In the SHO approximation, $v_x^{(0)} = \sqrt{15\epsilon_c}$, $v_y^{(0)} = \sqrt{15(\epsilon_c - \epsilon_b)}$, with $\{\epsilon_b, \epsilon_c\}$ the dimensionless measures of elongation defined in equation (4.143). More generally, integrable motion will still be
representable as uniform motion on the torus but the frequencies and the relations
between ℓ and the angles will be different. Finally, the dimensionless time, τ, in
equation (6.147) is defined as in equation (4.141a) (i.e., $\tau = v_0 t$), where $v_M = 3\ell v_0$ and v_M is the precession frequency due to the spherical part of the potential,
equation (4.88).

Stars are lost when $\ell(\theta_1, \theta_2) \leq \ell_{lc}$. Consider the loss region centered at $(\theta_1, \theta_2) = (0, 0)$. This is one of four such regions, of equal size and shape, that correspond
to the four corners of the base of the pyramid. For small ℓ_{lc}, the loss region is
approximately an ellipse,

$$\frac{\ell_{x0}^2}{\ell_{lc}^2}\theta_1^2 + \frac{\ell_{y0}^2}{\ell_{lc}^2}\theta_2^2 \lesssim 1. \tag{6.148}$$

The area enclosed by this "loss ellipse" is

$$\pi \frac{\ell_{lc}^2}{\ell_{x0}\ell_{y0}}. \tag{6.149}$$

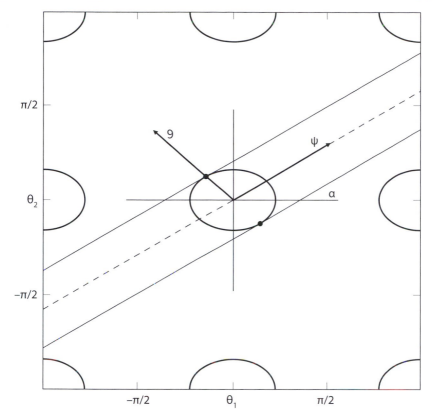

Figure 6.14 Two-torus describing oscillations of (e_x, e_y) for a pyramid orbit [378]. The ellipses correspond to regions near the four corners of the pyramid's base where
$\ell \leq \ell_{lc}$. In the orbit-averaged approximation, trajectories proceed smoothly
along lines parallel to the solid lines, with slope $\tan \alpha = v_y/v_x$. In reality, successive periapsis passages occur at discrete intervals, once per radial period.

There are four such regions on the torus; together, they constitute a fraction

$$\mu = \frac{1}{\pi} \frac{\ell_{lc}^2}{\ell_{x0}\ell_{y0}} \tag{6.150}$$

of the torus.

In the orbit-averaged approximation, stars move in the θ_1–θ_2 plane along lines
with slope $\tan \alpha = v_y^{(0)}/v_x^{(0)}$, at a constant angular rate of

$$\sqrt{\left(v_x^{(0)}\right)^2 + \left(v_y^{(0)}\right)^2}.$$

But since periapsis passages occur only once per radial period, a star will move a
finite step in the phase plane between close encounters with the SBH. The dimensionless time between successive periapsis passages is $\Delta\tau = v_0(2\pi/v_r)$. The angle

traversed during this time is

$$\Delta\theta = 2\pi\frac{v_0}{v_r}\sqrt{\left(v_x^{(0)}\right)^2 + \left(v_y^{(0)}\right)^2}.$$

The rate at which stars move into one of the four loss ellipses is given roughly by the number of stars that lie an angular distance $\Delta\theta$ from one side of a loss ellipse, divided by $\Delta\tau$.

This is not quite correct however, since a star satisfying this condition may precess past the loss ellipse before it has had time to reach periapsis. We carry out a more exact calculation by assuming that the torus is uniformly populated at some initial time, with unit total number of stars. To simplify the calculation, we first transform to a new phase plane defined by

$$\psi = \frac{v_x^{(0)}\ell_{x0}^2\theta_1 + v_y^{(0)}\ell_{y0}^2\theta_2}{\sqrt{\left(v_x^{(0)}\ell_{x0}\right)^2 + \left(v_y^{(0)}\ell_{y0}\right)^2}}, \qquad \vartheta = \frac{-v_y^{(0)}\ell_{x0}\ell_{y0}\theta_1 + v_x^{(0)}\ell_{x0}\ell_{y0}\theta_2}{\sqrt{\left(v_x^{(0)}\ell_{x0}\right)^2 + \left(v_y^{(0)}\ell_{y0}\right)^2}}. \qquad (6.151)$$

With this transformation, the phase velocity becomes

$$\dot\psi = \left[\left(v_x^{(0)}\ell_{x0}\right)^2 + \left(v_y^{(0)}\ell_{y0}\right)^2\right]^{1/2}, \qquad \dot\vartheta = 0, \qquad (6.152)$$

and the loss regions become circles of radius $\ell_{\rm lc}$. The angular displacement in one radial period is

$$\Delta\psi = 2\pi\frac{v_0}{v_r}\left[\left(v_x^{(0)}\ell_{x0}\right)^2 + \left(v_y^{(0)}\ell_{y0}\right)^2\right]^{1/2}. \qquad (6.153)$$

The density of stars is $(4\pi^2\ell_{x0}\ell_{y0})^{-1}$.

At any point in the ψ–ϑ plane, stars have a range of radial phases. Assuming that the initial distribution satisfies Jeans's theorem, stars far from the loss regions are uniformly distributed in χ where

$$\chi = P^{-1}\int_{r_p}^{r}\frac{dr}{v_r}; \qquad (6.154)$$

here $P \equiv 2\pi/v_r$ is the radial period, r_p is the periapsis distance and v_r is the radial velocity. The integral is performed along the orbit, hence χ ranges between 0 and 1 as r varies from r_p to apoapsis and back to r_p.

Figure 6.15 shows how stars move in the χ–ψ plane at fixed ϑ. The loss region extends in ψ a distance $2\left(\ell_{\rm lc}^2 - \vartheta^2\right)^{1/2}$, from $\psi_{\rm in}$ to $\psi_{\rm out}$. Stars are lost to the SBH if they reach periapsis while in this region.

Two regimes must be considered, depending on whether $\Delta\psi$ is less than or greater than $\psi_{\rm out} - \psi_{\rm in}$:

1. $\Delta\psi < \psi_{\rm out} - \psi_{\rm in}$ (figure 6.15a). In one radial period, stars in the orange region are lost. One half of this region lies *within* the loss ellipse; these are stars with $\ell < \ell_0$ but which have not yet attained periapsis. The persistence of stars inside the "loss cone" is similar to what occurs in the case of diffusive loss-cone repopulation in the spherical geometry; in that case, the width of the boundary layer depends on the ratio of the relaxation time to the radial period

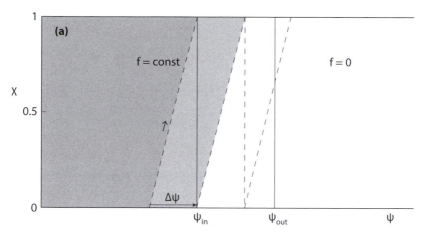

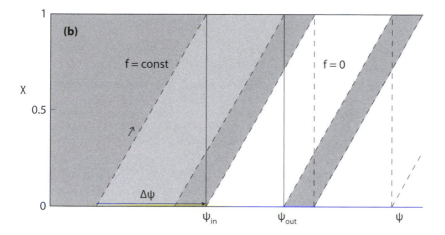

Figure 6.15 Trajectories of stars in the $\psi-\chi$ plane [378]. Stars encounter a loss region from
left to right, defined as $\psi_{in} \leq \psi \leq \psi_{out}$. χ increases from 0 at periapsis, to $1/2$
at apoapsis, to 1 at subsequent periapsis. Trajectories are indicated by dashed
lines. Stars are lost if they reach periapsis while inside the loss region. Stars
within the lightly shaded region are lost in one radial period. (a) $\Delta\psi < \psi_{in} - \psi_{out}$; (b) $\Delta\psi > \psi_{in} - \psi_{out}$.

(section 6.1.2). The other half consists of stars that have not yet entered the
loss region. The area of the orange region is equal to the area of a rectangle of
unit height and width $\Delta\psi$; since stars are distributed uniformly on the $\chi-\psi$
plane, the number of stars lost per radial period is equal to the total number
of stars, of any radial phase, contained within $\Delta\psi$.

2. $\Delta\psi > \psi_{out} - \psi_{in}$ (figure 6.15b). In this case, some stars manage to cross the
loss region without being captured. The area of the orange region is equal to
that of a rectangle of unit height and width $\psi_{out} - \psi_{in}$. The number of stars

lost per radial period is therefore equal to the number of stars, of arbitrary radial phase, contained within $\psi_{\text{out}} - \psi_{\text{in}} = 2 \left(\ell_{\text{lc}}^2 - \vartheta^2 \right)^{1/2}$.

To compute the total loss rate, we integrate the loss per radial period over ϑ. It is convenient to express the results in terms of q, where

$$q \equiv \frac{\Delta\psi}{2\ell_{\text{lc}}} = \pi \frac{v_0}{v_r} \frac{1}{\ell_{\text{lc}}} \sqrt{\left(v_x^{(0)} \ell_{x0} \right)^2 + \left(v_y^{(0)} \ell_{y0} \right)^2}. \tag{6.155}$$

A value of $q \ll 1$ corresponds to an "empty loss cone" and $q \gg 1$ to a "full loss cone." However, we note that, for any $q < 1$, there are values of ϑ such that the width of the loss region, $\psi_{\text{out}} - \psi_{\text{in}}$, is less than $\Delta\psi$. In terms of the integral W defined in equation (4.157), q becomes simply

$$q = \frac{P v_0}{6\ell_{\text{lc}}} \sqrt{W}. \tag{6.156}$$

Unlike the case of collisional loss-cone refilling, where $q = q(E)$ is a function only of energy, here q is also a function of a second integral W. Pyramid orbits with small opening angles will have small W and small q.

The area on the $\psi-\vartheta$ plane that is lost, in one radial period, into one of the four loss regions is

$$2 \int_0^{\vartheta_c} \Delta\psi \, d\vartheta + 2 \int_{\vartheta_c}^{\ell_{\text{lc}}} (\psi_{\text{out}} - \psi_{\text{in}}) d\vartheta, \tag{6.157}$$

where

$$\vartheta_c \equiv \ell_{\text{lc}} \sqrt{1 - q^2} \tag{6.158}$$

is the value of ϑ where $\Delta\psi = \psi_{\text{out}} - \psi_{\text{in}}$; for $q \geq 1$, $\vartheta_c = 0$. For $q \leq 1$, the area integral becomes

$$4q\ell_{\text{lc}} \int_0^{\vartheta_c} d\vartheta + 4 \int_{\vartheta_c}^{\ell_{\text{lc}}} \sqrt{\ell_{\text{lc}}^2 - \vartheta^2} d\vartheta$$

$$= 4q\ell_{\text{lc}}^2 \sqrt{1 - q^2} + 4\ell_{\text{lc}}^2 \int_{\sqrt{1-q^2}}^1 dx \sqrt{1 - x^2}$$

$$= \ell_{\text{lc}}^2 \left(\pi + 2q\sqrt{1 - q^2} - 2 \arcsin \sqrt{1 - q^2} \right)$$

$$= 4q\ell_{\text{lc}}^2 f(q),$$

$$f(q) = \frac{1}{2}\sqrt{1 - q^2} + \frac{1}{2q} \arcsin(q), \tag{6.159}$$

and for $q > 1$ it is $\pi \ell_{\text{lc}}^2$. The function $f(q)$ varies from $f(0) = 1$ to $f(1) = \pi/4 \approx 0.785$.

Considering that there are four loss regions, the instantaneous total loss rate $\mathcal{F}$, in dimensionless units, is

$$\mathcal{F} = f(q) \frac{2\ell_{lc}}{\pi^2 \, \ell_{x0}\ell_{y0}} \sqrt{\left(v_x^{(0)}\ell_{x0}\right)^2 + \left(v_y^{(0)}\ell_{y0}\right)^2}$$

$$= \frac{\mu}{Pv_0} \frac{4q \, f(q)}{\pi} \quad \text{for } 0 \le q \le 1, \tag{6.160a}$$

$$\mathcal{F} = q^{-1} \frac{\ell_{lc}}{2\pi \, \ell_{x0}\ell_{y0}} \sqrt{\left(v_x^{(0)}\ell_{x0}\right)^2 + \left(v_y^{(0)}\ell_{y0}\right)^2}$$

$$= \frac{1}{2\pi^2} \frac{\ell_{lc}^2}{\ell_{x0}\ell_{y0}} \frac{v_r}{v_0} = \frac{\mu}{Pv_0} \quad \text{for } q > 1. \tag{6.160b}$$

The second expression for the loss rate, equation (6.160b), can be called the "full-loss-cone" loss rate, since it corresponds to completely filling and emptying the loss regions in each radial step. Note that the loss rate for $q < 1$ is $\sim q$ times the full-loss-cone loss rate; a similar relation was found to hold in the case of collisionally repopulated loss cones in the spherical geometry (equation 6.70).

The inverse of the loss rate $\mathcal{F}$ gives an estimate of the (dimensionless) time τ_{drain} required to drain an orbit, or equivalently the time for a single star, selected randomly on the torus, to go into the SBH. This time is

$$\tau_{\text{drain}} = \frac{1}{f(q)} \frac{3\pi^2}{2\sqrt{W}} \frac{\ell_{x0}\ell_{y0}}{\ell_{lc}} \quad \text{for } 0 \le q \le 1, \tag{6.161a}$$

$$= \frac{6\pi q}{\sqrt{W}} \frac{\ell_{x0}\ell_{y0}}{\ell_{lc}} \quad \text{for } q > 1. \tag{6.161b}$$

If we consider a "typical" pyramid orbit with $\ell_{x0} \approx \ell_{y0} \equiv \ell_0$ and $v_x^{(0)} \approx v_y^{(0)} \equiv v^{(0)}$, the (dimensionless) precessional period in either x or y is $\tau_{\text{pyr}} = 2\pi/v^{(0)}$ (equation 4.156), and

$$\sqrt{W} = 6\sqrt{2}\pi \ell_0 \tau_{\text{pyr}}^{-1}. \tag{6.162}$$

The (physical) times for orbital draining can then be written approximately in terms of the (physical) precession time as

$$t_{\text{drain}} \approx \frac{1}{f(q)} \frac{\pi\sqrt{2}}{8} \frac{\ell_0}{\ell_{lc}} t_{\text{pyr}} \quad \text{for } 0 \le q \le 1, \tag{6.163a}$$

$$\approx q \frac{\sqrt{2}}{2} \frac{\ell_0}{\ell_{lc}} t_{\text{pyr}} \quad \text{for } q > 1. \tag{6.163b}$$

We recall from section 4.4.3 that the maximal ℓ_0 for pyramids is $\sim (\epsilon_{b,c})^{1/2}$ (equation 4.159) where $\epsilon_{b,c}$ is given by equation (4.143). Thus, in the "empty-loss-cone" regime ($q \ll 1$), the draining time is a factor $\sim \sqrt{\epsilon}/\ell_{lc}$ longer than the pyramid precessional period, or $\sim \ell_{lc}^{-1}$ longer than the typical mass precession time $PM_\bullet/M_\star$. These inequalities reflect the fact that capture only occurs near the corners of the pyramid, when oscillations in both x and y are simultaneously near their peaks, and much more rarely than once per precessional period in either x or y.

If $\eta_{pyr}(E)$ is the fraction of stars at energy E that are on pyramid orbits, the differential loss rate can be written approximately as

$$\dot{N}(E) \approx \eta_{pyr} N(E) t_{drain}^{-1}$$

$$\approx \eta_{pyr} \frac{M_\star}{M_\bullet} \ell_{lc} \frac{N}{P} \quad \text{for } 0 \leq q \leq 1, \tag{6.164a}$$

$$\approx q^{-1} \eta_{pyr} \frac{M_\star}{M_\bullet} \ell_{lc} \frac{N}{P} \quad \text{for } q > 1. \tag{6.164b}$$

Recall that in a *spherical* galaxy, the full-loss-cone capture rate is $\sim \ell_{lc}^2 N/P$. It is clear from equation (6.164) that the loss rate due to draining of the pyramids can be comparable to this. Even though the time to drain one pyramid orbit is much longer than P, the number of stars available to be captured in one draining time, $\eta_{pyr} N$, can be much larger than the number of stars on loss-cone orbits in a spherical galaxy, $\sim \ell_{lc}^2 N$.

At what distance from the SBH do we expect the transition from empty to full loss cones? Using equation (6.156) with $W = (15\epsilon_c)^2$, the maximum value for a pyramid, and equations (6.6) and (4.141a), the condition $q = 1$ becomes

$$1 = \frac{40\pi^2}{3} \frac{\alpha_2}{(2-\gamma)(3-\gamma)} \frac{\rho_0 r_0^3}{M_\bullet} \left(\frac{r_{crit}}{r_0}\right)^{3-\gamma} \epsilon_c(r_{crit}) \sqrt{\frac{r_{crit}}{\Theta r_g}}. \tag{6.165}$$

Without loss of generality, we can set $r_0 = r_m$, the radius containing a mass in stars twice the mass of the SBH. Then using equation (4.143),

$$\frac{r_{crit}}{r_m} = \left(\frac{3}{20\pi} \frac{2-\gamma}{\alpha_2}\right)^{2/7} \left[\frac{\sqrt{\Theta}}{\epsilon_c(r_m)} \sqrt{\frac{r_g}{r_m}}\right]^{2/7} \tag{6.166a}$$

$$\approx 0.5 \left[\frac{\sigma}{c} \frac{\sqrt{\Theta}}{\epsilon_c(r_m)}\right]^{2/7}, \tag{6.166b}$$

where $r_m \approx r_h$ has been used in the second line. The giant elliptical galaxies for which feeding by pyramid orbits is most relevant have $\sigma \approx 200$–300 km s^{-1} and $\Theta \approx$ a few, so that

$$\frac{r_{crit}}{r_m} \approx 0.1 \left[\epsilon_c(r_m)\right]^{-2/7}. \tag{6.167}$$

For $\epsilon_c(r_m) = 0.1(0.01)$, $r_{crit}/r_m \approx 0.2(0.4)$. To a reasonable approximation, we can say that feeding by stars inside r_m occurs via an empty loss cone. Recall that the pyramids are essentially replaced by chaotic centrophilic orbits when $a \approx r_m$; to the extent that these orbits behave similarly to pyramids, their feeding to the SBH would take place in the full-loss-cone regime.

The expressions just derived for the flux and for t_{drain} assumed a fully populated torus. In reality, after ~ 1 precessional periods, some parts of the torus that are entering the loss regions will be empty and the loss rate will drop below equation (6.160). For $\Delta\psi \geq \psi_{in} - \psi_{out}$, the downstream density in figure 6.14, integrated over the radial phase, is easily shown to be $1 - q^{-1}\sqrt{1 - \vartheta^2/\ell_{lc}^2}$ times the upstream density while for $\Delta\psi < \psi_{in} - \psi_{out}$ the downstream density is zero. Integrated over ϑ, the downstream depletion factor becomes

$$1 - \frac{\pi}{4q} - \sqrt{1-q^2}(1+q) + \frac{1}{2q}\sin^{-1}\sqrt{1-q^2} \qquad (6.168)$$

for $q \leq 1$ and $1 - \pi/4q$ for $q > 1$; it is 0 for $q = 0$, ~ 0.215 for $q = 1$ and 1 for $q \to \infty$. For small q, the torus will become striated, containing strips of nearly zero density interlaced with undepleted regions; the loss rate will exhibit discontinuous jumps whenever a depleted region encounters a new loss ellipse and the time to totally empty the torus will depend in a complicated way on the frequency ratio v_x/v_y and on ℓ_{lc}. For large q, the loss rate will drop more smoothly with time, roughly as an exponential law with time constant $\sim t_{drain}$.

Next, consider the case of pyramids with arbitrary opening angles, that is, for which $\{e_{x0}, e_{y0}\}$ are not necessarily small. For each orbit one can compute μ, the fraction of the torus occupied by the loss cone (equation 6.150), by numerically integrating the equations of motion and analyzing the probability distribution for instantaneous values of ℓ^2: $P(\ell^2 < X) \propto X - \ell^2_{min}$, where ℓ^2_{min} allows for a nonzero lower bound on ℓ^2. Almost all pyramids have $\ell_{min} = 0$, but some of them happen to be resonances (commensurable v_x and v_y) and hence avoid approaching $\ell = 0$. This linear character of the distribution of ℓ^2 near its minimum corresponds to a linear probability distribution of periapsis radii ($P(r_{peri} < r) \propto r$), which is natural to expect if we combine a quadratic distribution of impact parameters at infinity with gravitational focusing [371].

The coefficient μ for each orbit is calculated as $P(\ell^2 < \ell^2_{lc})$. As seen from equation (6.150), the smaller the extent of a pyramid in any direction, the greater μ—this is true even for orbits with large e_{x0} or e_{y0}. While μ varies greatly from orbit to orbit, its overall distribution over the entire ensemble of pyramid orbits follows a power law:

$$P_\mu(\mu > Y) \approx \left(\frac{Y}{\mu_{min}}\right)^{-2}, \qquad \mu_{min} \approx \frac{\ell^2_{lc}}{2\tilde{\eta}}; \qquad (6.169)$$

P_μ is the probability of having μ greater than a certain value and $\tilde{\eta}$ is the fraction of pyramids among all orbits. The average μ for all pyramid orbits is therefore $\overline{\mu} = 2\mu_{min}$, and the average fraction of time that an orbit of any ℓ spends inside the loss cone is $\overline{\mu}\tilde{\eta} \simeq \ell^2_{lc}$ (almost independent of the potential parameters ϵ_b and ϵ_c)—the same number that would result from an isotropic distribution of orbits in a spherically symmetric potential. In other words, until such a time as the centrophilic orbits have been substantially depleted, loss-cone feeding rates should be roughly equal to full-loss-cone rates in the equivalent spherical model.

This is about as far as it is possible to go in terms of analytically computing feeding rates in triaxial galaxies. As discussed in section 3.5, a "zone of chaos" exists in triaxial potentials, starting at a few times r_m and extending outward to a radius containing a mass in stars of $\sim 10^2 M_\bullet$. Figure 6.16 illustrates some of the important properties of orbits that extend into this region. There are, broadly speaking, two types of centrophilic orbit: regular orbits that avoid passing through the very center, and chaotic orbits. The former orbits lie near to a "thin" orbit, that

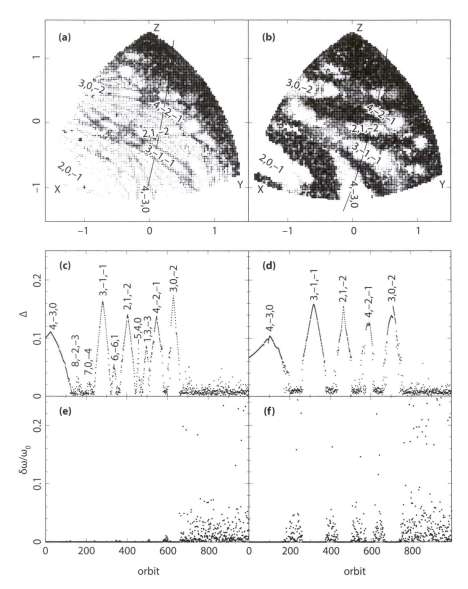

Figure 6.16 Properties of centrophilic orbits in triaxial galaxies, with (right) and without
(left) central SBHs [377]. The top panels show one octant of an equipotential surface located just inside the half-mass radius of the model. Orbits were
started on this surface with zero velocity. The top, left, and right corners correspond to the z- (short), x- (long), and y- (intermediate) axes. The gray scale is
proportional to the logarithm of the diffusion rate of orbits in frequency space,
computed in the same way as in section 3.1.2.1; initial conditions corresponding to regular orbits are white. The most important resonance zones are labeled
with their defining integers (m_1, m_2, m_3). Panels (c) and (d) show the distance
of closest approach, Δ, of orbits whose starting points lie along the heavy lines

Figure 6.16 Continued. in (a) and (b). The most important stable resonances are again
            labeled. Panels (e) and (f) show the degree of stochasticity of the orbits, as mea-
            sured by the change $\delta\omega$ in their "fundamental frequencies"; ω_0 is the frequency
            of the long-axis orbit and regular orbits have $\delta\omega/\omega_0 = 0$.

is, an orbit that respects a resonance between the fundamental frequencies:

$$m_1 \nu_1 + m_2 \nu_2 + m_3 \nu_3 = 0 \qquad (6.170)$$

with the m_i integers (equation 3.13). If the parent, resonant orbit avoids the center,
orbits that lie close to the resonant torus will do so as well, passing no closer to the
center than some minimum distance Δ. As the initial conditions move farther from
the resonant torus, the orbit broadens, causing it to approach more closely to the
destabilizing center. At some critical Δ—typically much larger than r_{lc}—the orbit
becomes chaotic. To a good approximation, *all* orbits that pass through the very
center and that extend outward into the "zone of chaos" are chaotic.

The complexity of the orbits in this region mandates a brute-force, numerical treatment of SBH feeding. Figure 6.17 shows the results of such an analysis,
starting from self-consistent models of triaxial nuclei [371]. Such calculations are
model-dependent, but they suggest feeding rates in triaxial galaxies of order

$$\dot{M} \approx 10^{-5} \eta \left(\frac{r_h}{100\,\mathrm{pc}} \right)^{-5/2} \left(\frac{M_\bullet}{10^8\,M_\odot} \right)^{5/2} M_\odot\,\mathrm{yr}^{-1} \qquad (6.171)$$

with a weak dependence on the degree of triaxiality; here η is the fraction of orbits
that are centrophilic. Equation (6.171) is based on a nuclear model in which $\rho \sim r^{-1}$, not too different from what is observed near the centers of bright elliptical
galaxies. Feeding rates due to collisional loss-cone refilling are very long in such
galaxies and even a modest fraction of centrophilic orbits could result in loss rates
that are far larger than predicted in collisional, spherical models (figure 6.18).

6.3 BINARY AND HYPERVELOCITY STARS

Interaction of binary stars with an SBH might seem to fit equally well into chapter 5
on gravitational encounters, or chapter 7 on collisional evolution of nuclei. The
topic is included here, since the continued supply of binary stars to the SBH is
inherently a loss-cone problem [578].

6.3.1 Basic concepts

It is believed that most stars, and in particular most massive stars, form in binary or
multiple systems. Studies of the multiplicity of solar-type stars in the solar neighborhood find that about half are in binary or triple systems [2]; the observed distribution of periods, or semimajor axes, is approximately log–normal with a peak
around 30 AU [126][9]. Upper main sequence (O, B) stars appear to almost always

---

[9]AU $=$ astronomical unit, the distance from the Earth to the Sun; 1 AU $= 1.50 \times 10^{13}$ cm $= 4.85 \times 10^{-6}$ pc.

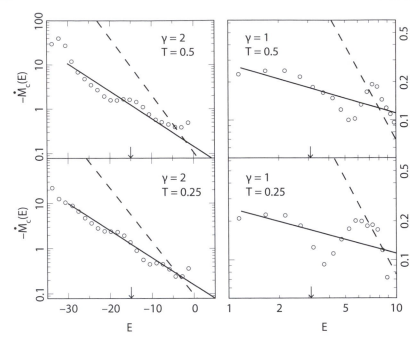

Figure 6.17 Energy-dependent capture rates due to centrophilic orbits in four triaxial models
at $t = 0$, i.e., before any orbital depletion has taken place [371]. Open circles
show $\dot{N} = m\dot{M}$ computed using the actual chaotic orbit populations in the self-
consistent models (the same models illustrated in figure 3.18). Dashed lines are
capture rates predicted for a spherical, full-loss-cone model. Solid lines are an
analytic approximation. The parameter γ is the exponent in $\rho \propto r^{-\gamma}$, and T is
the triaxiality index defined in equation (3.127). Arrows indicate E_h.

form in close binary systems [290]; typical semimajor axes are less than one parsec.
A lower limit on binary separations is set by the distance at which the components
of the binary would overlap; this is roughly 0.01 AU for stars of type F or G and
roughly five times larger for upper-main-sequence stars. Binaries with large separations are easily disrupted by interactions with field stars, as discussed in more
detail below.

In stellar systems containing a modest number of stars—globular clusters, for
instance—the gravitational energy of a single stellar binary can be comparable with
that of the cluster as a whole. Exchange of energy between binary and field stars
can play an important role in mediating processes like core collapse (chapter 7) in
these systems. No such role is played by binary stars in galactic nuclei, since the
internal energy of a binary is small compared with that of a single star on a tightly
bound orbit, $a \ll r_h$, around an SBH. It is for this reason that the existence of
binary stars can generally be ignored when discussing nuclear dynamics.

Binary stars are nevertheless interesting dynamically because they provide an
additional set of ways for *single* stars to interact closely with an SBH. The basic

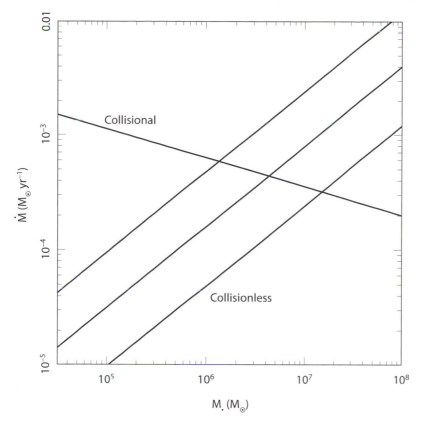

Figure 6.18 Comparison between collisional loss-cone feeding rates in spherical galaxies ("collisional") and rates of orbital draining in triaxial nuclei ("collisionless") [371]. All curves assume a nuclear density profile $\rho \propto r^{-2}$. Line labeled "collisional" is equation (6.86). In these models, the "collisionless" feeding rates scale with $\eta t_{10}^{-1/2}$, where η is the fraction of centrophilic orbits and t_{10} is the age of the galaxy in units of 10 Gyr; the three lines were computed assuming values of (0.3, 1, 3) for this parameter.

idea originated with J. G. Hills in 1988 [238]. A star that approaches closely to an SBH will have a velocity

$$v(r) = \sqrt{2(E - \Phi(r))} \approx (2GM_\bullet/r)^{1/2} \tag{6.172}$$

$$= 2.9 \times 10^3 \left(\frac{M_\bullet}{10^6 \, M_\odot}\right)^{1/2} \left(\frac{r}{1 \, \text{mpc}}\right)^{-1/2} \text{km s}^{-1}$$

(mpc = milliparsec = 10^{-3} pc $\approx$ 206 AU). Suppose that the star experiences a velocity change, $\delta v \ll v$, near its time of closest approach to the SBH. The perturbation could come from field stars, but it can also be due to the presence of the second star in a binary. The change in energy corresponding to the velocity change δv is $\delta E = \frac{1}{2}(v + \delta v)^2 - \frac{1}{2}v^2 \approx v \, \delta v$. If δE is much larger than $|E|$, the star can escape from the SBH, with velocity

$$(2v\,\delta v)^{1/2} \approx 1.5 \times 10^3 \left(\frac{M_\bullet}{10^6\,M_\odot}\right)^{1/4} \left(\frac{r}{1\,\mathrm{mpc}}\right)^{-1/4} \left(\frac{\delta v}{400\,\mathrm{km\,s^{-1}}}\right)^{1/2} \mathrm{km\,s^{-1}}.$$

(6.173)

Equation (6.173) is the velocity at the moment of departure; after climbing through the potential well of the SBH, and (perhaps) the galaxy, the ejected star will be moving more slowly. Nevertheless, this argument shows that **hypervelocity stars** are a likely consequence of the interaction of binary stars with an SBH; and their detection would constitute indirect evidence for the presence of an SBH. As discussed in more detail below, a handful of stars are observed, in the outskirts of the Milky Way, whose kinematical properties and ages are consistent with their having been ejected from the center of the Galaxy.

The same argument also implies that one member of a binary can *lose* energy; indeed (in the absence of dissipation, or additional stellar perturbations) the total energy of the three-body system—SBH plus two stars—is conserved, and the ejection of one star in an unbound orbit necessarily implies that the other star will move onto a more tightly bound orbit around the SBH. This mechanism has been proposed as a way to place the Galactic center S-stars onto their current orbits [206], and also as a way to inject compact objects onto relativistic orbits very near an SBH [385].

Binary stars in the environment of a galactic nucleus are "soft": the relative velocity of the two components of the binary is small compared with typical field-star velocities; that is, $V_{bin} < \sigma$. An equivalent statement is

$$\frac{|E_{bin}|}{m_{12}\sigma^2} \ll 1,$$

(6.174)

where $E_{bin} \equiv -Gm_1 m_2/(2a_{bin})$ is the total energy of the binary (equation 4.27) and $m_{12} = m_1 + m_2$ is the binary mass. Soft binaries tend to acquire smaller binding energies in interactions with passing stars ("soft binaries become softer" [236, 231]). A straightforward calculation yields the characteristic time over which a soft binary is disrupted by repeated encounters with field stars [532]:

$$t_{evap} \approx 0.1 \frac{\sigma}{G\rho a_{bin} \ln \Lambda},$$

(6.175)

where ρ is the field-star mass density and a_{bin} is the semimajor axis of the binary; equation (6.175) assumes that the binary components and the field stars have the same mass. Figure 6.19 shows estimates of the evaporation time for $2\,M_\odot$ binaries near the center of the Milky Way. Two different assumptions were made for $\rho(r)$: an approximately "isothermal" density, $\rho \sim r^{-2}$, as in equation (6.73); and a much flatter density profile, $\rho \sim r^{-1/2}$, motivated by the number counts of old stars at the Galactic center (figure 7.1). The figure suggests that—inside roughly 1 pc—essentially all binaries would be subject to evaporation over the age of the Galaxy. In the low-density model, the evaporation time exceeds 1 Gyr only for binaries with $a_{bin} \lesssim 0.03$ AU, while in the high-density model, only binaries with $a_{bin} \lesssim 0.01$ AU, that is, contact binaries, would survive for 1 Gyr inside ~ 0.1 pc.

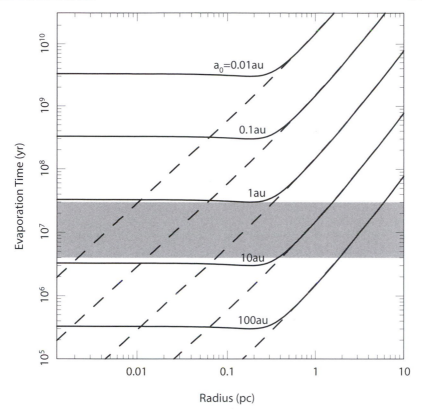

Figure 6.19 Evaporation time of binary stars as a function of distance from the center of the Milky Way [7]. Dashed curves were computed from equation (6.175) assuming a high density of field stars near the SBH, equation (6.73). Solid curves are based on a low-density model, motivated by the number counts in figure 7.1. The gray region shows the estimated ages of the S-stars [137], which might have been placed onto their current orbits following an exchange interaction between a primordial binary star and the SBH.

However, as we will see, consideration of the loss-cone problem suggests that most binaries destined to interact with the Milky Way SBH would originate on orbits with semimajor axes beyond one parsec, where the density of field stars is lower and evaporation times are much longer.

6.3.2 The binary loss cone

A binary that is tight enough to survive evaporation by field stars can nevertheless be disrupted if it passes sufficiently close to the SBH. The distance at which that occurs is given approximately by equation (6.2b), the radius of tidal disruption of a *single* star, if $R_\star$ is replaced by $a_{\rm bin}$ and $m_\star$ is replaced by $m_1 + m_2 \equiv m_{12}$:

$$r_{\rm t,bin} \approx a_{\rm bin} \left(\frac{M_\bullet}{m_{12}}\right)^{1/3} \approx 10 \left(\frac{a_{\rm bin}}{0.1\,{\rm AU}}\right) \left(\frac{M_\bullet}{10^6 m_{12}}\right)^{1/3}\,{\rm AU}. \qquad (6.176)$$

The velocity change experienced by either star is given roughly by its orbital velocity relative to the binary center of mass at the time of disruption. For m_1 this is

$$\delta v \approx \sqrt{\frac{G(m_{12})}{a_{\rm bin}}\frac{m_2}{m_{12}}}$$

$$\approx 67 \left(\frac{2m_2}{m_{12}}\right)^{1/2} \left(\frac{m_2}{1\,M_\odot}\right)^{1/2} \left(\frac{0.1\,{\rm AU}}{a_{\rm bin}}\right)^{1/2}\,{\rm km\,s}^{-1}, \qquad (6.177)$$

and similarly for m_2.

What happens next depends on the details: the relative orientation of the two orbital planes, the location of the two stars on their relative orbit at the time of closest approach to the SBH, etc. The outcome can be expressed statistically, based on the results of extensive numerical experiments [238]. If $m_1 \approx m_2$ and if the velocity of the binary at closest approach is similar to the value given by equation (6.172) (i.e., a marginally bound orbit with respect to the SBH), the probability of an "exchange interaction"—one star ejected, one remaining bound to the SBH—is a function mainly of

$$\lambda \equiv \frac{r_{\rm min}}{r_{\rm t,bin}}, \qquad (6.178)$$

where $r_{\rm min}$ is the periapsis of the binary's orbit around the SBH. Exchange interactions are found to occur in approximately 80% of encounters with $\lambda = 0.3$ and in 50% of encounters with $\lambda = 1$ [238]. If an exchange interaction occurs, the ejection velocity, defined as the velocity at infinity of the ejected star in the absence of the Galactic potential, is given approximately by [238]

$$v_{\rm ej} \approx 1800 \left(\frac{a_{\rm bin}}{0.1\,{\rm AU}}\right)^{-1/2} \left(\frac{m_{12}}{2\,M_\odot}\right)^{1/3} \left(\frac{M_\bullet}{4\times10^6\,M_\odot}\right)^{1/6}\,{\rm km\,s}^{-1}. \qquad (6.179)$$

We are interested in ejections with velocities high enough to escape the Galactic bulge, and perhaps even the Galactic halo. Ignoring the contribution to the gravitational potential from the SBH itself, the escape velocity from the center of the Galaxy, $v_{\rm esc}(0)$, is believed to be about $800\,{\rm km\,s}^{-1}$ [70]. Recalling that $v_{\rm ej}$ is the velocity of an ejected star after it has climbed out through the SBH potential well, energy conservation gives for its velocity at infinity,

$$v_\infty = \left[v_{\rm ej}^2 - v_{\rm esc}(0)^2\right]^{1/2}. \qquad (6.180)$$

Requiring $v_{\rm ej} > v_{\rm esc}(0)$ for exchange interactions with $r_{\rm min} \approx r_{\rm t,bin}$ then implies, via equation (6.179), $a_{\rm bin} \lesssim 0.5\,{\rm AU}$ if $m_1 = m_2 = 1\,M_\odot$. Based on figure 6.19, the survival time of binaries with separations of 0.5 AU exceed 10 Gyr beyond a radius of a few parsecs from Sgr A*.

Computing the rate of production of high-velocity stars (HVSs) from these formulas requires a number of additional pieces of information. We would need to know the fraction of stars that are in binary systems and the distribution of binary semimajor axes, and both quantities should preferably be expressed as functions of m_1 and m_2.

Perhaps most important, we need to specify the mechanism that is responsible for placing binary stars onto orbits that come close to the SBH; for the Milky Way, that means orbits with $r_{min} \lesssim r_{t,bin} \lesssim 125 a_{bin}$. This is, of course, a loss-cone problem. It differs from the loss-cone problems treated elsewhere in this chapter in one important respect: the radius of the tidal disruption sphere for a binary, $r_{t,bin}$, is much larger than that for a single star, r_t:

$$r_{lc} \approx r_{t,bin} \approx \frac{a_{bin}}{R_\star} r_t \approx 21 \left(\frac{a_{bin}}{0.1\,\text{AU}} \right) \left(\frac{R_\star}{R_\odot} \right)^{-1} r_t. \tag{6.181}$$

If we suppose that loss-cone orbits are resupplied by gravitational encounters, an important quantity is $q(\mathcal{E})$ (equation 6.37):

$$q(\mathcal{E}) \equiv \frac{P(\mathcal{E})\mathcal{D}(\mathcal{E})}{\mathcal{R}_{lc}(\mathcal{E})} \approx \frac{P(\mathcal{E})}{T_r(\mathcal{E})} \frac{L_c^2(\mathcal{E})}{2GM_\bullet r_{lc}}. \tag{6.182}$$

Recall that $q \approx 1$ defines the transition between the "full-" ($q \gg 1$) and "empty-" ($q \ll 1$) loss-cone regimes, and that for the loss cone corresponding to tidal disruption of single stars, this transition occurs roughly at energies corresponding to the SBH influence radius in a galaxy like the Milky Way (figure 6.5). The much larger value of r_{lc} in the binary problem implies that q will be much smaller at a given energy, and therefore that the loss cone will be emptier much farther out. In fact, it is reasonable to simply ignore any contribution to the feeding rate from the full-loss-cone regime, and define the encounter-driven, differential flux into the loss cone as

$$F(\mathcal{E}) \approx \eta \frac{F^{\max}(\mathcal{E})}{|\ln \mathcal{R}_{lc}(\mathcal{E})|} \approx \eta \frac{N(\mathcal{E})\mathcal{D}(\mathcal{E})}{|\ln \mathcal{R}_{lc}|} \tag{6.183}$$

(equations 6.70, 6.68), where η is the fraction by number of "stars" that are binary. Calculations based on an equation like (6.183) [578, 244] find a total production rate of

$$\dot{N}_{HVS} \approx 10^{-5} \left(\frac{\eta}{0.1} \right) \text{yr}^{-1}, \tag{6.184}$$

weakly (logarithmically) dependent on the assumed value of a_{bin}. The orbits of the binaries that dominate the flux into the loss cone have semimajor axes of a few parsecs, large enough to avoid disruption from field stars (figure 6.19).

6.3.3 Observed populations

At $1000\,\text{km s}^{-1}$, a star travels a distance of $100\,\text{kpc}$ in a time of $\sim 10^8\,\text{yr}$. If the production rate of HVSs is $\sim$ one per $10^5\,\text{yr}$, as suggested by equation (6.184), one would predict the existence of many such objects at any given time, distributed with a range of distances from the Galactic center.

The first candidate HVS was discovered in 2005 [64], with a galactocentric velocity of $709\,\text{km s}^{-1}$. The star's spectrum suggested either a type B9 main-sequence star, or a blue horizontal branch star; in the former case its photometric distance from the Galactic center would be about 100 kpc, in the latter case somewhat less. Subsequent observations [174] supported the main-sequence interpretation. The main-sequence lifetime of a B9 star is $\sim$ a few hundred Myr, consistent with the time required to reach such a great distance; the identification of this star as an HVS is further strengthened by the fact that B-type stars are rare so far from the Galactic center.

Figure 6.20 summarizes the current evidence for HVSs in the Milky Way [63]. The figure is based on a survey that targeted main-sequence stars of spectral type A; earlier (O, B) spectral types are more luminous but are easily confused with white dwarfs. The figure shows 14 stars having radial velocities and distances that place them beyond the curve of escape, computed using a standard model for the Galactic potential. Equating the measured (line-of-sight) velocity with the space velocity—likely to be approximately correct if these stars actually were ejected from the Galactic center—the inferred travel times range from $\sim$60 Myr to $\sim$240 Myr, comfortably shorter than main-sequence lifetimes. The observed set of travel times is consistent with a constant rate of production of HVSs, as opposed, say, to a burst.[10]

At first sight, the observed numbers of HVSs appear to be roughly consistent with the rate predicted by equation (6.184) [65]. However, it has been argued [422] that a careful accounting of the upper-main-sequence stars in binary systems in the inner one or two parsecs of the Milky Way yields much lower numbers than are implied by setting $\eta \approx 0.1$ in equation (6.184), and hence a much lower rate of production of the types of stars that would be picked up in the surveys (6.184). Alternative models, with possibly higher production rates, have been proposed and are discussed elsewhere in this book: scattering of binary stars by "massive perturbers" into the SBH loss cone (section 7.4); or ejection of stars by a binary SBH at the Galactic center (section 8.6.2).

Corresponding to each HVS, there should be a second star—the other component of the primordial binary—that becomes more tightly bound to the SBH. We can estimate the properties of the bound star's orbit as follows. Consider an equal-mass binary that is disrupted at a distance $r_{\mathrm{t,bin}} = a_{\mathrm{bin}}(M_\bullet/m_{12})^{1/3}$. The point of tidal disruption becomes the periapsis of the new orbit. The orbital energy extracted by the work of the tidal field on the binary,

$$|\delta E| \approx \left(\frac{GM_\bullet}{r_{\mathrm{t,bin}}^3}\right) a_{\mathrm{bin}} \times r_{\mathrm{t,bin}} \approx \frac{GM_\bullet^{1/3}}{a_{\mathrm{bin}}} m_{12}^{5/3}, \tag{6.185}$$

is carried away by the ejected star, so that the orbit of the captured star has energy $-|\delta E|$ and semimajor axis $a = -GM_\bullet m_{12}/(4|\delta E|)$, or

$$\langle a \rangle \approx \left(\frac{M_\bullet}{m_{12}}\right)^{2/3} a_{\mathrm{bin}} \tag{6.186a}$$

$$\approx 4 \left(\frac{M_\bullet}{4 \times 10^6 \, M_\odot}\right)^{2/3} \left(\frac{m_{12}}{5 \, M_\odot}\right)^{-2/3} \left(\frac{a_{\mathrm{bin}}}{0.1 \, \mathrm{AU}}\right) \, \mathrm{mpc}. \tag{6.186b}$$

---

[10] An alternate model for the production of HVSs involving an intermediate-mass black hole would predict a burst; see section 8.6.

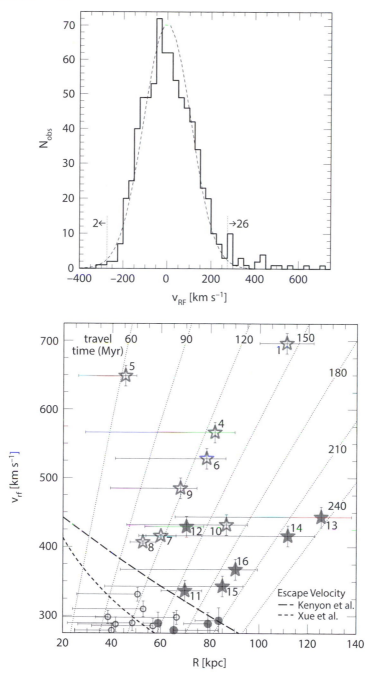

Figure 6.20 Evidence for hypervelocity stars in the Milky Way [63]. The top panel is the distribution of radial velocities for 759 stars in the survey. The dashed curve is the best-fit Gaussian distribution. Note the presence of outliers at positive (outgoing) velocities. The bottom panel plots radial velocity versus Galactocentric

Figure 6.20 Continued. distance R for the 26 stars with $v_{rf} > +275\,\mathrm{km\,s^{-1}}$. Distances were computed assuming that the stars are on the main sequence. The objects labeled with star symbols have distances and velocities that imply escape orbits; the dashed lines are estimates of the escape velocity curve based on models for the Galactic potential. The dotted lines are curves of constant travel time from the Galactic center computed from the same model, assuming that the observed velocity v_{rf} is the full space motion of the stars.

In effect, the distribution of binary separations is mapped into the distribution of orbital semimajor axes about the SBH. The eccentricity is very high and independent of a_{bin}:

$$\langle e \rangle \approx 1 - \left(\frac{m_{12}}{M_\bullet} \right)^{1/3} \gtrsim 0.95. \tag{6.187}$$

These properties are similar to those of the S-stars (table 4.1), the young, apparently normal main-sequence B-stars in the inner $\sim 50\,\mathrm{mpc}$ of the Milky Way, and it is natural to wonder whether the S-stars were placed onto their tightly bound, eccentric orbits by binary exchange interactions [206]. A special origin for the S-stars seems implied by the significant systematic differences between them and the other young stars in the inner parsec. For instance, stars in the two, parsec-scale stellar disks are much more massive, short-lived O-stars and their orbits are nearly circular. These differences would seem to preclude a model in which the two populations of young stars formed at the same time and in the same way [420].

The tidal capture hypothesis is supported by several lines of evidence. The luminosity function of the S-stars is close to the steep, "universal" luminosity function that is observed in the field, and quite different from the apparently flat ("top heavy") luminosity function of the disk stars [29]. The eccentricity distribution of the S-stars is close to "thermal," $N(e)de \sim e\,de$, although not as biased to high eccentricities as predicted by equation (6.187). However, the timescale for resonant relaxation to change orbital eccentricities at these small distances from the SBH may be short enough to convert an initial, highly nonthermal distribution into what is observed (figure 6.21). The distribution of semimajor axes of the captured stars is harder to predict since it depends on the poorly known, primordial distribution of binary separations. Perhaps the biggest problems with the capture hypothesis are the predicted rate of formation of HVSs (equation 6.184)—too low to deflect massive binaries from the field at a high enough rate to maintain a steady-state population of ~ 40 S-stars—and the necessity of postulating a continuous source of young binaries at distances of a few parsecs from Sgr A*.

6.4 RELATIVISTIC LOSS CONES AND EXTREME-MASS-RATIO INSPIRALS

So far in our discussion of the loss-cone problem, we have assumed standard, or "nonresonant," relaxation (NRR) as the mechanism that scatters stars onto

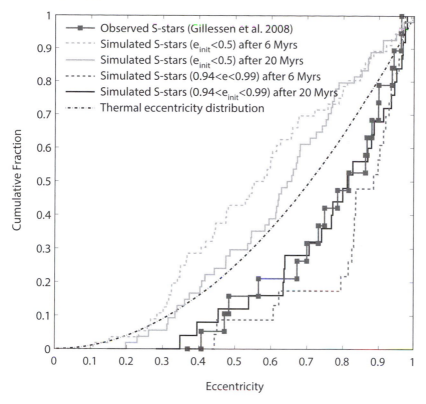

Figure 6.21 The evolution of the S-star eccentricity distribution due to resonant relaxation,
assuming the existence of a dense cluster of stellar-mass BHs around the Milky
Way SBH [421]. Two different assumptions were made about the initial eccentricity distribution of the S-stars: a distribution peaked around $e \approx 1$, as predicted by the binary exchange model; and a distribution with $e \leq 0.5$, as might
be expected if the S-stars formed in a thin disk. The former model is most consistent with the data, evolving, after ~ 20 Myr, into an $N(e)$ that is consistent
with the observed distribution.

loss-cone orbits. But as discussed in chapter 5, very near the center of a nucleus
containing an SBH, the mechanism of "resonant relaxation" (RR) can be more efficient than NRR at changing orbital angular momenta.

Recall from that discussion that RR is driven by torques from field-star orbits that
remain essentially fixed in their orientation for a time $\sim t_{\mathrm{coh}}$, the "coherence time."
For times $t \lesssim t_{\mathrm{coh}}$, the angular momentum of a test star changes roughly linearly
with time; this is the "coherent regime." On longer timescales, angular momenta
undergo a random walk, similar in character to the random walk due to NRR. In
this "incoherent regime," we define the RR time T_{RR} as the time for RR to change
the angular momentum of a test star by of order L_c, the angular momentum of

a circular orbit of the same energy (equation 5.239):

$$T_{RR} = \left(\frac{L_c}{|\Delta L|_{coh}} \right)^2 t_{coh},$$

where $|\Delta L|_{coh}$ is the change in L over one coherence time. The coherence time associated with Newtonian, or mass, precession was given by equation (5.202):

$$t_{coh,M}(a) \approx \frac{M_\bullet}{Nm} P(a). \tag{6.188}$$

Recall that equation (6.188) defines the *average* time for field-star orbits to precess; it was computed as an average over eccentricity. A second source of coherence breaking is geodetic, or Schwarzschild, precession due to the lowest-order (1PN) effects of relativity. The corresponding coherence time was given by equation (5.204) as

$$t_{coh,S}(a) \approx \frac{1}{12} \frac{a}{r_g} P(a), \tag{6.189}$$

which again is an average over eccentricity. In section 5.6.2, it was argued that $T_{RR} < T_{NRR}$ only inside $\sim 0.1 r_h$ for a galaxy like the Milky Way. This is much smaller than the radii from which most (normal) stars would be scattered into the SBH (figure 6.5).

Nevertheless, there are interesting regimes in which RR could make an important —even dominant—contribution to the feeding rate. One example would be an SBH that had been forcibly removed from its host galaxy, perhaps by a gravitational slingshot interaction involving three SBHs (section 8.7). Stars would only remain bound to the ejected SBH if their orbital velocities at the moment of the kick were higher than the kick velocity; for sufficiently large kicks, there would be few stars remaining bound at radii as large as $\sim r_h$, meaning that essentially all gravitational encounters would be in the RR regime [293].

Another, less speculative, example is the capture of compact remnants—white dwarfs, neutron stars, or stellar-mass BHs—by an SBH. Neutron stars and stellar BHs are not tidally disrupted, even by passing just outside the event horizon of an SBH. They can of course be directly captured. But a **plunge**—as these events are called by the gravitational-wave community—is considered less interesting than an **inspiral**: the gradual decay of an orbit due to emission of gravitational radiation.

Inspiral of a stellar-mass black hole or neutron star into an SBH is called an **extreme-mass-ratio inspiral**, or **EMRI** [240, 496]. While no gravitational-wave telescope yet exists that could detect the low frequency ($\sim 10^{-3}$ Hz) gravitational waves from such inspirals,[11] there is hope that such an instrument will be built in the near future. If so, an EMRI would provide a gravitational-wave train that could be continuously detected over thousands or tens of thousands of orbits, allowing the signal-to-noise ratio of the signal to be built up over time. The information so obtained might permit tests of theories of gravity in the strong-field limit, as

---

[11]LISA, the Laser Interferometer Space Antenna described in section 2.7, was to have been such a telescope [102]. Its status as of this writing is uncertain.

well as encoding more prosaic information like the mass, spin, and distance of the SBH [21].

In discussing the dynamics of EMRIs, it is useful to switch from (E, L) as orbital parameters to (a, e): the semimajor axis and eccentricity. (Since we are guaranteed to be in a mildly relativistic regime, an even better choice might be the 1PN generalizations of a and e derived in section 4.5.1.) Figure 6.22 shows the evolution of orbits plotted in this space. That figure is based on the first, direct N-body simulation of EMRI formation [359]. The simulation consisted of 50 particles representing stellar-mass black holes (BHs), distributed initially between 0.1 mpc $\leq a \leq 10$ mpc as $n(r) \propto r^{-2}$ around an $M_{\bullet} = 10^6 \, M_{\odot}$ SBH. In these simulations, the Newtonian and relativistic coherence times were roughly equal at $a \approx 1$ mpc. If we adopt $t_{\mathrm{coh,M}}$ as the coherence time, then equation (5.240) gives for the incoherent RR time

$$T_{\mathrm{RR}}(a) \approx 10^5 \left(\frac{a}{\mathrm{mpc}} \right)^{3/2} \mathrm{yr}, \tag{6.190}$$

while the NRR time is substantially longer,

$$T_{\mathrm{NRR}} \approx 5 \times 10^6 \left(\frac{a}{\mathrm{mpc}} \right)^{1/2} \mathrm{yr}. \tag{6.191}$$

The effect of incoherent RR is to scatter objects in angular momentum. The effects of this scattering can be seen in figure 6.22: the evolutionary tracks are mostly horizontal, since changes in L at fixed E correspond to changes in e at fixed a. In the absence of any other mechanisms, RR would be expected to scatter almost all the BHs into the SBH by 1 Myr. However, very few of the trajectories in figure 6.22 manage to reach the high eccentricities needed to become EMRIs. Instead, there is an apparent barrier in orbital eccentricity which inhibits the evolution to high e. BHs that "strike" this barrier are "reflected" back to smaller eccentricities.

The barrier is a consequence of the relativistic (Schwarzschild) precession [359]. As noted above, for orbits of *average* eccentricity, the rates of Newtonian and Schwarzschild precession are similar at these distances from the SBH. But the timescale associated with Schwarzschild precession of a single orbit,

$$t_{\mathrm{S}} = \frac{P(a)}{6} \frac{c^2 a}{GM_{\bullet}} (1 - e^2), \tag{6.192}$$

tends to zero as $e \to 1$. For a very eccentric orbit, the effective time over which background torques can act coherently is given by its precession time, and not the average (and much longer) precession time of the other orbits. The angular momentum barrier can be identified, in a qualitative way, with the value of e at which RR becomes ineffective due to relativistic precession.

The residual torque produced by an otherwise-spherical distribution of stars, at $r \approx a$, is given roughly by

$$T \approx \frac{Gm}{a} \sqrt{N(a)} \approx \frac{1}{\sqrt{N(a)}} \frac{GM_{\star}(a)}{a}. \tag{6.193}$$

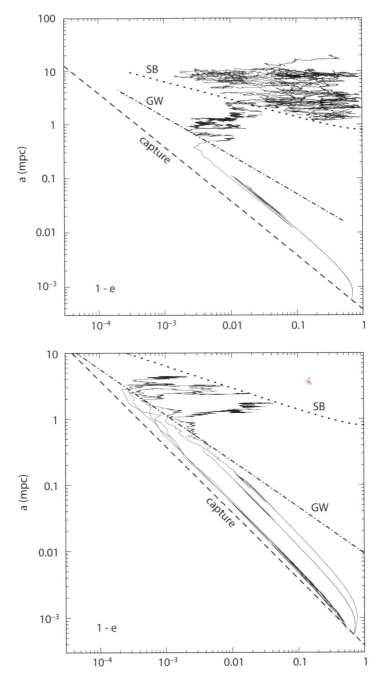

Figure 6.22 Simulation of EMRIs [7]. The top panel shows the trajectories, over a time interval of 2 Myr, of stellar-mass black holes orbiting a $10^6 \, M_\odot$ SBH as they undergo gravitational encounters with each other. Motion in the (a, e) plane is mostly horizontal due to the fact that resonant relaxation causes changes in

Figure 6.22 Continued. angular momentum (i.e., e) on a timescale that is much shorter than non-resonant relaxation causes changes in energy (i.e., a). The dashed line marked "capture" is the capture radius around the SBH; the dotted line marked "SB" is the Schwarzschild barrier, equation (6.195); and the dot-dashed line marked "GW" indicates the locus in the a–e plane where angular momentum loss due to gravitational radiation dominates changes due to gravitational encounters. Only one object manages to cross the latter line and become an EMRI; most of the other objects are reflected by the Schwarzschild barrier before reaching the gravitational-wave regime. There are no "plunges." The bottom panel, a montage from several independent N-body integrations, shows a number of EMRI events.

Writing $L = \left[GM_\bullet a(1 - e^2)\right]^{1/2}$ for the angular momentum of a test body, the timescale over which this fixed torque changes L is

$$\left|\frac{1}{L}\frac{dL}{dt}\right|^{-1} \approx \sqrt{N(a)}\frac{M_\bullet}{M(a)}\left[\frac{a^3(1 - e^2)}{GM_\bullet}\right]^{1/2}. \tag{6.194}$$

(Note that we are comparing changes in L to its own value, and not to L_c.) The condition that this time be shorter than the relativistic precession time is

$$(1 - e^2) > (1 - e^2)_{\mathrm{SB}} \approx \frac{r_g}{a}\frac{M_\bullet}{M_\star(a)}\sqrt{N(a)}. \tag{6.195}$$

This relation between a and e—the **Schwarzschild barrier**—is plotted as the dotted lines on figure 6.22.

Intuitively, one might expect that orbits would "hang up" after striking this barrier, since their precession is so rapid that the torques have become ineffective. What actually happens is a bit more interesting [359]. An orbit near the barrier has an eccentricity that varies over one precessional cycle. For instance, if the torquing potential is approximated by equation (5.215), and if the rate of precession in ω is determined by equation (6.192), then the equations of motion for ω, the argument of periastron, and $\ell \equiv (1 - e^2)^{1/2}$, the dimensionless angular momentum, are

$$\frac{d\omega}{d\tau} \approx \ell^{-2}, \qquad \frac{d\ell}{d\tau} \approx -Ae\cos\omega, \tag{6.196}$$

where

$$\tau \equiv 6\pi\frac{t}{P(a)}\frac{r_g}{a}. \tag{6.197}$$

The first equation of motion assumes that the precession in ω is driven entirely by relativity—a good assumption if ℓ is small. In the second equation, the dimensionless factor A describes the asymmetry in the background potential due to the finite value of N at a radius $r \approx a$:

$$A \approx \frac{1}{\sqrt{N}}\frac{M_\star(a)}{M_\bullet}\frac{a}{r_g}. \tag{6.198}$$

Equations (6.196) have approximate solution

$$1 - \frac{\ell(t)}{\langle\ell\rangle} \approx \langle\ell\rangle A\cos(\nu t), \tag{6.199}$$

where

$$\nu \equiv \frac{3}{c^2} \frac{(GM_\bullet)^{3/2}}{\langle \ell^2 \rangle a^{5/2}}. \tag{6.200}$$

As an orbit nears the Schwarzschild barrier, its angular momentum oscillates between the values ℓ_- and ℓ_+, where

$$\ell_+ - \ell_- \approx 2\langle \ell \rangle^2 A. \tag{6.201}$$

According to equation (6.196), it "lingers" at values of ω corresponding to large ℓ, that is, low eccentricity. Now, if we consider times long enough that the background potential is changing—that is, longer than the *mean* coherence time for all the orbits at $r \approx a$—then changes in the potential are most likely to catch a test body when it is at the low-e part of its precessional cycle. This is one reason why orbits tend to "bounce" after striking the Schwarzschild barrier: there is a bias toward changes that decrease e. As a result of the bounce, and because RR is so rapid to the right of the Schwarzschild barrier, the angular momentum distribution in this region remains close to that associated with an isotropic phase-space density; that is, $N(\ell)\,d\ell \approx \text{constant} \times \ell\,d\ell$. This contrasts with the logarithmic decrease in N with respect to ℓ associated with NRR.

If EMRIs are to occur, BHs must sometimes find their way out of this region, to the left of the Schwarzschild barrier. Here, we are free to invoke NRR, which is not quenched by relativistic precession. However, the fact that the angular momentum of a test body near the Schwarzschild barrier is changing on a timescale much less than T_{NRR}—due to the relativistic precession, and less rapidly, due to RR—must be taken into account. Nonresonant relaxation must change ℓ by an amount greater than $\delta\ell = \ell_+ - \ell_- \approx \ell_+ - \ell_{\text{SB}}$, in a time that is less than the coherence time of the background potential; the latter time limits how long a BH lingers near the barrier. These two conditions can be shown [359] to imply a critical value of a above which NRR is able to push objects past the barrier:

$$\left(\frac{a}{\text{mpc}}\right)_{\text{penetrate}} \approx 15 \left(\frac{M_\bullet}{10^6\,M_\odot}\right)^{5/2} \left(\frac{m}{10\,M_\odot}\right)^{-3/2} \left(\frac{N}{10^2}\right)^{-1/2}. \tag{6.202}$$

This prediction has been verified in direct N-body integrations (figure 6.23).

Assuming a BH does make it past the barrier, it will no longer be subject to RR, and its angular momentum will undergo a random walk due to nonresonant perturbations. If its eccentricity should become large enough, it will either undergo a plunge, or else the timescale for gravitational-wave emission to change a,

$$t_{\text{GW}} \equiv \left| \frac{1}{a} \frac{da}{dt} \right| \tag{6.203}$$

$$= \frac{5}{64} \frac{c^5 a^4}{G^3 M_\bullet^2 m} \left(1 - e^2\right)^{7/2} \left(1 + \frac{73}{24}e^2 + \frac{37}{96}e^4\right)^{-1}$$

$$\approx 1.2 \times 10^{14} \left(\frac{m}{50\,M_\odot}\right)^{-1} \left(\frac{M_\bullet}{10^6\,M_\odot}\right)^{-2} \left(\frac{a}{\text{mpc}}\right)^4 (1-e)^{7/2} \text{ yr},$$

becomes shorter than $\sim 2(1-e)t_{\text{NRR}}$, the time for the angular momentum to change by of order itself due to NRR. As discussed in chapter 4, in this high-eccentricity

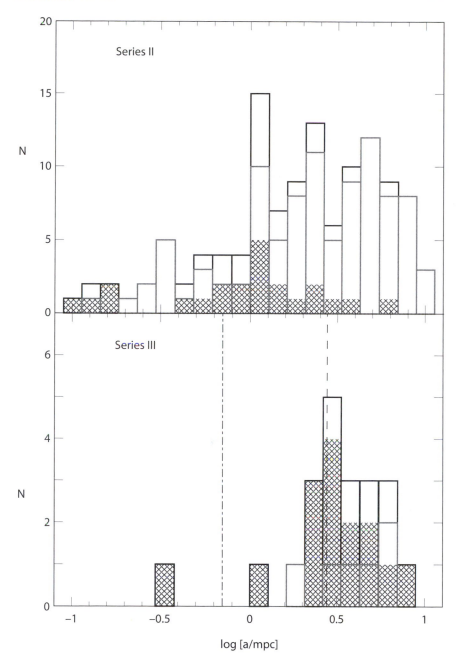

Figure 6.23 The distribution of semimajor axes for black holes that are captured, in a
set of N-body integrations [359]. "Series II" (top) consists of integrations in
which only the 2.5PN terms were added to the Newtonian equations of motion;
"Series III" integrations (bottom) included the 1PN and 2PN terms also, per-
mitting relativistic (Schwarzschild) precession. Unfilled histograms show the

Figure 6.23 Continued. plunges; (cross-hatched) histograms show the EMRIs; the totals are
indicated in black. In the upper panel the initial value of a is used; in the lower
panel, the value of a during the final crossing of the Schwarzschild barrier is
used. In both panels, the elapsed time is 2×10^6 yr. To the left of the dashed
vertical line in the lower panel, NRR is predicted to be ineffective at pushing
stars past the Schwarzschild barrier. To the left of the dot-dashed vertical line,
the Schwarzschild barrier does not exist.

regime, inspiral driven by gravitational-wave emission occurs along lines of fixed
slope in the a–$(1 - e)$ plane:

$$\frac{\Delta(1 - e)}{1 - e} \approx -\frac{\Delta a}{a}, \tag{6.204}$$

such that $r_p = (1 - e)a$ is approximately constant. The locus in the a–e plane
where these two timescales become equal is shown as the dot-dashed (blue) curves
in figure 6.22.

There is generally a rather small range of a values from which EMRIs can
form: small enough that GW emission can overcome stellar perturbations, but large
enough that NRR can push stars past the Schwarzschild barrier. Interestingly, for
sufficiently dense clusters, this range can go to zero, implying essentially no EM-
RIs; however, it appears that the required densities are one or two orders of magnitude larger than expected for real galactic nuclei (figure 6.24).

The existence of the Schwarzschild barrier implies yet another characteristic radius associated with galactic nuclei. Setting $e = 0$ in equation (6.195) yields

$$\frac{a}{r_g} \approx \frac{M_\bullet}{M_\star(a)} \sqrt{N(a)}. \tag{6.205}$$

The radius $a = a_{\mathrm{SB}}$ that satisfies equation (6.205) is the maximum radius for which
the Schwarzschild barrier exists. It is interesting to evaluate a_{SB} in the two models
presented in section 5.6.2 for the distribution of mass near the center of the Milky
Way. The first, "relaxed" (i.e., high-density) model, equation (5.245), had

$$N(a) \approx 10^5 \left(\frac{m_\star}{M_\odot}\right)^{-1} \left(\frac{a}{0.2\,\mathrm{pc}}\right) \tag{6.206}$$

yielding

$$a_{\mathrm{SB}} \approx 5 \left(\frac{m_\star}{10\,M_\odot}\right)^{-1/3} \mathrm{mpc}. \tag{6.207}$$

The second, "core" (i.e., low-density) model, equation (5.247), had

$$N(a) \approx 10^5 \left(\frac{m_\star}{M_\odot}\right)^{-1} \left(\frac{a}{0.2\,\mathrm{pc}}\right)^{5/2} \tag{6.208}$$

so that

$$a_{\mathrm{SB}} \approx 30 \left(\frac{m_\star}{M_\odot}\right)^{-2/9} \mathrm{mpc}. \tag{6.209}$$

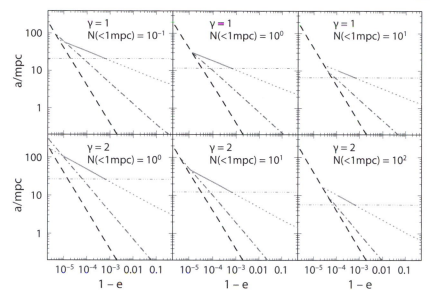

Figure 6.24 Critical curves for EMRI formation in models of nuclear star clusters [359]. $M_{\bullet} = 10^6 \, M_{\odot}$ and $m = 10 \, M_{\odot}$ were assumed, and the density of stellar BHs was assumed to obey $n(r) \propto r^{-\gamma}$, with various slopes and normalizations. Dashed (black) line: capture radius; dot-dashed line: radius at which GW emission dominates stellar perturbations. The Schwarzschild barrier is shown as the dotted line; Below the horizontal line, NRR is expected to be inefficient at pushing stars past the barrier.

Interestingly, these radii are comparable with the semimajor axes of the S-stars, which means that Nature is quite capable of depositing stars in this region of the (a, e) diagram. A deeper understanding of the evolution of orbits "below the barrier" is clearly to be desired.

Chapter Seven

Collisional Evolution of Nuclei

In chapter 3, a distinction was made between collisionless and collisional nuclei. The former were defined as nuclei in which the relaxation time T_r—the time required for random gravitational encounters between stars to deflect them from their otherwise fixed orbits—is long compared with the age of the universe. As we saw in chapter 3, the allowed equilibrium states of collisionless galaxies are very diverse, including models with essentially any radial distribution of matter around the supermassive black hole (SBH), as well as many different morphologies (spherical, axisymmetric, triaxial) and velocity distributions (isotropic, highly rotating, chaotic). The only a priori requirement that can be placed on the stellar distribution function f of a collisionless galaxy is that it satisfies Jeans's theorem in the combined gravitational potential, Φ, of the stars and the SBH; but this requirement turned out to place only very weak constraints on f and Φ. As discussed in detail in chapters 2 and 3, the degeneracy in f in collisionless nuclei is the biggest impediment to inferring SBH masses from kinematical data, since in only a handful of galaxies is the sphere of influence well-enough resolved that a Keplerian rise in velocities near the center is seen.

A collisionless stellar system is like a gas in which the molecules have not yet experienced a single collision: the particle positions and velocities still reflect in large measure the "initial conditions." By analogy, one expects that collisional nuclei— nuclei that are older than one relaxation time—might be much simpler than collisionless nuclei, with phase-space densities that are determined by just a handful of parameters: the "temperature" (i.e., the velocity dispersion), the mean density, the distribution of stellar masses.

That expectation will turn out to be correct. But before exploring the consequences of collisionality for the structure of galactic nuclei, it is interesting to ask whether collisional nuclei exist, and if so, which galaxies are likely to host them.

For many years, the nuclear star cluster (NSC) of the Milky Way was believed to be a prototypical example of a collisional nucleus. This belief was based on what appeared to be a steeply rising density of stars inside r_h—a density "cusp" [183]. The inferred density was so high—roughly $10^5 \, M_\odot \, \mathrm{pc}^{-3}$ at $r = r_h$ and even higher at smaller radii—that the implied relaxation time, which depends inversely on stellar density, fell well below 10^{10} yr at $r \ll r_h$. But careful observations around 2009 by three groups [66, 121, 29] revealed that the density cusp was present only in the handful of young, or early-type, stars that dominate the total light. The old, or late-type, stars have a very different radial distribution, consistent with a number density that is flat, or even declining, toward the center (figure 7.1a).

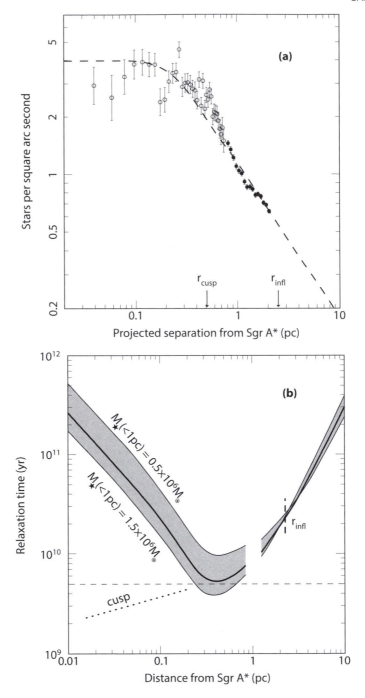

Figure 7.1 (a) Density of late-type (i.e., old) stars at the center of the Milky Way. Open
circles are binned counts of late-type stars brighter than apparent magnitude

Figure 7.1 Continued. $m_K = 15.5$ [66]. Filled circles show the density of all stars with $m_K \leq 15.5$ mag and $R \geq 20''$ after corrections for crowding and completeness [480]. The dashed line is a broken-power-law model with $\Sigma \propto R^{-0.8}$ at large radii and inner slope of zero; the core radius, defined as the radius at which the surface density falls to 1/2 of its central value, is 0.49 pc. Arrows show the SBH influence radius and the expected outer radius of the Bahcall–Wolf cusp. (b) Estimates of the relaxation time, assuming a single-mass population of 1 $M_\odot$ stars [357]. The dashed horizontal line indicates the mean age of stars assuming a constant rate of formation over the last 10 Gyr.

One needs to be careful in drawing conclusions from observations like these, since they are technically very difficult, and since they can only tell us about the tip of the iceberg—about the fraction of old stars (mostly 1–3 $M_\odot$ red giants) that are bright enough to be distinguished from the background light and to have their spectral types assigned (figure 7.2). But it is a reasonable assumption (and one commonly made in observations of external galaxies) that the brightest, late-type stars are distributed in the same way as the old stellar population generally, and therefore that their number counts trace the total stellar density. If one accepts this ansatz, it follows that the relaxation time is rather long, of order 10^{10} yr, everywhere in the central parsec of the Milky Way (figure 7.1b).[1]

There is an independent line of argument that leads to the same conclusion [357]. Suppose the Galactic center were older than one relaxation time; for instance, there might be a population of undetected stars or stellar remnants with a density much higher than the density inferred from the bright giants. If this were true, it can be shown that every old stellar population in this region (including the bright giants) would have a density that rises steeply toward the center at radii $r \lesssim 0.2r_h \approx 0.5$ pc. (The justification for this statement is presented later in this chapter.) This is very different from the data plotted in figure 7.1a, which reveal a core inside 0.5 pc, not a cusp. We are led again to the conclusion that the relaxation time at the center of our galaxy is longer than the time since the formation of the observed giants.

If we are to find galaxies with collisional nuclei, figure 3.1 suggests that we need to look in galaxies with spheroids less luminous than that of the Milky Way. As discussed in chapter 2, low-luminosity spheroids often contain compact nuclei, or NSCs, with bulk properties (size, luminosity) similar to those of the Milky Way's nucleus.[2] A few galaxies with NSCs are known to be active [164, 203], but the fraction of NSCs that contain SBHs can only be guessed at; beyond the Local Group, none is near enough that a kinematical detection of an SBH would be possible unless its mass far exceeded the mass predicted by the $M_\bullet$–σ relation. Typically, the only structural information that can be derived for NSCs beyond the Local Group is their half-light radii, r_{eff}, and total luminosities; the latter can be converted into masses by assuming a reasonable mass-to-light ratio. From these numbers,

---

[1] Assuming a typical stellar mass of ~ 1 $M_\odot$. It is argued in section 7.1.2 that a better choice for this mass would be ~ 0.5 $M_\odot$, implying a relaxation time that is twice as long.

[2] In fact, the existence of NSCs in external galaxies was recognized for some years before a definite connection was made with the nucleus of the Milky Way [479].

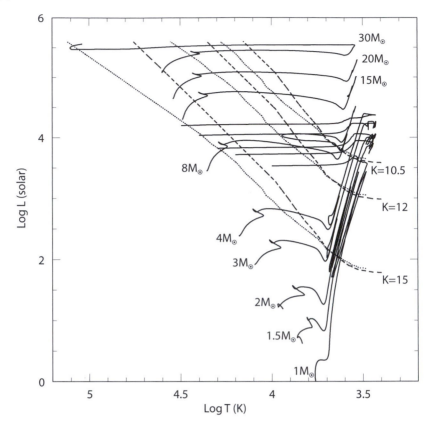

Figure 7.2 Theoretical evolutionary tracks of stars on a Hertzsprung–Russell diagram, com-
        pared with contours of constant K-band apparent magnitude as seen from the
        distance of the Galactic center, taking into account the known extinction due to
        dust [97]. The dotted lines show contours appropriate for main-sequence stars
        and the dashed lines are contours appropriate for giants. The number counts on
        which figure 7.1 was based were complete roughly to magnitude $m_K = 15.5$;
        based on this figure, most of those stars are expected to be red giants.

the half-mass relaxation time [502] can be computed:

$$T_{rh} = \frac{1.7 \times 10^5 \left[r_{\text{eff}} \ (\text{pc}) \right]^{3/2} N^{1/2}}{\left[m_\star / M_\odot \right]^{1/2}} \ \text{yr.} \qquad (7.1)$$

Equation (7.1) is the relaxation time, equation (3.2), if ρ is set to the average density inside r_{eff} (assumed to be the radius containing half the total mass), and σ is
identified with the velocity dispersion of all the stars in the NSC; N is the total
number of stars and the effects of an SBH on σ have been ignored. Figure 7.3
shows masses, radii and relaxation times for a sample of NSCs, all in galaxies that
belong to the Virgo Cluster [96]. Least-squares fits of log T_{rh} to M_B—the absolute

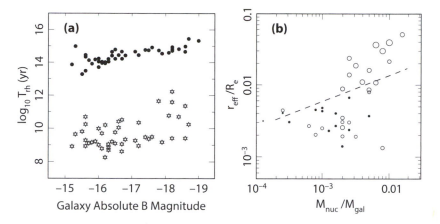

Figure 7.3 Properties of nuclear star clusters (NSCs) in galaxies belonging to the Virgo Galaxy Cluster [356]. The plotted points represent all Virgo galaxies, among the 100 brightest, that have compact nuclei. (a) Half-mass relaxation times of NSCs (stars) and their host galaxies (filled circles) plotted against absolute blue magnitude of the galaxy. Relaxation times were computed from equation (7.1) assuming $m_\star = M_\odot$. (b) The vertical axis is the ratio of half-light radii of NSCs ($r_{\rm eff}$) to the half-light radius of the host galaxies (R_e); the horizontal axis is the ratio of nuclear mass to galaxy mass. Symbol size is proportional to the logarithm of the nuclear relaxation time in (a). Open circles have $\log_{10}(T_{rh}/{\rm yr}) \geq 9.5$ and filled circles have $\log_{10}(T_{rh}/{\rm yr}) < 9.5$. The dashed line is an estimate of the critical value of $r_{\rm eff}/R_e$ above which NSCs expand (section 7.5.3).

blue magnitude of the galaxy (not the NSC)—are

$$\text{galaxies:}\quad \log_{10}(T_{rh}/{\rm yr}) = 14.2 - 0.336(M_B + 16), \tag{7.2a}$$

$$\text{NSCs:}\quad \log_{10}(T_{rh}/{\rm yr}) = 9.38 - 0.434(M_B + 16). \tag{7.2b}$$

Nuclear half-mass relaxation times appear to fall below $10\,{\rm Gyr}$ in galaxies with absolute magnitudes fainter than $M_B \approx -17$, or luminosities less than $\sim\!4 \times 10^8\,L_\odot$. The NSCs in these galaxies have masses below $\sim\!10^7\,M_\odot$ and half-light radii $\lesssim\!10\,{\rm pc}$. The NSC in the Milky Way has properties that are close to these limiting values, consistent with the fact that its relaxation time is $\gtrsim\!10\,{\rm Gyr}$.

We are led to the interesting hypothesis that the only spheroids likely to contain collisional nuclei are fainter than the spheroid of the Milky Way. While some of these systems are known to contain SBHs, it is unclear how many do, nor are the SBH masses well constrained in any of them. Indeed, if SBHs with masses below that of the Milky Way are rare—which is entirely consistent with what little is known of SBH demographics—it is possible that collisionality is a property that is associated primarily with galaxies lacking SBHs. The collisional evolution of such NSCs is discussed toward the end of this chapter.

Perhaps the safest conclusion to be drawn from this discussion is that many galaxies—including, most likely, the Milky Way—have nuclei that are in an intermediate state: not fully collisionless, but not yet old enough that random gravitational encounters have been able to establish a "collisional steady state" with easily predictable properties. Another way to state this is to say that the "initial conditions" of galactic nuclei should only have been partially erased by the effects of random gravitational encounters.[3] The title of this chapter—collisional *evolution* of nuclei—reflects this point of view.

As noted in the introduction to chapter 6, there are two timescales associated with the approach to a collisional steady state for stars near an SBH: the time for changes in L and the time for changes in E. If the initial conditions are far from a collisionally relaxed state, the two timescales may be comparable. On the other hand, there are some situations in which it makes sense to assume a steady state with regard to L even if the energy distribution is still evolving. This is the case, for instance, if the only significant changes in L are occurring near the loss cone. Another example is very near to the SBH, where resonant relaxation dominates the changes in L and acts much more quickly than nonresonant relaxation (chapter 5). Yet another reason for focusing on the energy distribution is the fact that the radial density profile is determined essentially by $N(E)$. Finally, it turns out that computing the evolution of $f(E, L, t)$ for stars near an SBH is so demanding that few attempts have been made to solve the full time-dependent problem; the great majority of published studies focuses on the approach to a steady state of the energy distribution alone, or on finding steady-state solutions $f(E, L)$. Throughout much of this chapter therefore, "collisional evolution" will be synonymous with "evolution in the energy distribution."

7.1 EVOLUTION OF THE STELLAR DISTRIBUTION AROUND A SUPERMASSIVE BLACK HOLE

7.1.1 Nuclei with a single mass group

In discussing the collisional evolution of nuclei, a natural starting point is a spherical nucleus containing stars of a single mass orbiting around an SBH. Until otherwise indicated, we will assume that gravitational encounters are uncorrelated—in other words, that resonant relaxation is unimportant—and that stars are far enough from the SBH that relativistic effects are also unimportant. Even in regimes where resonant relaxation is effective, it does not directly affect the energy distribution, which is the main focus of our attention here.

Suppose as well that f is nearly constant with respect to L at some initial time—in other words, that the velocity distribution is isotropic. As noted in the introduction, this is not a very reasonable assumption, since the timescales for establishing a steady state with respect to E and L are often similar. But writing $f = f(E, t)$ so greatly simplifies the evolution equations that it makes sense to investigate this problem before considering the more general case.

---

[3]Continuous star formation, which appears to be common in the nuclei of low-luminosity galaxies like the Milky Way, also implies that many stars will be young enough that their current distribution reflects the details of their formation.

Given these assumptions, the evolution equation for f is the isotropic, orbit-averaged Fokker–Planck equation (5.176):

$$\frac{\partial N}{\partial t} = -\frac{\partial \mathcal{F}_{\mathcal{E}}}{\partial \mathcal{E}}, \tag{7.3a}$$

$$\mathcal{F}_{\mathcal{E}} = -D_{\mathcal{E}\mathcal{E}} \frac{\partial f}{\partial \mathcal{E}} - m_{\star} D_{\mathcal{E}} f. \tag{7.3b}$$

Here $\mathcal{E} \equiv -E = -v^2/2 - \Phi(r) = -v^2/2 + \psi(r)$ is the binding energy, $N(\mathcal{E}) = 4\pi^2 p(\mathcal{E}) f(\mathcal{E})$ is the number of stars per unit energy, $p(\mathcal{E})$ is the phase-space volume element given by equation (5.178), $\mathcal{F}_{\mathcal{E}}$ is the flux of stars in energy space, and $D_{\mathcal{E}}$ and $D_{\mathcal{E}\mathcal{E}}$ are diffusion coefficients that depend on f as given in equations (5.176c,d).

A natural, inner boundary condition is $f(\mathcal{E}_t) = 0$, where $\mathcal{E}_t$ is the energy above which stars are captured or disrupted by the SBH. Of course, the real capture condition is only expressible in terms of a radius r_t; whether a star passes within r_t depends on both $\mathcal{E}$ and L. But since we are ignoring the L dependence of f, setting $f = 0$ at $\mathcal{E}_t = GM_{\bullet}/r_t$ is the best we can do. As for the outer boundary condition, we assume that at some sufficiently small $\mathcal{E} = \mathcal{E}_{\infty}$ (large r), the relaxation time is so long that f is unchanging: in other words, $f(\mathcal{E}_{\infty}, t) = f_{\infty}$.

Bahcall and Wolf [14] first presented numerical solutions to equations (7.3). They simplified the problem even further by ignoring the contribution to the gravitational potential from the stars, that is, by setting $\psi(r) = GM_{\bullet}/r$. This might seem questionable—after all, unless the stars act on each other gravitationally, there can be no relaxation! But the effects of gravitational encounters are described entirely by the expressions for the flux; if the contribution of the stars to the potential is negligible, their density normalization serves only to set the timescale for changes in f. In this approximation, the expressions for $p(\mathcal{E})$ and the related function $q(\mathcal{E})$ take on the simple forms given in equations (5.182). And for this assumed potential, the outer boundary condition becomes $f(\mathcal{E} = 0) = f_h$ where f_h can be identified with the phase-space density in the galaxy at energy $\mathcal{E} \approx GM_{\bullet}/r_h$.

Given these additional approximations, the evolution equation for $f(\mathcal{E}, t)$ becomes

$$\frac{\partial f}{\partial t} = 4\pi \Gamma \mathcal{E}^{5/2} \frac{\partial}{\partial \mathcal{E}} \left[-f \int_{\mathcal{E}}^{\mathcal{E}_t} d\mathcal{E}' \mathcal{E}'^{-5/2} f \right.$$
$$\left. + \frac{2}{3} \frac{\partial f}{\partial \mathcal{E}} \left(\mathcal{E}^{-3/2} \int_0^{\mathcal{E}} d\mathcal{E}' f + \int_{\mathcal{E}}^{\mathcal{E}_t} d\mathcal{E}' \mathcal{E}'^{-3/2} f \right) \right], \tag{7.4}$$

where $\Gamma \equiv 4\pi G^2 m_{\star}^2 \ln \Lambda$. It is clear from this equation that changing the normalization of f is equivalent to rescaling the time. We can make the equation dimensionless by defining $\mathcal{E} = [\mathcal{E}] \mathcal{E}^*$, $t = [t] t^*$, $f = [f] f^*$, where

$$[\mathcal{E}] = \sigma_0^2, \quad [t] = \frac{1}{4\pi\Gamma} \frac{(2\pi\sigma_0^2)^{3/2}}{n_0}, \quad [f] = \frac{n_0}{(2\pi\sigma_0^3)^{3/2}}. \tag{7.5}$$

Here, n_0 and σ_0 are understood to be the stellar number density and velocity dispersion just beyond the SBH influence radius. The unit of time can be rewritten as

$$[t] = \frac{1}{8}\sqrt{\frac{2}{\pi}} \frac{\sigma_0^3}{G^2 m_\star \rho_0 \ln \Lambda}, \tag{7.6}$$

where $\rho_0 \equiv m_\star n_0$; aside from a constant factor, this is the relaxation time defined in equation (5.61). Substituting these expressions into equation (7.4), and removing the '$\star$'s, gives us the evolution equation in terms of dimensionless variables:

$$\frac{\partial f}{\partial t} = \mathcal{E}^{5/2} \frac{\partial}{\partial \mathcal{E}} \left[-f \int_{\mathcal{E}}^{\mathcal{E}_t} d\mathcal{E}' \mathcal{E}'^{-5/2} f \right. \tag{7.7}$$

$$\left. + \frac{2}{3} \frac{\partial f}{\partial \mathcal{E}} \left(\mathcal{E}^{-3/2} \int_0^{\mathcal{E}} d\mathcal{E}' f + \int_{\mathcal{E}}^{\mathcal{E}_t} d\mathcal{E}' \mathcal{E}'^{-3/2} f \right) \right].$$

Bahcall and Wolf solved this differential equation and found that a steady state is reached after roughly one relaxation time at $r_{\rm h}$. If $\mathcal{E}_t \gg GM_\bullet / r_{\rm h}$, that is, if $r_t \ll r_{\rm h}$, the steady-state solution was found to be close to a power law in both energy and configuration space:

$$f(\mathcal{E}) \propto \mathcal{E}^{1/4}, \quad \rho(r) \propto r^{-7/4}, \tag{7.8}$$

for $\sigma^2 \ll \mathcal{E} \ll \mathcal{E}_t, r_t \ll r \ll r_{\rm h}$. Equations (7.8) were derived in chapter 5; they represent a "zero-flux" solution, that is, a solution for which $\mathcal{F}_\mathcal{E} = 0$. The $\rho \propto r^{-7/4}$ density profile around an SBH is commonly called a **Bahcall–Wolf cusp**.

Figure 7.4 illustrates the evolution of $f(\mathcal{E}, t)$ in a galaxy model that extends far beyond the SBH's influence radius. The initial density follows $\rho \sim r^{-1/2}$ near the SBH. As in Bahcall and Wolf's numerical solution, the evolution in figure 7.4 assumes a fixed potential. However, this potential included the contribution from the stars:

$$\psi(r) = \frac{GM_\bullet}{r} + \psi_\star(r), \tag{7.9}$$

with $\psi_\star$ evaluated from $\rho(r)$ at $t = 0$. Assuming a fixed potential is often an excellent approximation in the context of galactic nuclei, at least in cases where the only significant changes in ρ take place well inside $r_{\rm h}$. (This is not the case in the absence of an SBH; examples are presented later in this chapter.) Including the stellar potential is important because it allows us to see how the Bahcall–Wolf cusp merges with the stellar distribution beyond $r_{\rm h}$.

Figure 7.4 shows that the Bahcall–Wolf cusp grows "from the outside in." A nearly steady state has been reached by a time of $T_r(r_{\rm m})$, the relaxation time at $r = r_{\rm m}$; however, even after half this time, the cusp is well established at all but the smallest radii. The *largest* radius to which the cusp extends is $\sim 0.2 r_{\rm m}$.

In numerical solutions like these, the steady-state flux is found to be small but nonzero, of order

$$\frac{n(r_t) r_t^3}{T_r(r_t)} \propto r_t. \tag{7.10}$$

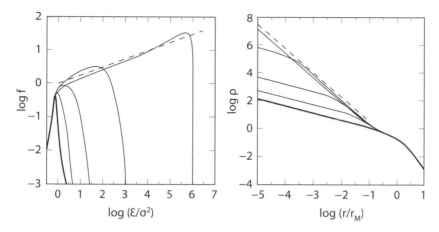

Figure 7.4 Evolution of the stellar distribution around an SBH due to energy exchange between stars. These curves were computed from the isotropic, orbit-averaged Fokker–Planck equation (7.3) with boundary condition $f = 0$ at $\log_{10}(\mathcal{E}/\sigma^2) = 6$. Left: Phase-space density f. Right: Configuration-space density ρ. The initial distribution (shown in bold) had $\rho \propto r^{-0.5}$ near the SBH; thin curves show f and ρ at times of $(0.2, 0.4, 0.6, 1.0)$ in units of the relaxation time at the SBH's initial influence radius r_m. Dashed lines have the slope of the "zero-flux" solution $f \propto \mathcal{E}^{1/4}$, $\rho \propto r^{-7/4}$. The steady-state density is well approximated by the zero-flux solution at $r \lesssim 0.2 r_m$.

In other words, the flux is limited by the rate at which stars can diffuse into the disruption sphere at r_t. As r_t is reduced, the flux approaches zero and the numerical solution approaches the strict power-law forms of equation (7.8).

Scaled to a galaxy like the Milky Way, the flux implied by the Bahcall–Wolf solution would be of order $\sim 10^{-12}$ stars per year—far too small to be physically interesting. But soon after Bahcall and Wolf published their 1976 paper, Frank and Rees [168] pointed out that the actual loss rate to an SBH would be dominated by changes in angular momentum, not energy, implying a much larger flux. In a second paper, Bahcall and Wolf [15] included these loss-cone effects heuristically, by adding a term to their equation:

$$\frac{\partial N}{\partial t} = -\frac{\partial \mathcal{F}_{\mathcal{E}}}{\partial \mathcal{E}} - F(\mathcal{E}, t). \qquad (7.11)$$

Even though F represents the loss of stars due to scattering in angular momentum, angular momentum does not appear in this equation; rather, it is assumed that the distribution over L is known and is independent of time. As discussed in chapter 6, that is not necessarily a bad assumption: the time required for f to reach a steady state, near the loss cone, can be much shorter than the time for changes in energy. In chapter 6, we derived several approximate expressions for this term. For instance, if f respects the Cohn–Kulsrud boundary layer solution, then equations (6.28), (6.64) state

$$F(\mathcal{E}, t) \approx \frac{4\pi^2 P(\mathcal{E}) L_c^2(\mathcal{E}) D(\mathcal{E}, t)}{\ln(1/R_0(\mathcal{E}))} \overline{f}(\mathcal{E}, t),$$

$$D(\mathcal{E}, t) = \lim_{\mathcal{R} \to 0} \frac{\langle (\Delta \mathcal{R})^2 \rangle_t}{2\mathcal{R}}. \tag{7.12}$$

Recall that $\mathcal{R} \equiv L^2/L_c^2(\mathcal{E})$ is a scaled angular-momentum variable, $L_c(\mathcal{E})$ is the angular momentum of a circular orbit of energy $\mathcal{E}$, P is the period of a radial orbit, and $\overline{f}$ is an angular-momentum-averaged f. $\mathcal{R}_0$ is the value of $\mathcal{R}$ at which $f(\mathcal{E}, \mathcal{R})$ drops to zero due to the competing effects of capture and diffusion (equation 6.65).

It is often the case that the scattering rate, for stars of energy $\mathcal{E}$, is dominated by stars with energies less than $\mathcal{E}$, allowing two of the three integrals that appear in the expression (6.31) for $D(\mathcal{E})$ to be ignored. In addition, the transition between the "pinhole" and "full-loss-cone" regimes (figure 6.3) is fairly sharp with respect to energy. Invoking these approximations, the loss term can be written [323, 395]

$$F(\mathcal{E}, t) \approx 4\pi^2 f(\mathcal{E}, t) \left[\frac{1 - \mathcal{R}_{\text{lc}}}{L_{\text{lc}}^2} + \frac{3\mathcal{E}}{5\pi^2} \frac{\ln\left(\mathcal{R}_{\text{lc}}^{-1} + \mathcal{R}_{\text{lc}} - 1\right)}{\Gamma' p(\mathcal{E}) \mathcal{M}_0(\mathcal{E})} \right]^{-1}, \tag{7.13a}$$

$$\mathcal{M}_0(\mathcal{E}) = 4\pi \int_0^{\mathcal{E}} \mu(\mathcal{E}') d\mathcal{E}'. \tag{7.13b}$$

Here $\Gamma' = 4\pi^2 G \ln \Lambda$ and L_{lc} is defined in equation (6.6). The expression for $\mathcal{M}_0$ allows for a distribution of masses in the field-star (scattering) population, with μ the second moment over mass of f (equation 5.181); for a single mass group, $\mu = m_\star^2 f$. (A more general treatment of the multimass case is given in the next section.) In equation (7.13a), the first term in the denominator represents the contribution from the pinhole regime, the second from the empty-loss-cone regime.

In their 1977 paper [15], Bahcall and Wolf adopted an expression similar to (7.13) for F. They found that inclusion of the loss term had only a small effect on the steady-state forms of $f(E)$ and $\rho(r)$, even though it substantially increased the implied loss rate. Figure 7.5 reproduces one of their steady-state solutions including the loss term.

Only a handful of attempts have been made to solve the full, anisotropic Fokker–Planck equation for stars orbiting near a massive black hole. Figure 7.5 shows the steady-state solutions from two such studies [491, 91]; plotted are angular-momentum averages of $f(\mathcal{E}, \mathcal{R})$. As in Bahcall and Wolf's 1977 paper, the parameters in these studies were chosen to describe stars orbiting a $10^3 \, M_\odot$ black hole in a globular cluster. For energies near $\mathcal{E}_t$, the isotropized f can be seen to differ rather strongly from the form calculated using the isotropic equation (7.11). Even more striking is the discrepancy between the inferred loss rates: in spite of having virtually identical steady-state f's, the two solutions based on the anisotropic equation had loss rates differing by a factor of two, presumably a consequence of different treatments of the boundary layer [91]. In these solutions, the logarithmic derivative of the steady-state density was found to be $d \ln \rho / (d \ln r) \approx -1.65$ for

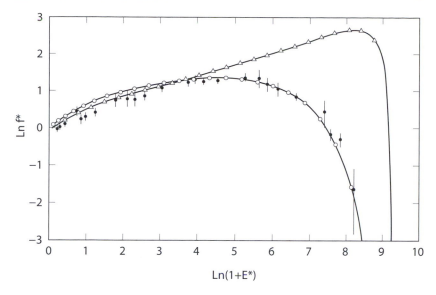

Figure 7.5 Steady-state solutions to the orbit-averaged Fokker–Planck equation describing
stars of a single mass around a massive black hole [91]. The upper curve (triangles) is from a calculation based on equation (7.11), with a loss-cone term
similar to equation (7.13) [15]. The lower curve (filled and open circles) shows
two, nearly indistinguishable solutions based on the anisotropic Fokker–Planck
equation (5.168); the plotted functions are angular-momentum-averaged f's
[491, 91]. The latter two solutions were computed using different expressions
for the angular-momentum dependence of f near the loss cone. All of these
models adopted parameters appropriate for a $10^3\, M_\odot$ black hole at the center of
a globular cluster.

r in the range $10^{-3} < r/r_h < 10^{-1}$, compared with the Bahcall–Wolf value of
-1.75. The velocity anisotropy was found to be close to zero for $r \gtrsim 10^{-3} r_h$.

Time-dependent solutions to equations like (7.11) need to be interpreted with
caution. If the initial conditions are far from a steady state with respect to L, the
early evolution will be strongly influenced by changes in angular momentum, and
this is particularly true with regard to the feeding rate, as discussed in chapter 6.
Perhaps the most appropriate application of equations like (7.11) is to the time-
independent problem, $\partial N/\partial t = 0$: inclusion of the loss term yields steady-state
f's that reflect, at least approximately, the modifications in f at large $\mathcal{E}$ resulting
from the loss cone. In principle, estimates of the loss rate based on such solutions
should be more accurate than if a loss-cone boundary condition is applied to an
ad hoc f, as in several examples from chapter 6. However, the comparisons of
figure 7.5 suggest that such estimates may still only be order-of-magnitude correct.

The same approach can be used to compute, in an approximate way, the steady-
state $f(\mathcal{E})$ very near to the SBH, where the capture rate is determined by resonant
relaxation (RR) rather than by normal, or nonresonant relaxation (NRR) [245].
Since RR leaves orbital energies unchanged, its effects can be approximated by

including an additional loss term on the right-hand side of equation (7.11):

$$\frac{\partial N}{\partial t} = -\frac{\partial \mathcal{F}_{\mathcal{E}}}{\partial \mathcal{E}} - F_{\mathrm{NRR}}(\mathcal{E}, t) - F_{\mathrm{RR}}(\mathcal{E}, t). \tag{7.14}$$

Here, F_{NRR} is the same quantity that was called simply F in equation (7.11), while F_{RR} approximates the loss rate due to RR. For the latter term we can write

$$F_{\mathrm{RR}}(\mathcal{E}, t) = \chi \frac{N(\mathcal{E}, t)}{\overline{T}_{\mathrm{RR}}}, \tag{7.15}$$

where $\overline{T}_{\mathrm{RR}}$ is an angular-momentum-average of the timescale T_{RR} associated with incoherent RR, and the dimensionless factor χ parametrizes the uncertainties in the efficiency of RR and the neglect, in equation (7.15), of the partial depletion of phase space near the loss-cone boundary. Combining equation (5.239),

$$T_{\mathrm{RR}} = \left(\frac{L_c}{|\Delta L|_{\mathrm{coh}}}\right)^2 t_{\mathrm{coh}}, \tag{7.16}$$

with equation (5.226a),

$$\frac{\Delta L}{L_c} = \beta_s N^{1/2} \frac{m}{M_{\bullet}} \frac{\Delta t}{P}, \tag{7.17}$$

allows us to express T_{RR} in terms of an arbitrary coherence time as

$$T_{\mathrm{RR}} = \frac{1}{\beta_s^2} \frac{M_{\bullet}^2}{m^2} \frac{1}{N} \frac{P^2}{t_{\mathrm{coh}}}. \tag{7.18}$$

Defining an angular momentum average of this expression is problematic, for the reasons discussed in section 6.4. The authors chose to replace t_{coh} in equation (7.18) by

$$\frac{1}{t_{\mathrm{coh}}} \rightarrow \left| \frac{1}{t_{\mathrm{coh,M}}} - \frac{1}{t_{\mathrm{coh,S}}} \right| \tag{7.19}$$

with the understanding that $t_{\mathrm{coh,S}}$ is a function of L; the minus sign accounts for the fact that mass precession is retrograde and relativistic precession is prograde.

Figure 7.6 shows steady-state $f(\mathcal{E})$'s computed in this way, for various choices of the factor χ [245]. Plotted there is $g(X)$, where g is a normalized phase-space density, and the dimensionless energy variable, X, is defined as $X = \mathcal{E}/\sigma_h^2$ with σ_h the velocity dispersion at $r = r_{\mathrm{h}}$. Setting $\chi = 0$ reproduces the Bahcall–Wolf [14] solution; at high energies, there is a depletion that increases with χ. However, for the likely value $\chi = 1$, relativistic precession limits the effectiveness of RR at high energies, and the differences with the $\chi = 0$ case are modest.

The calculations leading to figure 7.6 did not take into account the presence of the "Schwarzschild barrier" discussed in section 6.4, which strongly mediates the evolution of low-angular-momentum orbits near an SBH [359]. The effect of the Schwarzschild barrier on the steady-state distribution of stars near an SBH is a topic of current research interest, but unfortunately is not yet fully understood.

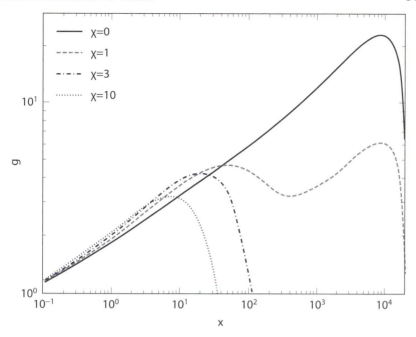

Figure 7.6 Steady-state solutions to equation (7.14) [245], which describes in an approximate way the depleting effects of resonant relaxation on the value of $f(\mathcal{E})$ at high energies. Symbols are defined in the text.

7.1.2 Nuclei with multiple mass groups

7.1.2.1 Stellar mass functions

Galaxies contain stars with a range of masses. The **stellar mass function** is defined such that $n(m, t)dm$ is the number of stars with masses in the range m to $m + dm$ at time t. Stellar mass functions can be constructed fairly easily for stars in the solar neighborhood, given measurements of apparent magnitude, distance, and multiplicity, for a large sample of individual stars. In distant galaxies, information about the stellar mass function is contained in just a few integrated quantities: colors, surface brightness, etc. Rather than attempt to infer $n(m)$ directly from such limited data, it is customary to express $n(m)$ in terms of two other quantities—the **initial mass function** (IMF), and the star formation history—and to vary the parameters defining those functions until a good fit to the data is obtained. Theoretical stellar evolution tracks are used to relate the initial mass to the observable properties of a star at time t.

The IMF, $\xi(m)$, is defined as $n(m)$ at the time when the stars were formed. The IMF is often approximated as a power law:

$$\xi(m)dm \propto (m/m_0)^{-\alpha}dm \qquad (7.20)$$

over some range in m. In 1955, E. Salpeter [471] proposed $\alpha = 2.35$, and a power-law IMF with this slope is called a **Salpeter IMF**.[4] More recent studies have suggested modifications to Salpeter's law at low and high stellar masses. The **Kroupa IMF** [302] is defined as a broken power law, $n(m) \propto m^{-\alpha_i}$, with

$$
\begin{aligned}
\alpha_1 &= 0.3, \quad 0.01 \leq m/M_\odot \leq 0.08, \\
\alpha_2 &= 1.3, \quad 0.08 < m/M_\odot \leq 0.5, \\
\alpha_3 &= 2.3, \quad 0.5 < m/M_\odot \leq 120.
\end{aligned}
\tag{7.21}
$$

The main difference between the Salpeter and Kroupa IMFs is the smaller number of low-mass stars predicted by the latter. IMFs like Kroupa's are often assumed, as a working hypothesis, to be "universal," that is, to characterize star formation at all locations and all times.

Stellar evolution causes masses to change with time. Massive stars evolve quickly to their final states: neutron stars or black holes, the remnants of the collapsing cores that accompany supernova explosions. Stars with initial masses less than about $10\,M_\odot$ are believed to end their lives as white dwarfs. The **initial-to-final mass relation**, $m_{\rm rem}(m_{\rm init})$, relates the mass of a star on the zero-age main sequence to the mass of its remnant. While there are many uncertainties, the following simple form is consistent with what is currently known:

$$
m_{\rm rem} = \begin{cases}
m_{\rm init}, & 0 < m/M_\odot < m_{\rm to}, \\
0.109\,m_{\rm init} + 0.394\,M_\odot, & m_{\rm to} < m_{\rm init}/M_\odot < 8, \\
1.4\,M_\odot, & 8 < m_{\rm init}/M_\odot < 30, \\
0.1 m_{\rm init}, & 30 < m_{\rm init}/M_\odot < 120.
\end{cases}
\tag{7.22}
$$

Stars with masses below the main-sequence turnoff mass $m_{\rm to}(t)$ have not yet left the main sequence. The main-sequence lifetime, $T_{\rm MS}$, is approximately [227]

$$
T_{\rm MS} \approx 10^{10} \left(\frac{m_{\rm init}}{1\,M_\odot} \right)^{-2.5} \text{yr},
\tag{7.23}
$$

so for an old ($\sim 10\,\text{Gyr}$) stellar population, $m_{\rm to} \approx 1\,M_\odot$. Mass loss can generally be ignored for stars with $m < m_{\rm to}$. The masses of observed white dwarfs and their progenitors can be estimated via a number of techniques; the relation given above is based on these observational studies [268]. Progenitor masses of neutron stars and black holes are based primarily on theoretical modeling [569]; masses of a handful of neutron stars that are observed to be in binary systems can be dynamically determined, and they appear to span a quite narrow range, $1.35 \pm 0.04\,M_\odot$ [515]. Dynamical measurements of black-hole masses in about 20 X-ray binaries yield estimates in the range $3 \lesssim m/M_\odot \lesssim 14$ [74, 409]; these values are consistent with the (still highly uncertain) predictions of supernova models [569], which suggest that about one tenth of a star's initial mass ends up in the black hole.

By combining the IMF, equation (7.20) or (7.21), with the initial-to-final mass ratio, equation (7.22), it is straightforward to compute the fraction of the mass in an evolved stellar population that would be associated with stars and remnants of

---

[4]Salpeter defined the mass function as $dN/(d \log m) \propto m^{-\Gamma}$, so that $\Gamma = \alpha - 1 = 1.35$.

Table 7.1 An evolved stellar population model.

Initial mass	Number fraction	Mass fraction	Remnant	Mass fraction
$0.01 < m/M_\odot < 1$	0.94	0.47	–	0.80
$1 < m/M_\odot < 8$	5.7×10^{-2}	0.32	White dwarf	0.16
$8 < m/M_\odot < 30$	3.4×10^{-3}	0.12	Neutron star	2.1×10^{-2}
$30 < m/M_\odot < 120$	6.1×10^{-4}	8.5×10^{-2}	Black hole	1.5×10^{-2}
				$\tilde{m} = 0.55\,M_\odot$
				$\Delta \approx 0.05$

various types. Table 7.1 shows the results, based on a Kroupa IMF, and assuming a main-sequence turnoff mass of $1\,M_\odot$. The fraction of the total mass in remnants is a few percent. From the point of view of collisional dynamics, what is more important is the mass fraction in stellar black holes alone, since these have masses $\sim 10\,M_\odot$, while neutron stars, white dwarfs, and lower-main-sequence stars all have masses below $\sim 1.5\,M_\odot$. The predicted black-hole mass fraction is roughly 1%. Table 7.1 also gives the mass $\tilde{m}$ defined in equation (5.62b) that determines the relaxation time via equation (5.61); for the assumed Kroupa IMF, $\tilde{m} \approx 0.55\,M_\odot$.

As long as the last star-forming event took place more than $\sim 10^8$ yr ago, the main-sequence lifetime of stars with $m_{\mathrm{init}} \approx 8\,M_\odot$, the numbers of dark remnants (neutron stars, black holes) will not be changing. But if star formation is an ongoing process, some high-mass stars may still be sitting on the main sequence at any given time. This appears to be the case for the nuclear star cluster of the Milky Way, as discussed in chapter 4. The stellar luminosity function of stars in the inner ~ 50 pc of our galaxy is poorly fit by assuming that all stars formed at the same time; a constant star-formation rate over the last 10 Gyr gives a better fit, although models which combine constant star formation with an old, "starburst" population are also satisfactory [163]. NSCs in external galaxies cannot be resolved into individual stars, but spectral data often reveals the presence of young stars, particularly in the nuclei of late-type (disk) galaxies [553]. Even in these nuclei, most of the mass appears typically to be attributable to stars with ages on the order of 10 Gyr.

Modeling studies like these often assume that the IMF is universal, with a known functional form; all that is then left to vary when fitting the data is the star formation history. But evidence for departures from a universal IMF has surfaced from time to time, typically in studies of star formation in "extreme" environments [30]. One such environment is the Galactic center, where the presence of massive, main-sequence O and B stars in the inner parsec implies that one or more star formation events have taken place in the last 10^8 yr alone [415]. It has been suggested that the luminosity function of these young stars implies an IMF that is flatter at the high-mass end (i.e., that falls off less steeply with m) than a Salpeter or Kroupa IMF—a so-called **top-heavy IMF**—with a correspondingly larger fraction of high-mass stars [29]. Theoretical studies of star formation in the tidal field of an SBH also suggest that high masses would be preferred [318]. If one assumes that star formation in the inner parsec of the Milky Way has *always* obeyed a top-heavy IMF, the fraction of accumulated mass in dark stellar remnants could be much larger than the roughly 1% listed in table 7.1. This is an exciting possibility from many perspectives; for instance, it would imply a high rate of gravitational-wave inspiral events. But such models are constrained by the known mass-to-light ratio of the inner parsec, which probably limits the remnant fraction to a few percent [325].

A reasonable first approximation would be to represent the mass function of an evolved stellar population in terms of just two groups: a dominant population, consisting of stars of roughly one solar mass or less; and a smaller population of black holes of roughly ten solar masses each. For standard (Salpeter, Kroupa) IMFs, the fraction of mass in the black holes would be about 1%, while a top-heavy IMF would imply a larger fraction.

7.1.2.2 Equipartition and mass segregation

Locally, gravitational encounters tend to establish equipartition of kinetic energy between stars of different masses:

$$m_1 \overline{v_1^2} = m_2 \overline{v_2^2} \tag{7.24}$$

(section 5.3). The characteristic time to reach energy equipartition between two mass groups can be estimated as follows [502]. Suppose that both groups adhere to a Maxwellian velocity distribution. Combining equations (5.124), (5.25) and (5.57), we find for the local diffusion coefficient describing changes in the specific energy of a star of mass m_1

$$\langle \Delta E \rangle = v \langle \Delta v_\| \rangle + \frac{1}{2} \langle (\Delta v_\|)^2 \rangle + \frac{1}{2} \langle (\Delta v_\perp)^2 \rangle$$

$$= \Gamma n_2 v^{-1} \left[-\frac{m_1}{m_2} \mathrm{erf}(x) + \left(1 + \frac{m_1}{m_2} \right) x \, \mathrm{erf}'(x) \right], \tag{7.25}$$

where $x \equiv v/(2^{1/2}\sigma_2)$ and $\Gamma \equiv 4\pi G^2 m_2^2 \ln \Lambda$. The mean rate of kinetic energy change of all the stars of mass m_1 is

$$\frac{d}{dt} \frac{m_1 \overline{v_1^2}}{2} = \frac{2^{1/2}}{\pi^{1/2}\sigma_1^3} \int_0^\infty dv \, v^2 \exp \left[-v^2/(2\sigma_1^2) \right] m_1 \langle \Delta E \rangle. \tag{7.26}$$

Combining equations (7.25) and (7.26) and integrating by parts,

$$\frac{d}{dt} \frac{m_1 \overline{v^2}}{2} = \frac{4 (6\pi)^{1/2} G^2 m_1 \rho_2 \ln \Lambda}{\left(v_1^2 + v_2^2 \right)^{3/2}} \left(m_2 \overline{v_2^2} - m_1 \overline{v_1^2} \right). \tag{7.27}$$

It is clear from this expression that energy exchange stops when $m_1 \overline{v_1^2} = m_2 \overline{v_2^2}$.

Let $\epsilon_i = \frac{1}{2} m_i \overline{v_i^2}$, and suppose that $m_2 \ll m_1$. The time for stars of each mass group to reach energy equipartition with stars in the other group is

$$T_1 \equiv \left| \frac{1}{\epsilon_1} \frac{d\epsilon_1}{dt} \right|^{-1} \approx \frac{0.0814 v_{\mathrm{rms}}^3}{G^2 m_1 \rho_2 \ln \Lambda}, \qquad T_2 \equiv \left| \frac{1}{\epsilon_2} \frac{d\epsilon_2}{dt} \right|^{-1} \approx \frac{\rho_2}{\rho_1} T_1. \tag{7.28}$$

In writing these expressions we have assumed that $\overline{v_1^2} \approx \overline{v_2^2} \equiv v_{rms}^2$; this condition might hold, for instance, shortly after a galaxy forms. Consider first the case that the heavier stars dominate the total density and have a relaxation time $T_{r,1}$. Then $T_2 \approx T_{r,1}$ and the light stars reach equipartition on the same timescale that the heavy stars establish a collisional steady state. On the other hand, if the light stars dominate, with relaxation time $T_{r,2}$, then $T_1 \approx (m_2/m_1)T_{r,2}$, and the heavy stars lose energy to the light stars very rapidly compared with $T_{r,2}$. In either case, the time for the subdominant population to reach equipartition with the dominant population is, at most, of order the relaxation time defined for the dominant component.

So far in our discussion of equipartition we have ignored the fact that in the absence of perturbations, stars move along orbits in the smooth galactic potential. A change of energy due to encounters implies a change in the unperturbed orbit: as heavy stars lose energy in their interactions with lighter stars, they congregate closer to the center, and the lighter stars are pushed out. This is called **mass segregation**. In the case of our simple, two-component mass function, mass segregation implies that the heavy stars (stellar black holes) should cluster more strongly around the SBH than the light (solar-mass) stars. As a consequence, they will end up moving *faster*, on average, than the light stars—an example of the "negative specific heat" of gravitating systems [502].

The steady-state distributions near an SBH can be found using the orbit-averaged Fokker–Planck equation. As in the single-mass case, we begin by assuming isotropy, $f = f(\mathcal{E}, m, t)$. Equation (5.176) describes the evolution of f for an arbitrary distribution of masses. It is straightforward to rewrite that equation for the case of a discrete mass function. Let $f_i(\mathcal{E}, t)$ be the phase-space number density of the ith species, of mass m_i. The evolution equation for f_i is

$$4\pi^2 p(\mathcal{E})\frac{\partial f_i}{\partial t} = -\frac{\partial \mathcal{F}_i}{\partial \mathcal{E}}, \tag{7.29a}$$

$$\mathcal{F}_i = \sum_j \left(-D_{\mathcal{E}\mathcal{E}ij}\frac{\partial f_i}{\partial \mathcal{E}} - D_{\mathcal{E}ij} f_i\right), \tag{7.29b}$$

$$D_{\mathcal{E}\mathcal{E}ij} = 16\pi^3 \Gamma' m_j^2 \left[q(\mathcal{E})\int_0^{\mathcal{E}} f_j(\mathcal{E}', t)d\mathcal{E}' + \int_{\mathcal{E}}^{\infty} f_j(\mathcal{E}', t)q(\mathcal{E}')d\mathcal{E}'\right], \tag{7.29c}$$

$$D_{\mathcal{E}ij} = -16\pi^3 \Gamma' m_i m_j \int_{\mathcal{E}}^{\infty} f_j(\mathcal{E}', t)p(\mathcal{E}')d\mathcal{E}'. \tag{7.29d}$$

Returning again to the case of a nucleus containing just two mass groups: to avoid confusion, we will use the subscript H (heavy) for the first species and L (light) for the second species; as before, we assume $m_H \gg m_L$. Suppose first that the heavier stars dominate the local mass density, $\rho_H \gg \rho_L$. Then $m_H^2 f_H \gg m_L^2 f_L$, and the evolution equation for the light component is

$$\frac{\partial f_L}{\partial t} \approx \frac{1}{4\pi^2 p}\frac{\partial}{\partial \mathcal{E}}\left(D_{\mathcal{E}\mathcal{E}LH}\frac{\partial f_L}{\partial \mathcal{E}}\right). \tag{7.30}$$

A steady state is reached when

$$\frac{\partial f_L}{\partial \mathcal{E}} = 0, \tag{7.31}$$

that is, when $f_L(\mathcal{E}) = \text{constant}$ [368]. Remarkably, this result is independent of the distribution of the heavy stars: when phase space is uniformly populated, scattering leaves the light-star distribution unchanged.

The relation between $f(\mathcal{E})$ and $n(r)$ is given by equation (3.46):

$$n(r) = 4\pi \int_0^{\psi(r)} f(\mathcal{E}) \sqrt{2\left[\psi(r) - \mathcal{E}\right]} \, d\mathcal{E}. \tag{7.32}$$

If f is a power law in $\mathcal{E}$, then

$$n(r) \propto \int_0^{\psi} \mathcal{E}^\alpha (\psi - \mathcal{E})^{1/2} \, d\mathcal{E} \propto \psi^{3/2+\alpha}. \tag{7.33}$$

Near the SBH, this implies

$$f(\mathcal{E}) = f_0 \mathcal{E}^\alpha, \quad n(r) = n_0 r^{-\gamma}, \quad \gamma = \frac{3}{2} + \alpha. \tag{7.34}$$

Setting $\alpha = 0$ for the light population,

$$n_L(r) \propto r^{-3/2} \quad (\rho_H \gg \rho_L). \tag{7.35}$$

This is a (slightly) shallower radial dependence than the Bahcall–Wolf solution, $n \propto r^{-7/4}$. If we assume that the dominant (heavy) population has the Bahcall–Wolf form, then its density will increase more rapidly toward the center than that of the light population, and sufficiently close to the SBH, the assumptions made in deriving equation (7.35) will be satisfied. Exactly *how* close depends on the relative numbers in the two species; we return to this question shortly.

Before doing so, we consider the complementary case in which the light stars dominate the total density, that is, $\rho_L \gg \rho_H$; as before, $m_L \ll m_H$. These assumptions might be satisfied at sufficiently large distances from the SBH where the steeper radial falloff of the heavy component has resulted in a small density compared with the light component.

Returning to equation (7.29), we now seek steady-state solutions of the form $\partial f_H/\partial t = 0$. Since $\rho_L \gg \rho_H$, we can ignore the terms in that equation that result from interactions between the heavy stars. The energy-space flux for the heavy stars is given approximately by

$$\mathcal{F}_H \approx -D_{\mathcal{E}\mathcal{E}\,HL} \frac{\partial f_H}{\partial \mathcal{E}} - D_{\mathcal{E}\,HL} f_H. \tag{7.36}$$

Suppose that the dominant (light) population has

$$f_L(\mathcal{E}) = f_{L,0} \mathcal{E}^\alpha, \quad n_{L,0}(r) = n_{L,0} r^{3/2+\alpha}. \tag{7.37}$$

For instance, if the light population has reached a steady state with regard to self-interactions, then $\alpha = 1/4$, $\gamma = 7/4$. Inserting equation (7.37) for f_L into the expressions (7.29c), (7.29d) for the diffusion coefficients, the energy-space flux of the heavy population becomes

$$\mathcal{F}_H \approx 2^{7/2} \pi^4 \Gamma' f_{L,0} G^3 M_{\bullet}^3 m_L m_H$$
$$\times \left[-\frac{m_L}{m_H} \frac{1}{(\alpha + 1)(1 - 2\alpha)} \mathcal{E}^{\alpha - 1/2} \frac{\partial f_H}{\partial \mathcal{E}} + \frac{2}{3 - 2\alpha} \mathcal{E}^{\alpha - 3/2} f_H \right], \quad (7.38)$$

where equations (5.182) have been used for p and q and $\Gamma' = 4\pi G^2 \ln \Lambda$.

We are now faced with the same choice that confronted us when deriving the Bahcall–Wolf solution in chapter 5: whether to set the flux to zero, or to a constant, nonzero value. It turns out that the constant-flux solution is the physically more relevant one here. It is easily seen to have the form

$$f_H(\mathcal{E}) = f_{H,0} \mathcal{E}^\beta, \quad \beta = \frac{3}{2} - \alpha. \quad (7.39)$$

If the light component has reached the Bahcall–Wolf steady state, then $\alpha = 1/4$ and $\beta = 5/4$, and the density of the heavy component is

$$n_H(r) = n_{H,0} r^{-11/4} \quad (\rho_H \ll \rho_L), \quad (7.40)$$

a steeper radial falloff than that of the Bahcall–Wolf cusp.

This solution was obtained using equation (7.38), but it is equally valid if only the second term in that equation, corresponding to dynamical friction, is nonzero; and since $m_L \ll m_H$, that is in fact nearly the case. In other words, equation (7.39) corresponds to a constant (with respect to energy) flux in the heavy population due to dynamical friction against the light population [3]. This is the kind of solution we expect if the "supply" of heavy stars far from the SBH has not been seriously depleted: heavy stars will continue to spiral in from large to small radii, and their density will adjust in such a way that the inward flux is nearly constant. A zero-flux solution also exists, but it has a slope that is proportional to m_H/m_L, much steeper than the constant-flux solution. The zero-flux solution describes a case in which the supply of heavy stars at large radii is inadequate to maintain a constant inspiral rate.

The two steady-state solutions just derived are asymptotic, in the sense that the heavy stars either dominate the total density, or contribute negligibly to it. Under what conditions should we expect these solutions to be valid? What matters here is the relative strength of self-interactions between the heavy stars, on the one hand, and interactions between the heavy and light stars on the other. This ratio can be estimated using the velocity-space diffusion coefficients derived in chapter 5. As a measure of the rate at which the heavy stars are scattered, consider

$$\langle (\Delta v)^2 \rangle \equiv \frac{1}{2} \left(\langle \Delta v_\parallel^2 \rangle + \langle \Delta v_\perp^2 \rangle \right); \quad (7.41)$$

the right-hand side of this expression contains the same combination of scattering coefficients that appears in equation (7.25) for $\langle \Delta E \rangle$. We can furthermore break $\langle (\Delta v)^2 \rangle$ for the heavy stars into two pieces, describing self-interactions (H) and interactions with the light stars (L). Assuming Maxwellian velocity distributions for both populations,

$$\langle(\Delta v)^2\rangle_H = \frac{m_H \rho_H \Gamma'}{2^{1/2}\sigma} \frac{\mathrm{erf}(x)}{x}, \tag{7.42a}$$

$$\langle(\Delta v)^2\rangle_L = \frac{m_L \rho_L \Gamma'}{2^{1/2}\sigma} \frac{\mathrm{erf}(x)}{x}. \tag{7.42b}$$

Note that the subscripts refer to the scattering populations; $x \equiv v/\sqrt{2}\sigma$, and we have assumed that both populations have the same velocity dispersion σ— reasonable if we apply the relations to stars at a given distance from the SBH. As in the derivation of equation (7.27), we can average these expressions over the heavy-star velocity distribution, yielding

$$\overline{\langle(\Delta v)^2\rangle}_H = \frac{m_H \rho_H \Gamma'}{\pi^{1/2}\sigma}, \quad \overline{\langle(\Delta v)^2\rangle}_L = \frac{m_L \rho_L \Gamma'}{\pi^{1/2}\sigma}. \tag{7.43}$$

Referring again to equation (7.25), the effect of dynamical friction on the energy distribution of the heavy stars is described by $v\langle\Delta v_\parallel\rangle$. Breaking that quantity again into contributions from the light and heavy stars, and averaging over velocities as before, we find

$$\overline{v\langle\Delta v_\parallel\rangle}_H = -\frac{m_H \rho_H \Gamma'}{\pi^{1/2}\sigma}, \tag{7.44a}$$

$$\overline{v\langle\Delta v_\parallel\rangle}_L = -\frac{1}{2}\left(1 + \frac{m}{m_f}\right)\frac{m_L \rho_L \Gamma'}{\pi^{1/2}\sigma}. \tag{7.44b}$$

Equations (7.42) and (7.44) allow us to define a simple measure of the relative importance of heavy–heavy compared with heavy–light interactions [3]:

$$\Delta \equiv \frac{\left|\overline{v\langle\Delta v_\parallel\rangle}_H\right| + \left|\overline{\langle(\Delta v)^2\rangle}_H\right|}{\left|\overline{v\langle\Delta v_\parallel\rangle}_L\right| + \left|\overline{\langle(\Delta v)^2\rangle}_L\right|} = \frac{m_H \rho_H}{m_L \rho_L}\frac{4}{3 + m_H/m_L}. \tag{7.45}$$

When $\Delta \gg 1$, self-interactions dominate the evolution of the heavy component; their density near the SBH will follow $n \sim r^{-7/4}$, and the light stars will follow $n_L \propto r^{-3/2}$, as in the first of the two cases considered above. When $\Delta \ll 1$, interactions with the light stars dominate the evolution of the heavy stars, and their density falls off more steeply with radius, according to equation (7.39). If we set $m_H/m_L \approx 10$, then the condition $\Delta > 1$ becomes

$$\frac{\rho_H}{\rho_L} \gtrsim \frac{13}{40} \approx 0.3. \tag{7.46}$$

According to table 7.1, the mass fraction in black holes that is predicted by standard IMFs is $\rho_H/\rho_L \approx 0.015$, somewhat smaller than the value in equation (7.46). However, in the presence of mass segregation, ρ_H/ρ_L will increase as one approaches the SBH.

Figure 7.7 shows time-dependent solutions to equation (7.29) for two mass groups, representing solar-mass stars (MS) and ten-solar-mass black holes (BH). The initial densities of both species followed $n \sim r^{-1/2}$ near the SBH; recall that this is the flattest density profile consistent with a nonnegative $f(E)$. The relative

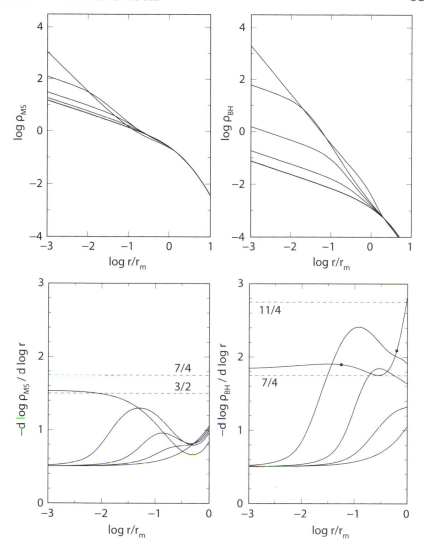

Figure 7.7 Evolution toward a steady state of two mass groups around an SBH. These are solutions to the orbit-averaged, isotropic Fokker–Planck equation (7.29). The component labeled "MS" represents stars of 1 $M_\odot$; the component labeled "BH" represents 10 $M_\odot$ black holes. Initially, the two components have similar radial density profiles, and the BH density is 0.005 times the MS density. Times shown are 0, 0.05, 0.1, 0.2, 0.5 in units of the initial MS relaxation time at the influence radius r_m. The MS slope grows initially toward 7/4, the Bahcall–Wolf single-component value, at small radii but asymptotes to a value closer to 3/2 as the BH density climbs. The BH slope tends initially toward 11/4, equation (7.40), then asymptotes to 7/4 after the BH density begins to dominate. The radii at which $\Delta = 1$ and $\Delta = 0.1$ are plotted as filled circles. Scaled to the Milky Way, the unit of time is $\sim 2.5 \times 10^{10}$ yr and the unit of length is ~ 2.5 pc.

numbers in the two populations were $n_{BH}/n_{MS} = 5 \times 10^{-4}$, that is, $\rho_{BH}/\rho_{MS} = 0.005$. After one-half relaxation time at the SBH influence radius the two distributions have reached a nearly steady state. The BHs dominate the mass density inside $\sim 10^{-2} r_m$, and at small radii, $n_{MS} \sim r^{-3/2}$, $n_{BH} \sim r^{-7/4}$ as expected. Farther from the SBH, at $r \approx r_m$, the stars dominate the mass density; equation (7.39) tells us that the density in BHs should be falling off faster than the MS density at these radii and this is clearly seen to be the case.

In models like these, the "observable" is the distribution of the stars, and it is interesting to ask how that distribution is affected by the presence of the black holes. Comparison of two-component solutions, like the one plotted in figure 7.7, with the single-component solution, figure 7.4, reveals the following differences:

1. The $n \sim r^{-7/4}$ Bahcall–Wolf cusp in the stars is replaced by a weaker cusp, $n \sim r^{-\gamma}$, $3/2 \lesssim \gamma \lesssim 7/4$.
2. The radial extent of the stellar cusp decreases somewhat with increasing black hole density: from $\sim 0.2 r_m$ when $n_{BH} = 0$, to $\sim 0.1 r_m$ when the black holes dominate the gravitational scattering.
3. The time required for growth of the stellar cusp is reduced somewhat by the presence of the black holes, although for physically relevant black hole numbers, this change is slight.

How do these predictions compare with the distribution of (old) stars near the center of the Milky Way? The influence radius for Sgr A* is estimated to be $2\,\mathrm{pc} \lesssim r_m \lesssim 3\,\mathrm{pc}$, and the relaxation time at this radius, assuming $\tilde{m} = 1\,M_\odot$, is 20–30 Gyr [357].[5] The final time plotted in figure 7.7 is half of this time, or roughly the age of the universe: perhaps twice the mean age of the stars. An immediate conclusion is that—if our estimate of the relaxation time is correct—we would not necessarily expect the stars at the Galactic center to have reached a collisionally relaxed steady state by now. On the other hand, if there *has* been time enough to form a steady-state cusp—perhaps because $\tilde{m}$ is greater than $1\,M_\odot$ and T_r correspondingly shorter— then the stellar density should rise as $n \sim r^{-3/2}$ inside a radius $\sim 0.1 r_m \approx 0.25\,\mathrm{pc} \approx 2''$. This radius is well resolved by the number counts, but as figure 7.1 shows, no cusp is observed; instead the counts level off, or even begin to decrease, inside roughly this radius. The most straightforward interpretation is that the nucleus of the Milky Way has not yet reached a collisional steady state [357].

Solutions to the isotropic Fokker–Planck equation that incorporate more than just two mass groups have also been constructed [395, 170, 245]. In studies like these, loss of stars to the SBH can be approximated by adding a term similar to equation (7.13) to the right-hand side of the evolution equation for each f_i, equation (7.29a). The only additional change required is to replace $\mathcal{R}_{lc}$ in the loss term by $\mathcal{R}_{lc,i}$, the appropriate value for the ith group. Figure 7.8 shows a steady-state solution for the stars near the Milky Way SBH that was computed in this way [245]. Four mass groups were included, with masses approximating those of main-sequence stars (MS), white dwarfs (WD), neutron stars (NS), and stellar-mass black

---

[5]Table 7.1 suggests that $\tilde{m}$ may be smaller than one solar mass, which would make T_r longer.

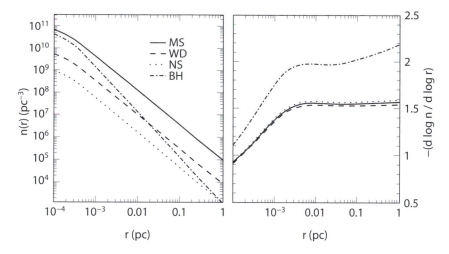

Figure 7.8 A steady-state solution to the isotropic Fokker–Planck equation (7.29) with four mass groups, and including a loss term similar to equation (7.13). The left panel shows the number density profiles $n(r)$ and the right panel shows logarithmic slopes. This model does not include the gravitational potential due to the stars themselves and so is only valid well inside the SBH influence radius of $\sim 2\,\mathrm{pc}$. Inclusion of a loss term is responsible for the flattening of the profiles inside $\sim 0.005\,\mathrm{pc}$. (Based on calculations of C. Hopman and T. Alexander.)

holes (BH):

$$m_{\mathrm{MS}} = 1\,M_\odot, \quad m_{\mathrm{WD}} = 0.6\,M_\odot, \quad m_{\mathrm{NS}} = 1.4\,M_\odot, \quad m_{\mathrm{BH}} = 10\,M_\odot. \tag{7.47}$$

The relative numbers in the four groups at the outermost radius ($r = 1\,\mathrm{pc}$) were fixed at $\mathrm{MS : WD : NS : BH} = 1 : 0.1 : 0.01 : 0.001$, consistent with the density ratios in the second column of table 7.1. In this collisionally relaxed solution, the stellar black holes dominate the mass density inside $r \approx 0.005\,\mathrm{pc}$; the three lighter species have logarithmic density slopes that are all close to $\gamma = 3/2$, while the black holes have $n \sim r^{-2}$. At very small radii, $r \lesssim 10^{-3}\,\mathrm{pc}$, the density of each of the species falls below the power-law solutions due to the inclusion of a loss term. The predicted number of black holes within $1\,\mathrm{pc}$ is 1.8×10^3, and within $0.1\,\mathrm{pc}$ is 150. But just as in the two-component models, the distribution of observable stars predicted by this model is seriously in conflict with observations of the Milky Way nucleus.

7.2 CUSP (RE)GENERATION

Several arguments have been presented in this book in support of the hypothesis that the nucleus of the Milky Way is not in a collisionally relaxed state. These include the long relaxation time computed for stars within the SBH influence radius (figure 7.1b), and the apparent absence of a Bahcall–Wolf cusp in the late-type stars (figure 7.1a). Given the empirical scaling of nuclear relaxation time with galaxy luminosity shown in figure 3.1, it is reasonable to conclude that galaxies brighter than the Milky Way also harbor unrelaxed nuclei. Galaxies *less* luminous than the Milky Way often contain dense nuclear star clusters, like that of the Milky Way, and these NSCs display a trend of decreasing relaxation time with decreasing galaxy luminosity (figure 7.3). But it is unclear how often NSCs contain SBHs. In the absence of an SBH, evolution of a nucleus over relaxation timescales will be fundamentally different than what has been described so far in this chapter; we reserve a detailed discussion of that case until later in this chapter.

In a collisionally unrelaxed nucleus, predicting the distribution of stars and stellar remnants at any given time becomes an initial-value problem. In other words, we need to say something about how the various components were distributed at some specified time, then integrate these initial conditions forward into the present epoch. The results of such a calculation can then be compared with the observations—assuming of course that there are any nuclei close enough that the relevant features can be resolved.

What should the initial conditions be? One could imagine going back to a time when both the stars and the SBH were being formed. Fortunately, in many galaxies, such an ambitious program is probably not necessary. It is likely that the nuclear "clock" was reset to zero much more recently, at the time of the last major merger that formed the stellar bulge. As discussed in detail in the next chapter, if the merging galaxies each contained SBHs, the massive binary that was created during the merger would efficiently eject passing stars via the gravitational slingshot, essentially erasing the preexisting nuclei and producing a low-density core comparable in size to the binary's semimajor axis. Such cores are in fact ubiquitous in galaxies close enough that features on scales of r_h can be well resolved (chapter 2).

Suppose that a preexisting binary SBH has ejected all stars that passed within some distance r_c of the galaxy center. We expect r_c to be of order the semimajor axis of the binary SBH at the time it first became "hard," in the sense defined in chapter 8: that is, sufficiently bound that it is capable of ejecting passing stars completely from the galaxy core. As shown in section 8.1, this separation is a fraction of r_h. Let $\mathcal{E}_c \equiv \psi(r_c) \approx G M_\bullet / r_c$ with $M_\bullet$ the mass of the single SBH that remains after the binary has coalesced. Approximating the galactic potential as spherical, stars will have been removed by the binary if their orbits satisfy either of the conditions

$$\mathcal{E} \gtrsim \mathcal{E}_c, \quad L^2 \lesssim L_c^2(\mathcal{E}) \equiv 2 r_c^2 \left(\mathcal{E}_c - \mathcal{E} \right). \tag{7.48}$$

L_c is the specific angular momentum of a star with periapsis at r_c. The configuration-space density corresponding to this "hole" in phase space will be zero at $r \leq r_c$; at radii beyond r_c, the density will have been lowered by the massive binary, though not all the way to zero, due to its removal of stars on low-angular-momentum orbits.

These "initial conditions" are the same ones that were explored in section 6.1.5, in our discussion of time-dependent loss cones around single SBHs. It was argued there that evolution would proceed on two timescales: a short timescale, $t_L \approx (r_c/r_h) T_r(r_h)$, associated with replenishment of low-angular-momentum orbits; and a longer timescale, $\sim T_r(r_{\rm h})$, over which orbital energies evolve. After a time $t_{\rm L}$, evolution in L will have "filled in" the angular-momentum gap created by the massive binary. Ignoring the evolution in $\mathcal{E}$ that takes place during this time, the resulting phase-space density can be approximated as

$$f(\mathcal{E}) = \begin{cases} f_0(\mathcal{E}), & \mathcal{E} \lesssim \mathcal{E}_c, \\ 0, & \mathcal{E} \gtrsim \mathcal{E}_c. \end{cases} \tag{7.49}$$

The configuration-space density corresponding to this f is

$$\rho(r) \approx \begin{cases} \rho_0(r), & r > r_c, \\ 4\sqrt{2}\pi \int_0^{\psi(r_c)} d\mathcal{E}\, f_0(\mathcal{E})\sqrt{\psi(r) - \mathcal{E}}, & r < r_c. \end{cases} \tag{7.50}$$

The latter expression is approximately

$$4\sqrt{2}\pi \sqrt{\psi(r)} \int_0^{\psi(r_c)} d\mathcal{E}\, f_0(\mathcal{E}) \propto r^{-1/2} \tag{7.51}$$

at $r \lesssim r_c$. Roughly speaking, this is what is meant by a "core": a region of nearly constant density inside r_c. In fact, the density in this "core" diverges weakly toward the center; but as seen in projection against the rest of the galaxy, such a weak density cusp is almost indistinguishable from a constant-density core.

We could have arrived at this model for the density via a less circuitous route. The luminosity profiles of elliptical galaxies are often observed to be flat, or slowly rising, at $r \lesssim r_{\rm h}$ (chapter 2). Regardless of how these cores formed, Eddington's formula, equation (3.47), can be used to infer $f(\mathcal{E})$ from the observed $\rho(r)$. If the density near the SBH is slowly rising, $f(\mathcal{E})$ will be similar in form to equation (7.49): it will increase with $\mathcal{E}$ until $\mathcal{E} \approx \mathcal{E}_c$, then drop off sharply for $\mathcal{E} \gtrsim \mathcal{E}_c$. Indeed, as noted in chapter 3, a density that rises as $r^{-1/2}$ is the shallowest profile consistent with an isotropic velocity distribution in a $1/r$ potential: it corresponds to an f that is zero near the SBH. Density profiles that are still flatter imply a deficit of eccentric orbits at large $\mathcal{E}$, that is, an anisotropic velocity distribution—as indeed might be created by a binary SBH, if it preferentially ejects stars on eccentric orbits.

Before exploring the evolution of models that start from initial conditions like these, there is an important point to be made about dynamical friction [8]. If the distribution function is sharply truncated as in equation (7.48), then the density at $r \leq r_c$ is zero, and Chandrasekhar's formula predicts zero frictional force. But it turns out that the same is true even if f has the form (7.49)—in spite of the fact that this f implies a nonzero *configuration*-space density everywhere!

Chandrasekhar's dynamical friction coefficient, in its standard form, is given by equation (5.23):

$$\langle \Delta v_\parallel \rangle = -\frac{4\pi G^2 \left(m_f + m\right) \ln \Lambda}{v^2} \rho\left(v_f < v\right), \tag{7.52}$$

where $\rho\left(v_f < v\right)$ is the mass density contributed by stars whose velocities at infinity are less than the test body's velocity v. If the field stars are described by an isotropic f, then

$$\rho(v_f < v; r) = 4\sqrt{2}\pi \int_{\psi(r)-v^2/2}^{\psi(r)} d\mathcal{E} f(\mathcal{E})\sqrt{\psi(r) - \mathcal{E}}. \tag{7.53}$$

If in addition, f is truncated at $\mathcal{E} > \mathcal{E}_c$, as in equation (7.49), then $\rho(v_f < v)$ falls to zero for orbits with

$$\psi(r) - \frac{v^2}{2} \geq \mathcal{E}_c \tag{7.54}$$

since there are *no* stars locally that move slower than v at these energies—regardless of whether the configuration-space density at r is nonzero. Assuming a circular orbit for the test body, and that the orbit lies inside the influence radius of the SBH, this condition becomes

$$r \lesssim \frac{r_c}{2}. \tag{7.55}$$

Thus, inside approximately half of the core radius, the frictional force drops precisely to zero.

What (if anything) went wrong? Aside from the assumption of isotropy, equation (7.52) is not based on any particular functional form for f. But its derivation was tainted at one point, when the logarithmic term was "taken out of the integral" (following equation 5.21). As discussed in chapter 5, this is the approximation that results in only the slow-moving stars contributing to the dynamical friction force. If the logarithmic term is left inside the integral, one obtains the more general expression for $\langle \Delta v_\parallel \rangle$ given by equation (5.20)—which indicates a contribution to the frictional force from field stars of all velocities, including $v_f > v$.

Clearly we need to worry about how important a contribution the fast-moving stars can make to the frictional force. Suppose that the stellar density near the SBH has the form $\rho \propto r^{-\gamma}$, $\gamma > 1/2$ at radii $r \lesssim r_h$, and consider a test mass moving in a circular orbit, $v = v_c(r)$. Using equation (5.20), the dynamical friction force can be computed and broken down into two contributions, due to field stars with $v_f > v$ and with $v_f < v$. Figure 7.9 shows the results, in a model scaled to match the density near the center of the Milky Way; the density inside the core (of radius 0.3 pc) has been allowed to have the form $\rho \propto r^{-\gamma}$, $\gamma > 1/2$. Inside the core, the fraction of the frictional force that comes from the fast-moving stars is substantial when $\gamma \lesssim 1$. Our initial conclusion that the frictional force should go to zero was wrong: even Chandrasekhar's treatment predicts a nonzero force, but only if the $\ln \Lambda$ term is left inside the integrand. Another useful comparison is made in the right panel of this figure, which plots the frictional force—including the fast-moving stars—as a fraction of the force that would be computed if one applied Chandrasekhar's formula in its standard form and assumed a locally Maxwellian velocity distribution, equation (5.23). The Maxwellian approximation can substantially overestimate the frictional force, whether the latter is computed from just the slow-moving stars, or from all of them.

The reader may be wondering at this point whether treatments like Chandrasekhar's are to be trusted in a case like this—even *if* the fast-moving stars are accounted for. That is a valid worry for a number of reasons: the inhomogeneity of

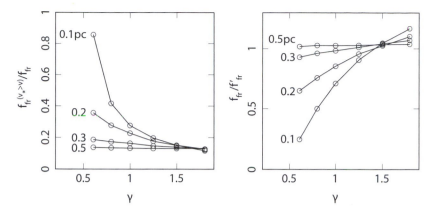

Figure 7.9 These figures illustrate how dynamical friction acts in a core around an SBH, in Chandrasekhar's formalism [8]. They are based on a density model for the center of the Milky Way that looks similar to the dashed line in figure 7.1a, except that the density inside the core (of radius $r_c = 0.3$ pc) is allowed to have the more general form $\rho \propto r^{-\gamma}$. Eddington's formula (3.47) was used to compute $f(\mathcal{E})$ from $\rho(r)$, and the test body was assumed to follow a circular orbit. The left panel shows the fraction of the frictional force that comes from stars with $v_f > v$ as a function of γ at different galactocentric radii: $r = 0.1, 0.2, 0.3$ and 0.6 pc. Equation (5.20) was used to compute these curves; N-body experiments verify that this equation accurately predicts the frictional force. The right panel compares the total frictional force computed by this equation with the force computed using equation (5.23), which assumes a Maxwellian distribution of velocities.

the models, the fact that the field stars are orbiting in the near-Keplerian field of the SBH, etc. One way to assess these concerns would be via a perturbative analysis that accounts properly for the orbits of the field stars. Unfortunately, such an analysis has never been carried out for models like these. But direct simulation via an N-body code is a viable alternative, and one which does not suffer from the limitations and ambiguities associated with a perturbative treatment. Such a study [8] confirms that Chandrasekhar's equation (5.20)—the version that includes the contribution from the fast-moving stars—does a very good job of predicting the orbital decay of a massive body in models with $\rho \propto r^{-\gamma}$ near an SBH.

Before returning to our discussion of cusp regeneration, we note that the velocity distribution inside a core with $n(r) \propto r^{-\gamma}$ is given approximately by

$$f(\mathcal{E}) = f_0 \mathcal{E}^{\gamma - 3/2}$$
$$\propto \left[\psi(r) - v^2/2 \right]^{\gamma - 3/2}$$
$$\propto \left(G M_\bullet / r - v^2/2 \right)^{\gamma - 3/2}$$
$$\propto \left(2 v_c^2 - v^2 \right)^{\gamma - 3/2}, \quad v \le 2^{1/2} v_c. \tag{7.56}$$

The first of these expressions is equation (3.49); note that f_0 in that equation becomes undefined for $\gamma \le 1/2$. The final expression replaces $2 G M_\bullet / r$ by v_c^2, the

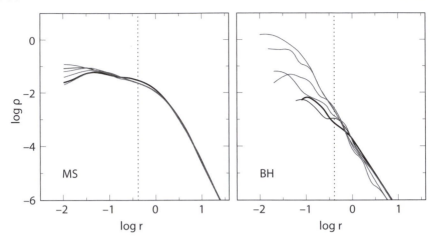

Figure 7.10 Regeneration of a mass-segregated cusp around an SBH following a galaxy
merger. The simulated galaxies contained four mass components, as in (7.47);
only two are shown here. Vertical (dotted) lines mark the initial influence radius
r_m. Times shown are $(0.2, 0.5, 1, 2, 3)$ in units of the initial relaxation time at
the influence radius, as defined by the MS stars. Other details are given in the
text. (Adapted from [220].)

square of the circular velocity at r. While it is not obvious from these expressions,
in the limit $\gamma \to 1/2$, $f(v)$ becomes a delta function at $v = \sqrt{2}v_c$; in other words,
all stars have zero energy with respect to the SBH. Here we have a concrete example of how non-Maxwellian the velocity distribution near an SBH is expected to
be.[6]

Armed with our insights into the workings of dynamical friction in cores, we
now consider two examples that illustrate the regeneration of cusps around SBHs.

The first example, illustrated in figure 7.10, is taken from a large-scale N-body
simulation of a galaxy merger [220]. The "initial conditions" in this plot consisted
of the N-body model at a time after the galaxy merger was essentially complete,
and the binary SBH—of mass ratio $M_2/M_1 = 1/3$—had created a core from the
preexisting, Bahcall–Wolf cusps in each of the two merging galaxies. These simulations included four mass groups, as in (7.47); before the merger, the nuclei of
the two galaxies were each in a collisionally relaxed, mass-segregated state, similar
to what is shown in figure 7.8. The galaxy merger, together with ejection of stars
by the binary SBH, have acted to create a large core around the single, coalesced
SBH. The radius of the core as defined by the "stars" (the component labeled MS)
is roughly twice the SBH influence radius r_m. The stellar black holes (BH) have a
smaller core initially—a relic of the pre-merger mass segregation, which was in-
completely erased by the massive binary. But because the core in the dominant
(MS) component is so large, the core radius scarcely evolves, even after a time of

[6]This is the same velocity distribution that was used in computing the dynamical friction wakes in
figure 5.3.

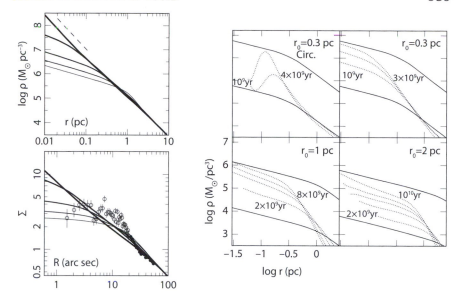

Figure 7.11 Left: Evolution of the surface density $\Sigma(R)$ and configuration-space density $\rho(r)$ for a population of $1\,M_\odot$ stars around the Milky Way SBH, assuming an initial core size of 1 pc. Increasing line thickness denotes increasing time, $t = (0, 0.2, 0.5, 1, 2) \times 10^{10}$ yr. The dashed line is the asymptotic form corresponding to a Bahcall–Wolf cusp, i.e., $\rho \propto r^{-7/4}$. (Adapted from [357].) Right: Evolution of the distribution of $10\,M_\odot$ black holes due to dynamical friction against the field stars. Evolution of the field-star distribution was ignored, and the frictional force was computed using equation (5.20), which accounts for the contribution of the fast-moving stars to the frictional force. (Adapted from [8].)

$3T_r(r_m)$. A Bahcall–Wolf cusp can be seen to grow in this component but only at radii $r \lesssim 0.2r_m \ll r_c$.

The BHs evolve against this essentially fixed background. But the frictional force inside the stellar core is small, and the BH distribution barely reaches a steady state by the end of the simulation, after ~3 MS relaxation times at r_m. At this time the BHs dominate the total density inside $\sim0.1r_m$.

A second example of cusp regeneration is shown in figure 7.11. Here, the initial conditions were designed specifically to represent the NSC of the Milky Way. The initial density was given by equation (7.50) with

$$\rho_0(r) = 1.5 \times 10^5 \left(\frac{r}{1\ \text{pc}}\right)^{-1.8} M_\odot\ \text{pc}^{-3} \tag{7.57}$$

and with a core of radius $r_c = 1$ pc. This functional form is a good match to what is observed in the Milky Way NSC, except that the adopted core radius is about a factor of two too large (figure 7.1). The initial core is still smaller than the SBH influence radius, however. Figure 7.11 (left), based on the single-component, isotropic Fokker–Planck equation, shows how the core gradually shrinks, as gravitational encounters drive the stellar density toward a Bahcall–Wolf cusp. Interestingly, the density falloff in the Milky Way NSC beyond the core is essentially identical with the Bahcall–Wolf form $\rho \sim r^{-7/4}$—even at radii much greater than r_h ($\approx$2–3 pc) where gravitational encounters are unlikely to have been effective over the age of the universe. Because of this coincidence, and because the initial core in our model lies completely inside r_h, the evolution has the effect of maintaining the slope of the density profile outside $\sim r_c(t)$, while gradually reducing r_c. After roughly 10 Gyr, the core size has decreased by about a half, to a size that is consistent with the size of the currently observed core.

Figure 7.11 (right) shows the results of a second set of integrations, this time of the orbits of 10 $M_\odot$ black holes, starting from a model in which both the BHs and the field (MS) stars had the same distribution as in figure 7.11 (left). The frictional force acting on the BHs was computed from equation (5.20), assuming that the field-star distribution was unevolving. These plots suggest that the spatial distribution of stellar-mass black holes near the Galactic center may not have reached a steady state under the influence of gravitational encounters—at least, if their initial conditions were as far from equilibrium as assumed in making the figure.

7.3 BLACK-HOLE-DRIVEN EXPANSION

The terms "equilibrium" and "steady-state" have been used freely throughout this chapter to describe time-independent models for stars around an SBH. But a little thought makes it clear that a true steady state can never exist, as long as the SBH is consuming or destroying stars. This is not so much because the supply of stars in a galaxy is finite; the number of stars is far greater than could be consumed by the SBH in any physically interesting time. But beyond a certain distance from the SBH, the relaxation time is so long that the encounter-driven flux of stars toward the galaxy center cannot keep up with losses near the SBH. As a result, the density near the SBH must drop. This gradual decrease in density can equally well be described as an expansion, in the sense that the radius of a sphere containing a fixed mass in stars will increase with time.

In steady-state models like those of Bahcall and Wolf [14, 15], as well as the solutions plotted in figure 7.5, this expansion was absent due to the choice of outer boundary condition: $f(E)$ was fixed at small $\mathcal{E}$, that is, far from the SBH. In effect, stars were resupplied at large radii at just the correct rate to counteract the losses to the SBH. The expansion also did not appear in the time-dependent solutions plotted in figures 7.4 or 7.7 since those integrations did not include a loss term.

We can estimate the approximate radius, in a spherical galaxy, beyond which gravitational encounters are unable to supply stars at a high enough rate to compensate for losses to the SBH. Equation (7.11) gives the rate of change of $N(\mathcal{E})$ due to the combined effects of diffusion in energy, and loss of stars to the SBH; recall that the latter process is driven almost entirely by changes in angular momentum. Integrating that equation in energy from $\mathcal{E}_t$ to $\mathcal{E}$ yields

$$\frac{\partial}{\partial t} N(> \mathcal{E}) \approx \mathcal{F}_{\mathcal{E}}(\mathcal{E}) - \dot{N}^{\text{lc}}. \qquad (7.58)$$

We have assumed that $\mathcal{E}$ is small compared with the energy of peak flux into the loss cone, hence the second term on the right-hand side can be identified with the total loss rate. Equation (6.86) gives $\dot{N}^{lc}$ in a nuclear star cluster like that of the Milky Way, with density $n(r) \sim r^{-2}$. We wish to find the largest r, that is, the smallest $\mathcal{E}$, for which $\mathcal{F}_{\mathcal{E}}$ can equal $\dot{N}^{lc}$. The most optimistic estimate of $\mathcal{F}_{\mathcal{E}}$ is $F^{max} \equiv N(\mathcal{E})\mathcal{D}(\mathcal{E})$ defined in equation (6.68). We can write that expression approximately in terms of radius by noting that $\mathcal{D}(\mathcal{E})$ is the inverse of an orbit-averaged relaxation time. The maximum inward flux of stars (number per unit time) at radius r is then

$$\dot{N}^{max}(r) \approx \frac{N(< r)}{T_r(r)}. \tag{7.59}$$

Using equation (5.61) for T_r, writing $N(< r) = 2(M_\bullet/m)(r/r_m)$, and setting $\sigma(r) = \sigma_0$, both appropriate for the singular isothermal sphere, we find

$$\dot{N}^{max}(r) \approx \frac{G^2 M_\bullet^2 \ln \Lambda}{\sigma_0^3 r_m^2 r} \tag{7.60a}$$

$$\approx \left(\frac{r_h}{r_m}\right)^2 \frac{\sigma_0}{r} \ln \Lambda \tag{7.60b}$$

$$\approx 9 \times 10^{-5} \left(\frac{\sigma}{90 \, \mathrm{km \, s^{-1}}}\right) \left(\frac{r}{\mathrm{pc}}\right)^{-1} \ln \Lambda \, \mathrm{yr^{-1}}, \tag{7.60c}$$

where r_m has been identified with r_h, again appropriately for the singular isothermal sphere. This energy diffusion rate falls below the total loss rate, equation (6.86), at radii greater than

$$r \approx 3 \left(\frac{\sigma}{90 \, \mathrm{km \, s^{-1}}}\right)^{-5/2} \left(\frac{M_\bullet}{4 \times 10^6 \, M_\odot}\right) \mathrm{pc}, \tag{7.61}$$

or roughly the influence radius in the case of the Milky Way.

Figure 7.12 illustrates the expansion. Equation (7.11) was integrated until a Bahcall–Wolf cusp had formed. The integration was then continued for an additional $\sim$20 Gyr, based on a scaling that identified the final density at 1 pc with the density currently observed in the Milky Way (equation 6.73). The right panel in the figure compares the density at 0.1 pc to the time-integrated captured mass. By the time the SBH has destroyed a mass in stars equal to its own mass, the density, and therefore the rate of supply of stars to the loss cone, has fallen by roughly a factor of three. Multimass models that include a loss term behave in a similar manner [395, 170]; in particular, the dominant mass component maintains an approximate, $n \propto r^{-7/4}$ dependence near the SBH as the density drops.

7.4 MASSIVE PERTURBERS

In section 5.2.2 it was pointed out that the characteristic time for scattering of a test star by a set of field stars having a range of different masses could be written as

$$T_r = \frac{0.34 \sigma^3}{G^2 \tilde{m} \rho \ln \Lambda}, \tag{7.62}$$

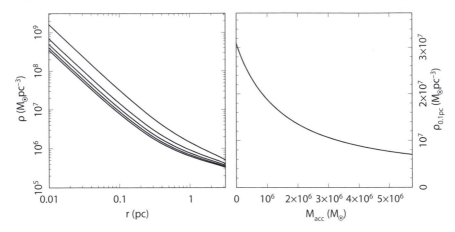

Figure 7.12 Expansion of a nucleus driven by loss of stars to the SBH. The left panel shows the density at constant time intervals after a Bahcall–Wolf cusp has been established; the final influence radius is $r_h \approx 3\,\mathrm{pc}$. The right panel shows the evolution of the density at 0.1 pc as a function of M_{acc}, the accumulated mass in tidally disrupted stars. As scaled to the Milky Way, the final time is roughly 2×10^{10} yr.

where $\tilde{m}$ is given by equation (5.62b):

$$\tilde{m} \equiv \frac{\int n(m)m^2 dm}{\int n(m)m\, dm} = \rho^{-1} \int n(m)m^2 dm, \tag{7.63}$$

and $n(m)dm$ is the number of field stars with masses in the interval dm centered on m. Equations (7.62) and (7.63) are appropriate if field stars of all masses have the same velocity distribution; a defensible assumption if the system is less than one relaxation time old.

Table 7.1 gave an estimate of $\tilde{m}$, $\tilde{m} \approx 0.55\, M_\odot$, in a stellar population with a "normal" initial mass function (IMF) and in which the most massive stars had evolved off the main sequence, leaving behind compact remnants. It is interesting to compute $\tilde{m}$ for other mass functions. For instance, consider a stellar cluster young enough that all its stars are still on the main sequence. In this case, $n(m)$ is equal to the IMF; writing the latter as $\xi(m) = \xi_0(m/m_0)^{-\alpha}$, we find

$$\rho\,\tilde{m} = \xi_0 \int_{m_1}^{m_2} m^2 \left(\frac{m}{m_0}\right)^{-\alpha} dm$$

$$= \frac{\xi_0 m_0^3}{3 - \alpha} \left[\left(\frac{m_2}{m_0}\right)^{3-\alpha} - \left(\frac{m_1}{m_0}\right)^{3-\alpha} \right]$$

$$\approx \frac{\xi_0 m_0^3}{3 - \alpha} \left(\frac{m_2}{m_0}\right)^{3-\alpha}. \tag{7.64}$$

The last line is valid if $m_2 \gg m_1$ and $\alpha \lesssim 3$. In fact, for a Salpeter IMF, $\alpha = 2.35$. For such a mass function, the value of $\tilde{m}$ depends sensitively on what is assumed

Table 7.2 Massive perturbers in the Galactic center [422].

Type	r (pc)	N	m ($M_\odot$)	$\tilde{m}$ ($M_\odot$)	R (pc)
GMCs	<100	~100	10^4–10^8	3×10^6–3×10^7	5
Clusters	<100	~10	10^2–10^5	4.8×10^3–2.4×10^4	1
Gas clumps	1.5–3	~25	10^2–10^5	3.7×10^3–4.1×10^4	0.25

for the maximum stellar mass. Setting $m_2 \approx 10^2\, M_\odot$ and $m_1 \approx 10^{-2}\, M_\odot$, one finds $\tilde{m} \approx 10^1\, M_\odot$, corresponding to a relaxation time that is roughly ten times shorter than in an evolved cluster with the same mass density.

In 1951, L. Spitzer and M. Schwarzschild [504] took this line of reasoning to a logical conclusion. They pointed out that the interstellar gas, whose mean density is similar to that of the stars in the Galactic disk, is distributed in a clumpy fashion, with individual clouds having masses as great as $10^5\, M_\odot$. They argued that the time for stars to be gravitationally scattered by gas clouds was therefore roughly 10^5 times shorter than the star–star relaxation time, short enough that star–cloud scattering could be responsible for the observed dependence of stellar scale height on age.

If we write the mass function in Spitzer and Schwarzschild's model as

$$n(m) = [n(m)]_{\text{star}} + [n(m)]_{\text{cloud}}, \qquad (7.65)$$

then equation (7.63) implies

$$\rho\tilde{m} = \left[\int n(m)m^2 dm \right]_{\text{star}} + \left[\int n(m)m^2 dm \right]_{\text{cloud}} \qquad (7.66a)$$

$$= \rho_{\text{star}}\tilde{m}_{\text{star}} + \rho_{\text{cloud}}\tilde{m}_{\text{cloud}}. \qquad (7.66b)$$

Now replacing the clouds by any collection of objects more massive than stars, the condition that the scattering time be dominated by the more massive objects is

$$(\rho\tilde{m})_{\text{MP}} \gg (\rho\tilde{m})_{\text{star}}. \qquad (7.67)$$

The subscript MP stands for **massive perturber** [422].

Near the center of a galaxy, massive perturbers can include gas clouds of various masses, up to and including giant molecular clouds (GMCs). Star clusters, both open and globular, also fall into this category. Table 7.2 gives estimates of the numbers and masses of massive perturbers near the center of the Milky Way. While mean number densities are very small—roughly 10^{-5} pc^{-3} in the case of GMCs— these objects are so much more massive than stars that they can easily dominate the gravitational scattering inside any region large enough to contain them. The dominant contribution to $\rho\tilde{m}$ comes from the GMCs; in the region $1.5\,\text{pc} \lesssim r \lesssim 5\,\text{pc}$, one finds [422]

$$(\rho\tilde{m})_{\text{GMC}} \approx (20\text{–}2000)\,(\rho\tilde{m})_{\text{star}}. \qquad (7.68)$$

The uncertainty in this estimate arises primarily from uncertainties in the measured masses of GMCs. In order to estimate the effective reduction in the timescale for gravitational scattering, we need to also take into account the large physical size of GMCs, which implies a lower effectiveness of close encounters. Recall from chapter 5 that the Coulomb logarithm can be written as

$$\ln \Lambda \approx \frac{1}{2} \ln \left[1 + \frac{p_{max}^2}{p_0^2} \right]$$

with

$$p_0 \equiv \frac{2Gm}{V^2} \approx 10^{-6} \left(\frac{m}{M_\odot} \right) \left(\frac{V}{10^2 \text{ km s}^{-1}} \right)^{-2} \text{ pc.}$$

Numerical experiments (section 5.2.3.3) suggest that p_{max} is roughly $1/4$ times the linear extent of the test star's orbit. Setting this size to 10 pc, and replacing p_0 by $1/2$ the physical size of a GMC, $R \approx 5$ pc (table 7.2), one finds that $\ln \Lambda$ near the center of the Milky Way decreases from ~ 15 in the case of star–star scattering to ~ 0.5 in the case of scattering by GMCs. Combined with equation (7.68), this result suggests that the effective timescale for gravitational scattering near the Galactic center might be reduced by a factor of $\sim 10^0$ to $\sim 10^2$ due to the presence of GMCs.

What would be the consequences of such a reduction? The effect on the distribution of stars inside the influence radius of Sgr A*, $r < r_h \approx r_m \approx 2$–3 pc, is likely to be small: at these radii, velocity perturbations are due mostly to objects within $\sim r_h$, a region that is not likely to contain a single massive perturber. But as we saw in section 6.1.3, the supply of stars to the SBH is likely to be dominated by gravitational encounters that take place at larger radii. In the absence of massive perturbers, the transition from empty- to full-loss-cone regimes takes place roughly at $r \approx r_h$ in a nuclear star cluster like that of the Milky Way (figure 6.5). Massive perturbers cannot increase the flux in the full-loss-cone regime, but they could convert an empty loss cone into a full loss cone, implying an increased rate of capture by the SBH.

Recall from equation (6.72) that the energy-dependent loss-cone flux is given in the two regimes as

$$F \approx F^{flc} \times \begin{cases} q |\ln \mathcal{R}_{lc}|^{-1}, & q \ll -\ln \mathcal{R}_{lc}, \\ 1, & q \gg -\ln \mathcal{R}_{lc}, \end{cases}$$

where F^{flc} is the full-loss-cone flux, $\mathcal{R}_{lc} \approx r_{lc}/r$, and $q \approx P/(T_r \mathcal{R}_{lc})$ with P the orbital period. A decrease in the effective value of T_r due to massive perturbers would imply a larger q, hence a smaller radius of transition to the full-loss-cone regime. But the effect on the net rate of stellar captures is likely to be modest, at least in a galaxy like the Milky Way, since the transition radius even in the absence of massive perturbers is $r_{crit} \approx r_h$, and since massive perturbers will not significantly decrease the effective value of T_r inside r_h.

These arguments are modified somewhat in the case of the interaction of *binary* stars with an SBH. As discussed in section 6.3, the distance from an SBH at which a binary star is tidally separated, $r_{t,bin}$, is larger than the tidal disruption radius for a single star, r_t, by a factor

$$\frac{r_{t,bin}}{r_t} \approx \frac{a_{bin}}{R_\star} \approx 21 \left(\frac{a_{bin}}{0.1 \text{ AU}} \right) \left(\frac{R_\star}{R_\odot} \right)^{-1}, \tag{7.69}$$

where a_{bin} is the binary semimajor axis and $R_\star$ is a stellar radius. Identifying $r_{t,bin}$ with r_{lc} allows us to define a "capture sphere" for binary stars; the rate of diffusion of binaries into this sphere determines the rate at which, for instance, hypervelocity stars are produced. Since the capture sphere for binaries is so much larger than r_t, the empty-loss-cone regime extends much farther out, and any mechanism that decreases the effective relaxation time can therefore have a substantial effect on the binary disruption rate. It has been argued that massive perturbers increase the rate of interaction of binary stars with the Milky Way SBH by a factor of 10^1–10^3, with corresponding increases in the rate of production of hypervelocity stars, and the rate of deposition of stars in tightly bound orbits near the SBH [422].

7.5 EVOLUTION OF NUCLEI LACKING MASSIVE BLACK HOLES

Throughout this book, the presence of an SBH at the center of a nucleus has been a default assumption. But there is no compelling reason to believe that every galaxy contains a massive black hole, and this is perhaps most true in the case of low-luminosity galaxies. For instance, the upper limit on the mass of any compact central object in the Local Group dwarf galaxy NGC 205 is a few times 10^4 $M_\odot$ [527]. But like many low-luminosity galaxies, NGC 205 does contain a compact nuclear star cluster (NSC), with a central relaxation time that is estimated at less than 10^8 yr. NSCs beyond the Local Group are not as well resolved as the one in NGC 205, but as discussed in the introduction to this chapter, half-mass relaxation times of NSCs show a trend with galaxy luminosity, falling below $\sim 10^9$ yr in the faintest NSCs whose structure can be reasonably well resolved (figure 7.3). Central relaxation times (assuming that no SBH is present) would be even shorter than half-mass relaxation times.

In the absence of a massive central object, nuclei evolve very differently in response to gravitational encounters. The evolution can be divided roughly into two phases. On timescales of order the initial, central relaxation time, the velocity distribution becomes approximately isothermal, $f(\mathcal{E}) \sim e^{\mathcal{E}/\sigma^2}$. The corresponding configuration-space density is flat, within a core of some radius that is determined by the initial conditions.

One might expect this configuration to represent a steady state. But suppose that the NSC is sufficiently dense that it can be modeled as an isolated system: in other words, gravitational interactions with stars in the larger galaxy can be ignored. In this case, gravitational encounters will gradually populate the high-velocity tails of the Maxwellian distribution, and stars will escape from the cluster. The binding energy of the remaining stars will grow, and this energy must be shared among a shrinking population. The cluster will therefore contract, and its central density will reach a (formally) infinite value in a finite time, given roughly by 10^2 central relaxation times. This process is called **core collapse**.

On the other hand, if an NSC is not so compact, there will be an additional source of evolution due to gravitational encounters between stars in the galaxy, and stars in the nucleus. This interaction can lead to the opposite outcome: the NSC becomes less gravitationally bound, until its density equals that of the galaxy.

We will call this process **core expansion**. Any further evolution due to gravitational encounters would then take place on the much longer timescale determined by the central relaxation time of the galaxy.

Whether a given NSC will expand or contract depends on its degree of compactness with respect to the larger galaxy. It turns out that the critical compactness (a term that will be defined more carefully below) lies well within the range of parameters that are observed to characterize real NSCs, suggesting that both types of evolution have occurred [356]: in other words, some NSCs may have undergone core collapse, while others may have reached their current states starting from denser initial conditions.

In the era before SBHs came to be seen as ubiquitous components of galaxies, theoretical studies often emphasized core collapse as the dominant mechanism driving the evolution of galactic nuclei [505, 506, 92, 474]. In these models, the increase in central density led to a significant rate of *physical* collisions between stars; when random velocities exceeded $\sim 10^3$ km s^{-1}—higher than the escape velocity from the surfaces of stars, and also much higher than the velocities observed in any NSC—collisions would liberate gas that fell to the center of the galaxy and condensed into new stars, generating further collisions. It was argued that the evolution of a dense nucleus would lead inevitably to the formation of a massive black hole at the center, either by runaway stellar mergers or by creation of a massive gas cloud which would collapse [37].

These models seem less relevant today. At least in the more luminous galaxies, SBHs are believed to have been present, with roughly their current masses, since very early times. Evolution of galactic nuclei during and after the era of peak quasar activity took place with the SBHs already in place, and processes like core collapse and the buildup of massive stars via collisions could not have occurred after this time due to the inhibiting effects of the SBH's gravitational field.

At least in the current universe, the only systems for which significant evolution toward a collapsed core may have occurred are the densest NSCs. In the Milky Way, if we imagine removing the SBH, the central relaxation time would be ~ 10 Gyr, its value at the SBH influence radius. In NGC 205, which appears to lack an SBH, this time is ~ 0.1 Gyr. If we suppose that these values define the range of central relaxation times of NSCs lacking SBHs, then evolution may have taken place for as little as ~ 1, and as many as $\sim 10^2$, relaxation times; the latter value is just long enough for core collapse to have occurred. We will take the conservative view here that few if any nuclei reach a state of post-core-collapse evolution in which close encounters and physical collisions between stars are important processes.

7.5.1 $t \lesssim T_r$

As we have seen, near an SBH, a time of $\sim T_r(r_h)$ is required for the distribution of stars to reach a nearly steady state under the influence of gravitational encounters. In the absence of an SBH, an NSC can also be expected to reach an approximately steady state in a time of order the relaxation time. Encounters drive the local velocity distribution toward a Maxwellian, $f(\boldsymbol{v}) \propto e^{-v^2/(2\sigma^2)}$. Invoking Jeans's theorem, a locally Maxwellian velocity distribution implies

$$f = f(\mathcal{E}) = f_0 e^{\mathcal{E}/\sigma^2} = f_0 \exp\left[\frac{\psi(r) - v^2/2}{\sigma^2}\right]. \tag{7.70}$$

The corresponding mass density is given by

$$\rho(r) = 4\pi f_0 m \int_{-\infty}^{\psi(r)} [2(\psi - \mathcal{E})]^{1/2} e^{\mathcal{E}/\sigma^2} d\mathcal{E} \tag{7.71a}$$

$$= 2^{5/2} \pi f_0 m \sigma^3 e^{\psi/\sigma^2} \tag{7.71b}$$

$$= \rho(0) e^{(\psi - \psi(0))/\sigma^2} \tag{7.71c}$$

with $\psi(r)$ the (still unspecified) potential. Combining equation (7.71c) with Poisson's equation (3.27b) gives

$$\nabla^2 \psi + 4\pi G \rho(0) e^{(\psi - \psi(0))/\sigma^2} = 0. \tag{7.72}$$

The solution to this differential equation, with appropriate boundary conditions, is called the **isothermal sphere** [77]. One class of solutions has been mentioned already several times in this book: the "singular isothermal sphere," for which $\rho(r) \propto r^{-2}$. The class of solution most relevant here has inner development

$$\rho(r) = \rho(0) \left(1 - Ar^2 + \cdots\right), \tag{7.73a}$$

$$\psi(r) = \psi(0) - Br^2 + \cdots, \tag{7.73b}$$

a constant-density core. Substituting (7.73a) into (7.72) yields

$$\rho(r) = \rho(0) \left(1 - \frac{3}{2}\frac{r^2}{r_c^2} + \cdots\right), \tag{7.74a}$$

$$\psi(r) = \psi(0) - \frac{2}{3}\pi G \rho(0) r^2 + \cdots \tag{7.74b}$$

with r_c, the core radius, defined by[7]

$$G\rho(0) = \frac{9}{4\pi}\frac{\sigma^2}{r_c^2}. \tag{7.75}$$

At large radii, $r \gg r_c$, it is easy to verify that

$$\psi(r) \to -2\sigma^2 \ln\left(\frac{r}{r_0}\right), \tag{7.76a}$$

$$\rho(r) \to \frac{\sigma^2}{2\pi G r^2}, \quad r \gg r_c. \tag{7.76b}$$

At intermediate radii the solution must be found numerically [77]. In the case of the singular isothermal sphere, the relations (7.76) hold exactly at all radii.

Our concern here is with NSCs that are roughly one relaxation time old: long enough to have reached a collisional steady state near the center, though not necessarily at $r \gg r_c$. It is interesting to consider the manner in which that approximate

---

[7]The numerical factor in equation (7.74a) is chosen so that equation (7.75) matches a widely used definition of the core radius [286].

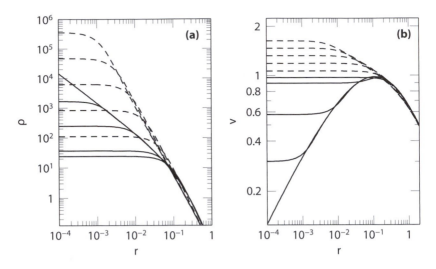

Figure 7.13 Evolution due to gravitational encounters of a nuclear star cluster lacking a central black hole [443]. The density and velocity dispersion are shown at various
times during the expansion (solid lines) and collapse (dashed lines) phases. The
initial conditions are given by the solid line with highest central value in the
case of ρ, and with the lowest central value in the case of σ. The unit of length
is approximately the half-mass radius.

steady state is reached. Figure 7.13a shows an example, based on integrations of
the orbit-averaged, isotropic Fokker–Planck equation [443]. The initial conditions
consisted of an NSC with a central density diverging as $\rho \sim r^{-1}$. Under the influence of gravitational encounters, the density near the center initially drops. This is
easy to understand. According to equation (3.39), the gravitational potential near
the center of a galaxy with $\rho(r) \propto r^{-\gamma}$ scales with radius as

$$\psi(r) \rightarrow \begin{cases} -r^{2-\gamma}, & \gamma \neq 2, \\ -\ln r, & \gamma = 2, \end{cases} \qquad (7.77)$$

and using the spherical Jeans equation (3.35), we find for the velocity dispersion
near the center,

$$\sigma^2(r) \rightarrow \begin{cases} r^{2-\gamma}, & 1 < \gamma < 3, \\ r^{\gamma}, & 0 \leq \gamma < 1. \end{cases} \qquad (7.78)$$

For values of γ less than or equal to 2, the velocity dispersion drops toward the
center, except in two special cases: $\gamma = 2$, the singular isothermal sphere; and $\gamma =
0$, which can be interpreted as describing a (nonsingular) isothermal sphere at $r \lesssim
r_c$. In all other cases, there is a "temperature inversion" near the center: the mean
square velocity rises outward, then (for reasonable models of the density profile at
large radii) reaches a peak before falling (figure 7.13b). If we think of the central

parts of the NSC as a "cool" subsystem, it is clear that gravitational encounters between these stars, and stars a little farther out, will tend to increase the mean square velocity of the inner stars (as described, for instance, by equation 7.27). In effect, heat flows into the center, causing it to expand.

This expansion continues until the temperature inversion has been erased, and the density near the center is approximately constant. At this time, the velocity distribution is well approximated by a Maxwellian, with constant σ, inside a radius r_c that is roughly equal to the radius of peak $\sigma(r)$ in the initial cluster. That radius can be large: comparable with the half-mass radius of the NSC. The time required to erase the temperature inversion is therefore comparable with the nuclear half-mass relaxation time; this is the same quantity that is plotted in figure 7.3a.

The evolution toward an isothermal core at times $t \lesssim T_r$ would not be very striking from an observational point of view, since it leaves observable quantities like the half-mass radius nearly unchanged. But it is important in that it determines the elapsed time required before the more interesting phases of evolution—for instance, core collapse—can get started. Figure 7.3 suggests that many NSCs will remain in this preliminary stage of evolution over their entire lifetimes.

7.5.2 $t \gg T_r$: Compact nuclear star clusters

By establishing a nearly isothermal core, the phase of nuclear expansion described in the previous section sets the initial conditions for the subsequent evolution. What happens next (assuming, of course, that the NSC exists for a time longer than a relaxation time) depends on how strongly the stars in the nucleus interact with the rest of the galaxy. We first consider the case in which those interactions can be ignored; the conditions under which this approximation is justified will be set out in detail in the next section, but the key requirement is that the NSC be much denser than the surrounding galaxy. This is the case, for instance, in the Milky Way.

Once its core is isothermal, an NSC can be expected to continue evolving toward the isothermal sphere solution described above, at a rate determined roughly by the local relaxation time at each radius. That is essentially correct; but something else happens too.

Beyond the core, the velocity dispersion is almost certain to decline with radius (e.g., figure 7.13b). This is because the $\rho \propto r^{-2}$ density profile implied by a constant velocity dispersion cannot extend arbitrarily far; that would imply a diverging total mass. Since the velocity dispersion falls off with radius, there will be a continuous transfer of heat from the inside to the outside. Roughly speaking, in each (local) relaxation time, the fraction of the stars in the tail of the Maxwellian velocity distribution that move faster than the local escape velocity (roughly 1%) will be lost [4]. This transfer of heat is in the opposite sense of the inward flow described in the previous section, and it has the opposite effect: the core must contract.

Up until now, we have largely ignored changes in the gravitational potential due to changes in the stellar distribution, based on the assumption that an SBH dominates the gravitational force; the only exception was our discussion, in chapter 3, of the adiabatic growth model, in which the SBH mass changes substantially over time. But in the absence of an SBH, changes in the gravitational potential due to a contracting core can become significant. To see this, we first need to derive an important property of self-gravitating systems.

The Jeans equation for an equilibrium spherical system is given by equation (3.31):

$$\frac{d\left(n\sigma_r^2\right)}{dr} + \frac{2n}{r}\left(\sigma_r^2 - \sigma_t^2\right) + n\frac{d\Phi}{dr} = 0, \qquad (7.79)$$

where σ_r and σ_t are the one-dimensional velocity dispersions, parallel and perpendicular, respectively, to the radius vector. Multiply both sides of this equation by $-4\pi r^3 m$ and integrate from $r = 0$ to infinity. Integrating the first term by parts, and assuming that $r^3\rho\sigma_r^2$ is zero both at very small and very large radii, yields

$$4\pi \int_0^\infty \rho\left(\sigma_r^2 + 2\sigma_t^2\right)r^2 dr - 4\pi \int_0^\infty \rho\frac{d\Phi}{dr}r^3 dr = 0 \qquad (7.80)$$

or

$$2K + W = 0, \qquad (7.81)$$

where K is the total kinetic energy and W is the total potential energy. This is a second form of the virial theorem that we first saw in chapter 3. (While we derived equation (7.81) under the assumption of spherical symmetry, it turns out to apply more generally [78].) Since the total energy is given by $E = K + W$, we can also write the virial theorem as

$$K + E = 0. \qquad (7.82)$$

The transfer of heat that occurs from the core of an NSC to its envelope corresponds to a loss of energy from the core. To the extent that we can consider the core to be a distinct stellar system, equation (7.82) tells us

$$\delta K = -\delta E. \qquad (7.83)$$

In other words, removal of energy from the core makes it hotter! The reason, of course, is that the system readjusts: its radius decreases (i.e., W becomes more negative) causing the stars to move, on average, faster. This property of self-gravitating systems is sometimes called a **negative specific heat**.

We seem to have established the conditions for a runaway: the core loses energy to the envelope; it contracts and becomes hotter; the rate of energy loss increases, etc. In such circumstances, we do not expect to find a steady-state solution. The most we might hope for is a **self-similar** solution: a solution in which the spatial distributions of the various properties (density, velocity dispersion, etc.), at different moments of time, can be obtained from one another by a similarity transformation—a rescaling of the axes. So, for instance, we might expect the density profile to achieve a fixed form that is close to the isothermal sphere profile, but the core radius and central density would change with time.

It is also clear that such self-similarity (if it exists) will not apply at all times and all radii; for instance, if the initial conditions are far from those of the isothermal sphere, self-similar behavior will require a finite time to be established. What we expect instead is what mathematicians call "intermediate-asymptotic" behavior:

self-similarity over some range in spatial and temporal scales [25]. In the case of core collapse, this means times long enough that the initial conditions have been "erased," but not so long that the core has shrunk to a size containing (say) just one star, or to a central density so high that physical collisions between stars would become important. Given that the runaway occurs near the center, we also do not expect self-similarity to apply at radii $r \gg r_c$.

Over such a limited range in time and radius, a self-similar solution would have the form

$$\rho(r, t) \approx \rho_c(t) \tilde{\rho}(\tilde{r}), \tag{7.84a}$$

$$\tilde{r} = r/r_c(t), \tag{7.84b}$$

where $\tilde{\rho}(\tilde{r})$ is a fixed function, and the time-dependent rescaling is determined by the functions $\rho_c(t)$ and $r_c(t)$. If we require that $\tilde{\rho}(0) = 1$, then ρ_c becomes the central density, and r_c can similarly be defined as the core radius.

Establishing self-similarity, and deriving the functional forms of $\tilde{\rho}(\tilde{r})$, $r_c(t)$ and $\rho_c(t)$, requires a numerical treatment. But one can make considerable progress based on simple physical arguments [230, 502]. In the self-similar regime, one expects

$$\xi \equiv T_{rc} \frac{1}{\rho_c} \frac{d\rho_c}{dt} \tag{7.85}$$

should be constant, where T_{rc} is the central relaxation time (itself a function of time). A similar relation will hold for other core parameters (r_c, σ_c), which implies that these parameters vary with each other as power laws. Writing $\rho_c \propto r_c^{-\alpha}$, substituting this relation into equation (7.84), and requiring that the time dependence vanish, we find

$$\rho(r) \propto r^{-\alpha} \tag{7.86}$$

which describes the (fixed) form of the density falloff at large radii. We also know from equation (7.75) that $\sigma_c^2 \propto r_c^{2-\alpha}$. Finally, if we assume that the time $t - t_{cc}$ remaining to a state of infinite central density is proportional to the current relaxation time,[8] then

$$\rho_c(t) = \rho_{c,0} \left(1 - t/t_{cc}\right)^\beta, \tag{7.87a}$$

$$r_c(t) = r_{c,0} \left(1 - t/t_{cc}\right)^\delta, \tag{7.87b}$$

with $\beta = -2\alpha/(6 - \alpha)$, $\delta = 2/(6 - \alpha)$.

There are various ways to proceed numerically; three important methods are summarized here:

1. **As an eigenvalue problem**. Mathematically, the assumption of self-similarity allows a set of partial differential equations that describe the time evolution of a system to be reduced to a set of ordinary differential equations that describe the time-independent, self-similar solution. In the case of core collapse, a starting point for this technique can be the orbit-averaged, isotropic Fokker–Planck equation.

---

[8]We ignore the dependence of the Coulomb logarithm on the core parameters.

We first need to generalize that equation slightly, to account for the fact that the gravitational potential is changing—like f, on a timescale long compared with orbital periods. This can be done [304, 235] by adding a term to the right-hand side of equation (7.3):

$$\frac{\partial f}{\partial t} = -\frac{1}{4\pi^2 p}\frac{\partial \mathcal{F}_\mathcal{E}}{\partial \mathcal{E}} - \frac{1}{p}\frac{\partial q}{\partial t}\frac{\partial f}{\partial \mathcal{E}}. \tag{7.88}$$

The first term on the right-hand side of equation (7.88) describes the diffusion of particles in phase space due to encounters. As a result of this diffusion, the density and potential slowly vary, so that the energy $\mathcal{E}$ associated with a given point in phase space changes with time. The second term compensates for this effect. Its physical meaning can be made clear by writing f as a function of the variable q, $f(\mathcal{E}, t) \equiv \zeta(q, t)$. Differentiating this function with respect to energy and time, and using the relation $p = -\partial q/\partial \mathcal{E}$,

$$\frac{\partial f}{\partial \mathcal{E}} = \frac{\partial \zeta}{\partial q}\frac{\partial q}{\partial \mathcal{E}} = -p\frac{\partial \zeta}{\partial q}, \tag{7.89a}$$

$$\frac{\partial f}{\partial t} = \frac{\partial \zeta}{\partial q}\frac{\partial q}{\partial t} + \frac{\partial \zeta}{\partial t}. \tag{7.89b}$$

Substituting these expressions into equation (7.88) and ignoring for the moment the energy diffusion term,

$$\frac{\partial \zeta}{\partial q}\frac{\partial q}{\partial t} + \frac{\partial \zeta}{\partial t} = \frac{\partial q}{\partial t}\frac{\partial \zeta}{\partial q}. \tag{7.90}$$

Equation (7.90) implies $\partial \zeta/\partial t = 0$, i.e., that the potential readjusts in such a way as to maintain f a fixed function of q. But q can be written as an integral over angular momentum of the radial action, $J_r \equiv \int_{r_-}^{r_+} v_r\, dr$:

$$q(\mathcal{E}) = \int_0^{L_c^2(\mathcal{E})} dL^2 J_r(\mathcal{E}, L) = \frac{4}{3}\int_0^{\psi^{-1}(\mathcal{E})} dr\, r^2 v^3(r, \mathcal{E}). \tag{7.91}$$

Forcing f to be a fixed function of q while the potential varies is therefore equivalent to assuming that the potential changes slowly enough, compared with orbital periods, that the radial adiabatic invariant J_r is conserved for each orbit.

With this addition, we are ready to convert the time-dependent Fokker–Planck equation into a set of time-independent equations by assuming self-similarity, this time for the functions $f(\mathcal{E})$ and $q(\mathcal{E})$:

$$f(\mathcal{E}, t) = f_c(t)\tilde{f}(\tilde{\mathcal{E}}), \tag{7.92a}$$

$$q(\mathcal{E}, t) = q_c(t)\tilde{q}(\tilde{\mathcal{E}}), \tag{7.92b}$$

$$\tilde{\mathcal{E}} = \mathcal{E}/\mathcal{E}_c(t). \tag{7.92c}$$

We can also write

$$\psi(r, t) = \psi_c(t)\tilde{\psi}(\tilde{r}), \tag{7.93a}$$

$$\tilde{r} = r/r_c(t). \tag{7.93b}$$

These expressions are then substituted into equations (7.88) and (7.91), and into Poisson's equation (3.27b), in the form

$$\frac{\partial^2 \psi}{\partial r^2} + \frac{2}{r}\frac{\partial \psi}{\partial r} = -16\pi^2 Gm \int_0^{\psi(r,t)} f(\mathcal{E})\,[2\,(\psi - \mathcal{E})]^{1/2}\,d\mathcal{E}. \qquad (7.94)$$

The result [235, 233] is a set of ordinary differential equations for $\tilde{f}$, $\tilde{q}$, and $\tilde{\psi}$, which depend on the constants

$$C_1 = \frac{1}{\Gamma f_c}\frac{d}{dt}\ln f_c, \quad C_2 = \frac{1}{\Gamma f_c}\frac{d}{dt}\ln \mathcal{E}_c. \qquad (7.95)$$

It is straightforward to show that the assumption of a power-law dependence of ρ on r at large radii, $\rho \propto r^{-\alpha}$, implies

$$\frac{C_1}{C_2} = \frac{6-\alpha}{2(\alpha-2)}, \qquad (7.96)$$

and the dimensionless collapse rate defined in equation (7.85) is

$$\xi = \frac{C_1 + \frac{3}{2}C_2}{0.167\pi^{1/2}}. \qquad (7.97)$$

Given appropriate boundary conditions, the set of differential equations can be solved, with the constants C_1, C_2 appearing as eigenvalues. The results are [233]

$$\alpha = 2.23, \quad C_1 = 9.1 \times 10^{-4}, \quad C_2 = 1.1 \times 10^{-4}, \quad \xi = 3.64 \times 10^{-3}. \qquad (7.98)$$

The numerical solution to $\tilde{\rho}(\tilde{r})$ is plotted as the filled circles in figure 7.14. The density falloff, $\rho \sim r^{-2.23}$, is seen to be slightly steeper than "isothermal," as expected. The computed value of ξ implies that, in the asymptotic limit, the time remaining until complete core collapse is

$$t - t_{cc} = \frac{2\alpha}{6-\alpha}\xi^{-1}T_{rc} \approx 330\,T_{rc}. \qquad (7.99)$$

2. **Via coarse dynamic renormalization.** The orbit-averaged Fokker–Planck equation is an approximation; some of the reasons for distrusting it were outlined in chapter 5. A fully general alternative is direct integration of the N-body equations of motion. That is the method of choice when N is small [1]. But values of N that are computationally tractable ($N \lesssim 10^6$) are smaller than the number of stars in a nucleus. During core collapse, the number of stars in the core decreases as collapse proceeds, and two-body and higher-order correlations, which are ignored in the Fokker–Planck equation, begin to dominate the evolution. In N-body simulations, this occurs at a time far earlier than would be the case in a real nucleus, and in fact no published N-body simulation has followed core collapse beyond a density contrast of $\sim 10^4$ (roughly where self-similar behavior first appears) before binary stars begin to lock up most of the gravitational energy.

A general framework exists for establishing self-similarity in problems like core collapse which can avoid these problems [276]. One starts from an algorithm for solving the exact, "microscopic" physics; in our case, an N-body code. A particular coarse-grained function—in our case, $f(\mathcal{E})$ or $\rho(r)$—is identified as the "macroscopic" quantity of interest. One then carries out short bursts of computation using

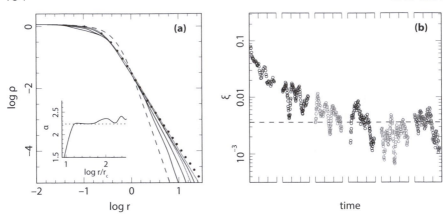

Figure 7.14 Core collapse via coarse-grained renormalization [510]. (a) Evolution of the rescaled density. The dashed line is the initial state and the points are the self-similar $\tilde{\rho}(r)$ derived by Heggie and Stevenson [233] from the Fokker–Planck equation. The inset shows $\alpha \equiv -d \log \rho / (d \log r)$ at the end of the final integration; the dashed line is $\alpha = 2.23$. (b) Evolution of the dimensionless collapse rate parameter $\xi \equiv (\dot{\rho}_c/\rho_c)/T_{rc}$. Each time interval corresponds to one "burst" of integration, after which the model was rescaled. The dashed line shows the asymptotic (self-similar) value $\xi = 3.6 \times 10^{-3}$ as computed via the Fokker–Planck equation.

the fine-scale model and extracts a smooth approximation of $\rho(r)$ at the end of the burst. The integration interval is chosen to be long enough that substantial evolution of the coarse-grained function occurs, but short enough that correlations due to the finite N are insignificant. The coarse-grained function, for example, $\rho(r)$, is then rescaled—shifted on a log–log plot—in such a way that the core properties are left unchanged. A new N-body model is then generated from the renormalized density, and another burst of integration is carried out; any correlations that do arise are eliminated during this step. By following this procedure, the number of particles in the core remains essentially constant as the core shrinks; in effect, the region being simulated shrinks along with the macroscopic observable (core radius). If self-similarity exists, the rescaling will be found, after a few interactions, to leave the macroscopic function unchanged. This method is able to establish self-similarity even if the equations describing the evolution of the macroscopic functions are unknown, or (as in our case) if they are approximations of unknown validity.

Figure 7.14 illustrates an application of this method to the core collapse problem [510]. The total number of particles was $N = 1.6 \times 10^4$, modest by N-body standards. Each integration interval was long enough for the central density to increase by roughly a factor of 5. After four or five such intervals, the renormalized density had attained a time-independent form, with $\rho \sim r^{-2.23}$ at $r \gg r_c$ (figure 7.14a). The other parameters that characterize the self-similar behavior are also derivable. For instance, if t_1 and t_2 are two distinct times in the self-similar regime, the exponent

β of equation (7.87) can be obtained as

$$\beta = (t_2 - t_1) \bigg/ \left(\frac{\rho_c(t_2)}{(d\rho_c/dt)|_{t_2}} - \frac{\rho_c(t_1)}{(d\rho_c/dt)|_{t_1}} \right) \qquad (7.100)$$

and similarly for δ. Figure 7.14b shows that the dimensionless collapse rate parameter ξ obtained by this method is consistent with the value obtained from the Fokker–Planck equation.

3. **As a solution to the time-dependent Fokker–Planck equation.** Integration of the time-dependent evolution equation for f provides another route to establishing self-similarity. A time-dependent solution also contains additional useful information: for instance, it can reveal over what range of spatial and temporal scales the self-similar solution is valid, and how the system behaves outside the self-similar regime. This is especially important in the context of NSCs, since in many cases, gravitational encounters may not have had time to reach the self-similar regime.

Integration of equations like (7.3) or (5.168) over space and time is in principle straightforward, except that questions of energy and number conservation become critical if the evolution is to be accurately followed far into the self-similar regime. Changes in the gravitational potential can be dealt with in one of two ways. If the independent variables are taken to be $\mathcal{E}$ and t, say, then changes in the potential require a readjustment of $f(\mathcal{E})$ so that f remains a fixed function of the q, as discussed above. This can be done via an iterative procedure at each time step [89, 90]. On the other hand, it is possible to recast the evolutionary equations in terms of the radial adiabatic invariant J_r rather than $\mathcal{E}$ [194]. This eliminates the need for iteration, and it also simplifies some computational tasks, for example, the orbit averaging; the cost is the extra effort required in converting from action-space variables to Cartesian coordinates.

7.5.3 $t \gg T_r$: Diffuse nuclear star clusters

So far, we have been discussing the evolution of NSCs without regard to the fact that they sit inside a larger galaxy. This idealization seems intuitively justified if the nucleus is sufficiently compact. But if the density of an NSC is low enough compared with that of the galaxy, the effects of encounters with stars belonging to the galaxy cannot be ignored [122, 269].

Consider a two-component system consisting of a galaxy and an NSC. Assume for simplicity that both components are homogeneous, with densities (ρ_{gal}, ρ_{nuc}), half-mass radii (r_{gal}, r_{nuc}), rms velocities (V_{gal}, V_{nuc}), and half-mass relaxation times (T_{gal}, T_{nuc}).

Define $\epsilon_{nuc} \equiv (1/2)\rho_{nuc} V_{nuc}^2$ to be the kinetic energy per unit volume of the nucleus. Assuming Maxwellian velocity distributions and a single stellar mass m, the rate of change of ϵ_{nuc} due to gravitational encounters is given by an equation similar to equation (7.27):

$$\frac{d\epsilon_{nuc}}{dt} = \frac{4\sqrt{6\pi}\, G^2 m \rho_{nuc} \rho_{gal} \ln \Lambda}{\left(V_{gal}^2 + V_{nuc}^2 \right)^{3/2}} \left(V_{gal}^2 - V_{nuc}^2 \right). \qquad (7.101)$$

A necessary and sufficient condition for the nucleus to be heated by the galaxy is $V_{\text{gal}} > V_{\text{nuc}}$. When this condition is satisfied, the nuclear heating time is

$$T_{\text{heat}} \equiv \left| \frac{1}{\epsilon_{\text{nuc}}} \frac{d\epsilon_{\text{nuc}}}{dt} \right|^{-1} \tag{7.102a}$$

$$= \frac{1}{48} \sqrt{\frac{6}{\pi}} \frac{V_{\text{nuc}}^3}{G^2 m \rho_{\text{nuc}} \ln \Lambda} \left(\frac{\rho_{\text{nuc}}}{\rho_{\text{gal}}} \right) \frac{\left((V_{\text{gal}}^2/V_{\text{nuc}}^2) + 1 \right)^{3/2}}{(V_{\text{gal}}^2/V_{\text{nuc}}^2) - 1}. \tag{7.102b}$$

In the limiting case $V_{\text{gal}} \gg V_{\text{nuc}}$ this becomes

$$T_{\text{heat}} = \left(\frac{\rho_{\text{nuc}}}{\rho_{\text{gal}}} \right) \left(\frac{V_{\text{gal}}}{V_{\text{nuc}}} \right) \frac{1}{48} \sqrt{\frac{6}{\pi}} \frac{V_{\text{nuc}}^3}{G^2 m \rho_{\text{nuc}} \ln \Lambda} \tag{7.103a}$$

$$\approx \left(\frac{\rho_{\text{nuc}}}{\rho_{\text{gal}}} \right) \left(\frac{V_{\text{gal}}}{V_{\text{nuc}}} \right) T_{\text{nuc}} \approx \left(\frac{V_{\text{gal}}}{V_{\text{nuc}}} \right)^2 T_{\text{gal}} \tag{7.103b}$$

$$\approx \left(\frac{\rho_{\text{nuc}}}{\rho_{\text{gal}}} \right)^{1/2} \left(\frac{V_{\text{nuc}}}{V_{\text{gal}}} \right)^{1/2} \left(T_{\text{nuc}} T_{\text{gal}} \right)^{1/2}. \tag{7.103c}$$

If $V_{\text{gal}} > V_{\text{nuc}}$, as assumed, then ρ_{nuc} cannot be large compared with ρ_{gal} (figure 7.15). Thus, the nuclear heating time is of the same order as, or somewhat less than, the geometric mean of T_{nuc} and T_{gal}. According to figure 7.3, this time is shorter than 10 Gyr in at least some galaxies.

Heating from the galaxy will reverse core collapse if T_{heat} is shorter than the nuclear core collapse time. One definition of the latter time is

$$T_{\text{cc}} \equiv \left| \frac{1}{\rho_{\text{nuc}}} \frac{d\rho_{\text{nuc}}}{dt} \right|^{-1} = \xi^{-1} T_{\text{nuc}}, \tag{7.104}$$

where ξ^{-1} varies from ~ 10 in the early stages of core collapse to an asymptotic value of ~ 300, as discussed above. The condition $T_{\text{heat}} < T_{\text{cc}}$ becomes

$$\frac{\rho_{\text{nuc}}}{\rho_{\text{gal}}} \frac{V_{\text{gal}}}{V_{\text{nuc}}} < \xi^{-1}, \tag{7.105}$$

where $V_{\text{gal}} \gg V_{\text{nuc}}$ has again been assumed. We can convert equation (7.105) into a relation between the quantities plotted in figure 7.3 by writing $\rho_{\text{nuc}} \approx M_{\text{nuc}}/r_{\text{nuc}}^3$ and by applying the virial theorem separately to both components, that is,

$$V_{\text{nuc}}^2 \sim \frac{G M_{\text{nuc}}}{r_{\text{nuc}}}, \quad V_{\text{gal}}^2 \sim \frac{G M_{\text{gal}}}{r_{\text{gal}}}; \tag{7.106}$$

the former expression will only be approximately true for low-density nuclei. With these substitutions, the condition (7.105) becomes

$$\frac{M_{\text{nuc}}}{M_{\text{gal}}} \lesssim 10^4 \left(\frac{\xi^{-1}}{100} \right)^2 \left(\frac{r_{\text{nuc}}}{r_{\text{gal}}} \right)^5. \tag{7.107}$$

A typical ratio of NSC mass to galaxy mass is $M_{\text{nuc}}/M_{\text{gal}} \approx 0.003$ [157, 557]. Using this value, equation (7.107) implies that $r_{\text{nuc}}/r_{\text{gal}}$ must be smaller than ~ 0.05

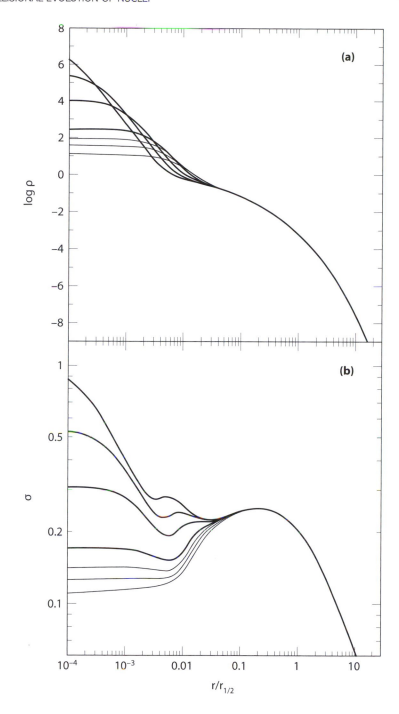

Figure 7.15 Models of galaxies with NSCs of various degrees of compactness [356]. Each of these models has $M_{nuc}/M_{gal} = 0.003$ but they differ in their degree of

Figure 7.15 Continued. nuclear concentration, from $r_{nuc}/r_{gal} \approx 0.0002$ to ~ 0.03. The top curves show the density while the bottom curves show the velocity dispersion. When $r_{nuc}/r_{gal} \gtrsim 0.003$, the central σ is lower than the peak value in the outer galaxy. "Temperature inversions" like these imply a flow of heat from galaxy to nucleus, which tends to counteract the outward flow that would otherwise drive the nucleus toward core collapse. Thick lines denote models in which the NSC undergoes "prompt" core collapse, while thin lines are models in which heat transfer from the galaxy causes the nucleus to expand initially, as shown in figure 7.16.

in order for core collapse to occur. Figure 7.3 suggests that some NSCs satisfy this condition.

A more accurate estimate of the critical degree of compactness can be obtained by integrating the time-dependent, isotropic Fokker–Planck equation. Figure 7.16 shows the results, starting from the initial conditions plotted in figure 7.15. As predicted, there is a critical value of r_{nuc}/r_{gal} above which evolution toward core collapse is halted and the nucleus expands. The initial evolution of these diffuse nuclei can be understood using the arguments in the previous section, except that now, the temperature inversion exists on a larger spatial scale: between the NSC as a whole, and the galaxy. Core collapse still occurs in these models but on a much longer timescale; reversing the gradient requires the creation of a large, flat core extending to roughly the half-mass radius of the galaxy, and the time required is roughly ten times the galaxy half-mass relaxation time—far longer than galaxy lifetimes; see figure 7.3. In the models with denser NSCs that contract, core collapse occurs in 15–20 times the initial nuclear half-mass relaxation time.

A large number of experiments like the ones illustrated in figure 7.16, starting from different models of the NSC and the galaxy, reveals that the critical compactness separating models that exhibit "prompt" core collapse (of the NSC) from models that undergo core collapse much later is given by a relation similar to equation (7.107):

$$\frac{M_{nuc}}{M_{gal}} = A \left(\frac{r_{nuc}}{r_{gal}} \right)^B \tag{7.108}$$

with $A \approx 350$ and $B \approx 2.5$. The exact values of A and B depend on the functional forms adopted for the density profiles of the NSC and galaxy [356]. Interestingly, figure 7.3 shows that observed NSCs almost all lie in the "prompt" core collapse regime; the only clear exceptions are nuclei with such long relaxation times ($\gtrsim 10\,\text{Gyr}$) that very little evolution would have occurred since their formation.

7.5.4 Rotating nuclei

As discussed in chapter 2, NSCs that are near enough for their internal structure and kinematics to be resolved are sometimes observed to be flattened and rotating; perhaps a consequence of star formation that took place in a gaseous disk, or perhaps an indication that the nucleus formed from the inspiral of star clusters. It is

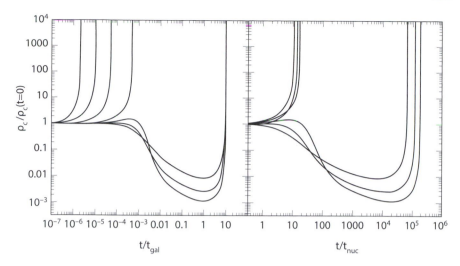

Figure 7.16 Long-term evolution of the galaxy models plotted in figure 7.15 [356]. The vertical axis is the core density normalized to its initial value. The horizontal axis is the time in units of the initial, half-mass relaxation time defined by the galaxy (left) or the NSC (right). The four models that undergo prompt core collapse are plotted with thick lines in figure 7.15.

interesting to ask how the shape and angular-momentum distribution of a such a nucleus would evolve under the influence of gravitational encounters.

Flattened nuclei are difficult to treat theoretically, for reasons that were discussed in chapters 3 and 5. Even assuming a gravitational potential with rotational symmetry, the only known integrals of motion are the energy E and the component L_z of the angular momentum parallel to the symmetry axis. In triaxial nuclei, the only known integral is typically E. Numerical integration of the equations of motion in axisymmetric potentials reveals that most orbits respect an additional isolating integral, I_3, but analytic expressions for $I_3(x, v)$ are almost never available. The techniques developed in the latter half of chapter 5 are essentially useless unless the functional forms of all the integrals are known. As a consequence, the few existing studies of relaxation in flattened systems have generally restricted f to the form $f = f(E, L_z)$; in other words, an axisymmetric nucleus in which the phase-space density is assumed, without real physical justification, to be constant with respect to the unknown I_3.

The equilibrium properties of these "two-integral" axisymmetric models were discussed in section 3.4.1. Such models are "isotropic" in a limited sense: the velocity dispersions σ_ϖ and σ_z in the meridional plane are equal. Writing $\sigma_\varpi = \sigma_z \equiv \sigma$, the dependence of σ on ϖ and z is determined uniquely by $n(\varpi, z)$ and $\Phi(\varpi, z)$ via the Jeans equation (3.119a). A similar relation, equation (3.119b), gives $\overline{v_\varphi^2}(\varpi, z)$ in terms of n and Φ. The only remaining freedom lies in how the φ motions are partitioned between streaming, that is, rotation, and dispersion: $\overline{v_\varphi^2} = \overline{v_\varphi}^2 + \sigma_\varphi^2$. Depending on how that partitioning is carried out, the velocity distribution in a locally "corotating" frame (a frame moving with the local, mean velocity) can be isotropic ($\sigma_\varphi = \sigma$) or anisotropic ($\sigma_\varphi \neq \sigma$). Another way to state this is in terms of f: the degree of streaming about the symmetry axis is determined by the odd part of f, $f_-(E, L_z) = (1/2)\left[f(E, L_z) - f(E, -L_z)\right]$, while the even part of f is fixed by n and Φ.

A little thought shows that the degree of rotation of a two-integral model need bear no relation to its shape. Even precisely spherical models can be made to "rotate" by selectively changing the sign of L for some fraction of the orbits.[9] But intuition suggests that the same dynamical processes that induce rotation in a stellar system will also cause it to be flattened. For instance, in the famous Maclaurin series of incompressible spheroids, the elongation of the fluid surface is expressible simply and uniquely in terms of the angular rotation rate. This is a consequence of the fact that the distribution of random velocities in an incompressible fluid is isotropic and independent of position. Stellar systems are certainly not homogeneous, but to the extent that their velocity ellipsoids are spherical, they should obey roughly similar relations between rotation and flattening.

Approximating a nucleus as an isolated system, with total mass M, energy E and angular momentum L,[10] we can define a dimensionless measure of its degree of rotation:

$$\lambda \equiv \frac{L\,|E|^{1/2}}{G\,M^{5/2}}. \tag{7.109}$$

In a self-gravitating system, λ is the *only* dimensionless combination of conserved quantities that is proportional to L. Let Ω be an average value of the angular velocity in the nucleus. Then, in an obvious notation, $\Omega \sim V/R$; and setting $E \sim G M^2/R$ via the virial theorem, we can write λ as

$$\lambda \approx \frac{\Omega}{\sqrt{G M/R^3}} \approx \frac{\Omega}{\sqrt{G\rho}} \tag{7.110}$$

with ρ a mean nuclear density. Since λ measures the dynamical importance of rotation, the nucleus should get flatter if λ increases and rounder if it decreases. A *constant* λ would imply

$$\Omega(t) \propto \rho(t)^{1/2}. \tag{7.111}$$

If Ω should be found to increase more steeply with time than $\rho^{1/2}$ during core collapse, this would correspond to the nucleus being "spun up." Such would be the case, for instance, if the total mass and angular momentum of the nucleus were conserved, since this would imply

$$L \propto MVR \propto M\Omega R^2 = \text{constant} \rightarrow$$
$$\Omega \propto R^{-2} \propto (M/R^3)^{2/3} \propto \rho^{2/3}.$$

But a constant L is unlikely, since we know that angular momentum in a differentially rotating system is redistributed by the viscosity. For instance, in a

---

[9]Spherical, rotating models were discussed in section 5.7.

[10]Note that E and L have the dimensions of energy and angular momentum, not specific energy and specific angular momentum.

self-gravitating fluid body, the **Navier–Stokes equations** state that

$$\rho \frac{Du}{Dt} = -\nabla P - \rho \nabla \Phi + \rho \nu \left(\nabla^2 u + \frac{1}{3} \nabla (\nabla \cdot u) \right), \tag{7.112}$$

where u is the mean velocity, P is the pressure, and ν is the kinematic viscosity. The viscosity has dimensions (length)^2/time. In a stellar system, dimensional analysis suggests that the quantity that plays the role of viscosity is

$$\nu \sim K_\nu \frac{(\sigma/\sqrt{G\rho})^2}{T_r} \tag{7.113}$$

with K_ν a dimensionless constant. The numerator of equation (7.113) is the square of the typical distance traversed by the star in its orbit, and the denominator is the time over which it exchanges energy with other stars.

Using equations (7.112) and (7.113), it is possible to search for self-similar solutions describing the collapse of a rotating nucleus. To equations (7.84) are added

$$v_r(r, t) = v_c(t) \, \tilde{v} \, (\tilde{r}), \tag{7.114a}$$

$$v(r, t) = v_c(t) \, \tilde{v} \, (\tilde{r}), \tag{7.114b}$$

$$\Omega(r, t) = \Omega_c(t) \, \tilde{\Omega} \, (\tilde{r}), \tag{7.114c}$$

where v_r is the radial component of the mean stellar velocity and Ω specifies the mean velocity via $\bar{v} = \Omega \varpi e_\varphi$. The result is [205]

$$\Omega_c \propto \rho^\delta, \quad 0.10 \lesssim \delta \lesssim 0.15. \tag{7.115}$$

The uncertainty in the exponent δ is due largely to uncertainties in the value of K_ν. Equation (7.115) states that the rate of rotation near the center increases with time during the late stages of core collapse, but it does so more gradually than would be needed to maintain a fixed ratio of rotational to gravitational energy (i.e., $\Omega_c \propto \rho_c^{1/2}$). The nucleus "spins down" as it collapses, becoming less flattened with time, due to the transfer of angular momentum outward.

Figure 7.17 shows results from a set of numerical integrations of the orbit-averaged Fokker–Planck equation for $f(E, L_z, t)$, equation (5.195), starting from flattened and rotating initial conditions [132]. As in most numerical treatments of the orbit-averaged equations, computation of the diffusion coefficients was simplified by assigning a simplified form to the field-star velocity distribution $f(v_f)$. In the study of figure 7.17, $f(v_f)$ was assumed to have the form

$$f(v_f) = \frac{\rho}{(2\pi\sigma^2)^{3/2}} \exp \left[-\frac{(v_f - \Omega \varpi e_\varphi)^2}{2\sigma^2} \right], \tag{7.116}$$

a "rotating Maxwellian"; the parameters $\{\rho, \Omega, \sigma\}$ were estimated at each time step by taking the appropriate moments over $f(v)$. With this ansatz, the diffusion coefficients are relatively straightforward to calculate using equations (5.198) from section 5.5.3; the reader will recall that the diffusion coefficients in that section were derived assuming a locally isotropic velocity distribution, as in equation (7.116). Initial conditions were taken to be

$$f_0(E, L_z) = f_0 \left(e^{-\beta E} - 1 \right) e^{\beta \Omega_0 L_z}. \tag{7.117}$$

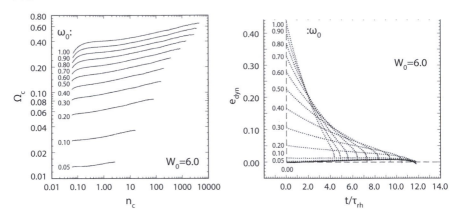

Figure 7.17 Results from the numerical integration of the orbit-averaged Fokker–Planck equation describing a rotating stellar system [132]. The distribution function was assumed to have the form $f(E, L_z; t)$, a "two-integral" model. Initial conditions were given by equation (7.117); the initial degree of rotation, or flattening, was determined by the dimensionless rotation parameter ω_0, where $\omega_0 = 0$ corresponds to a spherical, nonrotating model and $\omega_0 = 1$ to a maximally rotating model. The left panel shows the evolution of the central angular velocity, Ω_c, in terms of the central density n_c. The rotation rate increases as the core collapses, but much more slowly than would be required to maintain a fixed ratio of rotational kinetic energy to gravitational potential energy. The right panel shows the evolution of the mean shape of the model, defined in terms of a "dynamical ellipticity" e_{dyn}. The time is expressed in units of τ_{rh}, the initial, half-mass relaxation time; the curves terminate when the central density goes to infinity. The initially flattened models become rounder, in a typical time of $\sim\tau_{\mathrm{rh}}$. The thick curve near the bottom shows the evolution of the initially spherical model; due to the neglect of the third integral, this model's shape exhibits a slight, spurious evolution.

The parameters f_0 and β in this expression are scaling factors; the degree of initial rotation is fixed by the dimensionless angular velocity $\omega_0 = \sqrt{9/(4\pi G\rho_c)}\,\Omega_0$, which can vary between 0 (no rotation) and 1.

At late times, figure 7.17 shows that the central angular velocity Ω_c increases roughly as a power of the central density:

$$\frac{d\log\Omega_c}{d\log n_c} \approx \delta, \quad 0.06 \lesssim \delta \lesssim 0.08. \tag{7.118}$$

Consistent with the predictions of the more approximate, fluid dynamical model, the rate of rotation increases as the core collapses, but more slowly than the dependence ($\Omega_c \propto \rho^{1/2}$) that would be required to maintain a fixed degree of rotational support, and much more slowly than would be predicted if L were constant ($\Omega_c \propto \rho^{2/3}$). As a consequence, the central regions become rounder. The right panel of figure 7.17 plots the evolution of a global shape parameter, ϵ_{dyn}, defined as the ellipticity[11] of an oblate spheroid having the same partitioning of kinetic energy between mean and random motions as the numerical model. (For a Maclaurin spheroid, the relation between this shape parameter, and the parameter λ defined above, is $\lambda^2 \approx (24/625)e_{\text{dyn}}$ [205].) All of the models become essentially spherical after a few, half-mass relaxation times; in other words, in the time it takes the core to collapse.

Results like those of figure 7.17 have most often been applied to globular clusters. Globular clusters associated with the Milky Way are generically "old" (compared with their half-mass relaxation times), slowly rotating, and very round; few have ellipticities greater than about 0.2. On the other hand, globular clusters associated with the Large Magellanic Cloud, a dwarf companion galaxy to the Milky Way, exhibit a range of ages and flattenings: some appear to be as young as $\sim 10^7$ yr and there is a good correlation between shape and age, in the sense that younger clusters are more elongated. These observed regularities are well explained in terms of dynamical evolution [155, 326].

What about NSCs? The empirical trends shown in figure 7.3a suggest that the faintest NSCs might have relaxation times short enough for core collapse to have run to completion, implying that they could have evolved by now into nearly spherical, slowly rotating configurations. A complicating factor in the case of NSCs is the likelihood of ongoing star formation, from gas that accumulates at the bottom of the galaxy potential well; roughly speaking, the predictions made in this section would apply only to the oldest stellar population, and it is difficult to extract information about the old stars if there is also a population of bright young stars. The NSCs with the shortest relaxation times also tend to be the smallest (figure 7.3b) and therefore the most difficult to resolve. For these reasons, little evidence can currently be drawn from the observations about the influence of rotation on the dynamical evolution of NSCs.

---

[11]Ellipticity is defined as $\epsilon = 1 - b/a$ with b (a) the short (long) axis of the ellipse. It is related to eccentricity, e, via $\epsilon = 1 - \sqrt{1 - e^2}$.

Chapter Eight

Binary and Multiple Supermassive Black Holes

According to the currently accepted paradigm, galaxies grow through the agglomeration of smaller galaxies and protogalactic fragments—through **galaxy mergers**. If galaxies were no larger than implied by the sizes of their luminous components, mergers would be extremely rare. But many galaxies appear to be embedded in much larger systems: dark-matter halos that extend tens or hundreds of times farther than the stars or gas. According to large-scale simulations of the clustering of dark matter in the universe, the mean time between "major mergers"—mergers with mass ratios 3 : 1 or less[1]—varies from $\sim 0.2\,\mathrm{Gyr}$ at a redshift $z = 10$, to $\sim 10\,\mathrm{Gyr}$ at $z = 1$, with a weak dependence on halo mass [154]. These simulations do not contain baryonic matter; but mergers between halo-sized objects would be guaranteed to bring the central, luminous components together in a time comparable to the time required for the halos to merge, and this has been verified via detailed merger simulations of galaxies embedded in dark halos [28].

By the same reasoning, if the merging galaxies each contains a central supermassive black hole (SBH), the two SBHs will form a bound system in the merged galaxy—a **binary supermassive black hole**—shortly after the merger is complete [36, 463]. This idea has received considerable attention because the ultimate coalescence of such a binary would generate an observable outburst of gravitational waves [514]. Furthermore, to the extent that galaxies grow to their current sizes via mergers, so must their SBHs. The tight correlations found between SBH masses and the properties (luminosity, velocity dispersion) of their host galaxies could hardly be maintained otherwise.

The evolution of a binary SBH can be divided into three phases:

1. As the galaxies merge, the SBHs sink toward the center of the new galaxy via dynamical friction where they form a binary.

2. The binary interacts with nearby stars, ejecting them at velocities comparable to the binary's orbital velocity. The binary's binding energy increases as a result.

3. If the binary's separation decreases to the point where the emission of gravitational waves becomes efficient at carrying away the last remaining angular momentum, the SBHs coalesce.

---

[1] So defined, since early simulations suggested that this was the largest mass ratio capable of converting two disk galaxies into an elliptical galaxy.

The transition from (2) to (3) has long been seen as a potential bottleneck. At least in a spherical galaxy, the number of stars on orbits that intersect the massive binary is fairly small, and once these stars have been ejected from the galaxy's core, it is not clear that their orbits would be repopulated in a time shorter than the age of the universe. This has been called **the final-parsec problem** [387]; the name derives from the fact that the separation of a massive binary when it first forms at the center of a galaxy is roughly one parsec. Whether the massive binary continues to shrink beyond this separation is unclear, and probably depends in large measure on the details of its environment. One possibility is that the "loss cone" of orbits that intersect the binary is refilled by one of the mechanisms discussed in chapter 6: gravitational encounters between stars, torquing by a nonspherical galactic potential, massive perturbers, etc. For instance, a galaxy that formed via a major merger is likely to be nonspherical, even nonaxisymmetric, implying the existence of centrophilic orbits, like the saucers and pyramids. The amount of mass on such orbits can greatly exceed the mass of the central binary, and as shown in chapter 6, centrophilic orbits can maintain a "full loss cone" even in galaxies where timescales for collisional loss-cone repopulation are very long. There is some support for this idea in N-body merger simulations [277]. Interstellar gas could also play an important role in the dynamical evolution of binary SBHs. Any gas located close to a massive binary will be disturbed by the SBHs and exert gravitational torque on them, thereby affecting their orbit. Furthermore, if SBH coalescence is accompanied by the presence of gas, an observable electromagnetic "afterglow" might accompany the coalescence [248].

Of course, the final-parsec problem is a "problem" only from the point of view of those who would like to observe the final coalescence; there is probably an equally large community of scientists who would be happy to observe *un*coalesced binaries! As discussed in chapter 2, a handful of uncoalesced, binary SBHs have probably been observed, and there may be a great many more that have gone unnoticed. But there is circumstantial evidence that efficient coalescence is the norm. Jets in the great majority of radio galaxies do not show the wiggles expected if the SBH hosting the accretion disk were orbiting or precessing. The X-shaped radio sources [118] are probably galaxies in which SBHs have recently coalesced, causing jet directions to flip. The inferred production rate of the X-sources is comparable to the expected merger rate of bright ellipticals, suggesting that coalescence occurs relatively quickly following mergers [361]. If binary SBHs failed to merge efficiently, uncoalesced binaries would be present in many bright ellipticals, resulting in 3- or 4-body slingshot ejections when subsequent mergers brought in additional SBHs. This would produce off-center SBHs, which seem to be rare, as well as (perhaps) too much scatter in the $M_\bullet$–σ and $M_\bullet$–L relations [224]. Furthermore, the total mass density in SBHs in the local universe is consistent with that inferred from high-redshift AGN [362, 577], implying that only a modest fraction of SBHs could have been ejected from galaxies in the intervening period.

Whether or not a massive binary manages to coalesce, it will leave behind an imprint: a **mass deficit**—a lowered density of stars near the center of the galaxy. The displacement of matter takes place relatively quickly, as the separation between the two SBHs drops from $\Delta r \approx r_h$ to roughly one tenth this distance—the separation

at which slingshot ejections become efficient. Simulations suggest that this phase
of the evolution is unavoidable, and as discussed in chapter 2, observations confirm that low-density cores are ubiquitous in luminous elliptical galaxies; measured
mass deficits are roughly consistent with the predictions of the merger simulations.

Intermediate-mass black holes (IBHs), if they exist, could also form binary systems with SBHs. As summarized in chapter 2, the evidence for IBHs is circumstantial, and strong constraints can be put on the properties of a hypothetical IBH
near the center of our galaxy (figure 2.13). But an IBH at the Galactic center could
nicely explain a number of puzzling observations, including the parsec-scale core
recently discovered in the distribution of the old stars (figure 7.1), and the fact
that the S-stars in the inner tenth of a parsec have orbits that are so eccentric and
misaligned (figure 4.23). These possibilities are intriguing enough to motivate considerable work on the formation and evolution of IBHs near the centers of galaxies,
even in the absence of a secure detection.

8.1 INTERACTION OF A MASSIVE BINARY WITH FIELD STARS

Consider a binary system consisting of two SBHs of mass M_1 and M_2. Let $q \equiv M_2/M_1 \leq 1$ be the binary mass ratio and $M_{12} \equiv M_1 + M_2$ its total mass. If the two
SBHs are in a bound Keplerian orbit of semimajor axis a, the energy of the binary
is given by equation (4.27):

$$\mathsf{E}_{\text{bin}} = -\frac{G M_1 M_2}{2a} = -\frac{G \mu M_{12}}{2a}, \tag{8.1}$$

where $\mu = M_1 M_2/M_{12}$ is the reduced mass.[2] The binary's angular momentum is

$$\mathsf{L}_{\text{bin}} = \mu \left[G M_{12} a (1 - e^2) \right]^{1/2} \tag{8.2}$$

with e the eccentricity. The relative velocity of the two SBHs, assuming a circular
orbit, is

$$V_{\text{bin}} = \sqrt{\frac{G M_{12}}{a}} = 658 \left(\frac{M_{12}}{10^8 \, M_\odot} \right)^{1/2} \left(\frac{a}{1 \, \text{pc}} \right)^{-1/2} \, \text{km s}^{-1}; \tag{8.3}$$

note that V_{bin} is independent of the mass ratio.

From the point of view of a distant star, the binary appears almost as a single
mass. Suppose that the star, of mass $m_\star \ll M_{12}$, approaches the binary on an
unbound orbit. To a first approximation, the orbit of the star with respect to the
binary's center of mass is given by equations (5.13), after setting $m = M_{12}$ and
$m_f \approx 0$. The distance of closest approach, r_{min}, of the star to the binary center of
mass is

$$r_{\text{min}} \approx \frac{G M_{12}}{V^2} \left[\left(1 + \frac{p^2 V^4}{G^2 M_{12}^2} \right)^{1/2} - 1 \right] \approx \frac{p^2 V^2}{2 G M_{12}}, \tag{8.4}$$

---

[2]Note that E and L have the dimensions of energy and angular momentum, not specific energy and
specific angular momentum.

where p is the impact parameter and V is the velocity of the star with respect to the binary at infinity (figure 5.2); the second relation assumes $V^2 \ll GM_{12}/p$, appropriate for close encounters. Under the same approximation, the star's velocity at closest approach is

$$v_{\max} \approx \frac{2GM_{12}}{pV}. \tag{8.5}$$

We are interested in stars that come closer to the binary than a few times a; for such stars, the gravitational force from the binary will differ significantly from that of a single mass. Requiring $r_{\min} < Ka$ implies

$$\frac{p}{a} \lesssim (2K)^{1/2} \left(\frac{V}{V_{\text{bin}}} \right)^{-1} \tag{8.6}$$

and for trajectories that satisfy this condition,

$$v_{\max} \gtrsim \left(\frac{2}{K} \right)^{1/2} V_{\text{bin}}. \tag{8.7}$$

With very low probability, such a star can be captured onto a stable bound orbit about one or the other component of the binary. But it is much more likely that the star will escape again to infinity. The average velocity change can be determined via **scattering experiments** [237]: for a large number of different values of p and V, the orbit of a field star is integrated from some large starting distance until it has escaped again to a large distance from the binary. In the limit $m_\star \ll M_{12}$, the problem reduces to the **restricted three-body problem**: changes in the orbital motion of the binary due to the field star can be ignored. Use of this approximation greatly simplifies the integrations [383].

Figure 8.1 shows the results from a large set of such experiments, all with $V = 0.5V_{\text{bin}}$. In these experiments, the Keplerian elements describing the orientation and initial phase of the binary's orbit, (Ω, i, ω), were randomized from integration to integration, equivalent to assuming that stars approach from all directions with equal probability. Setting $K = 2$ in equation (8.6), we expect that stars with $p \lesssim 4a$ will feel the effects of the binary. Figure 8.1 confirms this prediction: for $p \gtrsim 4a$ the field star's velocity is nearly unchanged by the encounter. Furthermore, as p is reduced, there is an increasing bias toward positive changes in velocity: on average, the star *gains* energy from the binary.

The tendency for stars to gain energy after interaction with a massive binary is a manifestation of the **gravitational slingshot**.[3] This asymmetry is due in part to the prolonged character of binary-star interactions. After its first close encounter with the binary, a star may have more or less energy than before. If its energy is less, the star is likely to remain near the binary and interact with it again; this will continue until the star gains enough energy to escape. Such repeated interactions are common, and this means that the statistics of the "escapers" depends in part on how escape is defined. For instance, one can take the view that stars that remain near the binary for, say, 10 orbits or more have been "captured," even if those stars

[3] First described in the 1960s in the context of artificial satellites [390].

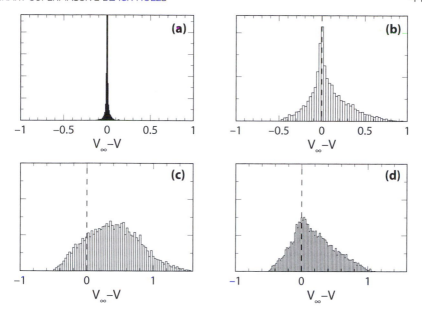

Figure 8.1 Distribution of velocity changes for stars that are scattered off a massive, circular-orbit binary with $M_1 = M_2$. Velocities are expressed in units of V_{bin}, the relative velocity of M_1 and M_2. The initial velocity of each star was $V = 0.5V_{\text{bin}}$; V_∞ is the velocity of the star after it has escaped far from the binary. Each panel includes stars with some range of impact parameters $[p_1, p_2]$; expressed in units of the binary semimajor axis a, these are (a) [4,10], (b) [2,4], (c) [1,2], (d) [0.2,0.4].

would eventually escape. In the context of galactic nuclei, what matters is the ratio of the time to escape to the time over which the orbital elements of the binary evolve. That ratio is typically small for all but a tiny fraction of the interactions.

The experiments illustrated in figure 8.1 were based on an equal-mass binary. In the case of binaries with $M_2 \ll M_1$, most of a star's energy gain results from interaction with the less massive component. An interaction force of $F \approx GM_2/a^2$ acts for a time $\Delta t \approx (a^3/GM_{12})^{1/2}$ to produce a velocity change $\Delta v \approx F \times \Delta t \approx (M_2/M_{12})V_{\text{bin}}$. The corresponding change in the star's (specific) energy is

$$\Delta E \approx \frac{1}{2}\left[(V + \Delta v)^2 - V^2\right]$$

$$\approx V \cdot \Delta v$$

$$\approx (M_2/M_{12})V_{\text{bin}}^2. \tag{8.8}$$

Of course, this energy is taken from the binary. Using the definitions of V_{bin} and E_{bin}, we can write the change in the binary's energy, $-m_\star \Delta E$, as

$$\frac{\Delta E_{\text{bin}}}{E_{\text{bin}}} \approx -2\frac{m_\star}{M_1}, \quad M_2 \ll M_1, \quad r_{\text{min}} \approx a. \tag{8.9}$$

This approximate result motivates the definition of a dimensionless energy change C [237]:

$$C \equiv \frac{M_{12}}{2m_\star} \frac{\Delta E_{\text{bin}}}{E_{\text{bin}}} = \frac{a \Delta E}{G \mu}, \tag{8.10}$$

where ΔE_{bin} and ΔE are computed using energies at a time when the field star has moved far from the binary. Evidently, C is of order unity for interactions that bring the field star close to the binary.

Let $\langle C \rangle$ be the value of C averaged over all orientations and initial phases of the binary, for star–binary interactions with a given p and V. The rate of change of the binary's energy due to interactions with a specified V is given by an integration over impact parameter:

$$\frac{dE_{\text{bin}}}{dt}\bigg|_V = 2\pi V \int_0^\infty dp \, p \Delta E_{\text{bin}} \tag{8.11a}$$

$$= -4\pi G^2 M_1 M_2 \rho V^{-1} \int_0^\infty dx \, x \langle C \rangle, \tag{8.11b}$$

where $\rho = m_\star n$ and

$$x = p/p_1, \quad p_1^2 = 2G M_{12} a / V^2; \tag{8.12}$$

p_1 is the approximate impact parameter corresponding to $r_{\text{min}} = a$. Here we note an important difference with the derivation of the single-particle diffusion coefficients in chapter 5: for large p, energy changes are negligible, and so there is no need to artificially truncate the integration over impact parameters. Expressed in terms of the binary's semimajor axis, equation (8.11) becomes

$$\frac{d}{dt}\left(\frac{1}{a}\right)\bigg|_V = \frac{8\pi G\rho}{V} \int_0^\infty dx \, x \langle C \rangle \tag{8.13a}$$

$$= \frac{G\rho}{V} H_1(V), \tag{8.13b}$$

where

$$H_1(V) = 8\pi \int_0^\infty dx \, x \langle C \rangle. \tag{8.14}$$

The final step is to perform an integration over the field-star velocity distribution $f(v_f)$. We define the **binary hardening rate**, H, that results from this integration as

$$H \equiv \frac{\sigma}{G\rho} \frac{d}{dt}\left(\frac{1}{a}\right), \tag{8.15}$$

where σ is the field-star velocity dispersion. Identifying $f(v_f)$ with the distribution of V,

$$H(\sigma) = 4\pi \int_0^\infty dv_f \, v_f^2 f(v_f/\sigma) \frac{\sigma}{V} H_1(V), \tag{8.16}$$

where it is understood that H is also a function of the binary orbital elements (a, e) and mass ratio q.

Table 8.1 Parameters for fits to H_1 (equation 8.17) assuming a circular binary [442].

q	H_0	λ
1	17.97	0.5675
1/4	20.54	0.4263
1/16	21.87	0.2228
1/64	22.78	0.1043
1/256	22.57	0.0573

Scattering experiments based on a circular-orbit binary yield a velocity-dependent hardening parameter H_1 that is well fit by [442]

$$H_1(V) = \frac{H_0}{\left[1 + V^4/(\lambda V_{\text{bin}})^4\right]^{1/2}}. \tag{8.17}$$

Table 8.1 gives numerically determined values of H_0 and λ for various binary mass ratios. H_0 is weakly dependent on q, while $\lambda \approx (M_2/M_{12})^{1/2}$. At low V, H_1 is nearly constant; it begins to drop off rapidly when $V \gtrsim \lambda V_{\text{bin}} \approx \sqrt{M_2/M_{12}} V_{\text{bin}}$.

Assuming a Maxwellian $f(v_f)$ in equation (8.16) yields

$$\frac{H}{H_1(\sqrt{3}\sigma)} \approx \left(\frac{2}{\pi}\right)^{1/2} + \ln\left[1 + \alpha \left(\frac{\sigma}{\lambda V_{\text{bin}}}\right)^{\beta}\right] \tag{8.18}$$

with $\alpha = 1.16$, $\beta = 2.40$. This relation shows that the binary's hardening rate is a function of its "hardness," V_{bin}/σ, that is, of the ratio of binary orbital velocity to typical stellar velocities. For "hard" binaries, binaries with $V_{\text{bin}} \gg \lambda^{-1}\sigma \approx q^{-1/2}\sigma$, the logarithmic term on the right-hand side of equation (8.18) is negligible and H_1 can be replaced by H_0, yielding

$$H \approx \left(\frac{2}{\pi}\right)^{1/2} H_0, \quad V_{\text{bin}} \gtrsim q^{-1/2}\sigma. \tag{8.19}$$

In other words, *hard binaries harden at a constant rate* [237]. Table 8.1 shows that this asymptotic hardening rate is $H \approx 16$, with a weak dependence on binary mass ratio.

It is worth dwelling for a moment on the meaning of the phrase "hard binary." In the context of star clusters, where all stars have roughly the same mass, a "hard binary" is one in which $V_{\text{bin}} \gtrsim \sigma$. Such binaries tend to acquire larger binding energies in interactions with passing stars ("hard binaries become harder" [236, 231]); they also harden at a nearly constant rate, that is, $(d/dt)(1/a)$ is independent of a. When the components of the binary are much more massive than any star, interactions *always* tend, on average, to harden the binary, regardless of the value of a. If we instead define a "hard" binary as one which hardens at a constant rate, equation (8.18) gives the condition

$$\frac{V_{\text{bin}}}{\sigma} \gg \lambda^{-1} \approx \left(\frac{M_{12}}{M_2}\right)^{1/2}. \tag{8.20}$$

Now, the fact that a binary SBH hardens at a constant rate is not particularly interesting. What is more significant is the typical velocity at infinity of a star ejected by such a binary. For stars that undergo close interactions with the binary, equation (8.8) implies

$$v_\infty \approx \left(\frac{2M_2}{M_{12}}\right)^{1/2} V_{\mathrm{bin}}. \tag{8.21}$$

A standard criterion for "escape" from a star cluster with velocity dispersion σ is $v \geq 2\sqrt{3}\sigma$ [502]. Substituting this for v_∞ in equation (8.21) gives

$$\frac{V_{\mathrm{bin}}}{\sigma} \gtrsim 6^{1/2} \left(\frac{M_{12}}{M_2}\right)^{1/2}, \tag{8.22}$$

similar to equation (8.20). In other words, a massive binary that hardens at a constant rate, also ejects stars with velocities high enough to escape from its vicinity.[4]

These arguments motivate the definition of the **hard binary separation**, a_{h}, as

$$a_{\mathrm{h}} \equiv \frac{G\mu}{4\sigma^2} = \frac{M_2}{M_{12}} \frac{r_{\mathrm{h}}}{4} \tag{8.23}$$

$$\approx 0.27 \,(1+q)^{-1} \left(\frac{M_2}{10^7 \, M_\odot}\right) \left(\frac{\sigma}{200 \,\mathrm{km\,s^{-1}}}\right)^{-2} \mathrm{pc},$$

where $r_{\mathrm{h}} = GM_1/\sigma^2$ is the influence radius of the larger SBH.[5] The numerical factor in equation (8.23) is somewhat arbitrary. In the next section, it will be shown that an expression like equation (8.23) predicts fairly well the value of a below which a massive binary at the center of a galaxy begins to "act" like a binary, in the sense of obeying the hardening equation (8.15). Prior to this time, the two SBHs interact with field stars in roughly the same way they would if the other SBH were not present [386].

A roughly equivalent definition of a hard binary is one for which the binding energy per unit mass, $|E_{\mathrm{bin}}|/M_{12} = G\mu/2a$, exceeds σ^2.

So far, we have considered only circular-orbit binaries. It turns out that the hardening rate is a weak function of the binary's eccentricity [383, 442], and in practice, the dependence of H on e can usually be ignored.

If we also ignore changes in the distribution of field stars due to the presence of the massive binary (an approximation that is justified only in certain circumstances—as discussed in detail in the next section), then the evolution of the binary's semimajor axis in the hard-binary regime is described approximately by

$$\frac{1}{a(t)} - \frac{1}{a_{\mathrm{h}}} \approx H\frac{G\rho}{\sigma}\,(t - t_h)\,, \quad t \geq t_h, \tag{8.24}$$

where t_h is the time at which $a = a_{\mathrm{h}}$. Let a_{GR} be the value of a at which gravitational radiation begins to dominate the loss of energy from the binary. If $a_{\mathrm{GR}} \ll a_{\mathrm{h}}$, the

---

[4]For a binary at the center of the deep potential well of a galaxy, this "escape" may be temporary; see the discussion of the "secondary slingshot" in section 8.3.1.1.

[5]The relation between a_{h} and r_{h} is unchanged if μ in equation (8.23) is replaced by M_2 and r_{h} by GM_{12}/σ^2. Both conventions, and some others, can be found in the literature; in fact, an alternative definition is used elsewhere in this book, in equation (8.71).

time to reach this separation is

$$\Delta t \equiv t(a_{\text{GR}}) - t(a_{\text{h}}) \approx \frac{\sigma}{HG\rho a_{\text{GR}}}$$

$$\approx 2.8 \times 10^8 \left(\frac{\sigma}{200\,\text{km s}^{-1}}\right)\left(\frac{\rho}{10^3\,M_\odot\,\text{pc}^{-3}}\right)^{-1}\left(\frac{a_{\text{GR}}}{10^{-2}\,\text{pc}}\right)^{-1}\,\text{yr.} \quad (8.25)$$

The choice of a_{GR} is somewhat arbitrary; a reasonable definition is the value of a at which da/dt due to stellar interactions is equal to da/dt due to gravitational-wave emission (equation 4.234). Assuming a circular-orbit binary, the result is

$$a_{\text{GR}}^5 = \frac{64G^2 M_1 M_2 M_{12}\sigma}{5Hc^5\rho}$$

$$\approx (1.65 \times 10^{-2}\,\text{pc})^5$$

$$\times \frac{q}{(1+q)^2}\left(\frac{M_{12}}{10^8\,M_\odot}\right)^3\left(\frac{\sigma}{200\,\text{km s}^{-1}}\right)\left(\frac{\rho}{10^3\,M_\odot\,\text{pc}^{-3}}\right)^{-1}. \quad (8.26)$$

Comparing equations (8.25) and (8.26), we see that a massive binary that hardens in a *fixed* stellar background can reach the gravitational-radiation-dominated regime in a time that is much shorter than galaxy lifetimes.

The "final-parsec problem" arises, in part, because the stellar background does not remain fixed. As noted above, a hard binary, $a \lesssim a_{\text{h}}$, ejects stars with velocities high enough to escape permanently from its vicinity. Suppose we define v_{ej} as the velocity at infinity of a star in the scattering experiments that can be considered to have "escaped," and M_{ej} as the total mass in such stars. We might guess that the binary needs to eject a mass comparable with its own mass in order to shrink by an appreciable factor, that is, that

$$\frac{\Delta M_{\text{ej}}}{M_{12}} \approx \frac{\Delta(1/a)}{1/a} \approx \Delta \ln(1/a). \quad (8.27)$$

This argument motivates the definition of a second dimensionless parameter, J, describing the rate of mass ejection:

$$J \equiv \frac{1}{M_{12}}\frac{dM_{\text{ej}}}{d\ln(1/a)}. \quad (8.28)$$

Comparing this expression with equation (8.15), we can write

$$J = \frac{\sigma}{GM_{12}\rho a H}\frac{dM_{\text{ej}}}{dt}. \quad (8.29)$$

If $F_{\text{ej}}(p, V)$ is the fraction of stars in the scattering experiments, with impact parameter p and initial velocity V, that satisfy $v_\infty \geq v_{\text{ej}}$, it is easy to show that

$$J(\sigma) = \frac{4\pi}{H}\int_0^\infty dv_f\, v_f^2 f(v_f/\sigma)\frac{\sigma}{V}4\pi\int_0^\infty dx\, x\, F_{\text{ej}}(x, V) \quad (8.30)$$

with x defined as in equation (8.12). For a hard binary, one finds [442] that $J \approx 1$, weakly dependent on q.

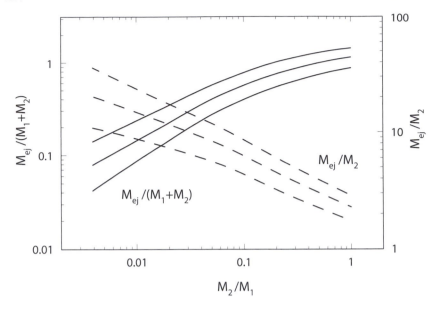

Figure 8.2 This figure shows the mass that would be ejected by a binary SBH in order to reach a value of a such that emission of gravitational waves would lead to co-alescence in a time of 10^{10} yr (lower), 10^9 yr (middle) and 10^8 yr (upper). This plot assumes a nuclear velocity dispersion $\sigma = 200\,\mathrm{km\,s^{-1}}$. Mass ejected by the binary before becoming "hard" is ignored; this figure also ignores the "secondary slingshot" (see section 8.3.1.1).

The mass ejected by the binary in decaying from $a = a_\mathrm{h}$ to $a = a_\mathrm{GR}$ is given by the integral of equation (8.28):

$$M_\mathrm{ej} = M_{12} \int_{a_\mathrm{GR}}^{a_\mathrm{h}} J(a) \frac{da}{a} \qquad (8.31a)$$

$$\approx M_{12} \ln\left(\frac{a_\mathrm{h}}{a_\mathrm{GR}}\right); \qquad (8.31b)$$

the latter expression uses $J(a < a_\mathrm{h}) \approx 1$. Figure 8.2 evaluates equation (8.31) using the results of scattering experiments [442]. The mass ejected in reaching coalescence is of order M_{12} for equal-mass binaries, and several times M_2 when $M_2 \ll M_1$. If this mass came mostly from stars that were originally in the nucleus, the density within $r \approx r_\mathrm{h}$ would drop drastically and the rate of binary evolution would go almost to zero. On the other hand, if the supply of stars is continuously re-plenished (via gravitational encounters, say), the change in density might be much smaller. Furthermore, as discussed in the next section, a binary can do considerable damage to a nucleus even before it reaches the hard-binary regime.

It was noted above that both H and J depend weakly on binary eccentricity. Nev-ertheless, the eccentricity of the binary can change in response to interactions with stars, and such changes are potentially important since the rate of orbital energy loss due to gravitational radiation grows steeply for $e \to 1$ (equation 4.234a).

The dimensionless parameter K, where

$$K \equiv \frac{de}{d \ln(1/a)}, \tag{8.32}$$

defines changes in the binary's eccentricity. The value of K can be derived from scattering experiments in much the same way as H and J [384, 442]. The change in the binary's eccentricity is expressed in terms of changes in its energy and angular momentum by differentiating equation (8.2):

$$\Delta e = -\frac{1 - e^2}{2e} \left(\frac{\Delta E_{bin}}{E_{bin}} + \frac{2 \Delta L_{bin}}{L_{bin}} \right). \tag{8.33}$$

Conservation of total angular momentum is used to relate ΔL_{bin} in a single scattering experiment to $\Delta L_\star$, the change in orbital angular momentum of the field star.

After integrating over an isotropic distribution of field-star velocities, net changes in binary eccentricity tend to be modest; they are due to the systematic differences between encounters that are direct or retrograde with respect to the binary's orbital motion. Except possibly in the case of soft, nearly circular binaries, evolution is always found to be in the direction of increasing eccentricity; that is, $K \geq 0$. Evolution rates tend to increase with increasing hardness of the binary, reaching maximum values of $K \approx 0.2$ for equal-mass binaries with $e \approx 0.75$ and falling to zero at $e = 0$ and $e = 1$. For an equal-mass binary, and in the limit of large binding energy, $V_{bin} \gg \sigma$, two approximate expressions have been derived for the dependence of K on e [384, 442]:

$$K_{MV}(e) \approx \frac{(1 - e^2)}{2e} \left[(1 - e^2)^m - 1 \right],$$
$$m = 0.3e^2 - 0.8, \tag{8.34}$$

and

$$K_Q(e) \approx e \left(1 - e^2 \right)^{k_0} (k_1 + k_2 e), \tag{8.35}$$

$$(k_0, k_1, k_2) = (0.731, 0.265, 0.230).$$

Figure 8.3 shows that the two expressions are in good agreement in spite of their disparate functional forms. Values of K have also been computed and tabulated for other binary mass ratios [488]. The implied changes in e as a binary decays from $a = a_h$ to a_{GR} are modest, $\Delta e \lesssim 0.2$.

On the other hand, if the nucleus is rotating, it is possible for the number of prograde and retrograde encounters with the binary to be very different, resulting in less "cancellation" and in a larger rate of change of the binary's eccentricity [487].

In order to understand this, we must consider the different ways in which stars that approach the binary on prograde and retrograde orbits (with respect to the binary orbit) end up interacting with it. It is clear that—all else being equal—a prograde encounter will result in larger changes to the orbit of the star, since the star

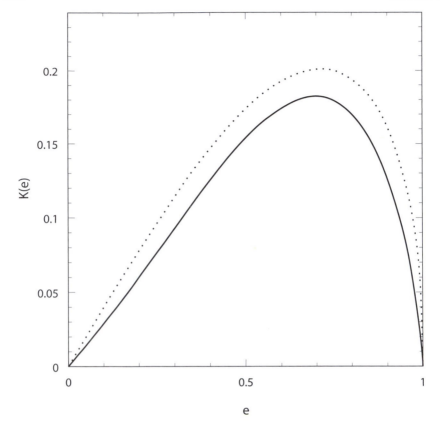

Figure 8.3 Two approximations, derived from scattering experiments, for the coefficient
K (equation 8.32) that describes the rate of eccentricity evolution of a massive binary in the limit of large binding energy, $V_{\mathrm{bin}} \gg \sigma$ [369]. Both assume
a Maxwellian distribution of field-star velocities with no net circulation. Solid
line: equation (8.35) [442]. Dashed line: equation (8.34) [384].

is moving in roughly the same sense, and with a similar velocity, as the (smaller
of the two) SBHs, and so the integrated momentum change can be larger. Stars on
retrograde orbits are likely to avoid ejection during their initial encounter with the
binary. But if the binary is even mildly eccentric, the torque that it exerts on such
a star can cause its orbit to radically change—in fact, the orbit can "flip," from retrograde to prograde (the "eccentric Kozai mechanism" [367]; section 8.6.4). When
such a star is finally ejected by the massive binary, it is likely to be on a prograde
orbit. The total change in the star's angular momentum might therefore be *greater*
than for a star that was on a prograde orbit initially, and the corresponding change
in the binary's eccentricity will also be greater. Since the star's angular momentum
has experienced a net increase in the direction of the binary's angular momentum,
the latter must decrease, implying an increase in the binary's eccentricity. This prediction is verified in N-body simulations (figure 8.4).

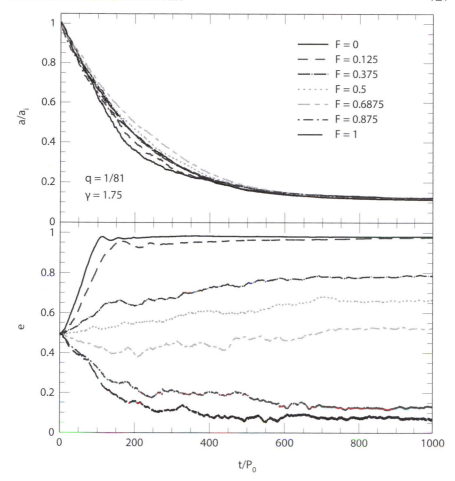

Figure 8.4 Evolution of binary SBHs in rotating nuclei [487]. The upper and lower panels
plot the changes in the binary's semimajor axis, a, and eccentricity, e, as it interacts with stars. The binary mass ratio is $1/81$ and time is given in units of P_0,
the initial period of the binary. The parameter F describes the degree of initial
rotation of the (spherical) nucleus: $F = 0.5$ is an isotropic nucleus, $F = 1$ is
a nucleus in which half of the orbits have been reversed so that all motion is
initially prograde with respect to the binary's orbit, and $F = 0$ has all orbits
initially retrograde. The initial eccentricity of the massive binary is $e = 0.5$.
For $F \gtrsim 0.7$, the eccentricity of the binary decreases, while for $F \lesssim 0.7$ the
eccentricity increases.

Angular momentum is a vector quantity, and its conservation during a single scattering event implies changes in both the magnitude, and direction, of the binary's
internal angular momentum, $\mathbf{L}_{bin}$. Changes in the magnitude of $\mathbf{L}_{bin}$ are equivalent to changes in eccentricity, as discussed above. Changes in the direction of $\mathbf{L}_{bin}$

imply a reorientation of the binary [353]. This process is similar to the reorientation of polar molecules, as studied by P. Debye in the context of dielectric theory [109, 110]. The polarization of a dielectric material is a competition between torques due to the imposed electric field, which tend to align the molecules, and collisions, which tend to destroy the alignment. In the case of binary SBHs, interactions with passing stars result both in a random walk of the binary's orientation, as well as a realignment if the passing stars are drawn from a velocity distribution with a net sense of circulation. It is reasonable to call the resulting evolution of the binary's orientation **rotational Brownian motion** by analogy with the use of this term in solid-state theory.

Following Debye [110], let $F(\theta, \phi, t)d\Omega$ be the probability that the spin axis of the binary is oriented within solid angle $d\Omega$ at time t. In the case that velocities in the stellar cluster are isotropic, the orientation of the axes that define (θ, ϕ) are arbitrary; if there is a net sense of cluster rotation, we define $\theta = 0$ in the direction of net rotational angular momentum. The evolution equation for F is

$$\frac{\partial F}{\partial t} = \frac{1}{\sin\theta} \frac{\partial}{\partial \theta} \left[\sin\theta \left(\frac{\langle \Delta\vartheta^2 \rangle}{4} \frac{\partial F}{\partial \theta} - \langle \Delta\theta \rangle \right) \right]. \tag{8.36}$$

In this equation, the evolution of the binary's orientation is determined by two diffusion coefficients. The second-order coefficient, $\langle \Delta\vartheta^2 \rangle$, is defined as

$$\langle \Delta\vartheta^2 \rangle = \int \Psi(d\Omega, d\Omega')\vartheta^2 d\Omega', \tag{8.37}$$

where $\Psi(d\Omega, d\Omega')d\Omega'$ is the probability that, during a unit interval of time, a binary whose angular momentum $\mathbf{L}_{bin}$ is directed toward $d\Omega$ will reorient itself such that $\mathbf{L}_{bin}$ lies within $d\Omega'$, and ϑ is the angular separation between $d\Omega$ and $d\Omega'$. The first-order coefficient, $\langle \Delta\theta \rangle$, is the rate of change of the angle between $\mathbf{L}_{bin}$ and the preferred axis.

Consider first the case $\langle \Delta\theta \rangle = 0$. In this case, gravitational encounters occur from random directions, and the massive binary responds by undergoing a random walk in its orientation (figure 8.5). We can define a dimensionless diffusion coefficient as

$$R_2 \equiv \frac{M_{12}}{m_\star} \frac{\sigma}{G\rho a} \langle \Delta\vartheta^2 \rangle. \tag{8.38}$$

The factor $\sigma/(G\rho a)$ is the hardening time defined above; the factor $M_{12}/m_\star$ accounts for the fact that the reorientation is a diffusive process. From its definition, $\langle \Delta\vartheta^2 \rangle$ is the sum, over a unit interval of time, of $(\delta\vartheta)^2$ due to encounters with field stars, and it can be computed from changes in the field-star orbital angular momentum obtained via scattering experiments [353]. After integrating over a Maxwellian distribution of field-star velocities, one finds that R_2 for an equal-mass, circular-orbit binary varies from ~ 30 for a binary with $a \approx a_h$ to ~ 60 in the hard-binary limit. The dependence of R_2 on binary mass ratio and eccentricity is approximately $R_2 \propto q^{-1}(1 - e^2)^{-1}$ [446].

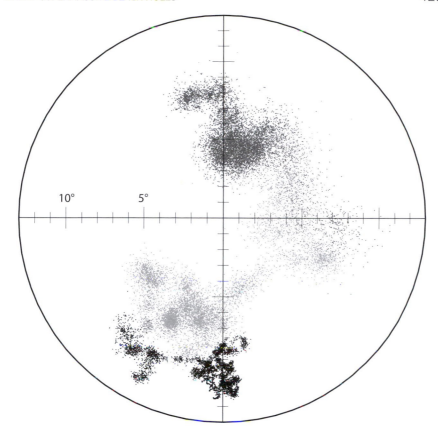

Figure 8.5 Rotational Brownian motion of an equal-mass, circular binary in three N-body
         simulations with different particle masses [386]. Plotted is the angular inclina-
         tion of the binary's axis of rotation. Medium dots: $M_{12}/m_\star = 164$; light dots:
         $M_{12}/m_\star = 328$; heavy dots: $M_{12}/m_\star = 655$. The amplitude of the angular
         changes scales approximately as $\sqrt{m_\star/M_{12}}$.

Changing the time variable from t to $x \equiv \ln(a/a_0)$ in equation (8.36), and using
equations (8.15) and (8.38), this can be written as

$$\frac{\partial F}{\partial x} = -\frac{1}{4}\frac{L}{H}\frac{m_\star}{M_{12}}\frac{\partial}{\partial \mu}\left[(1-\mu^2)\frac{\partial F}{\partial \mu}\right], \tag{8.39}$$

where $\mu \equiv \cos\theta$. In the hard-binary limit, the dependence of H and L on a can be
ignored. Defining $\overline{\mu}$ as the expectation value of μ, the solution to equation (8.39)
in this limit is

$$\overline{\mu}(a) = \overline{\mu}_0 \left(\frac{a}{a_0}\right)^n, \quad n = \frac{1}{2}\frac{L}{H}\frac{m_\star}{M_{12}}. \tag{8.40}$$

The exponent in this expression is of order $m_\star/M_{12} \ll 1$, hence

$$\bar{\mu}(a) \approx 1 + \frac{1}{2} \frac{L}{H} \frac{m_\star}{M_{12}} \ln\left(\frac{a}{a_0}\right), \tag{8.41}$$

where $\bar{\mu}_0$ has been set to unity, corresponding to an initial orientation parallel to the $\theta = 0$-axis. Let $\delta\theta \equiv \sqrt{2(1 - \bar{\mu})}$ be the rms change in the angle defined by the binary's spin axis. If we identify a_0 with a_h and set $L/2H \approx 60/(2 \times 16) \approx 2$, we find

$$(\delta\theta)^2 \approx \frac{2m_\star}{M_{12}} \ln\left(\frac{a_h}{a_{GR}}\right) \approx 10\frac{m_\star}{M_{12}}, \tag{8.42}$$

where, in the last expression, the argument of the logarithm has been set to $\sim 10^2$. For an equal-mass, circular binary of mass $10^6 \, M_\odot$ surrounded by $10 \, M_\odot$ black holes, this predicts $\delta\theta \approx 0.5°$. Larger reorientations would be produced by interaction with "massive perturbers" (section 7.4), or for binaries with large mass ratios or eccentricities.

Next consider the case that the stellar cluster has a net sense of rotation. A dimensionless, first-order diffusion coefficient can be defined as

$$R_1 \equiv \frac{\sigma}{G\rho a}\langle\Delta\theta\rangle, \tag{8.43}$$

where $\theta = 0$ defines the rotation axis of the stellar cluster. In a nonrotating nucleus, we expect $R_1 = 0$ by symmetry. In the case of rotation, the following argument suggests that the binary's angular momentum should align with that of the nucleus. Field stars that interact with the binary are ejected in nearly random directions; this is particularly true when the binary is eccentric. Assuming that the direction of ejection is completely random, the average change in angular momentum of stars that impinge on the binary is

$$\langle\delta\boldsymbol{L}_\star\rangle = \langle\boldsymbol{L}_{\star,\text{final}} - \boldsymbol{L}_{\star,\text{initial}}\rangle \tag{8.44a}$$

$$= -\langle\boldsymbol{L}_{\star,\text{initial}}\rangle. \tag{8.44b}$$

Since the change in the binary's angular momentum is opposite in sign to $\langle\delta\boldsymbol{L}_\star\rangle$, it follows that the binary's axis of rotation tends to align with that of the nucleus. One finds that to a good approximation, $\langle\Delta\theta\rangle \propto -\sin\theta$ [217].

Competing with this evolution is the tendency of encounters to randomize the binary's orientation, as described by the second-order coefficient. One expects the two effects to cancel, on average, when the binary angular momentum is inclined by a certain value with respect to $\boldsymbol{L}_{\text{tot}}$. In fact, the steady-state solution to equation (8.36) is

$$F(\theta) = F_0 e^{\alpha\cos\theta}, \tag{8.45}$$

where

$$\alpha \equiv -\frac{4}{\langle\Delta\vartheta^2\rangle} \frac{\langle\Delta\theta\rangle}{\sin\theta}. \tag{8.46}$$

The function (8.45) is peaked at $\theta = 0$ and falls off by a factor of order unity at an angle

$$\theta_{\text{crit}} = \sqrt{\frac{2}{\alpha}}. \tag{8.47}$$

Evaluating R_1 via scattering experiments [446] yields $R_1 \approx 3$ for hard, equal-mass, circular-orbit binaries; the dependence of R_1 on q and e is approximately $(1 - e^2)^{-1/2}$. The critical orientation works out to be

$$\theta_{\text{crit}} = 0.5 \sqrt{\frac{m_\star}{M_{12}}} \left(\frac{1}{\sqrt{q}} + \sqrt{q} \right) \left(1 - e^2\right)^{-1/4} (2a - 1)^{-1/2}. \tag{8.48}$$

This angle is again small, unless $m_\star$ refers to a "massive perturber." Nevertheless, there is likely to be a "big" effect associated with the reorientation from $\theta(t = 0)$ to θ_{crit} [217], with possibly observable consequences, since the spin axis of the coalesced binary will be determined by the angular momentum of the binary prior to coalescence.

Close encounters of field stars with the binary also contribute to the random walk of the binary's center of mass—to its translational Brownian motion. Brownian motion of single SBHs was considered in chapter 5. It was shown there that

$$V_{\text{rms}}^2 = \frac{3C}{2A} = \frac{m_\star}{M} v_{\text{rms}}^2, \tag{8.49}$$

where $v_{\text{rms}} = \sqrt{3}\sigma$ is the rms velocity of the field stars, and A and C define the behavior of the massive particle's diffusion coefficients at low velocity:

$$\langle \Delta v_\| \rangle = -Av + Bv^3 \cdots,$$
$$\langle (\Delta v_\|)^2 \rangle = C + Dv^2 \cdots.$$

Recall that both A and C are proportional to a term $\ln \Lambda' \approx \ln(p_{\text{max}}/p_{\text{min}})$ that plays the role of "Coulomb logarithm" for low test-particle velocities; however, this dependence drops out when taking the ratio in equation (8.49). If we now imagine replacing the single massive object by a binary, it is clear that the Brownian motion will be increased, since field stars gain kinetic energy on average from the binary, increasing the amplitude of the binary's recoil. This "superelastic scattering" will give the binary a larger random velocity than expected for a point particle in energy equipartition with background stars.

There is a second way in which close encounters contribute to the binary's Brownian motion. Translational Brownian motion represents a balance between dynamical friction and the random encounters that induce an acceleration. But as noted above, field stars that interact strongly with the binary are ejected in nearly random directions, and this reduces the dynamical friction force that they exert on the binary. The velocity change experienced by a field star in a low-impact-parameter collision with a point-mass perturber is $\sim -2V$, corresponding to a $180°$ change in its direction. When the point mass is replaced by a (hard) binary, the field star is ejected in a nearly random direction and its mean velocity change (averaged over many encounters with different phases and orientations of the binary) is there-fore $\sim -V$ in a direction parallel to $\mathbf{V}$. The drag force exerted on the massive object is proportional to the mean velocity change of the field stars and hence the contribution to the frictional force from close encounters is only $\sim 1/2$ as great in the case of a binary as in the case of a point mass.

These two effects can be evaluated, by using scattering experiments to compute the effective, single-particle diffusion coefficients for the binary [352]. The results can be expressed as

$$V_{\text{rms,bin}}^2 = \left(\frac{R_2}{R_1} \right) \left(\frac{m_\star}{M_{12}} \right) v_{\text{rms}}^2, \tag{8.50}$$

where R_1 and R_2 are the coefficients A and C computed for the binary, expressed in terms of the point-mass coefficients:

$$R_1 \equiv \frac{A_{\text{bin}}}{A}, \quad R_2 \equiv \frac{C_{\text{bin}}}{C}. \tag{8.51}$$

Now, the effects associated with super-elastic scattering are due almost entirely to field stars with low impact parameters. This means that both R_1 and R_2 will tend to unity if the integration over impact parameters is extended to large p_{max}, since the distant encounters will overwhelm the finite contribution from the close encounters. However, it was argued in chapter 5 that p_{max} is of order $r_{\text{h}} = GM/\sigma^2$ for a massive object at the center of a galaxy, and even smaller if the density profile around the SBH is steep. In this case, R_1 will be significantly less than one, and R_2 significantly greater than one. In fact, the results from the scattering experiments are well fit by

$$\frac{V_{\text{rms,bin}}}{V_{\text{rms}}} \approx \left(1 + \frac{0.18}{\ln \sqrt{1 + 2\Lambda^2}} \right)^{1/2}, \tag{8.52}$$

where $\Lambda \equiv p_{\text{max}}/(GM_{12}/\sigma^2)$, implying that the Brownian motion of a massive binary might be increased by a factor as great as ~ 2 compared with the motion of a single mass.

8.2 MASSIVE BINARY AT THE CENTER OF A GALAXY: I. EARLY EVOLUTION

The rates of binary evolution derived in the previous section can be applied to a binary SBH at the center of a galaxy, if the parameters ρ and σ that appear in equations like (8.15) and (8.38) are appropriately defined. But expressions like (8.24), which assumes a constant density of field stars as the binary evolves, are more problematic. By interacting with and ejecting stars, a massive binary inevitably changes the density of stars in its vicinity. Furthermore, quantities like M_{ej}, the mass in stars ejected by the binary, are not so clearly defined when the binary is embedded in a deep potential well.

Scattering experiments are a useful guide, but there is no substitute for fully self-consistent N-body integrations. Figure 8.6 shows the results from a set of such integrations, in which an SBH (i.e., a massive Newtonian particle) was placed on

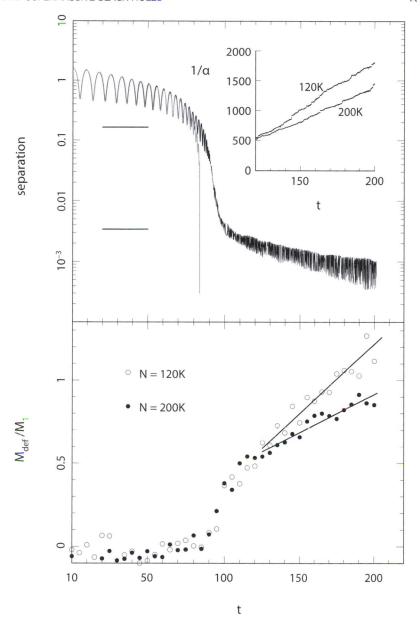

Figure 8.6 The three evolutionary phases of a binary SBH near the center of a spherical
galaxy [355]. These plots were derived from N-body integrations in which a
second SBH was added to a galaxy model that contained a larger SBH at its
center; the mass ratio was $M_2/M_1 = 0.1$ and the initial orbit of the smaller
SBH was eccentric. The unit of time is roughly the galaxy half-mass crossing
time. The top panel plots the distance between the two SBHs; the thin line is the
evolution predicted by the dynamical friction formula, equation (5.23), ignoring
changes in the galaxy. The two horizontal lines indicate r_h and a_h; the former is

Figure 8.6 Continued. the influence radius of the larger SBH and the latter is the hard-binary
          separation, equation (8.23). The inset shows the evolution of the inverse semima-
          jor axis of the binary in this integration, and in a second integration with roughly
          one half the number of particles; the latter curve lies above the former, i.e., the
          decay occurs more rapidly for smaller N when $a < a_h$, due to the higher rate
          at which star–star encounters repopulate orbits that were depleted by the binary.
          In the large-N limit of a spherical galaxy, the binary hardening rate would drop
          to zero at $a \approx a_h$. The lower panel shows evolution of the mass deficit in the
          same two N-body integrations; the solid lines show least-squares fits to the time
          interval $t \geq 120$.

an eccentric orbit ($e \approx 0.5$) near the center of a spherical galaxy containing a larger
SBH at its center; the mass ratio was $q = 0.1$.

The separation between the two SBHs, Δr, can be divided into the three, fairly
distinct regimes $\Delta r > r_h$, $r_h > \Delta r > a_h$, and $\Delta r < a_h$:

1. $\Delta r > r_h$: **dynamical-friction-driven inspiral**. At early times, the orbit of
   the smaller SBH decays due to dynamical friction from the stars. This phase
   ends when the separation between the two SBHs falls to $\sim r_h$, the gravitational influence radius of the larger SBH.

2. $r_h > \Delta r > a_h$: **formation of a hard binary**. When $\Delta r \lesssim r_h$, the two SBHs
   form a bound pair. The separation between the two SBHs drops rapidly in
   this phase, due first to dynamical friction acting on M_2, and later to ejection
   of stars by the binary. Energy input from the binary causes the stellar density
   to drop substantially. The motion of the smaller SBH around the larger is
   approximately Keplerian in this (and later) phases.

3. $\Delta r < a_h$: **continued hardening of the binary**. The rapid phase of binary evolution comes to an end when the binary's binding energy reaches
   $\sim M_{12}\sigma^2$, that is, when $a \approx a_h$, the "hard binary" separation. At this separation, the binary is ejecting nearby stars with high enough velocities that they
   can move far beyond its sphere of influence, effectively excluding them from
   further interactions. If the binary is to continue evolving, the depleted orbits
   must be replenished.

The first two of these evolutionary phases can be called "robust," in the sense that
the associated timescales, and the response of the galaxy to the binary, are fairly independent of details like the galaxy shape, the masses of the stars, etc. However, the
third phase, after the binary becomes hard, can be very dependent on such factors.
For instance, in a spherical galaxy, orbits conserve their angular momenta (except
for the short period of time when they are near the binary), and after roughly a
single galaxy crossing time, the hard binary will have ejected all stars that come
close enough to the galaxy center to interact with it. Continued evolution of the binary can only occur if these orbits are repopulated. In the simulations of figure 8.6,
that repopulation occurred via gravitational (star–star) encounters; the associated
timescale is the relaxation time, which is a function of the stellar mass.

In the remainder of this section, the first two evolutionary phases are discussed. We will proceed from a point in time after the galaxy merger has deposited the second SBH near the center of the merged galactic system, but before the two SBHs have formed a bound pair. Evolution in phase three, after the binary has become "hard," is treated in section 8.3.

8.2.1 $\Delta r > r_h$: Inspiral driven by dynamical friction

The most important question to be asked about evolution in this phase is, how quickly can the second SBH reach the center? We can address this question by adopting a simple model for the distribution of mass near the center of a galaxy and solving the equations of motion for M_2, including the acceleration $\langle \Delta v_\parallel \rangle$ from dynamical friction. (Recall that the dynamical friction force is independent of $m_\star$ as long as $M_2 \gg m_\star$.) Since the second SBH was brought in during the course of a galaxy merger, it will retain some fraction of its host galaxy's mass until late in the inspiral. We begin by ignoring the extra mass; this will yield a conservative upper limit on the inspiral time. We then consider a simple model that includes the time-dependent mass in stars around M_2 as it spirals in.

The relative orbit of the two *galaxies* that produced the binary SBH was probably eccentric. Dynamical friction tends to circularize orbits, and so as a first approximation, we assume that M_2 follows a circular orbit as it spirals in. The question of eccentricity evolution during this phase will be delayed until after a discussion in section 8.2.2 of how the inspiral affects the structure of the galaxy core.

Since the inspiral time is long compared with orbital periods, we can average over the angular motion of M_2 in its orbit, and equate the torque from dynamical friction with the rate of change of M_2's orbital angular momentum. Defining $r(t)$ as the orbital radius, we find

$$\frac{dL}{dt} = \frac{dL}{dr}\frac{dr}{dt} = r\langle \Delta v_\parallel \rangle \tag{8.53}$$

or

$$\frac{dr}{dt} = \frac{r\langle \Delta v_\parallel \rangle}{dL/dr}, \tag{8.54}$$

where

$$L^2(r) = r^2 v_c^2(r) = r^3 \frac{d\Phi}{dr}. \tag{8.55}$$

Near the center of a galaxy, the density profile can often be approximated as a power law, $\rho(r) = \rho_0(r/r_0)^{-\gamma}$. Equation (4.80) then tells us

$$\frac{dL}{dr} = (4-\gamma)(3-\gamma)^{-1/2} \left(\pi G\rho_0 r_0^2\right)^{1/2} \left(\frac{r}{r_0}\right)^{1-\gamma/2}. \tag{8.56}$$

If we identify r_0 with the galaxy half-mass radius r_e, it is not a bad approximation [111] to express ρ_0 in terms of the total galaxy mass via

$$\rho_0 = \frac{(3-\gamma)}{4\pi} \frac{M_{\text{gal}}}{r_e^3}. \tag{8.57}$$

Finally, using Eddington's formula (3.47), we can compute the isotropic $f(E)$ that generates the power-law density profile in the potential (4.80), and find the fraction of the density at each radius contributed by stars moving more slowly than v_c, which yields $\langle \Delta v_{\parallel} \rangle$. The result, after some algebra, is

$$
\frac{dr}{dt} = -\sqrt{\frac{G M_{\text{gal}}}{r_e}} \frac{M_2}{M_{\text{gal}}} \ln \Lambda \left(\frac{r}{r_e} \right)^{\gamma/2-2} F(\gamma),
$$

$$
F(\gamma) = \frac{2^{\beta+1}}{\sqrt{2\pi}} \frac{\Gamma(\beta)}{\Gamma(\beta - 3/2)} \frac{(3 - \gamma)}{4 - \gamma} (2 - \gamma)^{-\gamma/(2-\gamma)}
$$

$$
\times \int_0^1 dy \, y^{1/2} \left(y + \frac{2}{2 - \gamma} \right)^{-\beta},
$$

$$
\beta \equiv (6 - \gamma)/2(2 - \gamma).
$$

(8.58)

For $\gamma = (1.0, 1.5, 2.0)$, $F = (0.258, 0.362, 0.427)$.

Equation (8.58) implies that M_2 comes to rest at the center of the galaxy in a time

$$
\Delta t \approx 0.2 \sqrt{\frac{r_e^3}{G M_{\text{gal}}} \frac{M_{\text{gal}}}{M_2}} \left(\frac{r_i}{r_e} \right)^{(6-\gamma)/2}
$$

(8.59a)

$$
\approx 3 \times 10^9 \left(\frac{r_e}{1 \, \text{kpc}} \right)^{3/2} \left(\frac{M_{\text{gal}}}{10^{11} M_{\odot}} \right)^{1/2} \left(\frac{M_2}{10^7 M_{\odot}} \right)^{-1} \left(\frac{r_i}{r_e} \right)^{(6-\gamma)/2} \text{yr;}
$$

(8.59b)

the leading coefficient in equation (8.59a) turns out to depend weakly on γ. Here, r_i is the initial orbital radius, and following the discussion in chapter 5, $\ln \Lambda$ was set to 6.6. Evidently, an SBH can spiral in from $r_i = r_e$ to the center in a time less than 10 Gyr if

$$
M_2 \gtrsim 3 \times 10^6 \left(\frac{M_{\text{gal}}}{10^{11} M_{\odot}} \right)^{1/2} M_{\odot},
$$

(8.60)

a condition that is satisfied by all but the most massive galaxies—at least for SBHs that are more massive than the one in the Milky Way (which may describe *all* SBHs).

Next, we replace M_2 in the dynamical friction equation by $\mathcal{M}_2$, the mass that remains of the infalling SBH's host galaxy (assumed much larger than M_2). We assume that this mass is determined by the tidal field of the larger galaxy, hence it decreases with time as the inspiraling galaxy loses progressively more of its mass.

The radial acceleration per unit distance in a frame corotating with the smaller galaxy's orbit, of radius r, is given by [285]

$$
\frac{d F_T}{d \delta} = \frac{3}{r} \frac{d \Phi}{dr} - 4\pi G \rho,
$$

(8.61)

where δ is the distance measured from the position of M_2, and Φ and ρ refer to the larger galaxy. Equation (8.61) includes the tidal force due to the radial gradient in the larger galaxy's potential, as well as the centrifugal force from the smaller galaxy's orbit about the center of the larger galaxy, both of which act to remove stars from the smaller galaxy. Let δ_T be the tidally truncated outer radius of the smaller galaxy. The force holding a star onto the galaxy at its edge is

$$F_\star \approx \frac{G\mathcal{M}_2}{\delta_T^2}. \tag{8.62}$$

To make further progress, we need to relate $\mathcal{M}_2$ to δ_T. Suppose that the density of the smaller galaxy falls off with distance from its center as $\sim \delta^{-2}$—the singular isothermal sphere. This is a good description, for instance, of the density near the center of the Milky Way. Then

$$G\mathcal{M}_2 \approx \frac{1}{2}\alpha^2\sigma_2^2\delta_T, \tag{8.63}$$

where σ_2 is the (constant) velocity dispersion in the smaller galaxy, and $M_2 \ll \mathcal{M}_2$ has been assumed. The factor $\alpha \approx 1$ accounts for the fact that the smaller galaxy's density must fall below the assumed form near its edge; a sharp truncation would imply $\alpha = 2$. Substituting (8.63) into (8.62),

$$F_\star \approx \frac{1}{2}\frac{\alpha^2\sigma_2^2}{\delta_T}. \tag{8.64}$$

Equating F_T with $F_\star$ then yields

$$\delta_T \approx \frac{\alpha\sigma_2}{2^{1/2}}\left(\frac{3}{r}\frac{d\Phi}{dr} - 4\pi G\rho\right)^{-1/2}. \tag{8.65}$$

If we are willing to approximate the larger galaxy also as a singular isothermal sphere (an irresistible approximation, since it allows the galaxy's density to be specified by a single parameter, its velocity dispersion σ), then it is easy to show that equations (8.63) and (8.65) imply

$$G\mathcal{M}_2(r) \approx \frac{\sigma_2^3}{2\sigma}r. \tag{8.66}$$

At a distance

$$r = r_{\min} \approx \frac{2\sigma G M_2}{\sigma_2^3} \tag{8.67a}$$

$$\approx 2\left(\frac{M_2}{10^6\,M_\odot}\right)\left(\frac{\sigma}{200\,\text{km s}^{-1}}\right)\left(\frac{\sigma_2}{100\,\text{km s}^{-1}}\right)^{-3}\text{pc} \tag{8.67b}$$

from the center of the larger galaxy, the smaller galaxy has lost essentially all of its stars and its mass is $\sim M_2$.

Substituting $\mathcal{M}_2$ from equation (8.66) for the mass that appears in the dynamical friction formula, we find for the rate of orbital decay,

$$\frac{dr}{dt} = -0.30\frac{G\mathcal{M}_2}{\sigma r}\ln\Lambda \approx -0.15\frac{\sigma_2^3}{\sigma^2}\ln\Lambda, \quad r \gg r_{\min}. \tag{8.68}$$

Assuming $r_i \gg r_{min}$, the center is reached in a time

$$\Delta t \approx \frac{6.7}{\ln \Lambda} \frac{\sigma^2}{\sigma_2^3} r_i \tag{8.69}$$

$$\approx 1.3 \times 10^8 \left(\frac{\sigma}{200 \, \text{km s}^{-1}}\right)^2 \left(\frac{\sigma_2}{100 \, \text{km s}^{-1}}\right)^{-3} \left(\frac{\ln \Lambda}{2}\right)^{-1} \left(\frac{r_i}{1 \, \text{kpc}}\right) \, \text{yr.} \tag{8.70}$$

This second estimate of Δt is also likely to be shorter than 10 Gyr for all reasonable values of σ and σ_2.

8.2.2 $r_h > \Delta r > a_h$: Formation of a hard binary and the generation of cores

The second phase of binary evolution begins when the separation Δr between the two SBHs falls below $\sim r_h = GM_1/\sigma^2$, and it ends when $\Delta r \approx a_h$. Figure 8.7 illustrates this phase of the evolution, in a series of N-body integrations with various values of $q \equiv M_2/M_1$. In that figure, a slightly different definition was adopted for a_h:

$$a_h = \frac{\mu}{M_{12}} \frac{r_m}{4} = \frac{q}{(1+q)^2} \frac{r_m}{4} \tag{8.71}$$

with r_m defined as the radius containing a mass in stars equal to twice M_1. (Defining the "influence radius" in terms of an enclosed mass makes life computationally simpler—at least if the "galaxy" is an N-body model!) Over quite a wide span in binary mass ratios—from 1/2 down to 1/40—the condition $\Delta r \approx a_h$ is seen to accurately predict the point at which the binary's hardening drastically slows, announcing the end of phase two.

What drives the evolution during this phase? When the two SBHs are sufficiently far apart, deposition of energy into the stars by dynamical friction acting on the two SBHs individually is responsible for the orbital decay, while when they are sufficiently close, ejection of stars that interact with the binary is the dominant mechanism. Neither of these processes is well defined in the regime where the binary is neither very hard nor very soft. Nevertheless, we can write approximate expressions for the rate of energy loss from the two mechanisms, by assuming either that the two SBHs are moving independently of each other ($\Delta r \gtrsim r_h$), or as members of a tight binary ($\Delta r \approx a_h$). We focus here on changes in *energy*, since the energy lost by the binary is gained by the stars, and this energy gain turns out to imply substantial changes in the stellar density near the binary.

Dynamical friction causes the energy of M_2 to decrease at a rate given by equation (5.124):

$$\frac{dE}{dt} = M_2 v \langle \Delta v_\parallel \rangle = -4\pi G^2 M_2^2 \rho \ln \Lambda \frac{v}{\sigma^2} G(x) \tag{8.72}$$

with v the orbital velocity around M_1; $G(x) = G[v/(\sqrt{2}\sigma)]$ is defined in equation (5.25). When $\Delta r = r_h$, the mass enclosed within the orbit of M_2 is $\sim 3M_1$ and its orbital velocity is $v^2 \approx 3GM_1/r_h \approx 3\sigma^2$; thus $G(x) \approx G(\sqrt{3/2}) \approx 0.2$, and

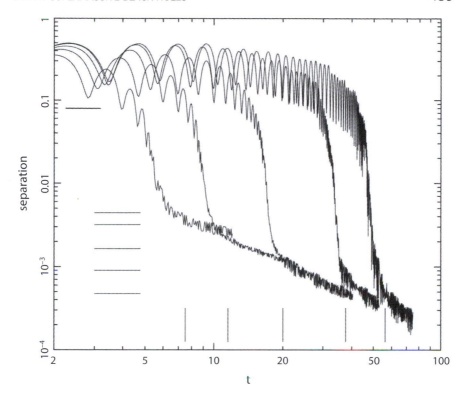

Figure 8.7 Evolution of the binary separation in five N-body integrations with initial conditions similar to those in figure 8.6 [355]. Binary mass ratios are, from left to right, 0.5, 0.25, 0.1, 0.05, 0.025. The upper horizontal line indicates r_m, the influence radius of the more massive SBH in the initial model. The lower horizontal lines show a_h as defined in equation (8.71). The vertical lines are estimates of the times at which hardening of the massive binary would "stall," in a galaxy containing a much larger number of stars than in the N-body model. Note that the rapid phase of binary hardening ("phase two") continues until $\Delta r \approx a_h \propto M_2$, with the result that the binary's binding energy at the end of this phase is nearly independent of M_2.

we find for the energy loss rate

$$\frac{dE}{dt} \approx -4.4 \frac{G^2 M_2^2 \rho \ln \Lambda}{\sigma}, \qquad \Delta r \approx r_h. \tag{8.73}$$

If instead we treat the two SBHs as members of a binary, equations (8.1) and (8.15) imply for the rate of change of the binary's energy,

$$\frac{dE}{dt} = -\frac{G^2 M_1 M_2 \rho H}{2\sigma} = -\frac{H(a)}{2q} \frac{G^2 M_2^2 \rho}{\sigma}. \tag{8.74}$$

Recall from section 8.1 that H is nearly independent of a and q when $a \lesssim a_h$: $H = H_\infty \approx 16$. When $a \approx r_h$, equations (8.17) and (8.18) imply $H \approx q H_\infty$. Thus, both mechanisms predict an energy loss rate that can be written as

$$\left|\frac{dE}{dt}\right| = C(a, q)G^2 M_2^2 \rho \sigma^{-1} \tag{8.75}$$

and when $\Delta r \approx r_h$,

$$C \approx \begin{cases} 9\,(\ln \Lambda/2)\,, & \text{dynamical friction,} \\ 8\,(H_\infty/16)\,, & \text{binary.} \end{cases} \tag{8.76}$$

Roughly speaking, dynamical friction, and slingshot ejection of stars, are equally responsible for the binary's evolution at the start of this phase. As the binary shrinks, the rate of energy generation by the binary increases, eventually by a factor $\sim q^{-1}$ (assuming a fixed ρ) when a has decreased to a_h.

The energy of the binary when $\Delta r \approx r_h$ is roughly $-M_2\sigma^2$, so the characteristic time over which either process extracts energy is approximately

$$T_E \equiv \left|\frac{1}{E}\frac{dE}{dt}\right|^{-1} \approx \frac{\sigma^3}{CG^2 M_2 \rho} \tag{8.77a}$$

$$\approx 4 \times 10^7 \left(\frac{C}{10}\right)^{-1} \left(\frac{\sigma}{200\,\text{km s}^{-1}}\right)^3 \left(\frac{M_2}{10^6\,M_\odot}\right)^{-1} \left(\frac{\rho}{10^3\,M_\odot\,\text{pc}^{-3}}\right)^{-1} \text{yr.} \tag{8.77b}$$

The transfer of energy from the SBHs into the surrounding stars is evidently a very rapid process—so rapid that it is difficult to see how a galaxy could avoid it.

The energy given up by the binary in shrinking from $\Delta r \approx r_h$ to $\Delta r \approx a_h$ is

$$\Delta E \approx -\frac{GM_1 M_2}{2r_h} - \left(-\frac{GM_1 M_2}{2a_h}\right) \tag{8.78a}$$

$$\approx -\frac{1}{2}M_2\sigma^2 + 2M_{12}\sigma^2 \approx 2M_{12}\sigma^2, \tag{8.78b}$$

roughly proportional to the *combined* mass of the two SBHs. The reason for this counterintuitive result is the dependence of a_h on M_2 (equation 8.23): smaller infalling SBHs form tighter binaries. An energy of $2M_{12}\sigma^2$ is comparable with the total energy of the stars within the sphere of influence of the binary. By absorbing such a large energy, the stars within r_h must undergo a substantial redistribution. If there was a density cusp or compact nuclear star cluster prior to formation of the binary, it will be replaced by a low-density core of radius $\sim r_h$. This process is called **cusp disruption**, and it is probably responsible for the almost ubiquitous presence of cores at the centers of luminous galaxies. Cusp disruption is illustrated in figure 8.8.

One would like to compare the structure of the cores produced in simulations with those of observed galaxies. One goal, of course, is to test the cusp-disruption model; another is to extract information about the merger histories of galaxies. This is a subtle and frustrating business. In a simulation, one knows the detailed distribution of matter both before and after the binary has done its work. In a real

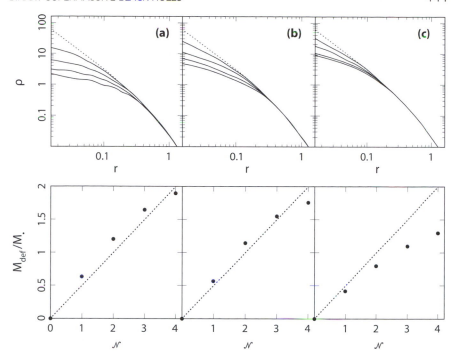

Figure 8.8 Results of N-body simulations that demonstrate the "scouring" effect of a binary SBH on a preexisting density cusp [355]. The binary mass ratio q is (a) 0.5, (b) 0.25, (c) 0.1. In the upper panels, dotted lines show the initial density profile and solid lines show $\rho(r)$ at the end of a set of successive inspiral events. In the lower panels, points show $M_{\text{def}}/M_\bullet$ and dotted lines show $M_{\text{def}}/M_\bullet = 0.5\mathcal{N}$, where $M_\bullet$ is the accumulated SBH mass and $\mathcal{N}$ is the number of inspiral events. The mass deficit depends weakly on the mass ratio of the binary, for reasons discussed in the text.

galaxy, one observes only the end result. A large core could indicate a substantial degree of cusp disruption; or it could simply mean that the galaxies from which the observed galaxy formed also contained low-density cores.

Most attempts to address this question start by adopting a simple parametrization for the distribution of mass (or light) near the center of a galaxy. A basic parameter is the core, or break, radius. If the surface density far from the center is well fit by some smooth function of radius—for instance, a Sérsic [486] law—one defines R_b as the radius where the observed density begins to depart from (fall below) the model density. An example of this procedure is shown in figure 2.2.

Since the binary is expected to inject a characteristic energy into the core, it is natural to parametrize its effects in terms of a displaced mass. In a spherical galaxy, the **mass deficit** [389] is defined as

$$M_{\text{def}} \equiv 4\pi \int_0^{r_{\max}} [\rho_{\text{init}}(r) - \rho(r)] \, r^2 dr. \qquad (8.79)$$

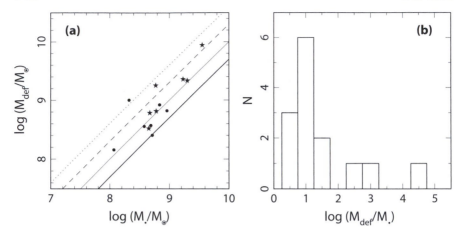

Figure 8.9 Mass deficits in galaxies with resolved cores [355]. (a) M_{def} in solar masses. Data
are from [207] (filled circles) and [158] (stars). Thick, thin, dashed, and dotted
lines show $M_{\text{def}}/M_\bullet = 0.5, 1, 2,$ and 4, respectively. (b) Histogram of $M_{\text{def}}/M_\bullet$
values in (a); $M_\bullet$ in most of the galaxies was computed from the $M_\bullet$–σ relation.

(The definition is straightforwardly generalized to the case of a nonspherical
galaxy.) Here, $\rho_{\text{init}}(r)$ and $\rho(r)$ are the density profiles before and after the binary
SBH has acted on the stars. As an integrated quantity, M_{def} might seem to be more
robust than R_b. But this is not really the case, for several reasons. First, the value
of the integral in equation (8.79) can depend sensitively on the upper integration
limit r_{max}. One can equate r_{max} with R_b, but R_b is typically also uncertain. Second, ρ_{init} for observed galaxies is not known. Assuming that a steep density cusp,
or dense nuclear star cluster, was present initially implies a larger M_{def} than if the
original galaxy contained a core. Finally, one is always faced with the problem of
converting the observable—the luminosity density—into a mass.

Perhaps because of these difficulties, there is some disagreement between published estimates of M_{def} in observed galaxies. However, most studies [95, 158, 207]
find that mass deficits correlate well with SBH masses, and that

$$M_\bullet \lesssim M_{\text{def}} \lesssim 2M_\bullet \tag{8.80}$$

for the majority of galaxies with resolved cores (figure 8.9). In other words, the typical mass deficit is comparable with the mass of the SBH (itself typically inferred
from the $M_\bullet$–σ relation.)

Computing mass deficits in simulations is somewhat less problematic. In the case
of N-body models like those illustrated in figures 8.6 and 8.7, in which a second
SBH was added gently to a galaxy already containing a larger SBH, one finds that
the mass deficit produced by inspiral of the binary to $\Delta r \approx a_{\text{h}}$ is [355]

$$M_{\text{def}} \approx 0.70 q^{0.2} M_{12} \tag{8.81}$$

for galaxies with initial nuclear density profiles $\rho \sim r^{-\gamma}$, $1 \lesssim \gamma \lesssim 1.5$.
Equation (8.81) implies a weak dependence on binary mass ratio, consistent with

our expectation that the energy injected by the binary into the stars depends essentially on M_{12} (equation 8.78). Assuming no other source of cusp disruption, equation (8.81) predicts

$$0.4 \lesssim \frac{M_{\text{def}}}{M_{12}} \lesssim 0.6, \quad 0.05 \lesssim q \lesssim 0.5. \tag{8.82}$$

This is reasonably consistent with the peak in the histogram of measured values, at $M_{\text{def}}/M_{\bullet} \approx 1$ (figure 8.9b). But some galaxies clearly have larger deficits, as large as several times $M_{\bullet}$.

There are a number of possible explanations for the existence of mass deficits greater than $\sim M_{\bullet}$. We list some of them briefly here, before discussing them in more detail later in this chapter:

1. Luminous elliptical galaxies are believed to have formed via a succession of mergers. Since the mass displaced by a binary SBH is nearly independent of q, the ratio of M_{def} to the (accumulated) SBH mass should increase with the number of mergers.

2. If a binary continues to harden beyond $a \approx a_{\text{h}}$, it can continue displacing stars.

3. Mechanisms that forcibly remove a single or binary SBH from a nucleus—for instance, a slingshot interaction involving three SBHs—can lower the density of stars.

Consider just the first of these. If the stellar mass displaced in a single merger is $\sim 0.5 M_{12}$, then—assuming that the two SBHs always coalesce before the next SBH falls in—the mass deficit following $\mathcal{N}$ mergers with $M_2 \ll M_1$ is $\sim 0.5 \mathcal{N} M_{\bullet}$. N-body simulations verify this prediction (figure 8.8). Mass deficits in the range $0.5 \lesssim M_{\text{def}}/M_{\bullet} \lesssim 1.5$ therefore imply $1 \lesssim \mathcal{N} \lesssim 3$, consistent with the number of gas-free mergers expected for bright galaxies.

The fact that there are potentially so many ways to create, or increase, mass deficits is disappointing: it is difficult to falsify a theory when the predictions of the theory are vague. On the other hand, the good observed correlation between M_{def} and $M_{\bullet}$ (figure 8.9), and the fact that observed mass deficits are at least approximately equal to $M_{\bullet}$, lends support to the view that galaxy cores are a consequence of binary SBHs.

Near the end of this evolutionary phase, the two SBHs are separated by a distance $\Delta r \approx a_{\text{h}}$. To reach this separation, the binary must have interacted with most or all stars on orbits having periapsis distances with respect to the galaxy center less than $\sim \Delta r(t)$. Those interactions were increasingly energetic as Δr decreased from $\sim r_{\text{h}}$ to $\sim a_{\text{h}}$; toward the end, sufficient energy was imparted to the stars to eject them from the galaxy core. What happens next?

There are many possibilities; some of these are discussed later in this chapter. But one possibility is that evolution of the binary essentially stalls at $a \approx a_{\text{h}}$. This would be the case, for instance, if the galaxy potential was accurately spherical, and the nuclear relaxation time long. Under these circumstances, the angular momenta

Table 8.2 Some of the properties that "stalled," binary SBHs would have if they were located at the centers of the brightest elliptical galaxies in the Virgo Cluster [355].
SBH masses are in units of $10^8 \, M_\odot$; excepting in the case of NGC 4486 (M87),
these were computed using the $M_\bullet$–σ relation. Influence radii r_m are given in pc
(arcsec); stalling radii a_h are in parsecs. v_∞ is the typical velocity with which a
star would be ejected (equation 8.83). In the last columns, v_{esc} is the velocity in
kilometers per second required to escape from the galaxy, starting from a distance
a_h from the center; the gravitational potential used in computing v_{esc} included the
contribution from the massive binary, modeled as a single body of mass $M_\bullet$.

Galaxy	$M_\bullet$	r_m	a_h $q = 0.5$	a_h $q = 0.1$	v_∞	v_{esc} $q = 0.5$	v_{esc} $q = 0.1$
NGC 4472	5.94	130. (1.6)	5.6	2.1	562.	1395.	1865.
NGC 4486	35.7	460. (5.7)	20.	7.6	733.	1480.	2175.
NGC 4649	20.0	230. (2.9)	10.	3.8	776.	1590.	2325.
NGC 4406	4.54	90. (1.1)	4.0	1.5	590.	1255.	1790.
NGC 4374	17.0	170. (2.1)	7.6	2.8	832.	1635.	2435.
NGC 4365	4.72	115. (1.4)	5.0	1.9	533.	1115.	1615.
NGC 4552	6.05	73. (0.91)	3.2	1.2	757.	1500.	2230.

of orbits with respect to the galaxy center would remain essentially fixed, and the
supply of stars on eccentric orbits capable of interacting with the binary would be
shut off.

Table 8.2 lists some properties of hypothetical, stalled binary SBHs at the centers of bright elliptical galaxies in the Virgo Galaxy Cluster. All of these galaxies
contain well-resolved cores [158]. The column labeled v_∞ lists the approximate
velocity with which a hard binary would eject stars. Recall from section 8.1 that
this velocity is $v_\infty \approx 2\sqrt{3}\sigma$; in terms of $r_h \equiv GM_\bullet/\sigma^2$,

$$v_\infty \approx 3.5 \left(\frac{GM_\bullet}{r_h} \right)^{1/2} \approx 725 \left(\frac{M_\bullet}{10^8 \, M_\odot} \right)^{1/2} \left(\frac{r_h}{10 \, \text{pc}} \right)^{-1/2} \text{km s}^{-1},$$

$$(8.83)$$

independent of mass ratio. (This property was built into the definition of a_h.) The
table also lists the escape velocity, defined as $\sqrt{-2\Phi(a_h)}$ with $\Phi(r)$ the galaxy's
gravitational potential including the contribution from the binary; the latter was
modeled as a point with mass equal to the inferred value of $M_\bullet$. Ejection velocities
are seen to be much smaller than v_{esc} in all cases, implying that essentially no
stars would be ejected beyond the galaxy. Production of such "hypervelocity stars"
(section 6.3) would require that the binary separation shrink well beyond $a \approx a_h$ in
these galaxies.

Assuming that evolution of the binary stalls, it is interesting to ask what the
phase-space distribution of the remaining stars would be. Figure 8.10, based on
Monte Carlo simulations of a binary SBH at the center of a galaxy [380], provides
a partial answer. As expected, a gap has been created in phase space, correspond-ing

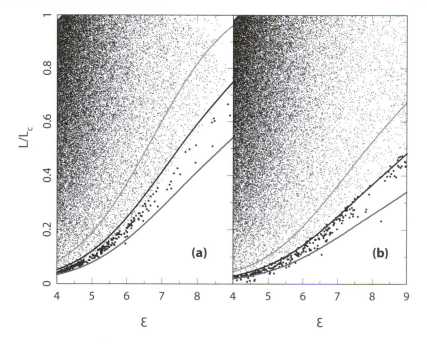

Figure 8.10 The gap in phase space created by a binary SBH at the center of a galaxy [380]. The binary mass ratios are (a) $q = 1$, and (b) $q = 1/8$. Monte Carlo simulations were used to track the evolution of stellar orbits as they interacted with the binary; the binary's semimajor axis was evolved in response to energy carried away by the stars. At the times shown, the binary has reached its "stalling" radius, $a \approx a_h$. Curves show the angular momenta of orbits with periapses of 0.5, 1, and 2 times a_h. The edge of the gap is approximately coincident with the middle curve, corresponding to orbits that graze the sphere $r = a_h$. The larger circles are stars that are still interacting with the binary, i.e., stars with periapses that lie within a few times a_h. These stars can experience a "secondary slingshot."

to orbits that intersected the binary at some point in its evolution from $\sim r_h$ to $\sim a_h$. The three curves in the figure are

$$L_{\mathrm{gap}} = K a_h \sqrt{2 \left[\psi(K a_h) - \mathcal{E} \right]}, \quad K = (0.5, 1, 2), \tag{8.84}$$

where $\mathcal{E} \equiv -E = -v^2/2 + \psi(r)$ and $\Phi(r) = -\psi(r)$ is the galaxy potential. A value $K = 1$ corresponds to an orbit that just grazes the sphere $r = a_h$. Figure 8.10 shows that there is a sharp drop in the phase-space density at the value of L corresponding to $K \approx 1$. A few stars linger within the gap; these are on orbits that intersect the (stalled) binary, but have not yet gained enough energy to escape. Models like the one in figure 8.10 provide the motivation for the initial conditions that were adopted in the time-dependent loss-cone calculations of section 6.1.5.

The binary preferentially ejects stars on eccentric orbits, but the strength of its interactions also depends on the sense of a star's orbital motion with respect to the binary's. Stars that orbit in a retrograde sense are "more stable," on average, than stars that orbit in a direct sense [347, 583]. This means that the phase-space gap extends to lower L for retrograde stars than for direct stars, and this effect is strongest for orbits that lay initially near the equatorial plane of the binary. The result is a toruslike structure in the region $a_h \lesssim r \lesssim 5a_h$: stars that remain bound move preferentially on retrograde, approximately spherical, orbits near the equatorial plane.

8.3 MASSIVE BINARY AT THE CENTER OF A GALAXY: II. LATE EVOLUTION

It was argued in the previous sections that a binary SBH is likely to reach the "hard binary" separation, $a = a_h$, in a time much shorter than galaxy lifetimes (equation 8.77). But a hard binary still has a long way to go before gravitational radiation can be an important source of energy loss. According to equation (4.241), the gravitational-wave inspiral time, starting from $a = a_h$, is

$$t_{GW} \approx 5.7 \times 10^{14} \frac{(1+q)^2}{q} \left(\frac{a_h}{1\,\text{pc}} \right)^4 \left(\frac{M_{12}}{10^8\,M_\odot} \right)^{-3} \text{yr} \qquad (8.85)$$

assuming a circular-orbit binary. Using the definition of a_h in equation (8.23), this can be written

$$t_{GW} \approx 3 \times 10^{16} \frac{q^3}{(1+q)^6} \left(\frac{M_{12}}{10^8\,M_\odot} \right) \left(\frac{\sigma}{200\,\text{km s}^{-1}} \right)^{-8} \text{yr.} \qquad (8.86)$$

Of course, the inefficiency of gravitational radiation at these parsec-scale separations is the origin of the "final-parsec problem."

We now discuss a variety of ways in which a binary SBH might be expected to evolve past $a \approx a_h$ due to interactions with stars. Of course, such mechanisms are only interesting if they act in a time that is reasonably short compared with galaxy lifetimes.

8.3.1 Spherical galaxies

8.3.1.1 The secondary slingshot

In the "hard" limit, $a \lesssim a_h$, a binary SBH imparts an energy $\Delta E \approx G\mu/a$ to stars that interact with it (equation 8.10 with $C = 1$). As noted in the previous section, even when $a \approx a_h$, this energy transfer is typically not large enough to put a star onto an escape orbit from its host *galaxy*; instead, the star is transferred to a different, bound orbit of greater energy, from which it may interact again with the binary. The result is a population of stars on orbits that pass inside $a(t)$, but with energies that have been boosted by previous interactions with the binary. The heavy dots in figure 8.10 belong to this population; the phase-space "gap" that was discussed in connection with that figure is only really empty at large $\mathcal{E}$ (i.e., very bound energies); as one moves to lower $\mathcal{E}$, the distribution of points can be seen to include stars with angular momenta less than L_{gap}—stars that have already interacted once with the binary but are still bound by the galactic potential.

Repeated interaction of stars with a massive binary is called the **secondary sling-shot** [388]. In spite of its name, there is nothing "secondary" about the importance of the mechanism; indeed, it is the *primary* way in which a massive binary, located at the bottom of a deep potential well, interacts with stars. Among other things, consideration of the secondary slingshot will lead to a sharper understanding of the concept of "escape" from a massive binary.

Let $L_{lc}(t)$ be the angular momentum of an orbit that grazes the sphere of radius $r = Ka(t)$ around the binary. Based on the discussion in section 8.1, a value $1 \lesssim K \lesssim 2$ should correspond to orbits that interact strongly with the binary. As in equation (8.84), we can write

$$L_{lc}(\mathcal{E}, t) = Ka(t)\sqrt{2[\psi(Ka) - \mathcal{E}]} \approx \sqrt{2GM_{12}Ka(t)}. \qquad (8.87)$$

By analogy with the single-SBH case, we define orbits having $L \leq L_{lc}$ as "loss-cone orbits." Of course, the physical scale associated with the loss cone of a massive binary is many orders of magnitude larger than that of the loss cones discussed in chapter 6. Furthermore, in the case of a binary, stars are not necessarily "lost" if $L < L_{lc}$. If $f(\mathcal{E}, L)$ is the stellar distribution function, we can use equation (5.159) to write for the energy distribution of loss-cone orbits,

$$N_{lc}(\mathcal{E})d\mathcal{E} = \left[\int_0^{L_{lc}^2} dL^2 4\pi^2 f_0(\mathcal{E}, L) P(\mathcal{E}, L) \right] d\mathcal{E}$$

$$\approx 8\pi^2 G M_{12} a f(\mathcal{E}) P(\mathcal{E}) d\mathcal{E}, \qquad (8.88)$$

where P is the radial period of a star in the galaxy's potential; in the final line, isotropy has been assumed, and P has been approximated by the period of a radial orbit of energy $\mathcal{E}$.

To understand how loss-cone orbits evolve in response to the secondary slingshot, it is instructive to look at an N-body simulation. Figure 8.11 shows the evolution of the energy distribution of stars that were initially within the binary's loss cone in such a simulation; here "initial" refers to a time when $a \approx a_h$. The initial energy distribution of these orbits is proportional to fP and is strongly peaked. As time progresses, the binary shrinks, and its interactions with stars become increasingly energetic. The peak in the energy distribution shifts toward $\mathcal{E} = \mathcal{E}_{eject}$, where

$$\mathcal{E}_{eject}(t) \approx \psi_{eject} - \frac{G\mu}{a(t)} \qquad (8.89)$$

and ψ_{eject} is the potential energy at a radius (roughly the binary's influence radius) from which the majority of ejections occur.

If the binary orbit decays slower than typical orbital periods, $(d \ln a/dt)^{-1} \gg P$, most stars inside the loss cone remain inside, encounter the binary at their next periapsis passage, and are "reejected." On the other hand, if the decrement in the binary separation is substantial, some stars that are ejected very near the loss-cone boundary finish just outside boundary as that boundary shifts inward. Both populations can be discerned in figure 8.11; the latter as the broad, secondary hump

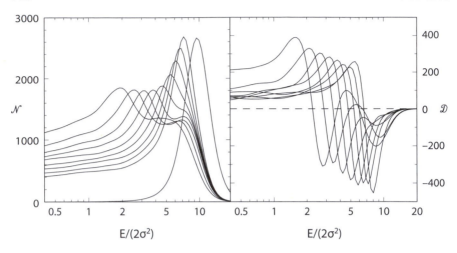

Figure 8.11 These plots show the evolution of $N_{lc}(\mathcal{E}, t)$, the distribution of orbital energies of stars that were initially within the binary loss cone, in an N-body simulation [388]. The binary mass ratio is $q = 1$ and the binary separation at $t = 0$ was roughly a_h. The binary separation decayed by a factor of about four during the integration. The distributions, from right to left, were recorded at logarithmically progressing times $t = (0, 1, 2, 4, 8, 16, 32, 64, 128, 256) \times 4 \times 10^{-4} P(0)$, with P defined as in equation (8.98). The left panel shows that the peak of the distribution travels from larger to smaller binding energies following the increase in the energy of ejection (equation 8.89). Note the late formation of the secondary hump at $|E|/(2\sigma^2) \approx 3.0$–$8.0$, leveling at $N_{lc} \sim 1400$; this hump consists of stars that had originally been inside the loss cone, but then finished just outside, beyond reach of the SBH binary. The right panel shows $\mathcal{D} \equiv 10^3 \partial N_{lc}(|E|, t)/\partial \ln t$ at $t = (1, 2, 4, 8, 16, 32, 64, 128) \times 4 \times 10^{-4} P(0)$.

that appears at late times. The stars in this hump are unable to interact further with the binary.

Returning to equation (8.88), let $N_{lc}(\mathcal{E}, t)$ be the time-dependent loss-cone population, defined by allowing a and f on the right-hand side to be functions of time. (Note that N_{lc} is determined by the number of stars actually within the loss cone at time t, and not by the initial loss-cone population as in figure 8.11.) To derive the time dependence of N_{lc}, we consider the evolution over a small interval Δt:

$$N_{lc}(\mathcal{E}, t + \Delta t) = \int N_{lc}(\mathcal{E}', t)\zeta(\mathcal{E}', \mathcal{E}, \Delta t)d\mathcal{E}'. \tag{8.90}$$

Here, $\zeta(\mathcal{E}', \mathcal{E}, \Delta t)$ is the transition probability between the energy $\mathcal{E}'$ at time t and the energy $\mathcal{E}$ at time $t + \Delta t$. The transition probability can be understood as a combination of the kicks from $\mathcal{E}$ into some other energy; the kicks from another energy $\mathcal{E}'$ into $\mathcal{E}$; and attrition due to decrease in the size of the loss cone as the binary shrinks. We can approximate these three terms as

$$\zeta(\mathcal{E}', \mathcal{E}, \Delta t) \approx \left[1 - \frac{\Delta t}{P(\mathcal{E}')}\right]\delta(\mathcal{E} - \mathcal{E}') + \frac{\Delta t}{P(\mathcal{E}')}\zeta_1(\mathcal{E}', \mathcal{E})$$
$$+ \left[\frac{L_{lc}^2(t + \Delta t)}{L_{lc}^2(t)} - 1\right]. \qquad (8.91)$$

Here, $\zeta_1(\mathcal{E}', \mathcal{E})$ is the probability that a star ejected from energy $\mathcal{E}'$ will end up on an orbit with energy E, normalized so that $\int \zeta_1(\mathcal{E}', \mathcal{E})d\mathcal{E}' = 1$. In writing these expressions, we have made a number of simplifying assumptions:

1. The probability for slingshot ejection is uniform with respect to time over an orbital period and totals to unity within one period.

2. Orbital phases of stars at a given energy are randomly distributed, even after one or more ejections.

3. f remains a function only of $\mathcal{E}$ in the loss cone, so that $N_{lc}(L)dL \propto L\,dL$.

Passing to the infinitesimal limit in equations (8.90) and (8.91), and using $(d/dt)\ln L_{lc}^2(t) = (d/dt)\ln a(t)$, we obtain an equation describing the evolution of the loss-cone orbital population:

$$\frac{\partial}{\partial t}N_{lc}(\mathcal{E}, t) = -\frac{N_{lc}(\mathcal{E}, t)}{P(\mathcal{E})} + \int \frac{N_{lc}(\mathcal{E}', t)}{P(\mathcal{E}')}\zeta_1(\mathcal{E}', \mathcal{E})d\mathcal{E}' + N_{lc}(\mathcal{E}, t)\frac{d\ln a(t)}{dt}. \qquad (8.92)$$

The total number of stars inside the loss cone evolves as

$$\frac{dN_{lc}(t)}{dt} \equiv \frac{d}{dt}\int N_{lc}(\mathcal{E}, t)d\mathcal{E} = N_{lc}(t)\frac{d\ln a(t)}{dt} \qquad (8.93)$$

and the total energy of these stars evolves as

$$\frac{d\mathcal{E}(t)}{dt} \equiv \frac{d}{dt}\int N_{lc}(\mathcal{E}, t)\mathcal{E}\,d\mathcal{E}$$
$$= \int \frac{N_{lc}(\mathcal{E}, t)}{P(\mathcal{E})}\left[-\mathcal{E} + \int \zeta_1(\mathcal{E}, \mathcal{E}')\mathcal{E}'d\mathcal{E}'\right]d\mathcal{E}$$
$$+ \int N_{lc}(\mathcal{E}, t)\frac{d\ln a(t)}{dt}\mathcal{E}\,d\mathcal{E}. \qquad (8.94)$$

The factor in brackets is the average energy change that a star originally at $\mathcal{E}$ experiences as a consequence of the gravitational slingshot:

$$\langle\Delta\mathcal{E}\rangle \equiv -\mathcal{E} + \int \zeta_1(\mathcal{E}, \mathcal{E}')\mathcal{E}'d\mathcal{E}'. \qquad (8.95)$$

The energy gained in the first term of equation (8.94) must be compensated by the change in the binary's binding energy:

$$\frac{d}{dt}\left(\frac{GM_1M_2}{2a}\right) = -m_\star \int \frac{N_{lc}(\mathcal{E}, t)}{P(\mathcal{E})}\langle\Delta\mathcal{E}\rangle d\mathcal{E}. \qquad (8.96)$$

The hardening of a binary SBH coupled to an evolving stellar population inside the loss cone is described by equations (8.92) and (8.96).

To make further progress, we must specify a model for the galactic potential. As elsewhere in this book, we adopt for simplicity the singular isothermal sphere,

$$\rho(r) = \frac{\sigma^2}{2\pi G r^2}, \quad \psi(r) = -2\sigma^2 \ln\left(\frac{r}{r_0}\right), \tag{8.97}$$

where σ is the galaxy velocity dispersion outside $\sim r_h$ and r_0 is an arbitrary radius. Note that we have ignored the contribution of the massive binary to the potential; this approximation is justified by the fact that essentially all stars interacting with the binary after it is hard are on orbits with $a > r_h$. The radial period of a star in this potential is

$$P(\mathcal{E}) = \oint \frac{dr}{v_r(\mathcal{E})} = \frac{\sqrt{\pi} r_0}{\sigma} e^{-\mathcal{E}/2\sigma^2}. \tag{8.98}$$

Substituting this into equation (8.96) gives

$$\frac{d}{dt}\left(\frac{G M_1 M_2}{2a}\right) = -\frac{m_\star}{P(0)} \int N_{\text{lc}}(\mathcal{E}, t) e^{\mathcal{E}/2\sigma^2} \langle \Delta \mathcal{E} \rangle d\mathcal{E}. \tag{8.99}$$

As noted above, the distribution of stars inside the loss cone peaks at the energy given by equation (8.89). In order to solve equation (8.99) analytically, we pretend that all stars inside the loss cone have the same energy, $\mathcal{E}_{\text{eject}}(t)$, at time t:

$$N_{\text{lc}}(\mathcal{E}, t) \to N_{\text{lc}}(t)\delta(\mathcal{E} - \mathcal{E}_{\text{eject}}). \tag{8.100}$$

Therefore

$$\frac{d}{dt}\left(\frac{G M_1 M_2}{2a}\right) \approx -\frac{m_\star}{P(0)} N_{\text{lc}}(t) \langle \Delta \mathcal{E} \rangle_{\mathcal{E}_{\text{eject}}} e^{\mathcal{E}_{\text{eject}}/2\sigma^2}. \tag{8.101}$$

Equation (8.101) can be simplified further by noting that the product $N(t)\langle \Delta \mathcal{E} \rangle_{\mathcal{E}_{\text{eject}}}$ is approximately constant with respect to time. Equation (8.93) implies that the total number of stars inside the loss cone decreases in proportion to the binary separation, $N(t) \propto a(t)$. After the two SBHs form a hard binary, stars that interact with the SBH for the first time receive a kick $\Delta \mathcal{E} \approx -G\mu/a \propto a^{-1}$. A similar result can be shown to hold late in the binary's evolution [388]. Making the approximation $N(t)\langle \Delta \mathcal{E} \rangle \propto a^1 a^{-1} = \text{constant}$, we can integrate equation (8.101) after substituting $\mathcal{E}_{\text{eject}}$ from equation (8.89):

$$\frac{1}{a(t)} = \frac{1}{a(0)} + \frac{2\sigma^2}{G\mu} \ln\left[1 + \frac{m_\star N_{\text{lc}} |\langle \Delta \mathcal{E} \rangle|}{M_{12}\sigma^2} \frac{t}{P(\mathcal{E}_0)}\right] \tag{8.102}$$

for $t > 0$, where $a(0)$ is the initial separation and $\mathcal{E}_0$ is the energy of ejected stars at the outset:

$$\mathcal{E}_0 = \mathcal{E}_{\text{eject}}\big|_{t=0} = \psi_{\text{eject}} - \frac{G\mu}{a(0)}. \tag{8.103}$$

Equation (8.102) tells us that, in the secondary slingshot, the binding energy of the binary increases as a logarithmic function of time. This result has been confirmed by Monte Carlo simulations [388].

How far beyond $a = a_h$ can the secondary slingshot bring the two SBHs? We can calculate this using equation (8.102), given an elapsed time t, and given an estimate for $m_* N_{lc} |\langle \Delta \mathcal{E} \rangle|$, the product of the initial mass in the loss cone and the typical energy change when $a \approx a_h$. From the definition of a hard binary we know that $|\langle \Delta \mathcal{E} \rangle| \approx G \mu / a_h \approx$ a few $\times \sigma^2$. The mass initially in the loss cone is given by integrating equation (8.88) over energy. If we continue to adopt the singular isothermal sphere model, and restrict the integration to energies beyond the sphere of influence, the result is $M_{lc} \approx$ a few $\times \mu$. Using these numbers, we find that a^{-1} will have increased by the factor

$$\frac{a_h}{a(t)} \approx 1 + \frac{1}{2} \ln \left[10 \times \frac{t}{P(\mathcal{E}_0)} \right]. \tag{8.104}$$

$P(\mathcal{E}_0)$ is the period of a star extending some way beyond r_h; in the Milky Way, $P(\mathcal{E}_0) \gtrsim r_h / \sigma \approx 10^{4.5}$ yr, and equation (8.102) implies $a_h / a(10 \, \text{Gyr}) \approx 8$. In galaxies with larger σ this factor will be smaller.

These estimates are probably too optimistic. Realistically, reejection becomes ineffective once a star gains too much energy: first because the density in many galaxies falls off more rapidly than r^{-2} beyond r_h, and second because a star with large apoapsis is easily perturbed from its nearly radial orbit on the way in or out. Furthermore, in many galaxies, other mechanisms (discussed below) can be expected to decrease a on timescales less than 10 Gyr, mitigating the importance of the secondary slingshot.

Finally, we return to the concept of "ejected mass" that was first discussed in section 8.1 in connection with the dimensionless parameter J, defined as

$$J \equiv \frac{1}{M_{12}} \frac{d M_{ej}}{d \ln(1/a)} = \frac{\sigma}{G M_{12} \rho a H} \frac{d M_{ej}}{dt}. \tag{8.105}$$

The reejection paradigm suggests a more general expression for the relation between the ejected mass and the change in a binary's energy. Equation (8.105) was motivated by scattering experiments in which stars are assumed lost if they exit with enough velocity to escape the binary's influence sphere. In the case of a binary embedded in a galactic potential, the critical quantity is the energy gained by a star between the time it first enters, and finally exits, the loss cone. Equating this with the change in the binary's absolute binding energy, we find

$$\frac{G M_1 M_2}{2} \left(\frac{1}{a_{final}} - \frac{1}{a_{initial}} \right) = M_{lost} \left(\bar{\mathcal{E}}_{enter} - \bar{\mathcal{E}}_{exit} \right). \tag{8.106}$$

Defining $\left(\bar{\mathcal{E}}_{enter} - \bar{\mathcal{E}}_{exit} \right) \equiv \Delta \Phi$ and passing to the infinitesimal limit $a_{initial} \to a_{final} \equiv a$ in equation (8.106), we obtain

$$\frac{1}{M_{12}} \frac{d M_{lost}}{d \ln(1/a)} = \frac{1}{(\Delta \Phi / 2 \sigma^2)(a / a_h)}. \tag{8.107}$$

Comparing this relation to equation (8.105), we identify the effective value of the mass-ejection parameter J:

$$J \approx \frac{1}{(\Delta \Phi / 2 \sigma^2)(a / a_h)}. \tag{8.108}$$

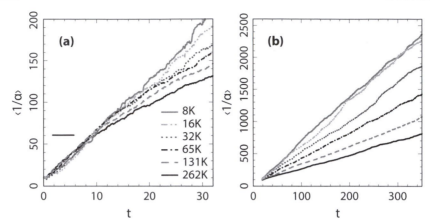

Figure 8.12 (a) Short-term, and (b) long-term evolution of the mean value of $1/a$ in a large
set of N-body integrations [369]. The sets differed in N, the number of "star"
particles making up the galaxy. The horizontal line in panel (a) is the hard-
binary radius; it also indicates approximately where the transition occurs between N-independent and N-dependent evolution.

To shrink the binary by one e-folding, a stellar mass of JM_{12} must be transported
from an energy marginally bound to the binary, to the galactic escape velocity.
These ideas will require further modification when we consider the additional effects of gravitational encounters between stars.

8.3.1.2 Collisional loss-cone repopulation

Until now, we have been treating the host galaxies of binary SBHs as collisionless systems, in the sense that gravitational encounters between stars have been
ignored.[6] That is not a bad approximation: nuclear relaxation times are long, even
in the densest of galaxies. But just as gravitational encounters tend to drive stars
into the loss cone of a single SBH, so can they drive stars into the larger loss cone
defined by a binary SBH. In this way, evolution of a massive binary can continue
indefinitely beyond $a \approx a_{\rm h}$, on a timescale of order the relaxation time, as stars
diffuse onto the eccentric orbits that were previously depopulated by the binary.

It is natural to address this kind of evolution using N-body simulations.
Figure 8.12 shows the results from a set of such simulations, in which the number N of stars representing the galaxy was varied, while quantities like the total
mass and the mass of the binary remained fixed. Evolution of the binary's binding
energy can be seen to exhibit two separate regimes. At early times, a^{-1} (i.e., the
energy of the binary) increases approximately linearly with time, at a rate that is
independent of N. This phase comes to an end when $a \approx a_{\rm h}$, and we identify it
with the second of the three phases discussed in section 8.2.2.

---

[6]Interactions of stars with the massive *binary* are certainly "collisional."

Once a has dropped below a_h, figure 8.12 shows that the linear dependence of a^{-1} on t continues, but the hardening rate is now seen to be a decreasing function of N. This is reasonable if we postulate that the binary's evolution is linked to gravitational encounters between stars, since the scattering rate scales in proportion with the mass of a single star, that is, as $\sim N^{-1}$. Evidently, for these values of N, evolution of the binary due to the secondary slingshot (which should be N independent) is swamped by the contribution from collisional loss-cone repopulation.

Inspection of the right panel of figure 8.12 reveals another interesting result: only for sufficiently large N is the binary's hardening rate N dependent; for the two smallest N values, the binary hardens at roughly the same rate. It turns out that this rate is consistent with equation (8.24), the rate of hardening of a binary in a fixed stellar background, if the quantities ρ and σ in that equation are assigned their values in the N-body models. A natural interpretation is that—for these low values of N—the timescale for gravitational encounters to scatter stars into the binary's loss cone is shorter than the time required for the binary to eject stars; in other words, we are in the "full-loss-cone" regime defined in chapter 6. Presumably, for the much larger N-values associated with real galaxies, a massive binary would be in the "empty-loss-cone" regime.

Armed with these insights, we can attempt to generalize the theory of collisional loss-cone repopulation presented in chapter 6 to the case of a massive binary. As in chapter 6, our model will be based on the orbit-averaged Fokker–Planck equation, and we will assume a fixed gravitational potential (though not a fixed density) for the galaxy. Changes that occur in the structure of the core prior to formation of a "hard binary" are not well treated by such a model and so we will assume $a \lesssim a_h$; evolution of the binary prior to its hard phase is best treated via N-body methods.

Let $F(\mathcal{E}, t)$ be the flux (number per unit energy per unit time) of stars that are driven via gravitational encounters into the loss cone of the massive binary. The relation between F and the rate of change of the binary's semimajor axis a is

$$\frac{d}{dt}\left(\frac{G M_1 M_2}{2a}\right) = -m_\star \int F(\mathcal{E}, t)\Delta\mathcal{E}\, d\mathcal{E}, \qquad (8.109)$$

where $\Delta\mathcal{E}$ is the mean specific-energy change of stars that interact with the binary. In the orbit-averaged approximation, it is appropriate to identify $\Delta\mathcal{E}$ with the total change in energy that occurs between the time a star first enters the loss cone, and the time that it ceases to interact with the binary: either because it has gained enough energy to escape from the galaxy, or because its angular momentum no longer satisfies $L > L_{lc}$.

The flux, F, will be a function of the constant K in equation (8.87) that defines the physical size of the interaction region around the binary; we expect $1 \lesssim K \lesssim 2$. Given a value of K, and ignoring for a moment the possibility of reejection, the average energy change $\Delta\mathcal{E}$ can be derived as follows [576].

Suppose that the stars beyond $\sim r_h$ define a constant-density core with velocity distribution

$$f(v) = \frac{\rho}{m_\star}\frac{1}{(2\pi\sigma^2)^{3/2}}\exp\left(-v^2/2\sigma^2\right); \qquad (8.110)$$

here ρ and σ are constants, which we identify with the ρ and σ that appear in expressions like equation (8.15) for H. The stars described by f pass within a distance Ka of the galaxy's center at a rate given by equation (6.10):

$$F^{\text{flc}}(\mathcal{E}) = 4\pi^2 L_{\text{lc}}^2(\mathcal{E}) f(\mathcal{E}) \approx 8\pi^2 G M_{12} K a f(\mathcal{E}) \qquad (8.111)$$

and so

$$\int F^{\text{flc}}(\mathcal{E}) d\mathcal{E} \approx 8\pi^2 G M_{12} K a \int_0^\infty f(v) v \, dv \approx \frac{2\sqrt{2\pi} G M_{12} K a \rho}{m_\star \sigma}. \qquad (8.112)$$

For stars that interact strongly with the binary, $\Delta\mathcal{E}$ depends weakly on $\mathcal{E}$ if the binary is hard, and equations (8.109) and (8.112) imply

$$\frac{d}{dt}\left(\frac{1}{a}\right) \approx -4\sqrt{2\pi} \frac{K a \rho}{\mu m_\star \sigma} \Delta\mathcal{E}, \quad K \gtrsim 1. \qquad (8.113)$$

Comparing this with equation (8.15) yields

$$\Delta\mathcal{E} \approx -\frac{H}{4\sqrt{2\pi}} \frac{G\mu}{Ka} \approx -1.6 \frac{G\mu}{Ka} \qquad (8.114)$$

which is the desired relation. Finally, for the binary hardening rate due to collisional loss-cone repopulation, equations (8.114) and (8.109) give

$$s(t) \equiv \frac{d}{dt}\left(\frac{1}{a}\right) \approx \frac{3.2}{Ka} \frac{m_\star}{M_{12}} \int F(\mathcal{E}, t) d\mathcal{E}. \qquad (8.115)$$

In general, F in this equation will not be linearly dependent on Ka, and some care must be put into selecting K. One way is by comparing Fokker–Planck models with N-body simulations; one finds $K \approx 1.2$ [42].

The physical size of a binary SBH, a, is many orders of magnitude larger than the radius of the tidal-disruption sphere or capture sphere around a single SBH. Since L_{lc}^2 is proportional to this size (equation 8.87), the time for a star to diffuse across the loss cone of a binary SBH will be much longer than in the case of a single SBH. We recall from chapter 6 that this implies a much smaller value for the quantity $q_{\text{lc}}(\mathcal{E})$;[7] in other words, we expect the binary's loss cone to be "empty": stars will execute many orbits just outside the loss cone before diffusing over the boundary and being lost. (The same conclusion follows from figure 8.12, where behavior corresponding to a "full loss cone" was only exhibited for values of N much smaller than in real galaxies.) When computing the flux into the loss cone of a binary SBH, it is therefore an excellent approximation to set $f(\mathcal{E}, L) \equiv 0$, $L \leq L_{\text{lc}}$.

A consistent choice for $F(\mathcal{E})$ [576] would be the steady-state, Cohn–Kulsrud flux derived in section 6.1.2. For $q_{\text{lc}} \ll 1$, the Cohn–Kulsrud f falls to zero just at L_{lc}:

$$f(\mathcal{R}) \approx \overline{f}(\mathcal{E}) \frac{\ln(\mathcal{R}/\mathcal{R}_{\text{lc}})}{\ln(1/\mathcal{R}_{\text{lc}})}, \quad \mathcal{R}_{\text{lc}} \leq \mathcal{R} \leq 1 \qquad (8.116)$$

---

[7]In this chapter, the quantity defined as $q(\mathcal{E})$ in chapter 6 will be written as $q_{\text{lc}}(\mathcal{E})$, to avoid confusion with the symbol for the binary mass ratio, $q = M_2/M_1$.

with $\mathcal{R} \equiv L^2/L_c^2(\mathcal{E})$. The flux would then be given by equation (6.64):

$$F(\mathcal{E}) \approx \overline{f}(\mathcal{E}) \frac{4\pi^2 q_{lc}(\mathcal{E}) L_{lc}^2(\mathcal{E})}{\ln(1/\mathcal{R}_{lc})}. \tag{8.117}$$

On the other hand, the Cohn–Kulsrud solution assumes a steady state with respect to L. But evolution of the massive binary prior to its becoming "hard" took place on a timescale much shorter than the relaxation time, and there is no reason to expect that f will have reached a form like equation (8.116) by the time $a \approx a_h$. A more likely form for f at this time is suggested by figure 8.10:

$$f(\mathcal{E}, \mathcal{R}) \approx \begin{cases} \overline{f}(\mathcal{E}), & \mathcal{R} > \mathcal{R}_{lc}(a_h), \\ 0, & \mathcal{R} < \mathcal{R}_{lc}(a_h), \end{cases} \tag{8.118}$$

a step function.

A similar argument was made in chapter 6, in the context of the feeding of single SBHs. In that case, the phase-space gap emptied by the binary needed to be refilled before stars could reach the much smaller capture sphere around the single SBH, implying a capture rate much lower than the steady-state value. In the present context, the binary "fills" the gap at the moment of its creation; since the collisional transport rate in phase space is proportional to the gradient of f with respect to $\mathcal{R}$ at $\mathcal{R} = \mathcal{R}_{lc}$, the steep gradients imply an *enhanced* flux into the loss cone, and a faster decay of the binary than if equation (8.117) were used for F.

If we artificially force the loss-cone boundary (i.e., a) to remain fixed, the evolution of f and F can be computed using the time-dependent machinery developed in section 6.1.5. Figure 8.13 shows the results, at one, arbitrarily chosen energy, assuming that the loss cone is empty within $\mathcal{R}_{lc}$ and that N is a constant function of $\mathcal{R}$ outside of $\mathcal{R}_{lc}$ at $t = 0$. The phase-space gradients $\partial N/\partial \mathcal{R}$ decay rapidly at first and then more gradually as they approach the equilibrium solution. It is evident that the total population $\int N d\mathcal{R}$ incurs a decrement of order unity before it has had time to reach the state of collisional equilibrium—the Cohn–Kulsrud solution. Of course, in reality, the loss of stars due to diffusion in L would be compensated for, in part, by their replacement via diffusion in E. But a decrease in N turns out to be a robust result, as discussed in more detail below. In other words, evolution of the binary past $a \approx a_h$ due to collisional loss-cone repopulation, is associated with a certain degree of continued "cusp destruction."

This model can be made one degree more realistic by allowing the binary, and hence the loss-cone boundary, to change with time. At the end of each time step, the binary semimajor axis is corrected for the energy that the binary has exchanged with the ejected stars via equation (8.115). The Fourier–Bessel coefficients in equation (6.127) are then recalculated using the new, advanced $N(\mathcal{E}, \mathcal{R}, t + \Delta t_i)$ and the corrected loss-cone boundary $\mathcal{R}_{lc}(\mathcal{E}, t + \Delta t_i)$. Figure 8.14 shows an example; the physical scalings are similar to those of figure 8.13. In the illustrated model, the nuclear relaxation time is roughly $10\,\mathrm{Gyr}$, and the binary shrinks by a substantial factor beyond its initial value ($\sim a_h$) in this time.

In the models of figures 8.13 and 8.14, the stellar distribution function evolved only with respect to L. Over times as long as nuclear relaxation times, orbital energies will change substantially as well. Qualitatively, we expect diffusion in $\mathcal{E}$ to

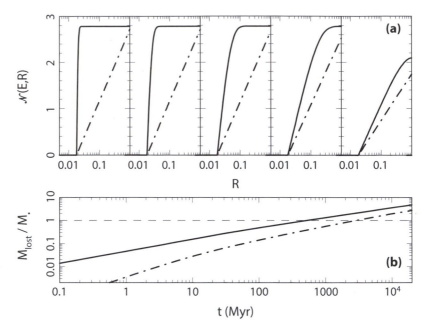

Figure 8.13 This figure shows how the empty loss cone created by a hard, binary SBH would refill due to gravitational encounters [388]. Physical scalings were based on the galaxy M32 assuming an initial separation between the SBHs of 0.1 pc. The inner loss-cone boundary was artificially fixed at $\mathcal{R} = 0.02$; in reality, this boundary would move inward as the binary shrinks. The panels in (a) show slices of the density $N(\mathcal{E}, \mathcal{R}, t)$ at one, arbitrary energy, at times, from left to right, of 10^0, 10^1, 10^2, 10^3, and 10^4 Myr. Also shown as the dotted curve is the equilibrium solution, equation (8.116). In (b), the integrated flux of stars into the loss cone as a function of time is shown as the solid curve, again compared with the steady-state curve. The horizontal dashed line indicates roughly the value of M_{lost} above which the assumption of a static loss-cone boundary would break down due to evolution of the binary.

accelerate the evolution of the binary, since orbits that were depleted by the angular-momentum diffusion can be repopulated by diffusion in energy.

As in chapter 7, we can model the effects of diffusion in energy space on $f(\mathcal{E})$ in an approximate way using equation (7.11):

$$\frac{\partial N}{\partial t} = -\frac{\partial \mathcal{F}_\mathcal{E}}{\partial \mathcal{E}} - F(\mathcal{E}, t). \tag{8.119}$$

This is the isotropic Fokker–Planck equation, with the loss-cone flux added as an energy-dependent "sink" term. Equation (8.119), coupled with an equation like (8.115) for the binary hardening rate, and a prescription for computing the loss-cone flux F given f, then defines the problem. The price we pay for this simplicity, of course, is that we cannot treat the detailed evolution of f with respect to L as in the previous examples.

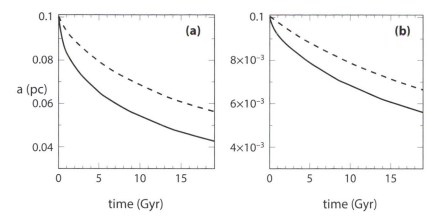

Figure 8.14 Solutions to equation (8.115) for a binary SBH at the center of a galaxy like M32 [388]. The dashed curves use the steady-state loss-cone flux, equation (8.117), while the solid curves are based on a model in which f evolves as in figure 8.13. These models ignore diffusion of stars in energy.

In deriving equation (8.115) for da/dt, the possibility of reejection of stars by the binary was ignored. The effects of the secondary slingshot can be included in an approximate way by modifying the prescription for computing $\Delta\mathcal{E}$, equation (8.114). If $\mathcal{E} + \Delta\mathcal{E}$ is less than zero, corresponding to escape from the galaxy, then no further interactions will occur, and equation (8.114) is correct. If $\mathcal{E} + \Delta\mathcal{E} > 0$, reejection will occur, and it is reasonable to assume that the star will eventually escape, with $\Delta\mathcal{E} = -\mathcal{E}$.

Figure 8.15 shows the results, assuming a binary with $M_{12} = 10^{-3} M_{\text{gal}}$, and two values of M_2/M_1. The loss-cone flux was computed from f and L_{lc} using the Cohn–Kulsrud prescription, which is likely to be valid for times greater than $\sim (M_2/M_1)T_r(r_h)$ (section 6.1.5). The basic physical time determining the rate of evolution of the binary is seen to be the relaxation time, expressed in figure 8.15 in terms of its value at r_m. There is an additional dependence on $N = M_{\text{gal}}/m_\star$, the number of stars in the galaxy. The N dependence appears through $q_{\text{lc}}(\mathcal{E})$ (equation 6.37), which we can write approximately in terms of the relaxation time as

$$q_{\text{lc}}(\mathcal{E}) \approx \frac{P(\mathcal{E})}{\mathcal{R}_{\text{lc}}(\mathcal{E})T_r(\mathcal{E})} \tag{8.120}$$

and so the flux of stars into the binary, equation (6.64), becomes

$$F(\mathcal{E}) = 4\pi^2 q_{\text{lc}} L_c^2 \mathcal{R}_{\text{lc}} \frac{\overline{f}(\mathcal{E})}{\ln(1/\mathcal{R}_0) - 1} \approx 4\pi^2 L_c^2 P T_r^{-1} \frac{\overline{f}(\mathcal{E})}{\ln(1/\mathcal{R}_0) - 1}. \tag{8.121}$$

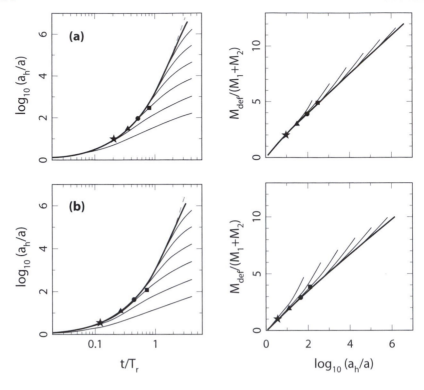

Figure 8.15 Joint evolution of a massive binary and its host galaxy due to gravitational encounters, as modeled via the isotropic Fokker–Planck equation including a sink
term, equation (8.119) [369]. (a) $q = 1$; (b) $q = 0.1$. The panels on the left show
the evolution of the binary semimajor axis a, starting from $a = a_h$. Time is reck-
oned from $a = a_h$ and is plotted in units of the relaxation time at the binary's
influence radius. The panels on the right show the mass deficit as a function of
binary separation, normalized to zero at $t = 0$. Different lines correspond to different values of $N \equiv M_{gal}/m_\star$: $N = 10^6, 10^7, \ldots, 10^{11}, 10^{12}$ (thick line). The
symbols mark the times t_{eq} at which the binary hardening rate equals the gravitational radiation evolution rate, assuming a binary mass of $10^5\, M_\odot$ (squares),
$10^6\, M_\odot$ (circles), $10^7\, M_\odot$ (triangles), and $10^8\, M_\odot$ (stars). Dashed lines in the
left-hand panels are the analytic model described in the text, equation (8.125).

Ignoring changes in the structure of the galaxy, the flux into the binary, integrated
over one relaxation time, scales approximately as

$$F(\mathcal{E})T_r(\mathcal{E}) \propto \left[\ln\left(1/R_0\right) - 1\right]^{-1} \qquad (8.122a)$$

$$\approx \begin{cases} \left[\ln R_{lc}^{-1}\right]^{-1}, & q_{lc} \ll 1, \\ q_{lc}^{-1}, & q_{lc} \gg 1. \end{cases} \qquad (8.122b)$$

The binary-hardening rate is fixed by F and a (equation 8.115), so these expressions imply that the binary's evolution over a specified number of relaxation times

will be smaller for smaller N, that is, larger q_{lc}; while in the large-N limit, the evolution rate at a given a will be determined solely by T_r. The latter regime is of most interest for real galaxies; in the large-N limit, we expect

$$\frac{1}{T_r}\left|\frac{\dot{a}}{a}\right| \equiv \frac{t_{\text{hard}}}{T_r} \propto \ln\left(\frac{a_h}{a}\right),\tag{8.123}$$

that is, the fractional change in a over one relaxation time is weakly dependent on a. The solutions plotted in figure 8.15 can in fact be well represented in terms of the slightly more general form,

$$\frac{1}{T_r}\left|\frac{\dot{a}}{a}\right| = A\ln\left(\frac{a_h}{a}\right) + B\tag{8.124}$$

with $A = 0.016$, $B = 0.08$. The weak dependence of the fitting parameters on binary mass ratio reflects the lack of a mass-ratio dependence in the evolution equations at a given a/a_h.

Integrating equation (8.124) gives a simple expression for the time dependence of the binary semimajor axis:

$$\ln\left(\frac{a_h}{a}\right) = -\frac{B}{A} + \sqrt{\frac{B^2}{A^2} + \frac{2}{A}\frac{t}{T_r(r_m)}},\tag{8.125}$$

where t is defined as the time since the binary first became hard, that is, the time since $a = a_h$. This function is plotted as the dashed curves in figure 8.15.

This model of binary evolution will remain valid until the two SBHs are so close together that energy losses due to gravitational radiation begin to dominate the binary's evolution. Using equation (8.86) together with the $M_\bullet$–σ relation, one finds for the value of $a = a_{eq}$ at which this first occurs,

$$\frac{a_h}{a_{eq}} \approx (315, 93, 27, 8.0)\tag{8.126}$$

for $M_\bullet = (10^5, 10^6, 10^7, 10^8)\, M_\odot$ and $q = M_2/M_1 = 1$, and

$$\frac{a_h}{a_{eq}} \approx (140, 40, 12, 3.5)\tag{8.127}$$

for $q = 0.1$. The corresponding times are

$$t_{eq} \approx \begin{cases} (0.73, 0.53, 0.35, 0.20) \times T_r(r_m), & q = 1, \\ (0.65, 0.54, 0.27, 0.13) \times T_r(r_m), & q = 0.1. \end{cases}\tag{8.128}$$

Evidently, the time required for a massive binary in a spherical galaxy to reach the gravitational radiation regime is of the same order as the relaxation time at the binary's influence radius.

The time to coalescence of the binary can be estimated by supposing that the rate of energy loss from the binary is the sum of the rates due to stellar interactions and gravitational radiation:

$$\frac{d}{dt}\left(\frac{1}{a}\right) = \frac{d}{dt}\left(\frac{1}{a}\right)_{\text{stars}} + \frac{d}{dt}\left(\frac{1}{a}\right)_{\text{GW}}\tag{8.129}$$

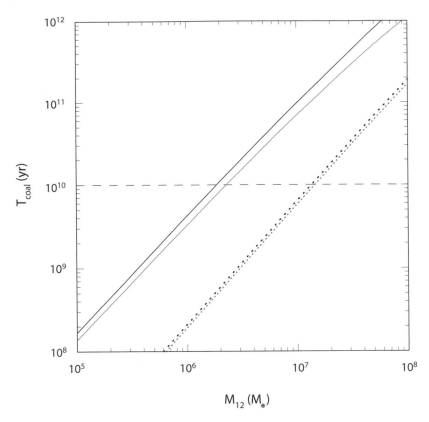

Figure 8.16 The time required for a massive binary in a spherical galaxy to reach coalescence, starting from the hard binary separation $a = a_h$, as a function of binary
mass [369]. The solid curves are derived from equation (8.130) after setting
$y_{max} = \infty$; the black curve is for a binary mass ratio $q = 1$, the gray curve for
$q = 0.1$. Dotted curves show the evolution time from $a = a_{eq}$ to $a = 0$, i.e.,
the time spent in the gravitational radiation regime only. Equation (8.132) gives
accurate analytic approximations to $T_{coal}(M_{12}; q)$.

The first term is given by equation (8.125) and the second by equation (8.86); for
the dependence of $T_r(r_m)$ on binary mass, we can use equation (3.6) after setting
$M_\bullet = M_{12}$. The time for the separation to drop from a_h to a is then given by

$$10^{10} \times \int_0^{y_{max}} \frac{Ay + B}{C + D\,(Ay + B)\,e^{4y}}\ \mathrm{yr}, \tag{8.130}$$

where

$$C = 1.25 M_6^{-1.54}, \tag{8.131a}$$

$$D = 1.75 \times 10^{-9} q^{-3}\,(1 + q)^6\,M_6^{0.65}, \tag{8.131b}$$

$y_{max} = \ln(a_h/a)$, and $M_6 \equiv M_{12}/10^6\,M_\odot$. The full time to coalescence, T_{coal}, starting from a_h is given by setting $y_{max} = \infty$ in this expression. Figure 8.16 shows T_{coal}

as a function of M_{12} for $q = (1, 0.1)$. Shown separately on this figure is the time to evolve from $a = a_{eq}$ to $a = 0$, that is, the time spent in the gravitational radiation regime alone. The latter time is a factor ~ 10 shorter than the total evolution time T_{coal}, which motivates fitting the following functional form to $T_{coal}(M_{12}; q)$:

$$Y = C_1 + C_2 X + C_3 X^2, \tag{8.132a}$$

$$Y \equiv \log_{10}\left(\frac{T_{coal}}{10^{10} \text{ yr}}\right), \tag{8.132b}$$

$$X \equiv \log_{10}\left(\frac{M_{12}}{10^6 \, M_\odot}\right). \tag{8.132c}$$

(This functional form is the integral of equation (8.125).) A least-squares fit to the curves in figure 8.16 gives

$$q = 1 \qquad C_1 = -0.372, \qquad C_2 = 1.384, \qquad C_3 = -0.025, \qquad \text{(8.133a)}$$
$$q = 0.1 \qquad C_1 = -0.478, \qquad C_2 = 1.357, \qquad C_3 = -0.041. \qquad \text{(8.133b)}$$

Based on figure 8.16, binary SBHs in spherical galaxies would be expected to reach gravitational-wave coalescence in 10 Gyr if $M_{12} \lesssim 2 \times 10^6 \, M_\odot$.

Before leaving the spherical paradigm, we consider how evolution of the binary past $a = a_h$ contributes to mass deficits. The right-hand panels of figure 8.15 show that M_{def} correlates well with binary hardness, a_h/a. This dependence is accurately described by

$$\frac{M_{def}}{M_{12}} \approx (1.8, 1.6) \log_{10}(a_h/a), \tag{8.134}$$

where the numbers in parentheses refer to $q = (1, 0.1)$, respectively. The mass deficits generated between formation of a hard binary, and the start of the gravitational radiation regime, are given by setting $a = a_{eq}$ in this expression; that is,

$$M_{def,eq} \approx \begin{cases} (4.5, 3.5, 2.6, 1.6) M_{12}, & q = 1, \\ (3.4, 2.6, 1.7, 0.9) M_{12}, & q = 0.1, \end{cases} \tag{8.135}$$

for $M_{12} = (10^5, 10^6, 10^7, 10^8) \, M_\odot$. These values are somewhat greater than the mass deficits $M_{def,h}$ generated *prior* to the hard-binary phase; that is, $M_{def,h} \approx 0.5 M_{12}$ (equation 8.82).

Why do mass deficits continue to grow? Diffusion in angular momentum causes stars to be lost to the binary, as represented by the term $-F(\mathcal{E}, t)$ on the right-hand side of equation (8.119). We saw earlier in this section how such a term causes the density of stars near the center of the galaxy to drop. But the depleted orbits can be repopulated by diffusion in energy, as represented by the first term on the right-hand side of equation (8.119). As mass deficits increase, so do the gradients in f, which tend to increase the energy flux and counteract the drop in density. In principle, these two terms could balance, at least over some range in energies, allowing the binary to harden without generating a mass deficit. This would require

$$\mathcal{F}_\mathcal{E}(\mathcal{E}) = \int_\mathcal{E}^\infty F(\mathcal{E}) d\mathcal{E}; \tag{8.136}$$

that is, the inward flux of stars due to energy diffusion at energy $\mathcal{E}$ must equal the *integrated* loss to the binary at all energies greater than $\mathcal{E}$. However, at sufficiently great distances from the binary, the relaxation time is so long that the local $\mathcal{F}_{\mathcal{E}}(\mathcal{E})$ drops below the integrated loss term. Growth of a mass deficit reflects the imbalance between these two terms.

It is tempting to associate the large mass deficits predicted by this evolutionary model—as great as several times M_{12} (equation 8.135)—with the large $M_{\text{def}}/M_\bullet$ values observed in a few galaxies (figure 8.9). That association is probably not justified, since the mass deficits plotted in that figure are in galaxies with long nuclear relaxation times. Aside from the Galactic center, no nucleus with $T_r(r_h) \lesssim 10^{10}$ yr is near enough that features on a scale of $r \lesssim r_h$ can be well resolved. Furthermore, assuming that the two SBHs eventually coalesce, evolution in such nuclei will continue on a relaxation timescale via the various mechanisms discussed elsewhere in this book: regeneration of a Bahcall–Wolf cusp at $r \lesssim 0.2r_m$ (section 7.2), nuclear heating by the SBH (section 7.3), and, if the nucleus is sufficiently diffuse, erasure of the "temperature inversion" (section 7.5.3). Galaxies that host SBHs of such low mass also tend to contain nuclear star clusters that are sites of ongoing star formation.

8.4 INTERACTION OF BINARY SUPERMASSIVE BLACK HOLES WITH GAS

Any model for how material is channeled into a single or binary SBH must describe the mechanism by which angular momentum is removed from the matter. In the case of stars, such mechanisms include random gravitational encounters with other stars or with massive perturbers, resonant relaxation, or torques from the mean gravitational field if the galaxy is nonspherical, as discussed in detail in chapters 4–6.

The dissipative nature of gas gives rise to behavior that can differ fundamentally from that of the stars. Possible mechanisms for the loss of angular momentum from gas are even more diverse than in the case of stars, and include torquing of gas flows by the rapidly fluctuating potential of merging galaxies [381] or by nested stellar bars [495], angular momentum transport by hydrodynamical turbulence driven by the onset of self-gravity [175, 494], magnetohydrodynamical turbulence [17], or magnetic braking [54], among many others.

Whatever the dominant mechanism for gas accretion may be, it is believed to operate universally during the epoch in which SBHs grew to their present masses, by driving rapid accretion of material onto preexisting black-hole "seeds." This is the same period in which the galaxy merger rate peaked [273]. While still largely inaccessible to observation due to obscuration [473], the nuclei of merging galaxies, which are also the sites for the formation of binary SBHs, are expected to contain the largest concentrations of dense gas anywhere in the universe. The inevitable abundance of gas suggests that gas dynamics be considered as a supplement to stellar dynamics during this early evolutionary phase.

It is natural to distinguish between two cases depending on whether or not the particles making up the gas are moving faster than the SBHs. The gas particles move at an average velocity given by

$$\frac{3}{2}kT_{\text{gas}} = \frac{1}{2}\mu m_p V_{\text{gas}}^2, \tag{8.137}$$

where μ is the mean particle mass in units of the proton mass m_p and k is Boltzmann's constant. The ratio of V_{gas} to V_{bin}, equation (8.3), is

$$\left(\frac{V_{\text{gas}}}{V_{\text{bin}}}\right)^2 = \frac{3kT_{\text{gas}}}{\mu m_p}\frac{a}{GM_{12}} = \frac{T_{\text{gas}}}{T_{\text{vir}}}, \tag{8.138}$$

where T_{vir} is the "virial temperature":

$$T_{\text{vir}} \equiv \frac{GM_{12}\mu m_p}{3ak} \tag{8.139a}$$

$$\approx 1.7 \times 10^9 \mu \left(\frac{M_{12}}{10^8\,M_\odot}\right)\left(\frac{a}{0.01\,\text{pc}}\right)^{-1} \text{K}. \tag{8.139b}$$

More generally, V_{bin} can be replaced in the definition of T_{vir} by the typical velocities of whatever objects are moving "virially" (i.e., in a steady state) in the local gravitational potential.

A "hot" flow is one in which $T_{\text{gas}} \gtrsim T_{\text{vir}}$, while in "cold" flows, $T_{\text{gas}} \ll T_{\text{vir}}$. The prototype of a hot flow is **Bondi accretion** [56], in which the accreting gas is supported by pressure against free infall toward the accretor. The prototype of a cold flow is a thin disk, in which the gas is rotationally supported against infall. Even in hot flows, gas with nonzero angular momentum that accretes sufficiently close to the central object can achieve rotational support. Astronomical observations offer abundant evidence for both hot and cold gas flows in the immediate vicinity of SBH candidates. The origin and the dynamical impact of the two classes of gas flow are distinct and are discussed here separately.

8.4.1 Interaction with hot gas

Hot gas permeates interstellar space in galaxies, and intergalactic space in groups of galaxies and galaxy clusters. Virial temperatures in these systems range from 10^6 K–10^8 K and the hot gas is almost completely ionized. Various sources contribute to the pool of hot gas. During the early stages of galaxy formation, intergalactic space contained partially ionized gas inherited from the pregalactic, early universe. Hydrogen recombines at redshifts $z \approx 1000$ and is reionized at redshifts $z \approx 10$ by the radiation emitted by the earliest structures. According to the current paradigm, this partially ionized gas cools within the confining gravitational potential of dark matter halos and filaments. Cold gas accelerates toward the halos' centers of gravity and is shock-heated to about the virial temperature. Cooling times at the centers of halos where the gas is the densest are short compared to the dynamical time and thus most of the primordial gas is consumed by star formation on a dynamical timescale.

The rate at which gas cools is determined by the **cooling function** $\Lambda(T)$; multiplying Λ by the gas density squared yields the rate of energy loss per unit volume.

The **cooling time**—the time for a given volume of gas to radiate an energy equal to $(3/2)nVkT$—is then

$$t_{\text{cool}} = \frac{3}{2}\frac{kT}{n\Lambda}. \tag{8.140}$$

Tenuous gas that remains after the cooling time has exceeded the dynamical time in the nascent galaxy might still be plentiful enough to feed a massive black hole growing at an Eddington-limited rate. Equating t_{cool} with r_h/σ yields, for the residual number density at the radius of influence of the SBH,

$$n \approx \frac{\sigma^3 kT_{\text{vir}}}{GM_\bullet\Lambda} \tag{8.141}$$

$$\approx 20\mu\left(\frac{M_\bullet}{10^8\,M_\odot}\right)^{5/(\alpha-1)}\left(\frac{\Lambda}{2\times10^{-23}\,\text{erg cm}^3\,\text{s}^{-1}}\right)^{-1}\text{cm}^{-3};$$

in the latter relation the $M_\bullet$–σ relation (2.33), $M_\bullet \propto \sigma^\alpha$, has been used to relate virial temperature to SBH mass.

This so-called "cooling flow model of quasar fueling" [87, 402] is, however, plagued by many problems. Most of the gas left over from star formation might be blown out by the mechanical feedback associated with the radiative and mechanical output of the accreting SBH, as discussed in section 2.4.5. A small amount of angular momentum in the gas results in circularization and settling into an accretion disk; this disk may be susceptible to fragmentation, thereby converting most of the gas mass into stars and effectively cutting off the supply of gas to the SBH [512].

The geometry of the flow of a hot, magnetized gas near a binary SBH is unknown. Assuming spherical, nonrotating accretion, the timescale on which the hot gas is captured by the SBH is

$$t_{\text{capt}} \equiv \left|\frac{M_\bullet}{\dot{M}}\right| \approx f_b\frac{\sigma^3}{G^2 M\mu m_p n}$$

$$\approx 10^8 f_b\mu^{-2}\left(\frac{M}{10^8\,M_\odot}\right)^{-2/\alpha}\left(\frac{\Lambda}{2\times10^{-23}\,\text{erg cm}^3\,\text{s}^{-1}}\right)\text{yr}, \tag{8.142}$$

where $f_b \approx 1$–10 is a numerical factor that depends on the equation of state of the gas. If a binary SBH is present, gravitational torques from the gas induce decay of the binary's semimajor axes on approximately the same timescale. This crude estimate is based on an analogy with the stellar interactions discussed in the previous sections: the binary must eject of order its own mass in stars to decay an e-folding in separation. Hot gas torquing the binary might be ejected in an outflow and thus the actual rate at which gas is accreting onto individual binary components might be severely suppressed compared to the accretion expected in an isolated SBH.

Galactic nuclei also contain hot gas produced by secondary sources. For example, observations with the Chandra X-ray Observatory have revealed tenuous ($n \approx 10$–$100\,\text{cm}^{-3}$), hot ($kT \approx 1\,\text{keV}$) plasma within a parsec of the Milky Way SBH [13]. This plasma is apparently being generated by the numerous massive, evolved stars in the Galactic center region through stellar wind and supernova activity. Since its temperature is higher than T_{vir}, most ($> 99\%$) of the plasma should escape the neighborhood of the SBH. While the hot gas densities in active galaxies might be transiently larger than that at the Galactic center, the tendency of the hot plasma to escape the neighborhood of the SBH reduces the likelihood that large quantities of virialized gas would remain enmeshed with the binary's orbit long enough to affect its dynamical evolution. If it did, however, the effect of the hot gas would be to induce a source of gravitational drag in addition to dynamical friction from the stars, thus assisting in the decay of the binary's orbit [146, 147].

8.4.2 Interaction with cold gas

The specific angular momentum of a cold gas flow might easily exceed that of a binary SBH. In this case, the gas tends to settle into rotationally supported, geometrically thin rings and disks. Such gas is "cold" in the sense $T_{gas} < T_{vir}$, but it can still be hot enough to be ionized; that is, $T_{gas} \gtrsim 10^4$ K.

Observations offer abundant evidence for the presence of dense gas in galactic nuclei. As discussed in chapter 2, thin molecular disks on scales 0.1–0.5 pc have been seen in the water maser emission in the nuclei of some Seyfert galaxies, most famously in NGC 4258 [216]. The Galactic SBH is surrounded by a molecular gas torus of mass $\sim 10^4$–$10^5 M_\odot$ at distances beyond ~ 1 pc from the SBH [255]. Compact stellar disks on scales $\gtrsim 20$ pc, which are most likely fossil relics of pre-existing gas disks, are seen in the nuclei of many galaxies [430]. Massive accretion disks must be present in quasars in order to account for what appears to be rapid accretion onto the central SBHs in these systems. However, the structure of these disks at radii comparable to the size of a hard SBH binary is unknown.

If a disk surrounding a binary SBH is initially inclined with respect to the binary's orbital plane, the quadrupole component of the binary's gravitational potential causes differential precession in the disk at the rate [307]

$$\Omega_{prec}(r) = \frac{3}{4} \frac{q}{(1+q)^2} \frac{(GM_{12})^{1/2}a^2}{r^{7/2}}, \quad q \equiv \frac{M_2}{M_1}, \tag{8.143}$$

which results in a warping of the disk. Depending on the details of the gas viscosity, the warp either dissipates, or smears around the binary, resulting ultimately in a nearly axisymmetric disk in the binary's orbital plane.

Interest in coplanar, circumbinary disks stems from their ability to extract a binary's angular momentum via a form of tidal coupling. Two interrelated questions might be posed: First, what is the response of a circumbinary disk to the binary's tidal forcing? Second, how does such a disk affect the evolution of the binary's orbit?

Existing attempts to answer these questions have employed rather ad hoc models for the form of the binary-disk torque coupling [438, 254], or have been restricted to binaries with components of very unequal mass [9], where an array of neighboring resonances facilitate binary-disk coupling [199], much like the coupling between a massive planet and its natal gas disk [200]. Early numerical simulations of circumbinary disks with nearly equal masses [10], however, suggested that the disks are truncated exterior to the resonances, which was interpreted as a consequence of a collisionless nonlinear parametric instability [467, 144]. Fluid dynamical theory of circumbinary disk truncation is still lacking.

The **outer Lindblad resonances** are radii in the disk where the natural, epicyclic frequency of radial oscillation is an integer multiple of the rate at which a packet of disk gas receives tidal "kicks" by the binary. In the case of a circular binary, these resonances are located at radii $r_{\rm m}/a = (1 + 1/m)^{2/3}$, where $m = 1, 2, \ldots$ is the order in the decomposition of the binary's gravitational potential into multipoles:

$$\varphi(r, \theta) = \sum_{m=0}^{\infty} \varphi_m(r) \cos\left[m\left(\theta - \Omega_{\rm bin}t\right)\right]. \tag{8.144}$$

The outermost resonance ($m = 1$) is located at $r \approx 1.6a$. The forcing near a resonance, as well as at a radius where surface density in the disk exhibits a large gradient, excites nonaxisymmetric propagating disturbances, or "density waves," in the disk.

The gravitational potential of eccentric binaries contains low-frequency components that are absent in circular binaries. These low-frequency components activate resonances located at larger radii than in the circular case, and might lead to mutual excitation and reinforcement of the binary and the disk eccentricities [413, 198]. Many extrasolar planets, which are thought to form in circumstellar disks, are notably eccentric, suggesting that dynamical coupling between a binary point mass (a star and a planet, or a pair of black holes) and a gas disk is conducive to eccentricity growth. The observed circumbinary disks in young stellar binaries are typically eccentric, are truncated at radii a few times the semimajor axis [343], which lends support to this hypothesis.

Density waves transport angular momentum outward through the circumbinary disk, and angular momentum carried by the waves is extracted from the binary's angular momentum. The binary experiences a negative torque equal and opposite to the total angular momentum flux transferred to the disk. The location of the inner edge of the disk reflects a balance between the angular momentum flux deposited into the disk, and the angular momentum flux transported through the disk by another, possibly viscous mechanism. Wave momentum is deposited into the disk material via a form of dissipative damping. The location in the disk where the waves are damped can be separated by many wavelengths from the location where they are excited. The damping could take place in the nonlinear steepening and the breaking of wave crests [477, 445]. In marginally optically thick disks, radiation damping might also play a role [75]. Yet another form of damping could be due to the dissipation of wave shear if the disk is strongly viscous [511]. The amplitude of the density waves is a steeply decreasing function of the radius of excitation. The amplitude is diminished if the waves are nonlinear at excitation and damp in situ, but then one expects the inner edge to recede where in situ damping shuts off.

The intricate and insufficiently understood nature of binary–disk interactions calls for grid-based hydrodynamical simulations with a shock-capturing capability. The necessity that the radial wavelength, which is smaller than the vertical scale height of the disk, be resolved by multiple cells, places severe demands on the computational resources, especially if a three-dimensional representation of the disk is required. It should also be noted that the radiative and thermal structure of accretion disks around *single* SBHs are not adequately understood on any radial scale.

As a binary's semimajor axis decreases due to stellar, gas dynamical, or gravitational radiation processes, a circumbinary disk's inner edge spreads inward viscously while maintaining constant edge-to-semimajor axis ratio, for example, $r_{edge}/a \sim 2$. In the final stages of the gravitational radiation-driven inspiral, however, the timescale on which the semimajor axis decays becomes shorter than the viscous timescale, and the disk can no longer keep up with the binary, resulting in binary–disk detachment. On the relevant length scales the disk might be dominated by radiation pressure and the electron scattering opacity; the structure and the stability of such disks is an active research area [525].

8.5 SIMULATIONS OF GALAXY MERGERS

In the stellar spheroids that exhibit evidence for the "scouring" effects of binary SBHs, nuclear relaxation times are always extremely long, $\gtrsim 10^{14}$ yr (figure 3.1). Repopulation of a massive binary's loss cone by star–star gravitational encounters, as discussed in section 8.3, would act very slowly in such galaxies: far too slowly to bring the binary to coalescence in 10 Gyr (figure 8.16). This is just another way of stating the "final-parsec problem."

On the other hand, the feeding rate of stars to *single* SBHs can be much higher in axisymmetric and triaxial galaxies than in spherical ones, and the same is presumably true for the feeding of stars to binary SBHs [576, 371]. Furthermore, galaxy mergers are rather asymmetrical events, known to be capable of converting spherical or axisymmetric galaxies into systems with more complicated shapes.

The first simulations of galaxy mergers including SBHs with sufficient resolution to follow the formation and evolution of a massive binary were only carried out in 2011 [277]. Figure 8.17 shows the results from such a simulation, in which the merging galaxies (of mass ratio 1 : 3) contained collisionally relaxed, multimass density cusps prior to the merger, intended to represent nuclear star clusters [220]. The binary SBHs created in these simulations continue to harden, long past the hard-binary stage, at a rate consistent with the "full-loss-cone" rate in a spherical galaxy, in spite of the fact that the number of particles in the simulations ($\sim 10^6$) implies a low rate of collisional loss-cone refilling. Partly because of the dense, preexisting nuclei, and partly because of the complex shape of the merger remnant, the supply of stars to the massive binary remains high.

Much more work along these lines needs to be carried out; in particular, it has not yet been established, using the torus-construction machinery described in section 3.1, precisely which orbits in the N-body models are responsible for the efficient loss-cone repopulation. Only after this is done will it be safe to extrapolate the N-body results to real galaxies. But a preliminary conclusion is that the final-parsec problem may not be as much of a problem as once believed.

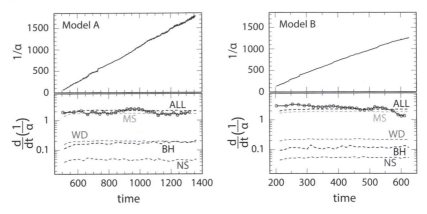

Figure 8.17 Evolution of the inverse binary semimajor axis (upper) and the binary hardening rate (lower) in N-body simulations of $3:1$ galaxy mergers [220]. The points
represent the N-body results while the dashed lines represent the semianalytic
estimates for the different mass groups, assuming a full loss cone. Model A
merged the galaxies on a circular orbit, and Model B from an eccentric orbit.
In spite of the large number of particles in the simulations, and the long relaxation time, the massive binary continues to harden as if its loss cone were
full—presumably due to the large number of centrophilic orbits. These simulations began from galaxies with collisionally relaxed nuclei containing four mass
groups around the SBH, as in figure 7.8.

The state of the art of simulation of gas-rich galaxy mergers is improving. As
of this writing, such simulations have resolutions of the order of $\sim 10^{-2}$ times
the half-mass radius of the merging systems, or $\sim 10\,\mathrm{pc}$ for a galaxy of half-mass
radius $1\,\mathrm{kpc}$ [341, 67]. This is far too coarse a resolution for treating the dynamics
of gas on the scale of a hardening binary SBH, and so such "subgrid physics" can
only be included in a very approximate way.

8.6 DYNAMICS OF INTERMEDIATE-MASS BLACK HOLES

As discussed in chapter 2, intermediate-mass black holes (IBHs) are objects whose
existence and mode of formation are still somewhat speculative. Even the range of
masses associated with this hypothetical class of object differs from author to author. From the point of view of nuclear dynamics, however, we can usefully define
an IBH as one whose mass satisfies

$$m_\star \ll M_{\mathrm{IBH}} \ll M_\bullet, \tag{8.145}$$

with $M_\bullet$, as always, the mass of the SBH at the galaxy's center. The condition
$M_{\mathrm{IBH}} \ll M_\bullet$ implies that the majority of stars moving within the SBH's sphere of
influence are essentially unaffected by the presence of the IBH, except when they
happen to come close to it. The condition $m_\star \ll M_{\mathrm{IBH}}$ means that perturbations

by stars of the motion of the IBH will take the form of a dynamical friction force; the effects of scattering by field stars will be negligible. These two approximations allow the interaction between an IBH and the surrounding stars to be computed in a simpler and more satisfying way than in the case of binary SBHs in which the two components are comparably massive.

8.6.1 Orbital decay

Since $M_{\text{IBH}} \gg m_\star$, the orbit of the IBH will gradually decay due to dynamical friction. Its eccentricity will also change; as we will see, that change is often expected to be in the direction of increasing eccentricity.

Since $M_{\text{IBH}} \ll M_\bullet$, we assume that the field-star distribution is unaffected by the presence of the IBH (aside, of course, from the dynamical-friction wake) and that changes in its orbit occur slowly enough that we can average the accelerations over an orbit. We also assume that the IBH is orbiting inside the sphere of influence of the SBH. Finally, we ignore for the moment the possibility of a significant mass in bound stars around the IBH, in spite of the fact that some models for IBH formation invoke stellar clusters.

The local diffusion coefficient describing changes in the scaled angular momentum variable $\mathcal{R} = L^2/L_c^2(E)$ is given by equation (5.167a). Near the SBH, $\mathcal{R} \approx 1 - e^2$, with e the eccentricity of the Keplerian orbit. Invoking our assumption $M_{\text{IBH}} \gg m_\star$, equation (5.167a) becomes

$$\langle \Delta \mathcal{R} \rangle = \frac{g'}{g} \mathcal{R} \langle \Delta E \rangle + 2gL \langle \Delta L \rangle \tag{8.146a}$$

$$= \langle \Delta v_\parallel \rangle \left(\frac{g'}{g} \mathcal{R} v + 2g \frac{L^2}{v} \right), \tag{8.146b}$$

where $g(E) = 1/L_c^2(E)$, and equations (5.124) and (5.125) have been used to express $\langle \Delta E \rangle$ and $\langle \Delta L \rangle$ in terms of the velocity-space diffusion coefficient $\langle \Delta v_\parallel \rangle$. Near the SBH, $L_c^2(E) = (GM_\bullet)^2/(2\mathcal{E})$ and

$$\frac{\langle \Delta \mathcal{R} \rangle}{\mathcal{R}} = \frac{\langle \Delta E \rangle}{E} + 2\frac{\langle \Delta L \rangle}{L} \tag{8.147a}$$

$$= \langle \Delta v_\parallel \rangle \left(\frac{v}{E} + \frac{2}{v} \right). \tag{8.147b}$$

If we adopt Chandrasekhar's first approximation, equation (5.23), to $\langle \Delta v_\parallel \rangle$, this becomes

$$\frac{\langle \Delta \mathcal{R} \rangle}{\mathcal{R}} = -4\pi G^2 M_{\text{IBH}} \ln \Lambda \frac{\rho(v_f < v)}{v^3} \left(\frac{v^2}{E} + 2 \right), \tag{8.148}$$

with $\rho(v_f < v)$ the density in field stars with velocities less than v.

Suppose that the field-star density is a power law in radius, $\rho(r) = \rho_0(r/r_0)^{-\gamma}$, and that the field-star velocity distribution is isotropic. Then $f(v_f)$ is given by equation (3.49), and

$$\frac{\rho(v_f < v, r)}{\rho(r)} = \frac{4}{\pi^{1/2}} \frac{\Gamma(\gamma + 1)}{\Gamma(\gamma - 1/2)} \int_0^{v/v_{\mathrm{esc}}(r)} dy \, y^2 \left(1 - y^2\right)^{\gamma - 3/2}, \tag{8.149}$$

where $v_{\mathrm{esc}}(r) = \sqrt{2GM_\bullet/r}$ is the escape velocity at r.

The final step is to orbit-average the local diffusion coefficient. Before doing so, we note an interesting special case [206]. If $\gamma = 3/2$, $f(\mathcal{E}) = f_0$, and equation (8.149) becomes simply

$$\frac{\rho(v_f < v)}{\rho(r)} = \left(\frac{v}{v_{\mathrm{esc}}}\right)^3 \tag{8.150}$$

and

$$\frac{\langle \Delta \mathcal{R} \rangle}{\mathcal{R}} = -4\pi G^2 M_{\mathrm{IBH}} \ln \Lambda \left[\frac{\rho(r)}{v_{\mathrm{esc}}^3(r)}\right] \left(\frac{v^2}{E} + 2\right), \quad \gamma = 3/2. \tag{8.151}$$

Since $v_{\mathrm{esc}} \propto r^{-1/2}$, the quantity in square brackets is constant. Furthermore, the time average of v^2 for a Keplerian orbit is $\overline{v^2} = -2E$. Hence $\langle \Delta \mathcal{R} \rangle_t = 0$ in this case: orbital eccentricity is conserved. Intuitively, we expect that for $\gamma > 3/2$, the greater density near periapsis will imply orbital circularization, and that for $\gamma < 3/2$, orbits should become more eccentric.

Figure 8.18 verifies this expectation. Plotted there is

$$\frac{\langle \Delta \mathcal{R} \rangle_t}{\mathcal{R}} \bigg/ \frac{\langle \Delta E \rangle_t}{E}, \tag{8.152}$$

the fractional change in $\mathcal{R} = 1 - e^2$ that occurs in one decay time of E or a. Except when γ is near 3/2, the fractional change is of order unity when e is large, falling to zero for circular orbits. In other words, substantial eccentricity increases should be expected for intermediate-mass-ratio inspirals in nuclei with density profiles flatter than $\sim r^{-3/2}$.

Now for a few caveats. As noted in section 7.2, the standard dynamical-friction formula, equation (5.23), predicts zero frictional force on a circular orbit in the $\gamma \to 1/2$ limit—an unphysical consequence of "taking the Coulomb logarithm out of the integral." (It is easy to show that the same is true for noncircular orbits when $\gamma = 1/2$.) It was argued there that field stars with $v_f > v$ become increasingly important as $\gamma \to 1/2$; these were ignored in the treatment just presented. Furthermore, if the core was created by a binary SBH, an isotropic velocity distribution may be unsuitable; instead there may be a genuine "hole" in configuration space corresponding to the ejection of stars that came close to the preexisting binary.

In spite of these caveats, the conclusion that eccentricities should increase during inspiral into cores is robust (figure 8.19).

8.6.2 Ejection of stars

A star that passes sufficiently close to an IBH can receive a velocity perturbation large enough for it to escape; in other words, large enough that the

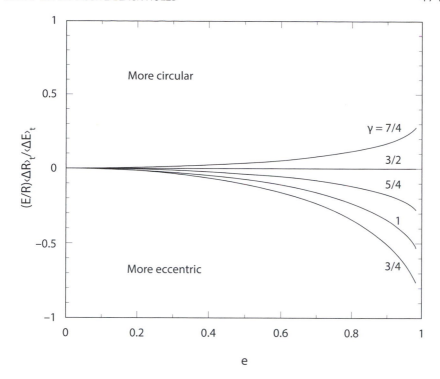

Figure 8.18 Changes in eccentricity for a test mass that inspirals toward an SBH, assuming a field-star density of $\rho \propto r^{-\gamma}$. The potential is assumed to be Keplerian, i.e., the distributed mass inside the orbit is much less than $M_\bullet$. The vertical axis is the fractional change in $\mathcal{R} = 1 - e^2$ in a time $E/\langle \Delta E \rangle_t$, i.e., the orbital decay time. For $\gamma = 3/2$, dynamical friction leaves the eccentricity unchanged; for larger γ, orbits circularize, and for smaller γ, they become more eccentric. These curves were computed as discussed in the text, using Chandrasekhar's diffusion coefficients in their standard forms.

star's velocity, after moving beyond the sphere of influence of the IBH, exceeds $v_{\rm esc} = (2GM_\bullet/r)^{1/2}$, the local escape velocity from the SBH. Escape was discussed in section 8.1 in the context of three-body scattering experiments; $v_{\rm ej}$ was defined as the final velocity of a star that was moving fast enough, after interacting with the massive binary, to escape from the galaxy core. In the scattering experiments, *all* stars are unbound from the massive binary, both before and after the interaction, and some consideration needed to be given to the precise definition of $v_{\rm ej}$. In the case of an IBH–SBH binary, we can compute the rate of escapers in a more definite way by invoking our assumption that $M_{\rm IBH} \ll M_\bullet$ [317, 406].

To simplify the notation, we use $\boldsymbol{v}$ to denote field-star velocities (rather than $\boldsymbol{v}_f$) and $\boldsymbol{v}_{\rm IBH}$ to denote the orbital velocity of the IBH around the SBH. Let $f(\boldsymbol{x}, \boldsymbol{v})$ be the steady-state distribution function describing stars that move in the fixed gravitational potential of the SBH at radius r; since $M_{\rm IBH} \ll M_\bullet$, we ignore the

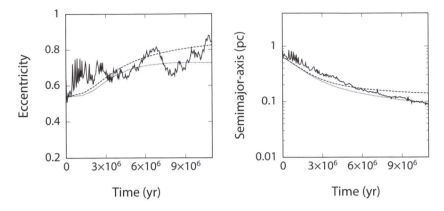

Figure 8.19 *N*-body simulation of the inspiral of a 5000 $M_\odot$ IBH into the center of the Milky Way [8]. The initial stellar density was $\rho \sim r^{-0.6}$ inside a core of radius $r_c \approx 0.3\,\mathrm{pc}$, falling off more steeply outside. The initial IBH orbit was eccentric, $e \approx 0.5$; as the orbit decays it becomes more eccentric, for the reasons discussed in the text. The dashed curves show the predictions of Chandrasekhar's dynamical-friction formula in its standard form, equation (5.23), which ignores the contribution from fast-moving stars; the dotted curve uses a more exact version that includes the contributions from these stars.

influence of the IBH on the field-star velocities when writing f, and $f = 0$ for $v > v_{\mathrm{esc}}(r) = (2GM_\bullet/r)^{1/2}$. The v that appears as an argument of f is understood to be the velocity of a star before it has come near enough to the IBH for its motion to be affected by it. Since stars will be affected by the IBH only when they are very close, the r that appears in f can be identified with the radius of the IBH's orbit.

Define v_{in} and v_{out} as the velocity of a star before and after it has passed near to the IBH, respectively (figure 8.20). The relative velocity vectors are

$$w_{\mathrm{in}} \equiv v_{\mathrm{in}} - v_{\mathrm{IBH}}, \qquad w_{\mathrm{out}} \equiv v_{\mathrm{out}} - v_{\mathrm{IBH}}. \tag{8.153}$$

Because the velocity change occurs very near to the IBH, it takes place in a time short compared with the time for v_{IBH} to change; hence we can write

$$|w_{\mathrm{in}}| = |w_{\mathrm{out}}| = w. \tag{8.154}$$

Stars of interest have initial velocities $v_{\mathrm{in}} < v_{\mathrm{esc}}$; that is,

$$|v_{\mathrm{IBH}} + w_{\mathrm{in}}| < v_{\mathrm{esc}}. \tag{8.155}$$

Let $v_{\mathrm{ej}} \geq v_{\mathrm{esc}}(r)$ be the minimum final velocity of a star that is considered to have escaped from the SBH. Then escapers satisfy

$$|v_{\mathrm{IBH}} + w_{\mathrm{out}}| > v_{\mathrm{ej}}. \tag{8.156}$$

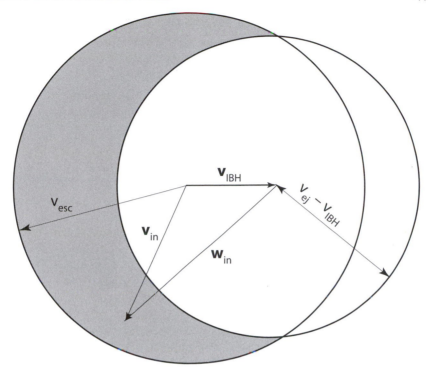

Figure 8.20 This figure shows, in velocity space, the vectors that are useful in describing the scattering of stars by an IBH. The initial stellar velocities, $\boldsymbol{v}_{\mathrm{in}}$, lie inside the sphere of radius v_{esc}, the local escape velocity from the SBH. Stars with velocities in the shaded region are able to be scattered by the IBH to velocities greater than v_{ej}.

This can be written as the joint conditions

$$w > v_{\mathrm{ej}} - v_{\mathrm{IBH}}, \tag{8.157a}$$

$$\cos \phi_{\mathrm{out}} > \cos \phi_0 = \frac{v_{\mathrm{ej}}^2 - v_{\mathrm{IBH}}^2 - w^2}{2 v_{\mathrm{IBH}} w}, \tag{8.157b}$$

where ϕ_{out} is the angle between $\boldsymbol{v}_{\mathrm{IBH}}$ and $\boldsymbol{w}_{\mathrm{out}}$.

However, not all stars satisfying the conditions (8.155), (8.157) will escape. Let $\Sigma(\boldsymbol{w}_{\mathrm{in}})$ be the cross section for a star with initial velocity $\boldsymbol{w}_{\mathrm{in}}$ to be scattered into the velocity-space volume defined by equation (8.157b). In terms of Σ, the rate at which stars are ejected with velocities greater than v_{ej} is

$$\dot{N}_{\mathrm{ej}} = \int f(\boldsymbol{v}_{\mathrm{in}}) w \, \Sigma(\boldsymbol{w}_{\mathrm{in}}) d^3 \boldsymbol{v}_{\mathrm{in}}. \tag{8.158}$$

The region of integration is shown as the shaded region in figure 8.20.

To calculate Σ, we first relate the initial and final velocities using the quantities defined in figure 5.2:

$$\boldsymbol{w}_{\text{out}} = \boldsymbol{w}_{\text{in}} \cos \chi - w \frac{\boldsymbol{p}}{p} \sin \chi, \qquad (8.159)$$

where $\boldsymbol{p}$ is a vector whose magnitude is p and which is directed from the IBH toward the star's initial trajectory as shown in the figure. Using equations (5.14) we can also write

$$\tan \left(\frac{\chi}{2} \right) = \frac{G M_{\text{IBH}}}{w^2 p}. \qquad (8.160)$$

From equation (8.159) we see that

$$\cos \phi_{\text{out}} = \cos \phi_{\text{in}} \cos \chi + \sin \phi_{\text{in}} \sin \chi \cos \phi_b, \qquad (8.161)$$

where ϕ_b is the polar angle in the $\boldsymbol{p}$ plane. Equations (8.160) and (8.161), together with equation (8.157b), define a finite domain in the $\boldsymbol{p}$ plane, the area of which is

$$\Sigma(\boldsymbol{w}_{\text{in}}) = \frac{\pi G^2 M_{\text{IBH}}^2}{w^4} \frac{\sin^2 \phi_0}{(\cos \phi_{\text{in}} - \cos \phi_0)^2}, \qquad (8.162)$$

where ϕ_{in} is the angle between $\boldsymbol{v}_{\text{IBH}}$ and $\boldsymbol{w}_{\text{in}}$.

In the previous section, we noted that a field-star density $n \propto r^{-3/2}$ has an isotropic distribution function at $r \ll r_{\text{h}}$ with the simple form $f = f_0 = \text{constant}$. Assuming this form for f, the ejection rate integral (8.158) is analytically tractable, and the rate of ejection at speeds greater than v_{ej} is

$$\dot{N}(v_{\text{out}} > v_{\text{ej}}) = \frac{3\pi}{2} \frac{G^2 M_{\text{IBH}}^2}{v_{\text{esc}}^3} n(r) R(\delta, \lambda)$$

$$\approx 1.4 \times 10^{-6} \left(\frac{M_{\text{IBH}}}{10^3 M_{\odot}} \right)^2 \left(\frac{M_{\bullet}}{4 \times 10^6 M_{\odot}} \right)^{-3/2}$$

$$\times \left(\frac{n_{1\text{pc}}}{10^5 \text{ pc}^{-3}} \right) R(\delta, \lambda) \text{ yr}^{-1}, \qquad (8.163a)$$

$$R(\delta, \lambda) \equiv \frac{-(\lambda - 1)^3 + 8\delta^3}{3(\lambda^2 - 1)\delta} - \ln \left(\frac{1 + \delta}{\lambda - \delta} \right),$$

$$1 \le \lambda \le 2\delta + 1, \qquad (8.163b)$$

where $\delta \equiv v_{\text{IBH}}/v_{\text{esc}} < 1$, $\lambda \equiv v_{\text{ej}}/v_{\text{esc}} > 1$. The differential ejection rate at v_{out} is

$$\left. \frac{d\dot{N}}{dv} \right|_{v=v_{\text{out}}} = \frac{3\pi}{2} \frac{G^2 M_{\text{IBH}}^2}{v_{\text{esc}}^4} n(r) S(\delta, v),$$

$$S(\delta, v) \equiv \frac{v(v - 2\delta - 1)^2 (v^2 + 2v - 4\delta^2 + 4\delta - 3)}{3\delta(v - \delta)(v^2 - 1)^2}, \qquad (8.164)$$

where $v \equiv v_{\text{out}}/v_{\text{esc}}$.

If the IBH is in a circular orbit about the SBH, $\delta = 2^{-1/2}$; figure 8.21, (left), plots $R(\lambda)$ for this case. The radial dependence of $\dot{N}$ comes entirely from the dependence of R on λ for this choice of density profile.

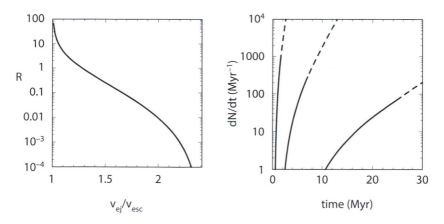

Figure 8.21 Left: The dimensionless quantity R, equation (8.163), that is proportional to the rate of ejection of stars with $v > v_{ej}$ by an IBH; $v_{esc} = (2GM_\bullet/r)^{1/2}$ is the escape velocity from the SBH. The IBH is assumed to be in a circular orbit around the SBH, and the field-star density is $\rho \propto r^{-3/2}$. Right: The rate of ejection of stars with velocities large enough to escape from the galaxy as a function of time, for inspiraling IBHs with three different masses: $10^3\,M_\odot$ (lower), $3 \times 10^3\,M_\odot$, and $10^4\,M_\odot$ (upper). In each case the inspiral began from a distance of 0.4 pc. Other parameters were chosen based on the Milky Way (see text). The curves change to dashed at the radius where the enclosed stellar mass is equal to the IBH mass; in this regime, changes in the stellar distribution due to the presence of the IBH could no longer be ignored. One expects the ejection rate to peak at roughly this time.

Writing $\rho(r) = \rho_0(r/r_0)^{-3/2}$ for the field-star density, and again assuming a circular orbit, equations (5.23), (8.54), and (8.150) give for the dependence of the IBH's orbital radius on time,

$$r(t) = r(0)e^{-t/T_{df}}, \tag{8.165a}$$

$$T_{df} = \frac{1}{8^{1/2}\pi} \frac{M_\bullet^{3/2}}{G^{1/2}M_{IBH}\ln\Lambda} \frac{1}{\rho_0 r_0^{3/2}} \tag{8.165b}$$

$$\approx 27 \left(\frac{M_\bullet}{4 \times 10^6\,M_\odot}\right)^{3/2} \left(\frac{M_{IBH}}{10^3\,M_\odot}\right)^{-1} \left(\frac{\rho(1\,pc)}{10^5\,M_\odot\,pc^{-3}}\right)^{-1}$$

$$\times \left(\frac{\ln\Lambda}{5}\right)^{-1} \text{Myr.} \tag{8.165c}$$

We can compute the time dependence of the ejection rate using equations (8.163)–(8.165), after specifying a value for v_{ej}, the minimum velocity of stars that are considered to have "escaped."

If we are interested in the stars that escape completely from the galaxy, it is appropriate to set

$$v_{ej}^2(r) = v_{esc}^2(r) + v_{galaxy}^2, \tag{8.166}$$

where v_{galaxy} is the escape velocity from the bottom of the galaxy's potential well. In the case of the Milky Way, an estimate of the latter is 10^3 km s^{-1}. Figure 8.21, (right), plots the result, assuming $\rho(1pc) = 5 \times 10^5 \, M_\odot \, pc^{-3}$, a starting radius of 0.4 pc, and three different values for M_{IBH}. Such stars have the correct kinematical properties to be associated with the "hypervelocity stars" discussed in section 6.3. However, it is currently unclear whether the inferred ejection times of these stars is consistent with the highly peaked rate of production predicted by this model [62].

8.6.3 Perturbation of stars by an IBH in a circular orbit

In addition to producing hypervelocity stars, an IBH can also affect stellar motions in a more gentle way, leaving them bound to the SBH but with altered orbits [44, 319]. Understanding this sort of perturbation has long been the bread and butter of celestial mechanicians; for instance, when predicting the motion of comets or asteroids that have been perturbed by Jupiter. In fact, the ratio between the mass of a $10^3 \, M_\odot$ IBH and a $10^6 \, M_\odot$ SBH is similar to the Jupiter/Sun mass ratio. If we assume that the orbit of the IBH remains fixed for many periods—in other words, that its inspiral has "stalled"—then the analogy is almost complete.

In the case of the solar system however, Jupiter's orbit is known to be nearly circular ($e \approx 0.05$), and a great deal of work has been devoted to this special case—the the **circular restricted three-body problem**. (Recall from chapter 4 that "restricted" means that the mass of the third body—in our case, a star—is negligible compared with the other two.) If the two massive bodies are moving in circular orbits, the distance between them is fixed, and each body moves about its common center of mass at a fixed angular velocity $n = 2\pi/P$, with P the binary period. It is well known that by transforming to a frame that rotates about the center of mass with frequency n, a conserved quantity appears in the motion of a test body, the **Jacobi constant**, given by

$$H_J = n^2 \left(\xi^2 + \eta^2\right) + 2\left(\frac{GM_\bullet}{r_1} + \frac{GM_{IBH}}{r_2}\right) - \dot{\xi}^2 - \dot{\eta}^2 - \dot{\zeta}^2, \tag{8.167}$$

where (ξ, η, ζ) are Cartesian coordinates in the rotating frame, (r_1, r_2) are, respectively, the distance of the test mass (star) from the SBH and the IBH, and the z-axis is identified with the binary's axis of rotation. The first term in equation (8.167) represents the centripetal acceleration. Transforming back to an inertial frame, H_J can be written

$$H_J = 2\left(\frac{GM_\bullet}{r_1} + \frac{GM_{IBH}}{r_2}\right) + 2nL_z - \dot{x}^2 - \dot{y}^2 - \dot{z}^2, \tag{8.168}$$

where $L_z = x\dot{y} - y\dot{x}$ is the specific angular momentum of the star about the binary center of mass. Note that L_z itself is not conserved; rather, only this particular combination of L_z and the energy.

Conservation of Jacobi's constant does not greatly restrict the motion of the star, which can in principle reach any point within the zero-velocity surface. But if we assume that $M_{IBH} \ll M_\bullet$, and that close encounters between star and IBH are rare, an alternate form for H_J can be derived that is more useful [516]. We first replace $\dot{x}^2 + \dot{y}^2 + \dot{z}^2$ in equation (8.168) by the approximation

$$v^2 = GM_\bullet \left(\frac{2}{r} - \frac{1}{a} \right); \tag{8.169}$$

in other words, we assume that the motion of the star is close to that of a Keplerian orbit about the dominant mass, with semimajor axis a, and we identify r_1 with r. We can also replace L_z by

$$L \cos i = \sqrt{GM_\bullet a(1 - e^2)} \cos i, \tag{8.170}$$

with i the inclination of the star's orbit with respect to the plane of the binary, and e its eccentricity. The Jacobi constant then becomes

$$H_J \approx 2 \left(\frac{GM_\bullet}{r_1} + \frac{GM_{IBH}}{r_2} \right)$$
$$+ 2n \cos i \sqrt{GM_\bullet a(1 - e^2)} - GM_\bullet \left(\frac{2}{r} - \frac{1}{a} \right). \tag{8.171}$$

Invoking our assumption of no close encounters allows us to neglect M_{IBH}/r_2 compared with $M_\bullet/r_1$, and again identifying r_1 with r, we find

$$H_J \approx \frac{2GM_\bullet}{a_{IBH}} \left\{ \frac{a_{IBH}}{2a} + \left[\frac{a}{a_{IBH}}(1 - e^2) \right]^{1/2} \cos i \right\}, \tag{8.172}$$

where n has been expressed in terms of the semimajor axis of the inner binary via $n^2 = GM_\bullet / a_{IBH}^3$.

The quantity in braces in equation (8.172) is called **Tisserand's parameter**, or T. "Tisserand's relation" is the statement that $T(a', e', i') = T(a, e, i)$, where (a, e, i) and (a', e', i') are the orbital elements of a comet (say) before and after an encounter with Jupiter (say). Tisserand's relation has been used to determine whether a "new" comet is actually a previously discovered one whose orbit has been altered by interaction with Jupiter.

In the context of galactic nuclei, Tisserand's relation can be used in the following way [44]. Imagine that an IBH is orbiting, undetected, somewhere near the Galactic center. Suppose in addition that this IBH was once associated with a cluster of stars, which were subsequently tidally dispersed, into a disk; and then perturbed again, by interactions with the IBH [226]. Finally, suppose that one knows which stars were brought into the center in this way. For instance, they might be the S-stars, which are young enough to be associated with a recent inspiral and whose current orbital elements are well determined (table 4.1).

When these stars were still in a disk, each had roughly the same $a = a_{IBH}$ and some eccentricity close to zero, yielding $T \approx 3/2$. In terms of the current orbital elements (a, e, i), Tisserand's relation for each star can therefore be written

$$\frac{a_{IBH}}{a} + 2 \left[\frac{a}{a_{IBH}}(1 - e^2) \right]^{1/2} \cos i \approx 3, \tag{8.173}$$

where the inclination is defined with respect to the (unknown) orientation of the $M_\bullet$–M_{IBH} orbital plane. One can then try to find a value for a_{IBH} and for the two angles that define the plane of the SBH–IBH orbit such that the relation (8.173) is most nearly satisfied for all the stars in question. (Note that the mass of the IBH does not enter into the problem.) So far, this procedure has not led to the identification of an unseen IBH. But the reader is encouraged to spend a few hours playing this game, using whatever are the best current estimates for the orbital elements of the young Galactic center stars.

Tisserand's relation allows for changes in the energy of the interacting star—that is, changes in its semimajor axis. If a is constant, Tisserand's relation simplifies to

$$\sqrt{1 - e^2}\cos i \approx \text{constant.} \tag{8.174}$$

Equation (8.174) states simply that the component of the star's angular momentum parallel to the SBH–IBH rotation axis is conserved. This is not too surprising: recall from section 4.8.2 that the same quantity was found to be conserved in the hierarchical three-body problem (HTBP), after averaging, when the lowest-order, or "quadrupole," approximation was made for the perturbing potential.

We can apply the results from our detailed discussion of the HTBP in section 4.8 to the motion of a star that is perturbed by an IBH, as long as we are careful to respect the assumptions that were made in that derivation. We considered two limiting cases, depending on the ratio m_1/m_2 between the masses of the inner and outer bodies: the "inner restricted problem" for which $m_1 \ll m_2$; and the "outer restricted problem" for which $m_2 \ll m_1$. In the current context, the inner problem refers to a star that orbits well inside the IBH orbit, and the outer problem to a star that orbits well outside. To summarize briefly what we found in chapter 4:

- **Inner problem.** The star's argument of periapsis ω either circulates ($0 \leq \omega \leq 2\pi$), or librates over a finite range; in either case, its eccentricity and inclination vary periodically with ω, in such a way as to conserve $(1 - e^2)\cos i$ ("Lidov–Kozai cycles").

- **Outer problem.** The star's eccentricity is constant, but its angular-momentum vector circulates, either about the angular momentum vector of the inner binary, or about its major axis.

At least in the case of a circular SBH–IBH binary, numerical experiments show that the HTBP is a reasonably good representation of the motion of nearby stars [219]. Figure 8.22 shows an example: the orbits of three of the Galactic center S-stars—S8, S12, and S27—as they evolve over the next 2 Myr, assuming the presence of a 4000 $M_\odot$ IBH in a circular orbit of radius 30 mpc. Referring to table 4.1, we see that apoapsides of these three stars lie at distances of 30.0, 23.6, and 35.7 mpc, respectively, from the SBH. Whereas all three stars spend much of their time inside the orbit of the hypothesized IBH, none can be said to strictly satisfy the conditions that define the "inner restricted problem." Nevertheless, each of the three orbits exhibits oscillations in eccentricity that are similar to those predicted

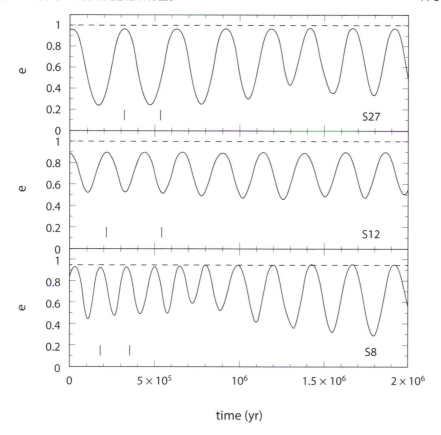

Figure 8.22 This figure shows how the orbits of three S-stars at the center of the Milky Way would evolve in eccentricity over the next two million years, in response to forcing from an IBH with mass $M_{\mathrm{IBH}} = 4000\,M_\odot$ and orbital radius $r = 30\,\mathrm{mpc}$. The initial relative inclinations were $i = 58°$ (S8), $86°$ (S12), and $79°$ (S27). The horizontal dashed lines show the predicted maximum value of the eccentricity, equation (8.174), and the space between the solid vertical lines indicates the predicted period, equation (8.175) [219].

by the HTBP, with period close to T_{Kozai}, the characteristic time for Lidov–Kozai oscillations (equation 4.314):

$$T_{\mathrm{Kozai}} = \frac{\sqrt{GM_\bullet}}{GM_{\mathrm{IBH}}} \frac{a_{\mathrm{IBH}}^3}{a^{3/2}} \left(1 - e_{\mathrm{IBH}}^2\right)^{3/2}$$

$$\approx 9 \times 10^5 \left(1 - e_{\mathrm{IBH}}^2\right)^{3/2} \left(\frac{M_{\mathrm{IBH}}}{10^3\,M_\odot}\right)^{-1} \left(\frac{a_{\mathrm{IBH}}}{0.1\,\mathrm{pc}}\right)^3$$

$$\times \left(\frac{a}{0.1\,\mathrm{mpc}}\right)^{-3/2} \mathrm{yr}. \qquad (8.175)$$

8.6.4 Perturbation of stars by an IBH in an eccentric orbit

As we saw in section 8.6.1, it is entirely possible that the orbit of an IBH–SBH binary would be substantially *non*circular. The treatment of the HTBP in section 4.8 allowed for nonzero eccentricities of both the test star and the perturber, but only under the quadrupole approximation to the perturbing potential, or equivalently, if $a_2 \gg a_1$. Adding the octopole and higher-order terms to the averaged Hamiltonian would be one way to improve the description of the motion, but the equations quickly become very complicated, and in any case the validity of the hierarchical approximation would always be questionable for stars that happen to come close to the IBH.

For these reasons, direct numerical integration of the equations of motion is probably the best way to investigate how an IBH on an eccentric orbit affects the motions of stars orbiting the SBH. The consequences of making the IBH orbit eccentric turn out to be dramatic [367]. Even a mild eccentricity ($e \gtrsim 0.5$) can cause the orbits of stars to "flip," that is, to change their sense of circulation around the SBH, from prograde (i.e., in the same sense as the IBH) to retrograde or vice versa. Furthermore, the eccentricities of the stellar orbits can be driven to very large values. Figure 8.23 shows an example in which the stellar orbits initially had small inclinations with respect to the SBH–IBH binary. Equation (8.174) would predict only modest changes in the eccentricity and inclination of these orbits, but after roughly 1 Myr, the eccentricity distribution is close to "thermal," $N(e)de \sim e\,de$, and the distribution of orbital inclinations is also essentially random.

Examination of the stellar orbits in these simulations reveals the following characteristics of the flipping phenomenon:

1. Large changes in inclination occur near the time when the eccentricity of the stellar orbit reaches large values (figure 8.24). It is intuitively obvious that changing the sign of L is "easiest" when the magnitude of L is small.

2. The timescale over which the eccentricity and inclination change is of a similar order to, but longer than, the time associated with Kozai–Lidov oscillations, equation (8.175), and much longer than orbital periods; in other words, this is a "secular" effect and not a result of (say) rare, close interactions with the IBH.

3. The large changes in e and $\cos i$ are not a result of close encounters between stars; evolution of the stellar orbits is nearly unchanged if the stars are replaced by test masses.

4. The flipping behavior "turns on" suddenly as the mass and the orbital eccentricity of the IBH are increased. For initial conditions like those of figure 8.23, flipping of the orbits requires an eccentricity of the IBH–SBH binary greater than ~ 0.5, and a mass ratio greater than $\sim 4 \times 10^{-4}$ [367].

Simulations like the ones shown in figures 8.23 and 8.24 are interesting because they provide a possible, albeit partial, solution to the S-star puzzle [367]. Recall that the S-stars—unlike, say, the young stars in the two nuclear disks—exhibit

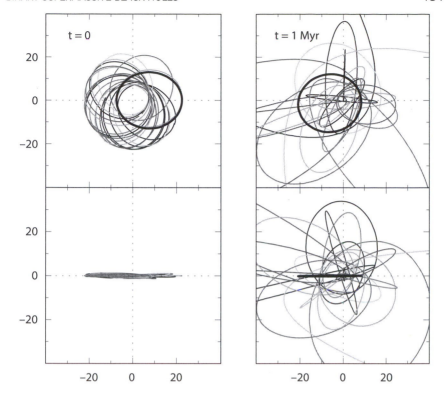

Figure 8.23 An IBH on an eccentric orbit can induce enormous changes in the orbits of nearby stars [367]. The two panels on the left show the initial conditions, which consisted of a 4000 $M_\odot$ IBH orbiting about the Galactic center SBH with $a_{IBH} = 15$ mpc and $e_{IBH} = 0.5$, and a set of stars on similar orbits. Such initial conditions might result from the tidal dispersion of a cluster of stars around the IBH [226]. The panels on the right show the same stars after 1 Myr. The IBH orbit is the heavy curve in all panels and the unit of length is milliparsecs. The initially disklike, corotating distribution of stars is converted, after 1 Myr, into an approximately isotropic distribution of orbits with a range of eccentricities, similar to what is observed for the S-stars. Many of the orbits "flip" in response to the perturbing force from the IBH; i.e., the direction of their angular momentum vector changes by roughly 180°. The flipping phenomenon is only observed when the IBH orbital eccentricity exceeds about 0.5, the value assumed in this figure.

an apparently random distribution of orientations and eccentricities, with no clear sense of rotation. Perhaps these stars were carried into their current location by an inspiraling IBH, and their orbits were then "randomized", as in figure 8.23, by perturbations from that same IBH, in a time much less than their current ages. The IBH responsible for this evolution might still be orbiting, unseen, somewhere inside the inner 10 mpc or so; or perhaps it managed to spiral in and merge with Sgr A*, contributing to the formation of the $\sim$ 0.5 pc core in the stellar distribution (figure 7.1).

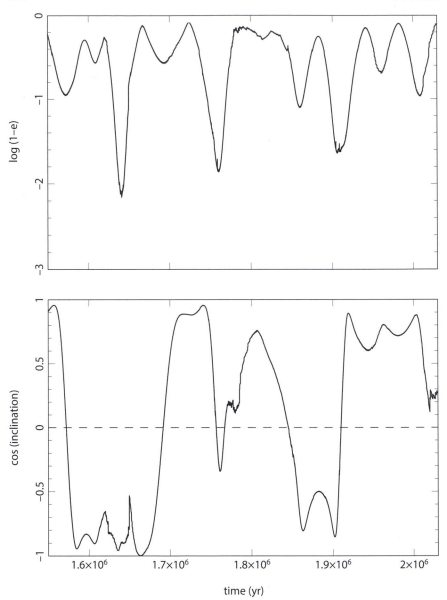

Figure 8.24 Flipping of orbits by a perturber in an eccentric orbit in the three-body (SBH–
           IBH–star) problem [367]. The figure shows the evolution of a single stellar
           orbit in the simulation of figure 8.23. The top and bottom panels show the star's
           eccentricity and inclination; the latter is defined with respect to the original
           orbital plane, which is also the plane of the SBH–IBH orbit. Large changes
           in inclination occur near the time when the eccentricity of the stellar orbit is
           largest.

8.7 TRIPLE SUPERMASSIVE BLACK HOLES AND THE FINAL-PARSEC PROBLEM

If a massive binary fails to coalesce, it can be present in a nucleus when a third SBH, or a second binary, is deposited there following a subsequent galaxy merger. The multiple SBH system that forms can then engage in three-body interactions. Possible outcomes include ejection of one or more of the SBHs from the nucleus by the gravitational slingshot; rapid coalescence of two SBHs that are placed onto a mutual orbit of high eccentricity or small semimajor axis; or both [476, 531]. This is one way to overcome the final-parsec problem.

Suppose that only one of the two merging galaxies contains a binary SBH, and that the larger of the two galaxies contains the largest of the three SBHs. Call the mass of this object M_1, and the mass of the primary (or single) SBH in the smaller galaxy M_3; the third SBH has mass M_2 and can be located either in the larger or smaller galaxy. Logically, there are three possibilities for the arrangement and relative masses of the three SBHs:

   I. The binary $M_1 + M_2$ is in the larger galaxy and $M_2 < M_3 < M_1$.

   II. The binary $M_1 + M_2$ is in the larger galaxy and $M_3 < M_2 < M_1$.

   III. The binary $M_3 + M_2$ is in the smaller galaxy and $M_2 < M_3 < M_1$.

Cases I and II have received the most attention in the literature and we focus on them here.

Recall from chapter 4 that gravitational-wave (GW) emission brings an isolated binary to coalescence in a time

$$t_{GW} \approx 6 \times 10^6 \frac{(1+q)^2}{q} \left(\frac{a_0}{10^{-2}\,\mathrm{pc}} \right)^4 \left(\frac{M_1 + M_2}{10^8\,M_\odot} \right)^{-3} \left(1 - e_0^2 \right)^{7/2}\,\mathrm{yr}$$

(equation 4.239), where $q \equiv M_2/M_1 \leq 1$ is the binary mass ratio and (a_0, e_0) are its initial semimajor axis and eccentricity. It is clear from this expression that even modest increases in eccentricity can greatly reduce t_{GW}. The presence of a third SBH can cause this to happen in one of two ways: via Lidov–Kozai oscillations when M_3 is relatively far from the binary; and via strong, three-body interactions should the orbit of M_3 decay (due to dynamical friction) so far that it is brought close to the central binary, before the latter has coalesced.

It is natural to consider these two regimes separately, since an inspiraling SBH will naturally transition from one to the other as its separation from the central binary decreases. In the language of the three-body problem, this transition corresponds roughly to a change from a "stable" to an "unstable" three-body system. "Stable" triples are defined as those that evolve in a manner similar to what we have seen in the hierarchical problem, with little exchange of energy between the inner and outer binaries. "Unstable" triples evolve chaotically, and the final outcome is ejection: one of the bodies extracts enough energy from the other two to escape, perhaps to infinity. The distinction is largely empirical; there is no fundamental constraint that would keep any three-body system from eventually evolving into a bound binary and a detached third body that escapes to infinity. Nevertheless, in numerical integrations, one often observes a sharp transition from one sort of behavior to another as a parameter (e.g., the initial relative separations) is changed.

Various, approximate criteria have been derived for the stability of hierarchical three-body systems based on numerical integrations. One widely used formula [339] predicts stability for

$$\frac{a_2}{a_1} > \Gamma \equiv \frac{2.8}{1-e_2}\left[\left(1+\frac{M_3}{M_1+M_2}\right)\frac{1+e_2}{(1-e_2)^{1/2}}\right]^{2/5}. \tag{8.176}$$

The notation here is the same as in the discussion of the hierarchical three-body problem in chapter 4: the subscript 1 on a or e refers to the inner (M_1, M_2) binary and the subscript 2 refers to the binary consisting of M_3 moving about the center of mass of the (M_1, M_2) system. Equation (8.176) was derived assuming coplanar, corotating orbits; more general configurations are expected to be more stable. Note that—for $M_3 \ll (M_1, M_2)$—equation (8.176) states roughly the same condition that was given in section 8.1 for a field star to interact strongly with a binary SBH: a distance of closest approach that is at most a few times the binary separation.

Ignoring the effects of the galaxy's confining potential, the evolution of a triple SBH in the "stable" regime, $a_2/a_1 \gtrsim \Gamma$, can be approximated using the hierarchical three-body equations of motion derived in section 4.8: either the inner restricted equations (case I) or the outer restricted equations (case II). Figure 8.25 shows the results of a set of such calculations, in which the averaged Hamiltonian included terms up to octupole order [50]. The effects of relativity were approximated by adding terms representing the averaged rate of Schwarzschild precession (1PN) and gravitational-wave energy loss (2.5PN) to the equations of motion of the inner binary. The figure shows that Lidov–Kozai oscillations can greatly accelerate the coalescence, particularly if the initial eccentricity is low.

Recall from chapter 4 that Lidov–Kozai oscillations are quenched by the Schwarzschild precession when

$$\left(1-e_1^2\right)\frac{a_1}{r_g} \lesssim \frac{3}{2\pi}\frac{M_1+M_2}{M_3}\frac{a_2^3}{a_1^3}\left(1-e_2^2\right)^{3/2}, \tag{8.177}$$

where $r_g \equiv GM_1/c^2$ (equation 4.332). Figure 8.26 shows the dependence of the coalescence time on the initial value of a_2/a_1 for triple systems with fixed initial inclination. For sufficiently large a_2/a_1, relativistic precession quenches the Lidov–Kozai cycles and the eccentricity of the inner binary is left almost unchanged. The dashed line in that figure was computed by using the merging time of equation (4.239) as a proxy for a_1, together with a relation similar to equation (8.177), yielding

$$t_{\mathrm{merge}} = 1.2 \times 10^6 \left(\frac{a_2/a_1}{10}\right)^{12}\left(\frac{2M_3}{M_1+M_2}\right)^{-4}\left(\frac{M_1}{10^6\,M_\odot}\right)^{-1}$$

$$\times \left(\frac{M_2}{10^6\,M_\odot}\right)^{-1}\left(\frac{M_1+M_2}{2\times 10^6\,M_\odot}\right)^3\frac{(1-e_2^2)^6}{(1-e_1^2)^{5/2}}f(e_1)\ \mathrm{yr}, \tag{8.178}$$

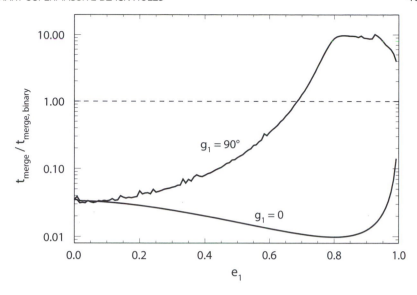

Figure 8.25 This figure shows how the time for a binary SBH to coalesce depends on its initial eccentricity e_1 in the presence of a third SBH [50]. These are integrations of the averaged, hierarchical three-body equations of motion, similar to equations (4.316) but derived from an octopole Hamiltonian and including terms corresponding to the 1PN and 2.5PN accelerations for the inner binary. The assumed masses were $M_1 = 2 \times 10^6 \, M_\odot$, $M_2 = M_3 = 10^6 \, M_\odot$, and the initial conditions of the triple system were $a_1 = 3.16 \times 10^{-3}$ pc, $a_2/a_1 = 10$, $e_2 = 0.1$, and $i = 80°$. The merger time is expressed as a multiple of the coalescence time for an isolated binary having the same initial parameters as the inner binary. Two initial values were chosen for the orientation of the inner binary, specified by the value of ω (indicated here by g_1). The first, $g_1 = 0$, corresponds to a circulating inner binary that starts out with an eccentricity that is at the minimum in the Lidov–Kozai cycle (as in figure 4.21a). The subsequent oscillations always result in higher eccentricities and a reduced time for GW energy loss. The second case, $g_1 = 90°$, corresponds to librating orbits for low e_1 and circulating orbits for high e_1. Irregularities in the latter curve reflect chaos in the motion.

with $f(e) \approx 1$ the same function defined in equation (4.239). Initial conditions to the right of this curve on figure 8.26 result in coalescence times that are almost unaffected by the Lidov–Kozai oscillations.

If the infalling SBH manages to reach a separation from the inner binary such that Γ in equation (8.176) is less than one, and does so before the inner binary has coalesced, the motion of the three SBHs will enter into a qualitatively different regime characterized by close encounters between the three massive bodies. If the infalling SBH is less massive than either of the components of the preexisting binary, $M_3 < (M_1, M_2)$, the subsequent evolution is similar to that of a star interacting with a massive binary, as discussed extensively in the first parts of this chapter. Each close interaction of the smaller SBH with the binary increases the

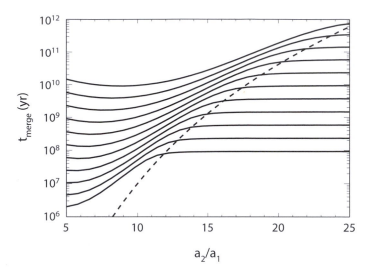

Figure 8.26 Illustrating how the coalescence time of the inner binary in a triple-SBH system
depends on the initial ratio of semimajor axes [50]. The component masses are
$M_1 = 2 \times 10^6\,M_\odot$, $M_2 = 10^6\,M_\odot$ and $M_3 = 10^6\,M_\odot$; the initial conditions
were $e_1 = 0.1$, $e_2 = 0.1$, $\omega_1 = 0$, $\omega_2 = 90°$, and $i = 80°$. The curves show
results for different initial values of a_1 spaced at equal logarithmic intervals:
$a_1 = \{1.00, 1.26, 1.58, 2.00, 2.51, 3.16, 3.98, 5.01, 6.31, 7.94, 10.0\} \times 10^{-3}$ pc.
The dashed curve is equation (8.178); to the right of this curve, Lidov–Kozai
cycles are quenched by relativistic precession.

latter's binding energy by $\langle |\Delta E/E| \rangle \approx 0.4 M_3/(M_1 + M_2)$ [239], and the ultimate
outcome is likely to be ejection of the smaller SBH with a velocity roughly $1/3$ the
relative velocity of the two SBHs in the inner binary [476]. Since the escaping SBH
carries with it a nonnegligible linear momentum, the binary also recoils, but with
a speed that is lower by $\sim M_3/(M_1 + M_2)$. The smaller SBH will escape entirely
from the galaxy if

$$V_{\text{bin}} \gtrsim 3 V_{\text{esc}},\qquad\qquad(8.179)$$

where V_{bin} is the relative velocity of the inner binary during the last close encounter
and V_{esc} is the escape velocity from the center of the galaxy. This condition can be
written

$$\left(\frac{M_1 + M_2}{10^8\,M_\odot}\right)^{1/2} \left(\frac{a_1}{10^{-2}\,\text{pc}}\right)^{-1/2} \gtrsim 0.5 \left(\frac{V_{\text{esc}}}{1000\,\text{km s}^{-1}}\right).\qquad(8.180)$$

Escape of the least massive of the three bodies from the galaxy is evidently to be
expected in many cases, and if the three masses are comparable, the binary may also
gain enough energy from the close interaction to escape. Even if it does not escape
entirely, its temporary displacement from the galaxy center will cause the inner
regions of the galaxy to expand, in a region $r \lesssim r_{\text{h}} = G(M_1 + M_2)/\sigma^2$, increasing
the size of the core that was created when the binary first formed. Subsequent decay

of the displaced binary's orbit due to dynamical friction against the stars will result in additional transfer of energy to the galaxy, causing the central stellar density to decrease still more. Figure 5.5 (which was based on a single displaced SBH, not a binary) illustrates this evolution.

If $M_3 > M_2$, the most likely outcome of the close three-body encounters will be an exchange interaction, with the lightest SBH ejected and the two most massive SBHs forming a new binary. Further interactions then proceed as in the case $M_3 < (M_1, M_2)$.

During these strong, three-body interactions, both the semimajor axis and eccentricity of the dominant binary will change stochastically, in a manner that is not well described by the hierarchical equations of motion. The timescale for GW coalescence can be enormously shortened, although to an extent that is difficult to estimate from the initial conditions [383, 531, 533].

One would obviously like to know how important such interactions are likely to be for the evolution of galaxies, and their central SBHs, over cosmological times. There is very little that can be said with certainty. If binary SBHs manage to coalesce in a time that is less than the typical time between galaxy mergers, triple SBH systems will almost never form. One study [549], based on a highly simplified model for binary/galaxy interactions, estimates that three-body recoils could result in 10% of SBHs having been ejected completely from galaxies. If this has occurred, a significant fraction of nuclei could be left with no SBH, with an offset SBH, or with an SBH whose mass is lower than expected based on the $M_{\bullet}-\sigma$ or $M_{\bullet}-L$ relations.

Suggestions for Further Reading

Chapter 1, "Introduction and historical overview"

Popular and semipopular books on supermassive black holes abound. For the Milky Way supermassive black hole in particular, see

> *The Black Hole at the Center of our Galaxy*, F. Melia. Princeton University Press, 2003,

and

> *The Black Hole at the Center of the Milky Way*, A. Eckart, R. Schoedel, and C. Straubmeier. Imperial College Press, 2005.

A more rigorous text on black holes is

> *Physics of Black Holes*, I. D. Novikov and V. P. Frolov. Kluwer, 1989.

Gravitational waves are the subject of

> *Einstein's Unfinished Symphony: Listening to the Sounds of Space-Time*, Marcia Bartusiak. Berkley Trade, 2003.

A broader, more speculative and more personal book about the general theory of relativity is

> *Black Holes and Time Warps: Einstein's Outrageous Legacy*, K. Thorne. W. W. Norton, 1994.

Chapter 2, "Observations of galactic nuclei and supermassive black holes"

An article on galaxy structure that pays close attention to the history of the subject is

> "A review of elliptical and disc galaxy structure, and modern scaling laws," A. Graham, in *Planets, Stars and Stellar Systems*, T. D. Oswalt (editor). Springer, 2012.

The nucleus of the Milky Way is treated in a number of review articles. For a mostly observational slant, see

> "The Galactic center massive black hole and nuclear star cluster," R. Genzel, F. Eisenhauer, and S. Gillessen, *Reviews of Modern Physics*, 82(4):3121–3195, 2010,

and for a more theoretical perspective,

"Stellar processes near the massive black hole in the Galactic center," T. Alexander, *Physics Reports*, 419(2–3):65–142, 2005.

The technique of reverberation mapping is reviewed in

"The central black hole and relationships with the host galaxy," B. Peterson, *New Astronomy Reviews*, 52(6):240–252, 2008.

For the history of the $M_\bullet$–σ relation, see

"Relationship of black holes to bulges," D. Merritt and L. Ferrarese, in I. Shlosman J. H. Knapen, J. E. Beckman and T. J. Mahoney, editors, *The Central Kiloparsec of Starbursts and AGN: The La Palma Connection*, volume 249 of *Astronomical Society of the Pacific Conference Series*, page 335, 2001.

Observational evidence for binary supermassive black holes is reviewed in

"Observational evidence for binary black holes and active double nuclei," S. Komossa, *Memorie della Società Astronomica Italiana*, 77:733–741, 2006.

A standard reference for theory and detection of gravitational waves is

Gravitational Waves: Theory and Experiments, M. Maggiore, Oxford University Press, 2007.

Chapter 3, "Collisionless equilibria"

Action-angle variables are discussed in many textbooks on classical mechanics. Nonintegrability is central to galactic dynamics however, and a comprehensive text that covers both sorts of motion is

Regular and Chaotic Dynamics, A. J. Lichtenberg and M. A. Lieberman. Springer, 1992.

A review of galactic dynamics that emphasizes chaos is

"Elliptical galaxy dynamics," D. Merritt, *Publications of the Astronomical Society of the Pacific*, 111(756):129–168, 1999.

A slightly dated text on galactic dynamics, of which the author is still very fond (perhaps because it was the text he learned from) is

Galactic Astronomy, D. Mihalas and P. M. Routly. W. H. Freeman and Company, 1968.

Chapter 4, "Motion near supermassive black holes"

Motion in the Schwarzschild and Kerr metrics is covered in many texts on relativity; one that places the subject in a broader astrophysical context is

> *Black Holes, White Dwarfs and Neutron Stars: The Physics of Compact Objects*, S. L. Shapiro and S. Teukolsky. Wiley, 1986.

A comprehensive review of the two-body problem in the post-Newtonian approximation—though excluding spin—is

> "Gravitational radiation from post-Newtonian sources and inspiralling compact binaries," L. Blanchet, *Living Reviews in Relativity*, 9(4), 2006.

Two texts that extensively use Lagrange's planetary equations to derive the post-Newtonian equations of motion, including spin (but in the context of the solar system, not black holes) are

> *Essential Relativistic Celestial Mechanics*, V. A. Brumberg. Adam Hilger, 1991,

and

> *Relativity in Astrometry, Celestial Mechanics and Geodesy*, M. H. Soffel. Springer, 1989.

The three-body problem is the subject of quite a number of books; one that includes applications to supermassive black holes and galactic nuclei is

> *Three-Body Problem*, M. Valtonen and H. Karttunen. Cambridge University Press, 2006.

Chapter 5, "Theory of gravitational encounters"

This topic is covered in the book by Mihalas and Routly cited above, and also in

> *Dynamical Evolution of Globular Clusters*, L. Spitzer. Princeton University Press, 1987,

and

> *Galaxy Dynamics*, 2nd edition, J. Biney and S. Tremaine. Princeton University Press, 2002.

Both texts contain a more extensive discussion of binary stars than this book; the latter contains some basic material relating to supermassive black holes and galactic nuclei. Chandrasekhar's text is still quite readable:

> *Principles of Stellar Dynamics*, S. Chandrasekhar. Dover Publications, 1943.

Chapter 6, "Loss-cone dynamics"

A short, qualitative review containing many of the important references is

"Loss cone: Past, present and future," S. Sigurdsson, *Classical and Quantum Gravity*, 20(10):S45–S54, 2003.

Observational evidence relating to tidal disruption events is reviewed in

"X-ray evidence for supermassive black holes at the centers of nearby, non-active galaxies," S. Komossa, *Reviews in Modern Astronomy*, 15:27–35, 2002,

and this article summarizes observational evidence for hypervelocity stars:

"Hypervelocity stars and the Galactic center," W. Brown, *Galactic Center Newsletter*, 28:7–12, 2008.

A review article on the detection of EMRIs and the (ever-changing) status of LISA is

"Low-frequency gravitational-wave science with eLISA/NGO," P. Amaro-Seoane et al., *Classical and Quantum Gravity*, 29(12):124016, 2012.

Chapter 7, "Collisional evolution of nuclei"

The book cited above for chapter 5 by L. Spitzer, as well as

The Gravitational Million-Body Problem: A Multidisciplinary Approach to Star Cluster Dynamics, D. Heggie and P. Hut. Cambridge University Press, 2003,

cover collisional evolution of stellar systems in detail, mostly in the context of star clusters; Spitzer's text is more rigorous, although succinct, while the second approaches the topic from a more intuitive point of view.

Chapter 8, "Binary and multiple supermassive black holes"

Two textbooks that deal with structure formation in a hierarchical universe are

Cosmological Physics, J. A. Peacock. Cambridge University Press, 1999,

and

Modern Cosmology, S. Dodelson. Academic Press, 2003.

Techniques of N-body simulation are comprehensively reviewed in

Gravitational N-body Simulations, S. J. Aarseth. Cambridge University Press, 1991.

Dynamics of gas in galactic nuclei is treated in

Accretion Power in Astrophysics, 3rd edition, J. Frenk, A. King, and D. J. Raine. Cambridge University Press, 2002.

References

[1] S. J. Aarseth. *Gravitational N-Body Simulations*. Cambridge University Press, Cambridge, 2003.

[2] H. A. Abt and S. G. Levy. Multiplicity among solar-type stars. *Astrophys. J. Suppl.*, 30:273–306, March 1976.

[3] T. Alexander and C. Hopman. Strong mass segregation around a massive black hole. *Astrophys. J.*, 697:1861–1869, June 2009.

[4] V. A. Ambartsumian. Title unknown. *Uchenniye Zapiskiy, Leningrad State University*, 22:19, 1938.

[5] J. Anderson and R. P. van der Marel. New limits on an intermediate-mass black hole in Omega Centauri. I. Hubble Space Telescope photometry and proper motions. *Astrophys. J.*, 710:1032–1062, February 2010.

[6] R. Angélil, P. Saha, and D. Merritt. Toward relativistic orbit fitting of Galactic center stars and pulsars. *Astrophys. J.*, 720:1303–1310, September 2010.

[7] F. Antonini, J. C. Lombardi Jr., and D. Merritt. Tidal breakup of binary stars at the Galactic center. II. Hydrodynamic simulations. *Astrophys. J.*, 731:128, April 2011.

[8] F. Antonini and D. Merritt. Dynamical friction around supermassive black holes. *Astrophys. J.*, 745:83, January 2012.

[9] P. J. Armitage and P. Natarajan. Accretion during the merger of supermassive black holes. *Astrophys. J. Lett.*, 567:L9–L12, March 2002.

[10] P. Artymowicz and S. H. Lubow. Dynamics of binary–disk interaction. 1: Resonances and disk gap sizes. *Astrophys. J.*, 421:651–667, February 1994.

[11] E. Athanassoula, J. C. Lambert, and W. Dehnen. Can bars be destroyed by a central mass concentration?- I. Simulations. *Mon. Not. R. Astron. Soc.*, 363:496–508, October 2005.

[12] R. Bacon, E. Emsellem, F. Combes, Y. Copin, G. Monnet, and P. Martin. The M 31 double nucleus probed with OASIS. A natural vec m = 1 mode? *Astron. Astrophys.*, 371:409–428, May 2001.

[13] F. K. Baganoff, Y. Maeda, M. Morris, M. W. Bautz, W. N. Brandt, W. Cui, J. P. Doty, E. D. Feigelson, G. P. Garmire, S. H. Pravdo, G. R. Ricker, and L. K. Townsley. Chandra X-ray spectroscopic imaging of Sagittarius A* and the central parsec of the galaxy. *Astrophys. J.*, 591:891–915, July 2003.

[14] J. N. Bahcall and R. A. Wolf. Star distribution around a massive black hole in a globular cluster. *Astrophys. J.*, 209:214–232, October 1976.

[15] J. N. Bahcall and R. A. Wolf. The star distribution around a massive black hole in a globular cluster. II. Unequal star masses. *Astrophys. J.*, 216:883–907, September 1977.

[16] J. Bailey. Detection of pre-main-sequence binaries using spectro-astrometry. *Mon. Not. R. Astron. Soc.*, 301:161–167, November 1998.

[17] S. A. Balbus and J. F. Hawley. Instability, turbulence, and enhanced transport in accretion disks. *Reviews of Modern Physics*, 70:1–53, January 1998.

[18] J. A. Baldwin, M. M. Phillips, and R. Terlevich. Classification parameters for the emission-line spectra of extragalactic objects. *Publ. Astron. Soc. Pacific*, 93:5–19, February 1981.

[19] L. Ballo, V. Braito, R. Della Ceca, L. Maraschi, F. Tavecchio, and M. Dadina. Arp 299: A second merging system with two active nuclei? *Astrophys. J.*, 600:634–639, January 2004.

[20] L. Barack and C. Cutler. LISA capture sources: Approximate waveforms, signal-to-noise ratios, and parameter estimation accuracy. *Phys. Rev. D*, 69(8):082005, April 2004.

[21] L. Barack and C. Cutler. Using LISA extreme-mass-ratio inspiral sources to test off-Kerr deviations in the geometry of massive black holes. *Phys. Rev. D*, 75(4):042003, February 2007.

[22] J. M. Bardeen. Kerr metric black holes. *Nature*, 226:64–65, April 1970.

[23] J. M. Bardeen and J. A. Petterson. The Lense–Thirring effect and accretion disks around Kerr black holes. *Astrophys. J. Lett.*, 195:L65–L68, January 1975.

[24] J. M. Bardeen, W. H. Press, and S. A. Teukolsky. Rotating black holes: Locally nonrotating frames, energy extraction, and scalar synchrotron radiation. *Astrophys. J.*, 178:347–370, December 1972.

[25] G. I. Barenblatt. *Scaling, Self-Similarity, and Intermediate Asymptotics*. Cambridge University Press, December 1996.

[26] B. M. Barker and R. F. O'Connell. Effect of the rotation of the central body on the orbit of a satellite. *Phys. Rev. D*, 10:1340–1342, August 1974.

[27] B. M. Barker and R. F. O'Connell. Gravitational two-body problem with arbitrary masses, spins, and quadrupole moments. *Phys. Rev. D*, 12:329–335, July 1975.

[28] J. E. Barnes. Merger time scales. In T. von Hippel, C. Simpson, and N. Manset, editors, *Astrophysical Ages and Times Scales*, volume 245 of *Astronomical Society of the Pacific Conference Series*, page 382, 2001.

[29] H. Bartko, F. Martins, S. Trippe, T. K. Fritz, R. Genzel, T. Ott, F. Eisenhauer, S. Gillessen, T. Paumard, T. Alexander, K. Dodds-Eden, O. Gerhard, Y. Levin, L. Mascetti, S. Nayakshin, H. B. Perets, G. Perrin, O. Pfuhl, M. J. Reid, D. Rouan, M. Zilka, and A. Sternberg. An extremely top-heavy initial mass function in the Galactic center stellar disks. *Astrophys. J.*, 708:834–840, January 2010.

[30] N. Bastian, K. R. Covey, and M. R. Meyer. A universal stellar initial mass function? A critical look at variations. *Ann. Rev. Astron. Astrophys.*, 48:339–389, September 2010.

[31] K. Basu. *The Three Body Problem*. West Chester University, 2009.

[32] D. Batcheldor. The $M_\bullet$–σ_* relation derived from sphere of influence arguments. *Astrophys. J. Lett.*, 711:L108–L111, March 2010.

[33] P. Batsleer and H. Dejonghe. The Kuzmin–Kutuzov two integral axisymmetric galaxy model revisited. *Astron. Astrophys.*, 271:104–108, April 1993.

[34] H. Baumgardt, P. Hut, J. Makino, S. McMillan, and S. Portegies Zwart. On the central structure of M15. *Astrophys. J. Lett.*, 582:L21–L24, January 2003.

[35] H. Baumgardt, J. Makino, P. Hut, S. McMillan, and S. Portegies Zwart. A dynamical model for the globular cluster G1. *Astrophys. J. Lett.*, 589:L25–L28, May 2003.

[36] M. C. Begelman, R. D. Blandford, and M. J. Rees. Massive black hole binaries in active galactic nuclei. *Nature*, 287:307–309, September 1980.

[37] M. C. Begelman and M. J. Rees. The fate of dense stellar systems. *Mon. Not. R. Astron. Soc.*, 185:847–860, December 1978.

[38] R. Bender. Unraveling the kinematics of early-type galaxies - Presentation of a new method and its application to NGC4621. *Astron. Astrophys.*, 229:441–451, March 1990.

[39] R. Bender, J. Kormendy, G. Bower, R. Green, J. Thomas, A. C. Danks, T. Gull, J. B. Hutchings, C. L. Joseph, M. E. Kaiser, T. R. Lauer, C. H. Nelson, D. Richstone, D. Weistrop, and B. Woodgate. HST STIS spectroscopy of the triple nucleus of M31: Two nested disks in Keplerian rotation around a supermassive black hole. *Astrophys. J.*, 631:280–300, September 2005.

[40] R. Bender, J. Kormendy, and W. Dehnen. Improved evidence for a 3 × $10^6 M_\odot$ black hole in M32: Canada–France–Hawaii telescope spectroscopy with FWHM = 0″.47 resolution. *Astrophys. J. Lett.*, 464:L123–L126, June 1996.

[41] M. C. Bentz, B. M. Peterson, H. Netzer, R. W. Pogge, and M. Vestergaard. The radius–luminosity relationship for active galactic nuclei: The effect of host-galaxy starlight on luminosity measurements. II. The full sample of reverberation-mapped AGNs. *Astrophys. J.*, 697:160–181, May 2009.

[42] P. Berczik, D. Merritt, and R. Spurzem. Long-term evolution of massive black hole binaries. II. Binary evolution in low-density galaxies. *Astrophys. J.*, 633:680–687, November 2005.

[43] F. Bertola and M. Capaccioli. Dynamics of early type galaxies. I - The rotation curve of the elliptical galaxy NGC 4697. *Astrophys. J.*, 200:439–445, September 1975.

[44] S. J. Berukoff and B.M.S. Hansen. Cluster core dynamics in the Galactic center. *Astrophys. J.*, 650:901–915, October 2006.

[45] J. Binney. On the rotation of elliptical galaxies. *Mon. Not. R. Astron. Soc.*, 183:501–514, May 1978.

[46] J. Binney and G. A. Mamon. M/L and velocity anisotropy from observations of spherical galaxies, or must M87 have a massive black hole. *Mon. Not. R. Astron. Soc.*, 200:361–375, July 1982.

[47] J. Binney and D. Spergel. Spectral stellar dynamics. *Astrophys. J.*, 252: 308–321, January 1982.

[48] J. Binney and S. Tremaine. *Galactic Dynamics*. Princeton, NJ, Princeton University Press, 2nd edition, 2008.

[49] J. J. Binney, R. L. Davies, and G. D. Illingworth. Velocity mapping and models of the elliptical galaxies NGC 720, NGC 1052, and NGC 4697. *Astrophys. J.*, 361:78–97, September 1990.

[50] O. Blaes, M. H. Lee, and A. Socrates. The Kozai mechanism and the evolution of binary supermassive black holes. *Astrophys. J.*, 578:775–786, October 2002.

[51] L. Blanchet and B. R. Iyer. Third post-Newtonian dynamics of compact binaries: equations of motion in the centre-of-mass frame. *Classical and Quantum Gravity*, 20:755–776, February 2003.

[52] R. Blandford and S. A. Teukolsky. Arrival-time analysis for a pulsar in a binary system. *Astrophys. J.*, 205:580–591, April 1976.

[53] R. D. Blandford and C. F. McKee. Reverberation mapping of the emission line regions of Seyfert galaxies and quasars. *Astrophys. J.*, 255:419–439, April 1982.

[54] R. D. Blandford and D. G. Payne. Hydromagnetic flows from accretion discs and the production of radio jets. *Mon. Not. R. Astron. Soc.*, 199:883–903, June 1982.

[55] N. N. Bogoliubov and Y. A. Mitropolski. *Asymptotic Methods in the Theory of Non-Linear Oscillations*. New York, Gordon and Breach, 1961.

[56] H. Bondi. On spherically symmetrical accretion. *Mon. Not. R. Astron. Soc.*, 112:195–204, 1952.

[57] T. R. Bontekoe and T. S. van Albada. Decay of galaxy satellite orbits by dynamical friction. *Mon. Not. R. Astron. Soc.*, 224:349–366, January 1987.

[58] A. H. Boozer. Establishment of magnetic coordinates for a given magnetic field. *Physics of Fluids*, 25:520–521, March 1982.

[59] M. Born, J. W. Fisher, and D. R. Hartree. *The Mechanics of the Atom.* International text-books of exact science. London, G. Bell and Sons, 1927.

[60] R. H. Boyer and R. W. Lindquist. Maximal analytic extension of the Kerr metric. *J. Math. Phys.*, 8:265–281, February 1967.

[61] A. Bressan, F. Fagotto, G. Bertelli, and C. Chiosi. Evolutionary sequences of stellar models with new radiative opacities. II - Z = 0.02. *Astron. Astrophys. Suppl.*, 100:647–664, September 1993.

[62] B. C. Bromley, S. J. Kenyon, W. R. Brown, and M. J. Geller. Runaway stars, hypervelocity stars, and radial velocity surveys. *Astrophys. J.*, 706:925–940, December 2009.

[63] W. R. Brown, M. J. Geller, and S. J. Kenyon. MMT hypervelocity star survey. *Astrophys. J.*, 690:1639–1647, January 2009.

[64] W. R. Brown, M. J. Geller, S. J. Kenyon, and M. J. Kurtz. Discovery of an unbound hypervelocity star in the Milky Way halo. *Astrophys. J. Lett.*, 622:L33–L36, March 2005.

[65] W. R. Brown, M. J. Geller, S. J. Kenyon, and M. J. Kurtz. A successful targeted search for hypervelocity stars. *Astrophys. J. Lett.*, 640:L35–L38, March 2006.

[66] R. M. Buchholz, R. Schödel, and A. Eckart. Composition of the galactic center star cluster. Population analysis from adaptive optics narrow band spectral energy distributions. *Astron. Astrophys.*, 499:483–501, May 2009.

[67] S. Callegari, S. Kazantzidis, L. Mayer, M. Colpi, J. M. Bellovary, T. Quinn, and J. Wadsley. Growing massive black hole pairs in minor mergers of disk galaxies. *Astrophys. J.*, 729:85, March 2011.

[68] N. Caon, M. Capaccioli, and M. D'Onofrio. On the shape of the light profiles of early-type galaxies. *Mon. Not. R. Astron. Soc.*, 265:1013–1021, December 1993.

[69] M. Cappellari, E. Emsellem, R. Bacon, M. Bureau, R. L. Davies, P. T. de Zeeuw, J. Falcón-Barroso, D. Krajnović, H. Kuntschner, R. M. McDermid, R. F. Peletier, M. Sarzi, R. C. E. van den Bosch, and G. van de Ven. The SAURON project – X. The orbital anisotropy of elliptical and lenticular galaxies: Revisiting the $(V/\sigma, \varepsilon)$ diagram with integral-field stellar kinematics. *Mon. Not. R. Astron. Soc.*, 379:418–444, August 2007.

[70] R. G. Carlberg and K. A. Innanen. Galactic chaos and the circular velocity at the Sun. *Astron. J.*, 94:666–670, September 1987.

[71] C. M. Carollo, M. Stiavelli, and J. Mack. Spiral galaxies with WFPC2. II. The nuclear properties of 40 objects. *Astron. J.*, 116:68–84, July 1998.

[72] B. Carter. Global structure of the Kerr family of gravitational fields. *Phys. Rev.*, 174:1559–1571, October 1968.

[73] B. Carter and J. P. Luminet. Pancake detonation of stars by black holes in galactic nuclei. *Nature*, 296:211–214, March 1982.

[74] J. Casares. Observational evidence for stellar-mass black holes. In V. Karas and G. Matt, editors, *Black Holes from Stars to Galaxies–Across the Range of Masses*, volume 238 of *IAU Symposium*, pages 3–12, April 2007.

[75] P. Cassen and D. S. Woolum. Radiatively damped density waves in optically thick protostellar disks. *Astrophys. J.*, 472:789, December 1996.

[76] D. Chakrabarty and P. Saha. A nonparametric estimate of the mass of the central black hole in the Galaxy. *Astron. J.*, 122:232–241, July 2001.

[77] S. Chandrasekhar. *An Introduction to the Study of Stellar Structure*. University of Chicago Press, 1939.

[78] S. Chandrasekhar. *Principles of Stellar Dynamics*. University of Chicago Press, 1942.

[79] S. Chandrasekhar. Dynamical friction. I. General considerations: The coefficient of dynamical friction. *Astrophys. J.*, 97:255–262, March 1943.

[80] S. Chandrasekhar. *Ellipsoidal Figures of Equilibrium*. Yale University Press, 1969.

[81] S. Chandrasekhar and J. von Neumann. The statistics of the gravitational field arising from a random distribution of stars. I. The speed of fluctuations. *Astrophys. J.*, 95:489–531, May 1942.

[82] S. Chapman, B. C. Garrett, and W. H. Miller. Semiclassical eigenvalues for nonseparable systems: Nonperturbative solution of the Hamilton–Jacobi equation in action-angle variables. *J. Chem. Phys.*, 64:502–509, January 1976.

[83] P. Charles. Black holes in X-ray binaries. In L. Kaper, E. P. J. van den Heuvel, and P. A. Woudt, editors, *Black Holes in Binaries and Galactic Nuclei*, page 27, 2001.

[84] P. Chatterjee, L. Hernquist, and A. Loeb. Brownian motion in gravitationally interacting systems. *Phys. Rev. Lett.*, 88(12):121103, March 2002.

[85] A. Chokshi and E. L. Turner. Remnants of the quasars. *Mon. Not. R. Astron. Soc.*, 259:421–424, December 1992.

[86] L. Ciotti. Stellar systems following the R exp 1/m luminosity law. *Astron. Astrophys.*, 249:99–106, September 1991.

[87] L. Ciotti and J. P. Ostriker. Cooling flows and quasars: Different aspects of the same phenomenon? I. Concepts. *Astrophys. J. Lett.*, 487:L105–L108, October 1997.

[88] R. S. Cohen, L. Spitzer, and P. M. Routly. The electrical conductivity of an ionized gas. *Phys. Rev.*, 80:230–238, October 1950.

[89] H. Cohn. Numerical integration of the Fokker–Planck equation and the evolution of star clusters. *Astrophys. J.*, 234:1036–1053, December 1979.

[90] H. Cohn. Late core collapse in star clusters and the gravothermal instability. *Astrophys. J.*, 242:765–771, December 1980.

[91] H. Cohn and R. M. Kulsrud. The stellar distribution around a black hole – numerical integration of the Fokker–Planck equation. *Astrophys. J.*, 226:1087–1108, December 1978.

[92] S. A. Colgate. Stellar coalescence and the multiple supernova interpretation of quasi-stellar sources. *Astrophys. J.*, 150:163–191, October 1967.

[93] S. Collin, T. Kawaguchi, B. M. Peterson, and M. Vestergaard. Systematic effects in measurement of black hole masses by emission-line reverberation of active galactic nuclei: Eddington ratio and inclination. *Astron. Astrophys.*, 456:75–90, September 2006.

[94] S. A. Cora, J. C. Muzzio, and M. M. Vergne. Orbital decay of galactic satellites as a result of dynamical friction. *Mon. Not. R. Astron. Soc.*, 289:253–262, August 1997.

[95] P. Côté, L. Ferrarese, A. Jordán, J. P. Blakeslee, C.-W. Chen, L. Infante, D. Merritt, S. Mei, E. W. Peng, J. L. Tonry, A. A. West, and M. J. West. The ACS Fornax cluster survey. II. The central brightness profiles of early-type galaxies: A characteristic radius on nuclear scales and the transition from central luminosity deficit to excess. *Astrophys. J.*, 671:1456–1465, December 2007.

[96] P. Côté, S. Piatek, L. Ferrarese, A. Jordán, D. Merritt, E. W. Peng, M. Haşegan, J. P. Blakeslee, S. Mei, M. J. West, M. Milosavljević, and J. L. Tonry. The ACS Virgo Cluster survey. VIII. The nuclei of early-type galaxies. *Astrophys. J. Suppl.*, 165:57–94, July 2006.

[97] J. E. Dale, M. B. Davies, R. P. Church, and M. Freitag. Red giant stellar collisions in the Galactic Centre. *Mon. Not. R. Astron. Soc.*, 393:1016–1033, March 2009.

[98] T. Damour. Gravitational radiation reaction in the binary pulsar and the quadrupole-formula controversy. *Phys. Rev. Lett.*, 51:1019–1021, September 1983.

[99] T. Damour and N. Deruelle. General relativistic celestial mechanics of binary systems. I. The post-Newtonian motion. *Annales de l'I.H.P. Physique théorique*, 43:107–132, 1985.

[100] T. Damour and G. Schafer. Higher-order relativistic periastron advances and binary pulsars. *Nuovo Cimento B Serie*, 101:127–176, 1988.

[101] J. M. A. Danby and G. L. Camm. Statistical dynamics and accretion. *Mon. Not. R. Astron. Soc.*, 117:50–71, 1957.

[102] K. Danzmann. LISA–laser interferometer space antenna, pre-phase. Technical report, MPQ 233, Max-Planck-Institut für Quantenoptic, 1988.

[103] K. Danzmann and LISA Study Team. LISA - an ESA cornerstone mission for a gravitational wave observatory. *Classical and Quantum Gravity*, 14:1399–1404, June 1997.

[104] W. De Sitter. Einstein's theory of gravitation and its astronomical consequences. *Mon. Not. R. Astron. Soc.*, 76:699–728, June 1916.

[105] W. De Sitter. On Einstein's theory of gravitation and its astronomical consequences. Second paper. *Mon. Not. R. Astron. Soc.*, 77:155–184, December 1916.

[106] G. de Vaucouleurs. Recherches sur les nebuleuses extragalactiques. *Annales d'Astrophysique*, 11:247–287, January 1948.

[107] G. de Vaucouleurs. On the distribution of mass and luminosity in elliptical galaxies. *Mon. Not. R. Astron. Soc.*, 113:134–161, 1953.

[108] T. de Zeeuw and D. Merritt. Stellar orbits in a triaxial galaxy. I – Orbits in the plane of rotation. *Astrophys. J.*, 267:571–595, April 1983.

[109] P. Debye. Zur Theorie der anomalen Dispersion im Gebiete der langweiligen elektrischen Strahlung. *Berichte der deutschen Physikalischen Gesellschaft*, 15:777, 1913.

[110] P. Debye. *Polare Molekeln*. Hirzel, 1929.

[111] W. Dehnen. A family of potential-density pairs for spherical galaxies and bulges. *Mon. Not. R. Astron. Soc.*, 265:250–256, November 1993.

[112] W. Dehnen. Modelling galaxies with $f(E, L_z)$: A black hole in M32. *Mon. Not. R. Astron. Soc.*, 274:919–932, June 1995.

[113] W. Dehnen and O. E. Gerhard. Three-integral models of oblate elliptical galaxies. *Mon. Not. R. Astron. Soc.*, 261:311–336, March 1993.

[114] A. T. Deibel, M. Valluri, and D. Merritt. The orbital structure of triaxial galaxies with figure rotation. *Astrophys. J.*, 728:128, February 2011.

[115] H. Dejonghe. Stellar dynamics and the description of stellar systems. *Physics Reports*, 133:217–313, February 1986.

[116] H. Dejonghe. A completely analytical family of anisotropic Plummer models. *Mon. Not. R. Astron. Soc.*, 224:13–39, January 1987.

[117] H. Dejonghe and D. Merritt. Inferring the mass of spherical stellar systems from velocity moments. *Astrophys. J.*, 391:531–549, June 1992.

[118] J. Dennett-Thorpe, P. A. G. Scheuer, R. A. Laing, A. H. Bridle, G. G. Pooley, and W. Reich. Jet reorientation in active galactic nuclei: two winged radio galaxies. *Mon. Not. R. Astron. Soc.*, 330:609–620, March 2002.

[119] S. Detweiler. Pulsar timing measurements and the search for gravitational waves. *Astrophys. J.*, 234:1100–1104, December 1979.

[120] P. Diener, A. G. Kosovichev, E. V. Kotok, I. D. Novikov, and C. J. Pethick. Non-linear effects at tidal capture of stars by a massive black hole – II. Compressible affine models and tidal interaction after capture. *Mon. Not. R. Astron. Soc.*, 275:498–506, July 1995.

[121] T. Do, A. M. Ghez, M. R. Morris, J. R. Lu, K. Matthews, S. Yelda, and J. Larkin. High angular resolution integral-field spectroscopy of the galaxy's nuclear cluster: A missing stellar cusp? *Astrophys. J.*, 703:1323–1337, October 2009.

[122] V. I. Dokuchaev and L. M. Ozernoi. Stellar systems fed by outside stars – the evolution of model galactic nuclei. *Soviet Astronomy Letters*, 11: 139–142, May 1985.

[123] J. L. Donley, W. N. Brandt, M. Eracleous, and T. Boller. Large-amplitude X-ray outbursts from galactic nuclei: A systematic survey using ROSAT archival data. *Astron. J.*, 124:1308–1321, September 2002.

[124] A. G. Doroshkevich. Variation of the moment of inertia of a star during accretion. *Astron. Zh.*, 43:105, 1966.

[125] J. Dubinski. The effect of dissipation on the shapes of dark halos. *Astrophys. J.*, 431:617–624, August 1994.

[126] A. Duquennoy and M. Mayor. Multiplicity among solar-type stars in the solar neighbourhood. II – Distribution of the orbital elements in an unbiased sample. *Astron. Astrophys.*, 248:485–524, August 1991.

[127] A. Eckart and R. Genzel. Observations of stellar proper motions near the Galactic Centre. *Nature*, 383:415–417, October 1996.

[128] A. S. Eddington. The distribution of stars in globular clusters. *Mon. Not. R. Astron. Soc.*, 76:572–585, May 1916.

[129] P. P. Eggleton. Why stars evolve to giant radii. In M. Livio, editor, *Unsolved Problems in Stellar Evolution*, page 172, 2000.

[130] P. P. Eggleton and R. C. Cannon. A conjecture regarding the evolution of dwarf stars into red giants. *Astrophys. J.*, 383:757–760, December 1991.

[131] E. Eilon, G. Kupi, and T. Alexander. The efficiency of resonant relaxation around a massive black hole. *Astrophys. J.*, 698:641–647, June 2009.

[132] C. Einsel and R. Spurzem. Dynamical evolution of rotating stellar systems – I. Pre-collapse, equal-mass system. *Mon. Not. R. Astron. Soc.*, 302:81–95, January 1999.

[133] A. Einstein. Über die von der molekularkinetischen Theorie der Wärme geforderte Bewegung von in ruhenden Flüssigkeiten suspendierten Teilchen. *Ann. Physik*, 322:549–560, 1905.

[134] A. Einstein. Zur Theorie der Brownschen Bewegung. *Ann. Physik*, 324: 371–381, 1906.

[135] A. Einstein. On a stationary system with spherical symmetry consisting of many gravitating masses. *Ann. Math.*, 4:922–936, 1940.

[136] A. Einstein, L. Infeld, and B. Hoffmann. The gravitational equations and the problem of motion. *Ann. Math.*, 39:65–100, January 1938.

[137] F. Eisenhauer, R. Genzel, T. Alexander, R. Abuter, T. Paumard, T. Ott, A. Gilbert, S. Gillessen, M. Horrobin, S. Trippe, H. Bonnet, C. Dumas, N. Hubin, A. Kaufer, M. Kissler-Patig, G. Monnet, S. Ströbele, T. Szeifert, A. Eckart, R. Schödel, and S. Zucker. SINFONI in the Galactic center: Young stars and infrared flares in the central light-month. *Astrophys. J.*, 628:246–259, July 2005.

[138] R. D. Ekers, R. Fanti, C. Lari, and P. Parma. NGC326 – A radio galaxy with a precessing beam. *Nature*, 276:588–590, December 1978.

[139] E. Emsellem, M. Cappellari, D. Krajnović, G. van de Ven, R. Bacon, M. Bureau, R. L. Davies, P. T. de Zeeuw, J. Falcón-Barroso, H. Kuntschner, R. McDermid, R. F. Peletier, and M. Sarzi. The SAURON project – IX. A kinematic classification for early-type galaxies. *Mon. Not. R. Astron. Soc.*, 379:401–417, August 2007.

[140] E. Emsellem and F. Combes. N-body simulations of the nucleus of M 31. *Astron. Astrophys.*, 323:674–684, July 1997.

[141] E. Emsellem, H. Dejonghe, and R. Bacon. Dynamical models of NGC 3115. *Mon. Not. R. Astron. Soc.*, 303:495–514, March 1999.

[142] R. Epstein. The binary pulsar – Post-Newtonian timing effects. *Astrophys. J.*, 216:92–100, August 1977.

[143] M. Eracleous, J. P. Halpern, A. M. Gilbert, J. A. Newman, and A. V. Filippenko. Rejection of the binary broad-line region interpretation of double-peaked emission lines in three active galactic nuclei. *Astrophys. J.*, 490:216–226, November 1997.

[144] P. Erwin and L. S. Sparke. Vertical instabilities and off-plane orbits in circumbinary disks. *Astrophys. J.*, 521:798–822, August 1999.

[145] P. Erwin and L. S. Sparke. Double bars, inner disks, and nuclear rings in early-type disk galaxies. *Astron. J.*, 124:65–77, July 2002.

[146] A. Escala, R. B. Larson, P. S. Coppi, and D. Mardones. The role of gas in the merging of massive black holes in galactic nuclei. I. Black hole merging in a spherical gas cloud. *Astrophys. J.*, 607:765–777, June 2004.

[147] A. Escala, R. B. Larson, P. S. Coppi, and D. Mardones. The role of gas in the merging of massive black holes in galactic nuclei. II. Black hole merging in a nuclear gas disk. *Astrophys. J.*, 630:152–166, September 2005.

[148] N. W. Evans. The power-law galaxies. *Mon. Not. R. Astron. Soc.*, 267: 333–360, March 1994.

[149] G. Fabbiano. Populations of X-ray sources in galaxies. *Ann. Rev. Astron. Astrophys.*, 44:323–366, September 2006.

[150] G. Fabbiano, J. Wang, M. Elvis, and G. Risaliti. A close nuclear black-hole pair in the spiral galaxy NGC3393. *Nature*, 477:431–434, September 2011.

[151] S. M. Faber, A. Dressler, R. L. Davies, D. Burstein, and D. Lynden-Bell. Global scaling relations for elliptical galaxies and implications for formation. In S. M. Faber, editor, *Nearly Normal Galaxies. From the Planck Time to the Present*, pages 175–183, 1987.

[152] S. M. Faber and R. E. Jackson. Velocity dispersions and mass-to-light ratios for elliptical galaxies. *Astrophys. J.*, 204:668–683, March 1976.

[153] A. C. Fabian and K. Iwasawa. The mass density in black holes inferred from the X-ray background. *Mon. Not. R. Astron. Soc.*, 303:L34–L36, February 1999.

[154] O. Fakhouri, C.-P. Ma, and M. Boylan-Kolchin. The merger rates and mass assembly histories of dark matter haloes in the two Millennium simulations. *Mon. Not. R. Astron. Soc.*, 406:2267–2278, August 2010.

[155] S. M. Fall and C. S. Frenk. The true shapes of globular clusters. *Astron. J.*, 88:1626–1632, November 1983.

[156] F. Farago and J. Laskar. High-inclination orbits in the secular quadrupolar three-body problem. *Mon. Not. R. Astron. Soc.*, 401:1189–1198, January 2010.

[157] L. Ferrarese, P. Côté, E. Dalla Bontà, E. W. Peng, D. Merritt, A. Jordán, J. P. Blakeslee, M. Haşegan, S. Mei, S. Piatek, J. L. Tonry, and M. J. West. A fundamental relation between compact stellar nuclei, supermassive black holes, and their host galaxies. *Astrophys. J. Lett.*, 644:L21–L24, June 2006.

[158] L. Ferrarese, P. Côté, A. Jordán, E. W. Peng, J. P. Blakeslee, S. Piatek, S. Mei, D. Merritt, M. Milosavljević, J. L. Tonry, and M. J. West. The ACS Virgo Cluster survey. VI. Isophotal analysis and the structure of early-type galaxies. *Astrophys. J. Suppl.*, 164:334–434, June 2006.

[159] L. Ferrarese and H. Ford. Supermassive black holes in galactic nuclei: Past, present and future research. *Space Science Reviews*, 116:523–624, January 2005.

[160] L. Ferrarese and D. Merritt. A fundamental relation between supermassive black holes and their host galaxies. *Astrophys. J. Lett.*, 539:L9–L12, August 2000.

[161] L. Ferrarese, R. W. Pogge, B. M. Peterson, D. Merritt, A. Wandel, and C. L. Joseph. Supermassive black holes in active galactic nuclei. I. The consistency of black hole masses in quiescent and active galaxies. *Astrophys. J. Lett.*, 555:L79–L82, July 2001.

[162] J. Fiestas, R. Spurzem, and E. Kim. 2D Fokker–Planck models of rotating clusters. *Mon. Not. R. Astron. Soc.*, 373:677–686, December 2006.

[163] D. F. Figer. Young massive clusters in the galactic center. In H. J. G. L. M. Lamers, L. J. Smith, and A. Nota, editors, *The Formation and Evolution of Massive Young Star Clusters*, volume 322 of *Astronomical Society of the Pacific Conference Series*, page 49, December 2004.

[164] A. V. Filippenko and L. C. Ho. A low-mass central black hole in the bulgeless Seyfert 1 Galaxy NGC 4395. *Astrophys. J. Lett.*, 588:L13–L16, May 2003.

[165] J. A. Fillmore and H. F. Levison. Dynamical models of highly flattened oblate elliptical galaxies with de Vaucouleurs' surface-brightness profiles. *Astron. J.*, 97:57–68, January 1989.

[166] G. Fiorentino, R. Contreras Ramos, E. Tolstoy, G. Clementini, and A. Saha. The ancient stellar population of M 32: RR Lyrae variable stars confirmed. *Astron. Astrophys.*, 539:A138, 2012.

[167] A. D. Fokker. Die mittlere Energie rotierender elektrischer Dipole im Strahlungsfeld. *Annalen der Physik*, 348:810–820, 1914.

[168] J. Frank and M. J. Rees. Effects of massive central black holes on dense stellar systems. *Mon. Not. R. Astron. Soc.*, 176:633–647, September 1976.

[169] M. Franx, G. Illingworth, and T. de Zeeuw. The ordered nature of elliptical galaxies - Implications for their intrinsic angular momenta and shapes. *Astrophys. J.*, 383:112–134, December 1991.

[170] M. Freitag, P. Amaro-Seoane, and V. Kalogera. Stellar remnants in galactic nuclei: mass segregation. *Astrophys. J.*, 649:91–117, September 2006.

[171] M. Freitag and W. Benz. A new Monte Carlo code for star cluster simulations. II. Central black hole and stellar collisions. *Astron. Astrophys.*, 394:345–374, October 2002.

[172] T. Fridman and D. Merritt. Periodic orbits in triaxial galaxies with weak cusps. *Astron. J.*, 114:1479–1487, October 1997.

[173] C. L. Fryer. Mass limits for black hole formation. *Astrophys. J.*, 522:413–418, September 1999.

[174] C. I. Fuentes, K. Z. Stanek, B. S. Gaudi, B. A. McLeod, S. Bogdanov, J. D. Hartman, R. C. Hickox, and M. J. Holman. The hypervelocity star SDSS J090745.0+024507 is variable. *Astrophys. J. Lett.*, 636:L37–L40, January 2006.

[175] C. F. Gammie. Nonlinear outcome of gravitational instability in cooling, gaseous disks. *Astrophys. J.*, 553:174–183, May 2001.

[176] K. Gebhardt, R. Bender, G. Bower, A. Dressler, S. M. Faber, A. V. Filippenko, R. Green, C. Grillmair, L. C. Ho, J. Kormendy, T. R. Lauer, J. Magorrian, J. Pinkney, D. Richstone, and S. Tremaine. A relationship between nuclear black hole mass and galaxy velocity dispersion. *Astrophys. J. Lett.*, 539:L13–L16, August 2000.

[177] K. Gebhardt, J. Kormendy, L. C. Ho, R. Bender, G. Bower, A. Dressler, S. M. Faber, A. V. Filippenko, R. Green, C. Grillmair, T. R. Lauer, J. Magorrian, J. Pinkney, D. Richstone, and S. Tremaine. Black hole mass estimates from reverberation mapping and from spatially resolved kinematics. *Astrophys. J. Lett.*, 543:L5–L8, November 2000.

[178] K. Gebhardt, T. R. Lauer, J. Kormendy, J. Pinkney, G. A. Bower, R. Green, T. Gull, J. B. Hutchings, M. E. Kaiser, C. H. Nelson, D. Richstone, and D. Weistrop. M33: A galaxy with no supermassive black hole. *Astron. J.*, 122:2469–2476, November 2001.

[179] K. Gebhardt, R. M. Rich, and L. C. Ho. An intermediate-mass black hole in the globular cluster G1: Improved significance from new Keck and Hubble Space Telescope observations. *Astrophys. J.*, 634:1093–1102, December 2005.

[180] K. Gebhardt, D. Richstone, S. Tremaine, T. R. Lauer, R. Bender, G. Bower, A. Dressler, S. M. Faber, A. V. Filippenko, R. Green, C. Grillmair, L. C. Ho, J. Kormendy, J. Magorrian, and J. Pinkney. Axisymmetric dynamical models of the central regions of galaxies. *Astrophys. J.*, 583:92–115, January 2003.

[181] K. Gebhardt and J. Thomas. The black hole mass, stellar mass-to-light ratio, and dark halo in M87. *Astrophys. J.*, 700:1690–1701, August 2009.

[182] R. Genzel, C. Pichon, A. Eckart, O. E. Gerhard, and T. Ott. Stellar dynamics in the Galactic centre: Proper motions and anisotropy. *Mon. Not. R. Astron. Soc.*, 317:348–374, September 2000.

[183] R. Genzel, R. Schödel, T. Ott, F. Eisenhauer, R. Hofmann, M. Lehnert, A. Eckart, T. Alexander, A. Sternberg, R. Lenzen, Y. Clénet, F. Lacombe, D. Rouan, A. Renzini, and L. E. Tacconi-Garman. The stellar cusp around the supermassive black hole in the Galactic center. *Astrophys. J.*, 594: 812–832, September 2003.

[184] R. Genzel, N. Thatte, A. Krabbe, H. Kroker, and L. E. Tacconi-Garman. The dark mass concentration in the central parsec of the Milky Way. *Astrophys. J.*, 472:153–172, November 1996.

[185] O. E. Gerhard. Line-of-sight velocity profiles in spherical galaxies: breaking the degeneracy between anisotropy and mass. *Mon. Not. R. Astron. Soc.*, 265:213–230, November 1993.

[186] O. E. Gerhard and J. Binney. Triaxial galaxies containing massive black holes or central density cusps. *Mon. Not. R. Astron. Soc.*, 216:467–502, September 1985.

[187] O. E. Gerhard and J. J. Binney. On the deprojection of axisymmetric bodies. *Mon. Not. R. Astron. Soc.*, 279:993–1004, April 1996.

[188] J. Gerssen, R. P. van der Marel, K. Gebhardt, P. Guhathakurta, R. C. Peterson, and C. Pryor. Hubble Space Telescope evidence for an intermediate-mass black hole in the globular cluster M15. II. Kinematic analysis and dynamical modeling. *Astron. J.*, 124:3270–3288, December 2002.

[189] A. M. Ghez, G. Duchene, K. Matthews, S. D. Hornstein, A. Tanner, J. Larkin, M. Morris, E. E. Becklin, S. Salim, T. Kremenek, D. Thompson, B. T. Soifer, G. Neugebauer, and I. McLean. The first measurement of spectral lines in a short-period star bound to the galaxy's central black hole: A paradox of youth. *Astrophys. J. Lett.*, 586:L127–L131, April 2003.

[190] A. M. Ghez, B. L. Klein, M. Morris, and E. E. Becklin. High proper-motion stars in the vicinity of Sagittarius A*: Evidence for a supermassive black hole at the center of our galaxy. *Astrophys. J.*, 509:678–686, December 1998.

[191] A. M. Ghez, S. Salim, N. N. Weinberg, J. R. Lu, T. Do, J. K. Dunn, K. Matthews, M. R. Morris, S. Yelda, E. E. Becklin, T. Kremenek, M. Milosavljevic, and J. Naiman. Measuring distance and properties of the Milky Way's central supermassive black hole with stellar orbits. *Astrophys. J.*, 689:1044–1062, December 2008.

[192] S. Gillessen, F. Eisenhauer, T. K. Fritz, H. Bartko, K. Dodds-Eden, O. Pfuhl, T. Ott, and R. Genzel. The orbit of the star S2 around SGR A* from very large telescope and Keck data. *Astrophys. J. Lett.*, 707:L114–L117, December 2009.

[193] S. Gillessen, F. Eisenhauer, S. Trippe, T. Alexander, R. Genzel, F. Martins, and T. Ott. Monitoring stellar orbits around the massive black hole in the Galactic center. *Astrophys. J.*, 692:1075–1109, February 2009.

[194] J. A. Girash. *A Fokker–Planck Study of Dense Rotating Stellar Clusters*. PhD thesis, Harvard University, 2010.

[195] A. Gnerucci, A. Marconi, A. Capetti, D. J. Axon, and A. Robinson. Spectroastrometry of rotating gas disks for the detection of supermassive black holes in galactic nuclei. I. Method and simulations. *Astron. Astrophys.*, 511:A19, February 2010.

[196] A. Gnerucci, A. Marconi, A. Capetti, D. J. Axon, A. Robinson, and N. Neumayer. Spectroastrometry of rotating gas disks for the detection of supermassive black holes in galactic nuclei. II. Application to the galaxy Centaurus A (NGC 5128). *Astron. Astrophys.*, 536:A86, December 2011.

[197] B. B. Godfrey. Mach's principle, the Kerr metric, and black-hole physics. *Phys. Rev. D*, 1:2721–2725, May 1970.

[198] P. Goldreich and R. Sari. Eccentricity evolution for planets in gaseous disks. *Astrophys. J.*, 585:1024–1037, March 2003.

[199] P. Goldreich and S. Tremaine. The excitation of density waves at the Lindblad and corotation resonances by an external potential. *Astrophys. J.*, 233:857–871, November 1979.

[200] P. Goldreich and S. Tremaine. Disk-satellite interactions. *Astrophys. J.*, 241:425–441, October 1980.

[201] H. Goldstein, C. Poole, and J. Safko. *Classical Mechanics*. San Francisco: Addison-Wesley, 3rd edition, 2002.

[202] P. Gondolo and J. Silk. Dark matter annihilation at the Galactic center. *Phys. Rev. Lett.*, 83:1719–1722, August 1999.

[203] R. M. González Delgado, E. Pérez, R. Cid Fernandes, and H. Schmitt. HST/WFPC2 imaging of the circumnuclear structure of low-luminosity active galactic nuclei. I. Data and nuclear morphology. *Astron. J.*, 135: 747–765, March 2008.

[204] J. Goodman and M. Schwarzschild. Semistochastic orbits in a triaxial potential. *Astrophys. J.*, 245:1087–1093, May 1981.

[205] J. J. Goodman. *Dynamical Relaxation in Stellar Systems*. PhD thesis, Princeton University, NJ., 1983.

[206] A. Gould and A. C. Quillen. Sagittarius A* companion S0-2: A probe of very high mass star formation. *Astrophys. J.*, 592:935–940, August 2003.

[207] A. W. Graham. Core depletion from coalescing supermassive black holes. *Astrophys. J. Lett.*, 613:L33–L36, September 2004.

[208] A. W. Graham. A review of elliptical and disc galaxy structure, and modern scaling laws. In T. D. Oswalt, editor, *Planets, Stars and Stellar Systems*. Springer, Berlin, 2012.

[209] A. W. Graham, P. Erwin, N. Caon, and I. Trujillo. A correlation between galaxy light concentration and supermassive black hole mass. *Astrophys. J. Lett.*, 563:L11–L14, December 2001.

[210] A. W. Graham, P. Erwin, I. Trujillo, and A. Asensio Ramos. A new empirical model for the structural analysis of early-type galaxies, and a critical review of the Nuker model. *Astron. J.*, 125:2951–2963, June 2003.

[211] A. W. Graham and R. Guzmán. HST photometry of dwarf elliptical galaxies in Coma, and an explanation for the alleged structural dichotomy between dwarf and bright elliptical galaxies. *Astron. J.*, 125:2936–2950, June 2003.

[212] J. Granholm. How to design a kludge. *Datamation*, 8:30–31, February 1962.

[213] J. E. Greene and L. C. Ho. Estimating black hole masses in active galaxies using the Hα emission line. *Astrophys. J.*, 630:122–129, September 2005.

[214] J. E. Greene and L. C. Ho. A new sample of low-mass black holes in active galaxies. *Astrophys. J.*, 670:92–104, November 2007.

[215] L. Greenhill. Extragalactic H_2O masers. In V. Migenes and M. J. Reid, editors, *Cosmic Masers: From Proto-Stars to Black Holes*, volume 206 of *IAU Symposium*, page 381, 2002.

[216] L. J. Greenhill, R. S. Booth, S. P. Ellingsen, J. R. Herrnstein, D. L. Jauncey, P. M. McCulloch, J. M. Moran, R. P. Norris, J. E. Reynolds, and A. K. Tzioumis. A warped accretion disk and wide-angle outflow in the inner parsec of the Circinus Galaxy. *Astrophys. J.*, 590:162–173, June 2003.

[217] A. Gualandris, M. Dotti, and A. Sesana. Massive black hole binary plane reorientation in rotating stellar systems. *Mon. Not. R. Astron. Soc.*, 420: L38–L42, February 2012.

[218] A. Gualandris and D. Merritt. Ejection of supermassive black holes from galaxy cores. *Astrophys. J.*, 678:780–797, May 2008.

[219] A. Gualandris and D. Merritt. Perturbations of intermediate-mass black holes on stellar orbits in the Galactic center. *Astrophys. J.*, 705:361–371, November 2009.

[220] A. Gualandris and D. Merritt. Long-term evolution of massive black hole binaries. IV. Mergers of galaxies with collisionally relaxed nuclei. *Astrophys. J.*, 744:74, January 2012.

[221] K. Gültekin, D. O. Richstone, K. Gebhardt, T. R. Lauer, S. Tremaine, M. C. Aller, R. Bender, A. Dressler, S. M. Faber, A. V. Filippenko, R. Green, L. C. Ho, J. Kormendy, J. Magorrian, J. Pinkney, and C. Siopis. The M–σ and M–L relations in galactic bulges, and determinations of their intrinsic scatter. *Astrophys. J.*, 698:198–221, June 2009.

[222] M. A. Gürkan, M. Freitag, and F. A. Rasio. Formation of massive black holes in dense star clusters. I. Mass segregation and core collapse. *Astrophys. J.*, 604:632–652, April 2004.

[223] M. A. Gürkan and C. Hopman. Resonant relaxation near a massive black hole: the dependence on eccentricity. *Mon. Not. R. Astron. Soc.*, 379:1083–1088, August 2007.

[224] M. G. Haehnelt and G. Kauffmann. Multiple supermassive black holes in galactic bulges. *Mon. Not. R. Astron. Soc.*, 336:L61–L64, November 2002.

[225] M. G. Haehnelt, P. Natarajan, and M. J. Rees. High-redshift galaxies, their active nuclei and central black holes. *Mon. Not. R. Astron. Soc.*, 300: 817–827, November 1998.

[226] B.M.S. Hansen and M. Milosavljević. The need for a second black hole at the Galactic center. *Astrophys. J. Lett.*, 593:L77–L80, August 2003.

[227] C. J. Hansen and S. D. Kawaler. *Stellar Interiors. Physical Principles, Structure, and Evolution.* Springer, 1994.

[228] R. S. Harrington. Dynamical evolution of triple stars. *Astron. J.*, 73:190–194, April 1968.

[229] H. Hasan, D. Pfenniger, and C. Norman. Galactic bars with central mass concentrations - Three-dimensional dynamics. *Astrophys. J.*, 409:91–109, May 1993.

[230] D. Heggie and P. Hut. *The Gravitational Million-Body Problem: A Multidisciplinary Approach to Star Cluster Dynamics.* Cambridge University Press, February 2003.

[231] D. C. Heggie. Binary evolution in stellar dynamics. *Mon. Not. R. Astron. Soc.*, 173:729–787, December 1975.

[232] D. C. Heggie and R. D. Mathieu. Standardised units and time scales. In P. Hut and S. L. W. McMillan, editors, *The Use of Supercomputers in Stellar Dynamics*, volume 267 of *Lecture Notes in Physics*, page 233. Springer : Berlin, 1986.

[233] D. C. Heggie and D. Stevenson. Two homological models for the evolution of star clusters. *Mon. Not. R. Astron. Soc.*, 230:223–241, January 1988.

[234] M. Henon. L'amas isochrone: I. *Annales d'Astrophysique*, 22:126–139, February 1959.

[235] M. Hénon. Sur l'évolution dynamique des amas globulaires. *Annales d'Astrophysique*, 24:369–419, February 1961.

[236] J. G. Hills. Possible power source of Seyfert galaxies and QSOs. *Nature*, 254:295–298, March 1975.

[237] J. G. Hills. The effect of low-velocity, low-mass intruders (collisionless gas) on the dynamical evolution of a binary system. *Astron. J.*, 88:1269–1283, August 1983.

[238] J. G. Hills. Hyper-velocity and tidal stars from binaries disrupted by a massive Galactic black hole. *Nature*, 331:687–689, February 1988.

[239] J. G. Hills and L. W. Fullerton. Computer simulations of close encounters between single stars and hard binaries. *Astron. J.*, 85:1281–1291, September 1980.

[240] D. Hils and P. L. Bender. Gradual approach to coalescence for compact stars orbiting massive black holes. *Astrophys. J. Lett.*, 445:L7–L10, May 1995.

[241] P. W. Hodge. The structure and content of NGC 205. *Astrophys. J.*, 182: 671–696, June 1973.

[242] L. Hoffman, T. J. Cox, S. Dutta, and L. Hernquist. Orbital structure of merger remnants. I. Effect of gas fraction in pure disk mergers. *Astrophys. J.*, 723:818–844, November 2010.

[243] M. Holman, J. Touma, and S. Tremaine. Chaotic variations in the eccentricity of the planet orbiting 16 Cygni B. *Nature*, 386:254–256, March 1997.

[244] C. Hopman. Binary dynamics near a massive black hole. *Astrophys. J.*, 700:1933–1951, August 2009.

[245] C. Hopman and T. Alexander. Resonant relaxation near a massive black hole: The stellar distribution and gravitational wave sources. *Astrophys. J.*, 645:1152–1163, July 2006.

[246] R. C. W. Houghton, J. Magorrian, M. Sarzi, N. Thatte, R. L. Davies, and D. Krajnović. The central kinematics of NGC 1399 measured with 14 pc resolution. *Mon. Not. R. Astron. Soc.*, 367:2–18, March 2006.

[247] S. A. Hughes and R. D. Blandford. Black hole mass and spin coevolution by mergers. *Astrophys. J. Lett.*, 585:L101–L104, March 2003.

[248] S. A. Hughes and D. E. Holz. Cosmology with coalescing massive black holes. *Classical and Quantum Gravity*, 20:565–572, May 2003.

[249] R. A. Hulse and J. H. Taylor. Discovery of a pulsar in a binary system. *Astrophys. J. Lett.*, 195:L51–L53, January 1975.

[250] C. Hunter. Determination of the distribution function of an elliptical galaxy. *Astron. J.*, 80:783–793, October 1975.

[251] C. Hunter, P. T. de Zeeuw, C. Park, and M. Schwarzschild. Prolate galaxy models with thin-tube orbits. *Astrophys. J.*, 363:367–390, November 1990.

[252] L. Infeld and J. Plebanski. *Motion and Relativity*. Pergamon Press, 1960.

[253] K. A. Innanen, J. Q. Zheng, S. Mikkola, and M. J. Valtonen. The Kozai mechanism and the stability of planetary orbits in binary star systems. *Astron. J.*, 113:1915–1919, May 1997.

[254] P. B. Ivanov, J. C. B. Papaloizou, and A. G. Polnarev. The evolution of a supermassive binary caused by an accretion disc. *Mon. Not. R. Astron. Soc.*, 307:79–90, July 1999.

[255] J. M. Jackson, N. Geis, R. Genzel, A. I. Harris, S. Madden, A. Poglitsch, G. J. Stacey, and C. H. Townes. Neutral gas in the central 2 parsecs of the Galaxy. *Astrophys. J.*, 402:173–184, January 1993.

[256] V. Jacobs and J. A. Sellwood. Long-lived lopsided modes of annular disks orbiting a central mass. *Astrophys. J. Lett.*, 555:L25–L28, July 2001.

[257] J. H. Jeans. On the theory of star-streaming and the structure of the universe. *Mon. Not. R. Astron. Soc.*, 76:70–84, December 1915.

[258] R. L. Jeffery. *The Theory of Functions of a Real Variable*. Toronto : University of Toronto Press, 1953.

[259] H. Jerjen and B. Binggeli. Are "dwarf" ellipticals genuine ellipticals? In M. Arnaboldi, G. S. Da Costa, and P. Saha, editors, *The Nature of Elliptical Galaxies*, volume 116 of *ASP Conference Series: 2nd Stromlo Symposium*, page 239, 1997.

[260] R. Jesseit, M. Cappellari, T. Naab, E. Emsellem, and A. Burkert. Specific angular momentum of disc merger remnants and the Lambda-R-parameter. *Mon. Not. R. Astron. Soc.*, 397:1202–1214, August 2009.

[261] D. H. Jones, J. R. Mould, A. M. Watson, C. Grillmair, J. S. Gallagher, III, G. E. Ballester, C. J. Burrows, S. Casertano, J. T. Clarke, D. Crisp, R. E. Griffiths, J. J. Hester, J. G. Hoessel, J. A. Holtzman, P. A. Scowen, K. R. Stapelfeldt, J. T. Trauger, and J. A. Westphal. Visible and far-ultraviolet WFPC2 imaging of the nucleus of the galaxy NGC 205. *Astrophys. J.*, 466:742–749, August 1996.

[262] C. L. Joseph, D. Merritt, R. Olling, M. Valluri, R. Bender, G. Bower, A. Danks, T. Gull, J. Hutchings, M. E. Kaiser, S. Maran, D. Weistrop, B. Woodgate, E. Malumuth, C. Nelson, P. Plait, and D. Lindler. The nuclear dynamics of M32. I. Data and stellar kinematics. *Astrophys. J.*, 550:668–690, April 2001.

[263] P. C. Joss, S. Rappaport, and W. Lewis. The core mass-radius relation for giants - A new test of stellar evolution theory. *Astrophys. J.*, 319:180–187, August 1987.

[264] A. Just, F. M. Khan, P. Berczik, A. Ernst, and R. Spurzem. Dynamical friction of massive objects in galactic centres. *Mon. Not. R. Astron. Soc.*, 411:653–674, February 2011.

[265] M. Kaasalainen. Torus construction in potentials supporting different orbit families. *Mon. Not. R. Astron. Soc.*, 268:1041–1050, June 1994.

[266] C. Kalapotharakos and N. Voglis. Global dynamics in self-consistent models of elliptical galaxies. *Celestial Mechanics and Dynamical Astronomy*, 92:157–188, April 2005.

[267] C. Kalapotharakos, N. Voglis, and G. Contopoulos. Chaos and secular evolution of triaxial N-body galactic models due to an imposed central mass. *Astron. Astrophys.*, 428:905–923, December 2004.

[268] J. S. Kalirai, B. M. S. Hansen, D. D. Kelson, D. B. Reitzel, R. M. Rich, and H. B. Richer. The initial–final mass relation: Direct constraints at the low-mass end. *Astrophys. J.*, 676:594–609, March 2008.

[269] H. E. Kandrup. The evolution of a system of stars in an external environment. *Astrophys. J.*, 364:100–103, November 1990.

[270] H. E. Kandrup and M. E. Mahon. Relaxation and stochasticity in a truncated Toda lattice. *Phys. Rev. E*, 49:3735–3747, May 1994.

[271] H. E. Kandrup and I. V. Sideris. Chaos in cuspy triaxial galaxies with a supermassive black hole: a simple toy model. *Celestial Mechanics and Dynamical Astronomy*, 82:61–81, January 2002.

[272] S. Kaspi, P. S. Smith, D. Maoz, H. Netzer, and B. T. Jannuzi. Measurement of the broad line region size in two bright quasars. *Astrophys. J. Lett.*, 471: L75–L78, November 1996.

[273] G. Kauffmann and M. Haehnelt. A unified model for the evolution of galaxies and quasars. *Mon. Not. R. Astron. Soc.*, 311:576–588, January 2000.

[274] W. C. Keel. Shock excitation, nuclear activity, and star formation in NGC 6240. *Astron. J.*, 100:356–372, August 1990.

[275] R. P. Kerr. Gravitational field of a spinning mass as an example of algebraically special metrics. *Phys. Rev. Lett.*, 11:237–238, September 1963.

[276] I. G. Kevrekidis, C. W. Gear, and G. Hummer. *Equationfree Modeling For Complex Systems*, page 1453. Springer Science+Business Media B. V. Springer, 2005.

[277] F. M. Khan, A. Just, and D. Merritt. Efficient merger of binary supermassive black holes in merging galaxies. *Astrophys. J.*, 732:89, May 2011.

[278] L. E. Kidder. Coalescing binary systems of compact objects to $(post)^{5/2}$-Newtonian order. V. Spin effects. *Phys. Rev. D*, 52:821–847, July 1995.

[279] E. Kim, I. Yoon, H. M. Lee, and R. Spurzem. Comparative study between N-body and Fokker–Planck simulations for rotating star clusters – I. Equal-mass system. *Mon. Not. R. Astron. Soc.*, 383:2–10, January 2008.

[280] S. C. Kim and M. G. Lee. Surface photometry of the dwarf elliptical galaxies NGC 185 and NGC 205. *J. Korean Astron. Soc.*, 31:51–65, April 1998.

[281] A. King. Black holes, galaxy formation, and the M_{BH}–σ relation. *Astrophys. J. Lett.*, 596:L27–L29, October 2003.

[282] A. King. The AGN-starburst connection, galactic superwinds, and M_{BH}–σ. *Astrophys. J. Lett.*, 635:L121–L123, December 2005.

[283] A. R. King. AGN have underweight black holes and reach Eddington. *Mon. Not. R. Astron. Soc.*, 408:L95–L98, October 2010.

[284] A. R. King and K. A. Pounds. Black hole winds. *Mon. Not. R. Astron. Soc.*, 345:657–659, October 2003.

[285] I. R. King. The structure of star clusters. I. An empirical density law. *Astron. J.*, 67:471–485, October 1962.

[286] I. R. King. The structure of star clusters. III. Some simple dynamical models. *Astron. J.*, 71:64–75, February 1966.

[287] H. Kinoshita and H. Nakai. General solution of the Kozai mechanism. *Celestial Mechanics and Dynamical Astronomy*, 98:67–74, May 2007.

[288] R. Kippenhahn and A. Weigert. *Stellar Structure and Evolution.* Springer, 1994.

[289] O. Klein. Zur statistischen Theorie der Suspensionen und Lösungen. *Arkiv für Matematik, Astronomi och Fysik*, 16:1–53, 1922.

[290] H. A. Kobulnicky and C. L. Fryer. A new look at the binary characteristics of massive stars. *Astrophys. J.*, 670:747–765, November 2007.

[291] S. Komossa. Ludwig Biermann Award Lecture: X-ray evidence for supermassive black holes at the centers of nearby, non-active galaxies. *Reviews of Modern Astronomy*, 15:27, 2002.

[292] S. Komossa, V. Burwitz, G. Hasinger, P. Predehl, J. S. Kaastra, and Y. Ikebe. Discovery of a binary active galactic nucleus in the ultraluminous infrared galaxy NGC 6240 using Chandra. *Astrophys. J. Lett.*, 582:L15–L19, January 2003.

[293] S. Komossa and D. Merritt. Tidal disruption flares from recoiling supermassive black holes. *Astrophys. J. Lett.*, 683:L21–L24, August 2008.

[294] S. Komossa, H. Zhou, and H. Lu. A recoiling supermassive black hole in the quasar SDSS J092712.65+294344.0? *Astrophys. J. Lett.*, 678:L81–L84, May 2008.

[295] J. Kormendy. Brightness profiles of the cores of bulges and elliptical galaxies. *Astrophys. J. Lett.*, 292:L9–L13, May 1985.

[296] J. Kormendy. Families of ellipsoidal stellar systems and the formation of dwarf elliptical galaxies. *Astrophys. J.*, 295:73–79, August 1985.

[297] J. Kormendy and R. D. McClure. The nucleus of M33. *Astron. J.*, 105: 1793–1812, May 1993.

[298] J. Kormendy and D. Richstone. Inward bound—the search for supermassive black holes in galactic nuclei. *Ann. Rev. Astron. Astrophys.*, 33:581, 1995.

[299] Y. Kozai. Secular perturbations of asteroids with high inclination and eccentricity. *Astron. J.*, 67:591, November 1962.

[300] A. Krabbe, R. Genzel, A. Eckart, F. Najarro, D. Lutz, M. Cameron, H. Kroker, L. E. Tacconi-Garman, N. Thatte, L. Weitzel, S. Drapatz, T. Geballe, A. Sternberg, and R. Kudritzki. The nuclear cluster of the Milky Way: Star formation and velocity dispersion in the central 0.5 parsec. *Astrophys. J. Lett.*, 447:L95–L99, July 1995.

[301] N. A. Krall and A. W. Trivelpiece. *Principles of Plasma Physics.* McGraw-Hill Kogakusha, 1973.

[302] P. Kroupa. On the variation of the initial mass function. *Mon. Not. R. Astron. Soc.*, 322:231–246, April 2001.

[303] C. Y. Kuo, J. A. Braatz, J. J. Condon, C. M. V. Impellizzeri, K. Y. Lo, I. Zaw, M. Schenker, C. Henkel, M. J. Reid, and J. E. Greene. The Megamaser Cosmology Project. III. Accurate masses of seven supermassive black holes in active galaxies with circumnuclear megamaser disks. *Astrophys. J.*, 727:20, January 2011.

[304] G. G. Kuzmin. Ehffekt sblizhenij zvezd i ehvolyuciya zvezdnyh skoplenij. The effect of star encounters and the evolution of star clusters. *Publications of the Tartu Astrofizica Observatory*, 33:75–102, 1957.

[305] G. G. Kuzmin. Quadratic integrals of motion and stellar orbits in the absence of axial symmetry of the potential. In *Dynamics of Galaxies and Star Clusters*, pages 71–75, 1973.

[306] P. Langevin. Sur la theorie du mouvement brownien. *Comptes Rendus de l'Académie des Sciences (Paris)*, 146:530–533, 1908.

[307] J. D. Larwood, R. P. Nelson, J. C. B. Papaloizou, and C. Terquem. The tidally induced warping, precession and truncation of accretion discs in binary systems: three-dimensional simulations. *Mon. Not. R. Astron. Soc.*, 282: 597–613, September 1996.

[308] J. Laskar. Frequency analysis for multi-dimensional systems. Global dynamics and diffusion. *Physica D Nonlinear Phenomena*, 67:257–281, August 1993.

[309] T. R. Lauer, S. M. Faber, E. J. Groth, E. J. Shaya, B. Campbell, A. Code, D. G. Currie, W. A. Baum, S. P. Ewald, J. J. Hester, J. A. Holtzman, J. Kristian, R. M. Light, C. R. Ligynds, E. J. O'Neil, Jr., and J. A. Westphal. Planetary camera observations of the double nucleus of M31. *Astron. J.*, 106:1436–1447, October 1993.

[310] R. Launhardt, R. Zylka, and P. G. Mezger. The nuclear bulge of the Galaxy. III. Large-scale physical characteristics of stars and interstellar matter. *Astron. Astrophys.*, 384:112–139, March 2002.

[311] J. P. Leahy and P. Parma. Multiple outbursts in radio galaxies. In *Extragalactic Radio Sources. From Beams to Jets*, page 307, 1992.

[312] M. G. Lee. Stellar populations in the central region of the dwarf elliptical galaxy NGC 205. *Astron. J.*, 112:1438, October 1996.

[313] M. H. Lee and S. J. Peale. Secular evolution of hierarchical planetary systems. *Astrophys. J.*, 592:1201–1216, August 2003.

[314] J. F. Lees and M. Schwarzschild. The orbital structure of galactic halos. *Astrophys. J.*, 384:491–501, January 1992.

[315] J. Lense and H. Thirring. Über den Einfluß der Eigenrotation der Zentralkörper auf die Bewegung der Planeten und Monde nach der Einsteinschen Gravitationstheorie. *Physikalische Zeitschrift*, 19:156, 1918.

[316] P. J. T. Leonard and D. Merritt. The mass of the open star cluster M35 as derived from proper motions. *Astrophys. J.*, 339:195–208, April 1989.

[317] Y. Levin. Ejection of high-velocity stars from the Galactic center by an inspiraling intermediate-mass black hole. *Astrophys. J.*, 653:1203–1209, December 2006.

[318] Y. Levin and A. M. Beloborodov. Stellar disk in the Galactic center: A remnant of a dense accretion disk? *Astrophys. J. Lett.*, 590:L33–L36, June 2003.

[319] Y. Levin, A. Wu, and E. Thommes. Intermediate-mass black hole(s) and stellar orbits in the Galactic center. *Astrophys. J.*, 635:341–348, December 2005.

[320] A. J. Lichtenberg and M. A. Lieberman. Regular and stochastic motion. *Applied Mathematics*, 38, 1983.

[321] M. L. Lidov. The evolution of orbits of artificial satellites of planets under the action of gravitational perturbations of external bodies. *Planetary and Space Science*, 9:719–759, October 1962.

[322] E. S. Light, R. E. Danielson, and M. Schwarzschild. The nucleus of M31. *Astrophys. J.*, 194:257–263, December 1974.

[323] A. P. Lightman and S. L. Shapiro. The distribution and consumption rate of stars around a massive, collapsed object. *Astrophys. J.*, 211:244–262, January 1977.

[324] X. Liu, Y. Shen, and M. A. Strauss. Cosmic train wreck by massive black holes: Discovery of a kiloparsec-scale triple active galactic nucleus. *Astrophys. J. Lett.*, 736:L7, July 2011.

[325] U. Löckmann, H. Baumgardt, and P. Kroupa. Constraining the initial mass function of stars in the Galactic Centre. *Mon. Not. R. Astron. Soc.*, 402: 519–525, February 2010.

[326] P.-Y. Longaretti and C. Lagoute. Rotating globular clusters. III. Evolutionary survey. *Astron. Astrophys.*, 319:839–849, March 1997.

[327] R. H. Lupton and J. E. Gunn. Three-integral models of globular clusters. *Astron. J.*, 93:1106–1113, May 1987.

[328] D. Lynden-Bell. Stellar dynamics: Exact solution of the self-gravitation equation. *Mon. Not. R. Astron. Soc.*, 123:447, 1962.

[329] D. Lynden-Bell. The invariant eccentricity of galactic orbits. *The Observatory*, 83:23–25, February 1963.

[330] D. Lynden-Bell. Statistical mechanics of violent relaxation in stellar systems. *Mon. Not. R. Astron. Soc.*, 136:101, 1967.

[331] D. Lynden-Bell and A. J. Kalnajs. On the generating mechanism of spiral structure. *Mon. Not. R. Astron. Soc.*, 157:1, 1972.

[332] F. Macchetto, A. Marconi, D. J. Axon, A. Capetti, W. Sparks, and P. Crane. The supermassive black hole of M87 and the kinematics of its associated gaseous disk. *Astrophys. J.*, 489:579, November 1997.

[333] W. Maciejewski and J. Binney. Kinematics from spectroscopy with a wide slit: detecting black holes in galaxy centres. *Mon. Not. R. Astron. Soc.*, 323:831–838, May 2001.

[334] J. Magorrian and S. Tremaine. Rates of tidal disruption of stars by massive central black holes. *Mon. Not. R. Astron. Soc.*, 309:447–460, October 1999.

[335] J. Magorrian, S. Tremaine, D. Richstone, R. Bender, G. Bower, A. Dressler, S. M. Faber, K. Gebhardt, R. Green, C. Grillmair, J. Kormendy, and T. Lauer. The demography of massive dark objects in galaxy centers. *Astron. J.*, 115:2285–2305, June 1998.

[336] C. Marchal. *The Three-Body Problem*. Elsevier Science, 1990.

[337] A. Marconi, D. J. Axon, A. Capetti, W. Maciejewski, J. Atkinson, D. Batcheldor, J. Binney, C. M. Carollo, L. Dressel, H. Ford, J. Gerssen, M. A. Hughes, D. Macchetto, M. R. Merrifield, C. Scarlata, W. Sparks, M. Stiavelli, Z. Tsvetanov, and R. P. van der Marel. Is there really a black hole at the center of NGC 4041? Constraints from gas kinematics. *Astrophys. J.*, 586:868–890, April 2003.

[338] A. Marconi and L. K. Hunt. The relation between black hole mass, bulge mass, and near-infrared luminosity. *Astrophys. J. Lett.*, 589:L21–L24, May 2003.

[339] R. A. Mardling and S. J. Aarseth. Tidal interactions in star cluster simulations. *Mon. Not. R. Astron. Soc.*, 321:398–420, March 2001.

[340] F. Martins, S. Gillessen, F. Eisenhauer, R. Genzel, T. Ott, and S. Trippe. On the nature of the fast-moving star S2 in the Galactic center. *Astrophys. J. Lett.*, 672:L119–L122, January 2008.

[341] L. Mayer, S. Kazantzidis, P. Madau, M. Colpi, T. Quinn, and J. Wadsley. Rapid formation of supermassive black hole binaries in galaxy mergers with gas. *Science*, 316:1874–1877, June 2007.

[342] A. Mazure and H. V. Capelato. Exact solutions for the spatial de Vaucouleurs and Sérsic laws and related quantities. *Astron. Astrophys.*, 383:384–389, January 2002.

[343] C. McCabe, G. Duchêne, and A. M. Ghez. NICMOS images of the GG Tauri circumbinary disk. *Astrophys. J.*, 575:974–988, August 2002.

[344] A. W. McConnachie, M. J. Irwin, A. M. N. Ferguson, R. A. Ibata, G. F. Lewis, and N. Tanvir. Determining the location of the tip of the red giant branch in old stellar populations: M33, Andromeda I and II. *Mon. Not. R. Astron. Soc.*, 350:243–252, May 2004.

[345] A. W. McConnachie, M. J. Irwin, A. M. N. Ferguson, R. A. Ibata, G. F. Lewis, and N. Tanvir. Distances and metallicities for 17 Local Group galaxies. *Mon. Not. R. Astron. Soc.*, 356:979–997, January 2005.

[346] R. J. McLure and J. S. Dunlop. The black hole masses of Seyfert galaxies and quasars. *Mon. Not. R. Astron. Soc.*, 327:199–207, October 2001.

[347] Y. Meiron and A. Laor. The stellar kinematic signature of massive black hole binaries. *Mon. Not. R. Astron. Soc.*, 407:1497–1513, September 2010.

[348] D. Merritt. The dynamical inverse problem for axisymmetric stellar systems. *Astron. J.*, 112:1085, September 1996.

[349] D. Merritt. Cusps and triaxiality. *Astrophys. J.*, 486:102, September 1997.

[350] D. Merritt. Recovering velocity distributions via penalized likelihood. *Astron. J.*, 114:228–237, July 1997.

[351] D. Merritt. Black holes and galaxy evolution. In F. Combes, G. A. Mamon, and V. Charmandaris, editors, *Dynamics of Galaxies: From the Early Universe to the Present*, volume 197 of *Astronomical Society of the Pacific Conference Series*, page 221, 1999.

[352] D. Merritt. Brownian motion of a massive binary. *Astrophys. J.*, 556: 245–264, July 2001.

[353] D. Merritt. Rotational Brownian motion of a massive binary. *Astrophys. J.*, 568:998–1003, April 2002.

[354] D. Merritt. A note on gravitational Brownian motion. *Astrophys. J.*, 628:673–677, August 2005.

[355] D. Merritt. Mass deficits, stalling radii, and the merger histories of elliptical galaxies. *Astrophys. J.*, 648:976–986, September 2006.

[356] D. Merritt. Evolution of nuclear star clusters. *Astrophys. J.*, 694:959–970, April 2009.

[357] D. Merritt. The distribution of stars and stellar remnants at the Galactic center. *Astrophys. J.*, 718:739–761, August 2010.

[358] D. Merritt, T. Alexander, S. Mikkola, and C. M. Will. Testing properties of the Galactic center black hole using stellar orbits. *Phys. Rev. D*, 81(6):062002, March 2010.

[359] D. Merritt, T. Alexander, S. Mikkola, and C. M. Will. Stellar dynamics of extreme-mass-ratio inspirals. *Phys. Rev. D*, 84(4):044024, August 2011.

[360] D. Merritt, P. Berczik, and F. Laun. Brownian motion of black holes in dense nuclei. *Astron. J.*, 133:553–563, February 2007.

[361] D. Merritt and R. D. Ekers. Tracing black hole mergers through radio lobe morphology. *Science*, 297:1310–1313, August 2002.

[362] D. Merritt and L. Ferrarese. Black hole demographics from the $M_\bullet$–σ relation. *Mon. Not. R. Astron. Soc.*, 320:L30–L34, January 2001.

[363] D. Merritt and L. Ferrarese. The $M_\bullet$–σ relation for supermassive black holes. *Astrophys. J.*, 547:140–145, January 2001.

[364] D. Merritt and L. Ferrarese. Relationship of black holes to bulges. In J. H. Knapen, J. E. Beckman, I. Shlosman, and T. J. Mahoney, editors, *The Central Kiloparsec of Starbursts and AGN: The La Palma Connection*, volume 249 of *Astronomical Society of the Pacific Conference Series*, page 335, 2001.

[365] D. Merritt, L. Ferrarese, and C. L. Joseph. No supermassive black hole in M33? *Science*, 293:1116–1119, August 2001.

[366] D. Merritt and T. Fridman. Triaxial galaxies with cusps. *Astrophys. J.*, 460:136, March 1996.

[367] D. Merritt, A. Gualandris, and S. Mikkola. Explaining the orbits of the Galactic center S-stars. *Astrophys. J. Lett.*, 693:L35–L38, March 2009.

[368] D. Merritt, S. Harfst, and G. Bertone. Collisionally regenerated dark matter structures in galactic nuclei. *Phys. Rev. D*, 75(4):043517, February 2007.

[369] D. Merritt, S. Mikkola, and A. Szell. Long-term evolution of massive black hole binaries. III. Binary evolution in collisional nuclei. *Astrophys. J.*, 671:53–72, December 2007.

[370] D. Merritt and S. P. Oh. The stellar dynamics of M87. *Astron. J.*, 113:1279–1285, April 1997.

[371] D. Merritt and M. Y. Poon. Chaotic loss cones and black hole fueling. *Astrophys. J.*, 606:788–798, May 2004.

[372] D. Merritt and G. D. Quinlan. Dynamical evolution of elliptical galaxies with central singularities. *Astrophys. J.*, 498:625, May 1998.

[373] D. Merritt and P. Saha. Mapping spherical potentials with discrete radial velocities. *Astrophys. J.*, 409:75–90, May 1993.

[374] D. Merritt, J. D. Schnittman, and S. Komossa. Hypercompact stellar systems around recoiling supermassive black holes. *Astrophys. J.*, 699:1690–1710, July 2009.

[375] D. Merritt, T. Storchi-Bergmann, A. Robinson, D. Batcheldor, D. Axon, and R. Cid Fernandes. The nature of the HE0450-2958 system. *Mon. Not. R. Astron. Soc.*, 367:1746–1750, April 2006.

[376] D. Merritt and M. Valluri. Chaos and mixing in triaxial stellar systems. *Astrophys. J.*, 471:82, November 1996.

[377] D. Merritt and M. Valluri. Resonant orbits in triaxial galaxies. *Astron. J.*, 118:1177–1189, September 1999.

[378] D. Merritt and E. Vasiliev. Orbits around black holes in triaxial nuclei. *Astrophys. J.*, 726:61, January 2011.

[379] D. Merritt and E. Vasiliev. Spin evolution of supermassive black holes and nuclear star clusters. *Phys. Rev. D*, December 2012.

[380] D. Merritt and J. Wang. Loss cone refilling rates in galactic nuclei. *Astrophys. J.*, 621:L101–L104, March 2005.

[381] J. C. Mihos and L. Hernquist. Gasdynamics and starbursts in major mergers. *Astrophys. J.*, 464:641, June 1996.

[382] S. Mikkola and P. Nurmi. Computing secular motion under slowly rotating quadratic perturbation. *Mon. Not. R. Astron. Soc.*, 371:421–423, September 2006.

[383] S. Mikkola and M. J. Valtonen. The slingshot ejections in merging galaxies. *Astrophys. J.*, 348:412–420, January 1990.

[384] S. Mikkola and M. J. Valtonen. Evolution of binaries in the field of light particles and the problem of two black holes. *Mon. Not. R. Astron. Soc.*, 259:115–120, November 1992.

[385] M. C. Miller, M. Freitag, D. P. Hamilton, and V. M. Lauburg. Binary encounters with supermassive black holes: Zero-eccentricity LISA events. *Astrophys. J. Lett.*, 631:L117–L120, October 2005.

[386] M. Milosavljević and D. Merritt. Formation of galactic nuclei. *Astrophys. J.*, 563:34–62, December 2001.

[387] M. Milosavljević and D. Merritt. The final parsec problem. In *The Astrophysics of Gravitational Wave Sources*, volume 686 of *AIP Conference Proceedings*, pages 201–210, October 2003.

[388] M. Milosavljević and D. Merritt. Long-term evolution of massive black hole binaries. *Astrophys. J.*, 596:860–878, October 2003.

[389] M. Milosavljević, D. Merritt, A. Rest, and F. C. van den Bosch. Galaxy cores as relics of black hole mergers. *Mon. Not. R. Astron. Soc.*, 331:L51–L55, April 2002.

[390] M. A. Minovitch. The invention that opened the solar system to exploration. *Planetary and Space Science*, 58:885–892, May 2010.

[391] J. Miralda-Escude and M. Schwarzschild. On the orbit structure of the logarithmic potential. *Astrophys. J.*, 339:752–762, April 1989.

[392] M. Miyoshi, J. Moran, J. Herrnstein, L. Greenhill, N. Nakai, P. Diamond, and M. Inoue. Evidence for a black-hole from high rotation velocities in a sub-parsec region of NGC4258. *Nature*, 373:127, January 1995.

[393] J. M. Moran, L. J. Greenhill, and J. R. Herrnstein. Observational evidence for massive black holes in the centers of active galaxies. *J. Astrophys. Astron.*, 20:165, September 1999.

[394] W. A. Mulder. Dynamical friction on extended objects. *Astron. Astrophys.*, 117:9–16, January 1983.

[395] B. W. Murphy, H. N. Cohn, and R. H. Durisen. Dynamical and luminosity evolution of active galactic nuclei - Models with a mass spectrum. *Astrophys. J.*, 370:60–77, March 1991.

[396] M. Nauenberg. Analytic approximations to the mass-radius relation and energy of zero-temperature stars. *Astrophys. J.*, 175:417, July 1972.

[397] S. Nayakshin, W. Dehnen, J. Cuadra, and R. Genzel. Weighing the young stellar discs around Sgr A*. *Mon. Not. R. Astron. Soc.*, 366:1410–1414, March 2006.

[398] J.-L. Nieto, F. D. Macchetto, M. A. C. Perryman, S. di Serego Alighieri, and G. Lelievre. UV observations of the nucleus of M31 with the ESA Photon Counting Detector (PCD). *Astron. Astrophys.*, 165:189–196, September 1986.

[399] C. A. Norman, A. May, and T. S. van Albada. Black holes and the shapes of galaxies. *Astrophys. J.*, 296:20–34, September 1985.

[400] I. Novikov and V. Frolov. *Physics of Black Holes.* Fundamental Theories of Physics. Kluwer, 1989.

[401] E. Noyola, K. Gebhardt, and M. Bergmann. Gemini and Hubble Space Telescope evidence for an intermediate-mass black hole in ω Centauri. *Astrophys. J.*, 676:1008–1015, April 2008.

[402] P. E. J. Nulsen and A. C. Fabian. Fuelling quasars with hot gas. *Mon. Not. R. Astron. Soc.*, 311:346–356, January 2000.

[403] S. Oh, S. S. Kim, and D. F. Figer. Mass distribution in the central few parsecs of our galaxy. *J. Korean Astron. Soc.*, 42:17–26, April 2009.

[404] A. Ollongren. Three-dimensional galactic stellar orbits. *Bull. Astron. Inst. Netherlands*, 16:241, October 1962.

[405] C. A. Onken, L. Ferrarese, D. Merritt, B. M. Peterson, R. W. Pogge, M. Vestergaard, and A. Wandel. Supermassive black holes in active galactic nuclei. II. Calibration of the black hole mass-velocity dispersion relationship for active galactic nuclei. *Astrophys. J.*, 615:645–651, November 2004.

[406] E. J. Opik. Collision probability with the planets and the distribution of planetary matter. *Proc. R. Irish Acad. Sect. A*, 54:165–199, 1951.

[407] J. R. Oppenheimer and H. Snyder. On continued gravitational contraction. *Phys. Rev.*, 56:455–459, September 1939.

[408] F. N. Owen, C. P. O'Dea, M. Inoue, and J. A. Eilek. VLA observations of the multiple jet galaxy 3C 75. *Astrophys. J. Lett.*, 294:L85–L88, July 1985.

[409] F. Özel, D. Psaltis, R. Narayan, and J. E. McClintock. The black hole mass distribution in the galaxy. *Astrophys. J.*, 725:1918–1927, December 2010.

[410] L. M. Ozernoy. Failure of the supermassive black hole concept? *The Observatory*, 96:67–69, April 1976.

[411] M. N. Özisik. *Heat Conduction*. New York : Wiley, 1993.

[412] P. L. Palmer and J. Papaloizou. On the interaction between a spherical stellar system and a companion. *Mon. Not. R. Astron. Soc.*, 215:691–699, August 1985.

[413] J. C. B. Papaloizou, R. P. Nelson, and F. Masset. Orbital eccentricity growth through disc-companion tidal interaction. *Astron. Astrophys.*, 366:263–275, January 2001.

[414] Y. Papaphilippou and J. Laskar. Global dynamics of triaxial galactic models through frequency map analysis. *Astron. Ap.*, 329:451–481, January 1998.

[415] T. Paumard, R. Genzel, F. Martins, S. Nayakshin, A. M. Beloborodov, Y. Levin, S. Trippe, F. Eisenhauer, T. Ott, S. Gillessen, R. Abuter, J. Cuadra, T. Alexander, and A. Sternberg. The two young star disks in the central parsec of the galaxy: Properties, dynamics, and formation. *Astrophys. J.*, 643:1011–1035, June 2006.

[416] P. J. E. Peebles. Gravitational collapse and related phenomena from an empirical point of view, or, black holes are where you find them. *General Relativity and Gravitation*, 3:63–82, June 1972.

[417] P. J. E. Peebles. Star distribution near a collapsed object. *Astrophys. J.*, 178:371–376, December 1972.

[418] I. C. Percival. Variational principles for the invariant toroids of classical dynamics. *J. Phys. A*, 7:794–802, May 1974.

[419] I. C. Percival. A variational principle for invariant tori of fixed frequency. *J. Phys. A*, 12:L57–L60, March 1979.

[420] H. B. Perets and A. Gualandris. Dynamical constraints on the origin of the young B-stars in the Galactic center. *Astrophys. J.*, 719:220–228, August 2010.

[421] H. B. Perets, A. Gualandris, G. Kupi, D. Merritt, and T. Alexander. Dynamical evolution of the young stars in the Galactic center: N-body simulations of the S-stars. *Astrophys. J.*, 702:884–889, September 2009.

[422] H. B. Perets, C. Hopman, and T. Alexander. Massive perturber-driven interactions between stars and a massive black hole. *Astrophys. J.*, 656:709–720, February 2007.

[423] P. C. Peters. Gravitational radiation and the motion of two point masses. *Phys. Rev. B*, 136:1224–1232, November 1964.

[424] B. M. Peterson. Reverberation mapping of active galactic nuclei. *Publ. Astron. Soc. Pac.*, 105:247–268, March 1993.

[425] B. M. Peterson. Toward precision measurement of central black hole masses. In *Co-evolution of Central Black Holes and Galaxies*, volume 267 of *IAU Symposium*, pages 151–160, May 2010.

[426] B. M. Peterson, M. C. Bentz, L.-B. Desroches, A. V. Filippenko, L. C. Ho, S. Kaspi, A. Laor, D. Maoz, E. C. Moran, R. W. Pogge, and A. C. Quillen. Multiwavelength monitoring of the dwarf Seyfert 1 Galaxy NGC 4395. I. A reverberation-based measurement of the black hole mass. *Astrophys. J.*, 632:799–808, October 2005.

[427] B. M. Peterson and A. Wandel. Evidence for supermassive black holes in active galactic nuclei from emission-line reverberation. *Astrophys. J. Lett.*, 540:L13–L16, September 2000.

[428] H. Pfister. On the history of the so-called Lense–Thirring effect. *General Relativity and Gravitation*, 39:1735–1748, November 2007.

[429] E. S. Phinney. Manifestations of a massive black hole in the Galactic Center. In M. Morris, editor, *IAU Symposium 136: The Center of the Galaxy*, page 543, 1989.

[430] A. Pizzella, E. M. Corsini, L. Morelli, M. Sarzi, C. Scarlata, M. Stiavelli, and F. Bertola. Nuclear stellar disks in spiral galaxies. *Astrophys. J.*, 573: 131–137, July 2002.

[431] M. Planck. Über einen Satz der statistischen Dynamik und seine Erweiterung in der Quantentheorie. *Sitzber. Preuß Akad. Wiss.*, page 324, 1917.

[432] H. C. K. Plummer. *An Introductory Treatise on Dynamical Astronomy*. Cambridge University Press, 1918.

[433] M. Y. Poon and D. Merritt. Orbital structure of triaxial black hole nuclei. *Astrophys. J.*, 549:192–204, March 2001.

[434] M. Y. Poon and D. Merritt. A self-consistent study of triaxial black hole nuclei. *Astrophys. J.*, 606:774–787, May 2004.

[435] S. F. Portegies Zwart and S. L. W. McMillan. The runaway growth of intermediate-mass black holes in dense star clusters. *Astrophys. J.*, 576:899–907, September 2002.

[436] M. C. Potter and J. F. Foss. *Fluid Mechanics*. New York : Ronald Press, 1975.

[437] W. H. Press and S. A. Teukolsky. On formation of close binaries by two-body tidal capture. *Astrophys. J.*, 213:183–192, April 1977.

[438] J. E. Pringle. The properties of external accretion discs. *Mon. Not. R. Astron. Soc.*, 248:754–759, February 1991.

[439] P. Prugniel and F. Simien. The fundamental plane of early-type galaxies: non-homology of the spatial structure. *Astron. Astrophys.*, 321:111–122, May 1997.

[440] T. Pursimo, L. O. Takalo, A. Sillanpää, M. Kidger, H. J. Lehto, J. Heidt, P. A. Charles, H. Aller, M. Aller, V. Beckmann, E. Benítez, H. Bock, P. Boltwood, U. Borgeest, J. A. de Diego, G. De Francesco, M. Dietrich, D. Dultzin-Hacyan, Y. Efimov, M. Fiorucci, G. Ghisellini, N. González-Pérez, M. Hanski, P. Heinämäki, R. K. Honeycutt, P. Hughes, K. Karlamaa, S. Katajainen, L. B. G. Knee, O. M. Kurtanidze, M. Kümmel, D. Kühl, M. Lainela, L. Lanteri, J. V. Linde, A. Lähteenmäki, M. Maesano, T. Mahoney, S. Marchenko, A. Marscher, E. Massaro, F. Montagni, R. Nesci, M. Nikolashvili, K. Nilsson, P. Nurmi, H. Pietilä, G. Poyner, C. M. Raiteri, R. Rekola, G. M. Richter, A. Riehokainen, J. W. Robertson, J.-M. Rodríguez-Espinoza, A. Sadun, N. Shakhovskoy, K. J. Schramm, T. Schramm, G. Sobrito, P. Teerikorpi, H. Teräsranta, M. Tornikoski, G. Tosti, G. W. Turner, E. Valtaoja, M. Valtonen, M. Villata, S. J. Wagner, J. Webb, W. Weneit, and S. Wiren. Intensive monitoring of OJ 287. *Astron. Astrophys. Suppl.*, 146:141–155, October 2000.

[441] E. E. Qian, P. T. de Zeeuw, R. P. van der Marel, and C. Hunter. Axisymmetric galaxy models with central black holes, with an application to M32. *Mon. Not. R. Astron. Soc.*, 274:602–622, May 1995.

[442] G. D. Quinlan. The dynamical evolution of massive black hole binaries I. Hardening in a fixed stellar background. *New Astronomy*, 1:35–56, July 1996.

[443] G. D. Quinlan. The time-scale for core collapse in spherical star clusters. *New Astronomy*, 1:255–270, November 1996.

[444] G. D. Quinlan, L. Hernquist, and S. Sigurdsson. Models of galaxies with central black holes: Adiabatic growth in spherical galaxies. *Astrophys. J.*, 440:554–564, February 1995.

[445] R. R. Rafikov. Nonlinear propagation of planet-generated tidal waves. *Astrophys. J.*, 569:997–1008, April 2002.

[446] A. Rasskazov and D. Merritt. Rotational Brownian motion of a massive binary. II. First and second order diffusion coefficients. Unpublished research note, January 2012.

[447] S. J. Ratcliff, K. M. Chang, and M. Schwarzschild. Stellar orbits in angle variables. *Astrophys. J.*, 279:610–620, April 1984.

[448] K. P. Rauch and B. Ingalls. Resonant tidal disruption in galactic nuclei. *Mon. Not. R. Astron. Soc.*, 299:1231–1241, October 1998.

[449] K. P. Rauch and S. Tremaine. Resonant relaxation in stellar systems. *New Astron.*, 1:149–170, October 1996.

[450] M. J. Rees. Tidal disruption of stars by black holes of 10^6–10^8 solar masses in nearby galaxies. *Nature*, 333:523–528, June 1988.

[451] M. J. Rees. 'Dead quasars' in nearby galaxies? *Science*, 247:817–823, February 1990.

[452] O. Regev and D. Portnoy. Tidal disruption of stars by a massive black hole. *Astrophysics and Space Science*, 132:249–256, April 1987.

[453] M. J. Reid and A. Brunthaler. The proper motion of Sagittarius A*. II. The mass of Sagittarius A*. *Astrophys. J.*, 616:872–884, December 2004.

[454] D. Richstone, E. A. Ajhar, R. Bender, G. Bower, A. Dressler, S. M. Faber, A. V. Filippenko, K. Gebhardt, R. Green, L. C. Ho, J. Kormendy, T. R. Lauer, J. Magorrian, and S. Tremaine. Supermassive black holes and the evolution of galaxies. *Nature*, 395:A14–19, October 1998.

[455] D. O. Richstone. Scale-free, axisymmetric galaxy models with little angular momentum. *Astrophys. J.*, 238:103–109, May 1980.

[456] D. O. Richstone. Scale-free models of galaxies. II - A complete survey of orbits. *Astrophys. J.*, 252:496–507, January 1982.

[457] D. O. Richstone and S. Tremaine. Maximum-entropy models of galaxies. *Astrophys. J.*, 327:82–88, April 1988.

[458] H. Risken. *The Fokker–Planck Equation. Methods of Solution and Applications*. Springer Series in Synergetics. Berlin, New York: Springer, 2nd edition, 1989.

[459] H.-W. Rix and S. D. M. White. Optimal estimates of line-of-sight velocity distributions from absorption line spectra of galaxies – Nuclear discs in elliptical galaxies. *Mon. Not. R. Astron. Soc.*, 254:389–403, February 1992.

[460] C. Rodriguez, G. B. Taylor, R. T. Zavala, A. B. Peck, L. K. Pollack, and R. W. Romani. A compact supermassive binary black hole system. *Astrophys. J.*, 646:49–60, July 2006.

[461] A. J. Romanowsky and C. S. Kochanek. Structural and dynamical uncertainties in modelling axisymmetric elliptical galaxies. *Mon. Not. R. Astron. Soc.*, 287:35–50, May 1997.

[462] G. E. Romero, L. Chajet, Z. Abraham, and J. H. Fan. Beaming and precession in the inner jet of 3C 273 — II. The central engine. *Astron. Astrophys.*, 360:57–64, August 2000.

[463] N. Roos. Galaxy mergers and active galactic nuclei. *Astron. Astrophys.*, 104:218–228, December 1981.

[464] N. Roos, J. S. Kaastra, and C. A. Hummel. A massive binary black hole in 1928 + 738? *Astrophys. J.*, 409:130–133, May 1993.

[465] M. N. Rosenbluth, W. M. MacDonald, and D. L. Judd. Fokker–Planck equation for an inverse-square force. *Phys. Rev.*, 107:1–6, July 1957.

[466] A. E. Roy. *Orbital Motion*. Adam Hilger, 2005.

[467] B. Rudak and B. Paczynski. Outer excretion disk around a close binary. *Acta Astronomica*, 31:13–24, 1981.

[468] G. B. Rybicki. Deprojection of galaxies – How much can be learned. In P. T. de Zeeuw, editor, *Structure and Dynamics of Elliptical Galaxies*, volume 127 of *IAU Symposium*, page 397, 1987.

[469] N. Sabha, A. Eckart, D. Merritt, M. Zamaninasab, G. Witzel, M. García-Marín, B. Jalali, M. Valencia-S., S. Yazici, R. Buchholz, B. Shahzamanian, C. Rauch, M. Horrobin, and C. Straubmeier. The S-star cluster at the center of the Milky Way. On the nature of diffuse NIR emission in the inner tenth of a parsec. *Astron. Astrophys.*, 545:A70, September 2012.

[470] L. Sadeghian and C. M. Will. Testing the black hole no-hair theorem at the galactic center: perturbing effects of stars in the surrounding cluster. *Classical and Quantum Gravity*, 28(22):225029, November 2011.

[471] E. E. Salpeter. The luminosity function and stellar evolution. *Astrophys. J.*, 121:161, January 1955.

[472] N. Sambhus and S. Sridhar. Stellar orbits in triaxial clusters around black holes in galactic nuclei. *Astrophys. J.*, 542:143–160, October 2000.

[473] D. B. Sanders, B. T. Soifer, J. H. Elias, B. F. Madore, K. Matthews, G. Neugebauer, and N. Z. Scoville. Ultraluminous infrared galaxies and the origin of quasars. *Astrophys. J.*, 325:74–91, February 1988.

[474] R. H. Sanders. The effects of stellar collisions in dense stellar systems. *Astrophys. J.*, 162:791, December 1970.

[475] W. C. Saslaw. The dynamics of dense stellar systems. *Publ. Astron. Soc. Pac.*, 85:5–23, February 1973.

[476] W. C. Saslaw, M. J. Valtonen, and S. J. Aarseth. The gravitational slingshot and the structure of extragalactic radio sources. *Astrophys. J.*, 190:253–270, June 1974.

[477] G. J. Savonije, J. C. B. Papaloizou, and D. N. C. Lin. On tidally induced shocks in accretion discs in close binary systems. *Mon. Not. R. Astron. Soc.*, 268:13–28, May 1994.

[478] G. Schäfer and N. Wex. Second post-Newtonian motion of compact binaries. *Phys. Lett. A*, 174:196–205, March 1993.

[479] R. Schödel. The Milky Way nuclear star cluster in context. In M. R. Morris, Q. D. Wang, and F. Yuan, editors, *The Galactic Center*, volume 439 of *Astronomical Society of the Pacific Conference Series*, page 222, May 2011.

[480] R. Schödel, A. Eckart, T. Alexander, D. Merritt, R. Genzel, A. Sternberg, L. Meyer, F. Kul, J. Moultaka, T. Ott, and C. Straubmeier. The structure of the nuclear stellar cluster of the Milky Way. *Astron. Astrophys.*, 469: 125–146, July 2007.

[481] R. Schödel, D. Merritt, and A. Eckart. The nuclear star cluster of the Milky Way: proper motions and mass. *Astron. Astrophys.*, 502:91–111, July 2009.

[482] M. Schwarzschild. A numerical model for a triaxial stellar system in dynamical equilibrium. *Astrophys. J.*, 232:236–247, August 1979.

[483] M. Schwarzschild. Triaxial equilibrium models for elliptical galaxies with slow figure rotation. *Astrophys. J.*, 263:599–610, December 1982.

[484] M. Schwarzschild. Self-consistent models for galactic halos. *Astrophys. J.*, 409:563–577, June 1993.

[485] J. L. Sérsic. Influence of the atmospheric and instrumental dispersion on the brightness distribution in a galaxy. *Boletin de la Asociacion Argentina de Astronomia La Plata Argentina*, 6:41, 1963.

[486] J. L. Sérsic. *Atlas de Galaxias Australes*. Cordoba, Argentina: Observatorio Astronomico, 1968.

[487] A. Sesana, A. Gualandris, and M. Dotti. Massive black hole binary eccentricity in rotating stellar systems. *Mon. Not. R. Astron. Soc.*, 415:L35–L39, July 2011.

[488] A. Sesana, F. Haardt, and P. Madau. Interaction of massive black hole binaries with their stellar environment. I. Ejection of hypervelocity stars. *Astrophys. J.*, 651:392–400, November 2006.

[489] A. C. Seth, R. D. Blum, N. Bastian, N. Caldwell, and V. P. Debattista. The rotating nuclear star cluster in NGC 4244. *Astrophys. J.*, 687:997–1003, November 2008.

[490] K. L. Shapiro, M. Cappellari, T. de Zeeuw, R. M. McDermid, K. Gebhardt, R. C. E. van den Bosch, and T. S. Statler. The black hole in NGC 3379: A comparison of gas and stellar dynamical mass measurements with HST and integral-field data. *Mon. Not. R. Astron. Soc.*, 370:559–579, August 2006.

[491] S. L. Shapiro and A. B. Marchant. Star clusters containing massive, central black holes - Monte Carlo simulations in two-dimensional phase space. *Astrophys. J.*, 225:603–624, October 1978.

[492] S. L. Shapiro and S. A. Teukolsky. *Black Holes, White Dwarfs, and Neutron Stars: The Physics of Compact Objects*. Wiley Interscience, 1983.

[493] M. Shaw, D. Axon, R. Probst, and I. Gatley. Nuclear bars and blue nuclei within barred spiral galaxies. *Mon. Not. R. Astron. Soc.*, 274:369–387, May 1995.

[494] I. Shlosman and M. C. Begelman. Evolution of self-gravitating accretion disks in active galactic nuclei. *Astrophys. J.*, 341:685–691, June 1989.

[495] I. Shlosman, J. Frank, and M. C. Begelman. Bars within bars - A mechanism for fuelling active galactic nuclei. *Nature*, 338:45–47, March 1989.

[496] S. Sigurdsson and M. J. Rees. Capture of stellar mass compact objects by massive black holes in galactic cusps. *Mon. Not. R. Astron. Soc.*, 284: 318–326, January 1997.

[497] J. Silk and M. J. Rees. Quasars and galaxy formation. *Astron. Astrophys.*, 331:L1–L4, March 1998.

[498] A. Sillanpaa, S. Haarala, M. J. Valtonen, B. Sundelius, and G. G. Byrd. OJ 287 - Binary pair of supermassive black holes. *Astrophys. J.*, 325:628–634, February 1988.

[499] A. Soltan. Masses of quasars. *Mon. Not. R. Astron. Soc.*, 200:115–122, July 1982.

[500] P. F. Spinnato, M. Fellhauer, and S. F. Portegies Zwart. The efficiency of the spiral-in of a black hole to the Galactic Centre. *Mon. Not. Royal Astron. Soc.*, 344:22–32, September 2003.

[501] L. Spitzer. Dynamical evolution of dense spherical star systems. In D. J. K. O'Connell, editor, *Pontificiae Academiae Scientiarum Scripta Varia, Proceedings of a Study Week on Nuclei of Galaxies, held in Rome, April 13–18, 1970*, page 443. Amsterdam: North Holland, and New York: American Elsevier, 1971.

[502] L. Spitzer. *Dynamical Evolution of Globular Clusters*. Princeton, NJ : Princeton University Press, 1987.

[503] L. Spitzer, Jr. and M. H. Hart. Random gravitational encounters and the evolution of spherical systems. I. Method. *Astrophys. J.*, 164:399–410, March 1971.

[504] L. Spitzer, Jr. and M. Schwarzschild. The possible influence of interstellar clouds on stellar velocities. *Astrophys. J.*, 114:385–397, November 1951.

[505] L. J. Spitzer and W. C. Saslaw. On the evolution of galactic nuclei. *Astrophys. J.*, 143:400–419, February 1966.

[506] L. J. Spitzer and M. E. Stone. On the evolution of galactic nuclei. II. *Astrophys. J.*, 147:519–528, February 1967.

[507] T. S. Statler, E. Emsellem, R. F. Peletier, and R. Bacon. Long-lived triaxiality in the dynamically old elliptical galaxy NGC 4365: a limit on chaos and black hole mass. *Mon. Not. R. Astron. Soc.*, 353:1–14, September 2004.

[508] H. Sudou, S. Iguchi, Y. Murata, and Y. Taniguchi. Orbital motion in the radio galaxy 3C 66B: Evidence for a supermassive black hole binary. *Science*, 300:1263–1265, May 2003.

[509] D. Syer and A. Ulmer. Tidal disruption rates of stars in observed galaxies. *Mon. Not. R. Astron. Soc.*, 306:35–42, June 1999.

[510] A. Szell, D. Merritt, and I. G. Kevrekidis. Core collapse via coarse dynamic renormalization. *Phys. Rev. Lett.*, 95(8):081102, August 2005.

[511] T. Takeuchi, S. M. Miyama, and D. N. C. Lin. Gap formation in protoplanetary disks. *Astrophys. J.*, 460:832–847, April 1996.

[512] J. C. Tan and E. G. Blackman. Star-forming accretion flows and the low-luminosity nuclei of giant elliptical galaxies. *Mon. Not. R. Astron. Soc.*, 362:983–994, September 2005.

[513] B. Terzić and A. W. Graham. Density-potential pairs for spherical stellar systems with Sérsic light profiles and (optional) power-law cores. *Mon. Not. R. Astron. Soc.*, 362:197–212, September 2005.

[514] K. S. Thorne and V. B. Braginskii. Gravitational-wave bursts from the nuclei of distant galaxies and quasars - Proposal for detection using Doppler tracking of interplanetary spacecraft. *Astrophys. J. Lett.*, 204:L1–L6, February 1976.

[515] S. E. Thorsett and D. Chakrabarty. Neutron star mass measurements. I. Radio pulsars. *Astrophys. J.*, 512:288–299, February 1999.

[516] F. F. Tisserand. *Traité de Mécanique Céleste IV*. Paris, Gauthier-Villars, 1896.

[517] F. Tombesi, R. M. Sambruna, J. N. Reeves, V. Braito, L. Ballo, J. Gofford, M. Cappi, and R. F. Mushotzky. Discovery of ultra-fast outflows in a sample of broad-line radio galaxies observed with Suzaku. *Astrophys. J.*, 719: 700–715, August 2010.

[518] J. L. Tonry. Evidence for a central mass concentration in M32. *Astrophys. J. Lett.*, 283:L27–L30, August 1984.

[519] J. L. Tonry. A central black hole in M32. *Astrophys. J.*, 322:632–642, November 1987.

[520] S. Tremaine. An eccentric-disk model for the nucleus of M31. *Astron. J.*, 110:628–633, August 1995.

[521] S. Tremaine and M. D. Weinberg. Dynamical friction in spherical systems. *Mon. Not. R. Astron. Soc.*, 209:729–757, August 1984.

[522] B. Tremblay and D. Merritt. Evidence from intrinsic shapes for two families of elliptical galaxies. *Astron. J.*, 111:2243–2247, June 1996.

[523] I. Trujillo, P. Erwin, A. Asensio Ramos, and A. W. Graham. Evidence for a new elliptical-galaxy paradigm: Sérsic and core galaxies. *Astron. J.*, 127:1917–1942, April 2004.

[524] R. B. Tully and J. R. Fisher. A new method of determining distances to galaxies. *Astron. Astrophys.*, 54:661–673, February 1977.

[525] N. J. Turner. On the vertical structure of radiation-dominated accretion disks. *Astrophys. J. Lett.*, 605:L45–L48, April 2004.

[526] G. E. Uhlenbeck and L. S. Ornstein. On the theory of the Brownian motion. *Phys. Rev.*, 36:823–841, September 1930.

[527] M. Valluri, L. Ferrarese, D. Merritt, and C. L. Joseph. The low end of the supermassive black hole mass function: Constraining the mass of a nuclear black hole in NGC 205 via stellar kinematics. *Astrophys. J.*, 628:137–152, July 2005.

[528] M. Valluri and D. Merritt. Regular and chaotic dynamics of triaxial stellar systems. *Astrophys. J.*, 506:686–711, October 1998.

[529] M. Valluri and D. Merritt. Orbital instability and relaxation in stellar systems. In V. G. Gurzadyan and R. Ruffini, editors, *The Chaotic Universe, Proceedings of the Second ICRA Network Workshop*, volume 10 of *Advanced Series in Astrophysics and Cosmology*, page 229. World Scientific, 2000.

[530] M. Valluri, D. Merritt, and E. Emsellem. Difficulties with recovering the masses of supermassive black holes from stellar kinematical data. *Astrophys. J.*, 602:66–92, February 2004.

[531] L. Valtaoja, M. J. Valtonen, and G. G. Byrd. Binary pairs of supermassive black holes - Formation in merging galaxies. *Astrophys. J.*, 343:47–53, August 1989.

[532] M. Valtonen and H. Karttunen. *The Three-Body Problem*. Cambridge University Press, 2006.

[533] M. J. Valtonen, S. Mikkola, P. Heinamaki, and H. Valtonen. Slingshot ejections from clusters of three and four black holes. *Astrophys. J. Suppl.*, 95:69–86, November 1994.

[534] T. S. van Albada. Dissipationless galaxy formation and the R to the 1/4-power law. *Mon. Not. R. Astron. Soc.*, 201:939–955, December 1982.

[535] S. van den Bergh. The stellar populations of M 33. *Publ. Astron. Soc. Pac.*, 103:609–622, July 1991.

[536] R. C. E. van den Bosch and P. T. de Zeeuw. Estimating black hole masses in triaxial galaxies. *Mon. Not. R. Astron. Soc.*, 401:1770–1780, January 2010.

[537] R. P. van der Marel. On the stellar kinematical evidence for massive black holes in galactic nuclei expected with the Hubble Space Telescope. *Astrophys. J.*, 432:L91–L94, September 1994.

[538] R. P. van der Marel. Velocity profiles of galaxies with claimed black-holes – Part Three – observations and models for M87. *Mon. Not. R. Astron. Soc.*, 270:271, September 1994.

[539] R. P. van der Marel. Relics of nuclear activity: Do all galaxies have massive black holes? In *Galaxy Interactions at Low and High Redshift*, volume 186 of *IAU Symposium*, page 333, 1999.

[540] R. P. van der Marel and J. Anderson. New limits on an intermediate-mass black hole in Omega Centauri. II. Dynamical models. *Astrophys. J.*, 710:1063–1088, February 2010.

[541] R. P. van der Marel, N. Cretton, P. T. de Zeeuw, and H.-W. Rix. Improved evidence for a black hole in M32 from HST/FOS spectra. II. Axisymmetric dynamical models. *Astrophys. J.*, 493:613–631, January 1998.

[542] R. P. van der Marel, N. W. Evans, H.-W. Rix, S. D. M. White, and T. de Zeeuw. Velocity profiles of galaxies with claimed black holes – II. $f(E, L_z)$ models for M 32. *Mon. Not. R. Astron. Soc.*, 271:99–117, November 1994.

[543] R. P. van der Marel and M. Franx. A new method for the identification of non-Gaussian line profiles in elliptical galaxies. *Astrophys. J.*, 407:525–539, April 1993.

[544] M. A. Vashkov'yak. Evolution of the orbits of distant satellites of Uranus. *Astron. Lett.*, 25:476–481, July 1999.

[545] S. Veilleux and D. E. Osterbrock. Spectral classification of emission-line galaxies. *Astrophys. J. Suppl.*, 63:295–310, February 1987.

[546] F. Verhulst. Discrete symmetric dynamical systems at the main resonances with applications to axi-symmetric galaxies. *Roy. Soc. London Phil. Trans. A*, 290:435–465, January 1979.

[547] E. K. Verolme, M. Cappellari, Y. Copin, R. P. van der Marel, R. Bacon, M. Bureau, R. L. Davies, B. M. Miller, and P. T. de Zeeuw. A SAURON study of M32: measuring the intrinsic flattening and the central black hole mass. *Mon. Not. R. Astron. Soc.*, 335:517–525, September 2002.

[548] M. Vestergaard. Determining central black hole masses in distant active galaxies. *Astrophys. J.*, 571:733–752, June 2002.

[549] M. Volonteri, F. Haardt, and P. Madau. The assembly and merging history of supermassive black holes in hierarchical models of galaxy formation. *Astrophys. J.*, 582:559–573, January 2003.

[550] M. von Smoluchowski. Zur kinetischen Theorie der Brownschen Molekular-bewegung und der Suspensionen. *Ann. Physik*, 21:756–780, 1906.

[551] M. von Smoluchowski. Drei Vorträge über Diffusion, Brownsche Molekularbewegung und Koagulation von Kolloidteilchen. *Physik. Zeits.*, 17:557–571, 1916.

[552] R. V. Wagoner and C. M. Will. Post-Newtonian gravitational radiation from orbiting point masses. *Astrophys. J.*, 210:764–775, December 1976.

[553] C. J. Walcher, T. Böker, S. Charlot, L. C. Ho, H.-W. Rix, J. Rossa, J. C. Shields, and R. P. van der Marel. Stellar populations in the nuclei of late-type spiral galaxies. *Astrophys. J.*, 649:692–708, October 2006.

[554] A. Wandel. The black hole-to-bulge mass relation in active galactic nuclei. *Astrophys. J. Lett.*, 519:L39–L42, July 1999.

[555] J. Wang and D. Merritt. Revised rates of stellar disruption in galactic nuclei. *Astrophys. J.*, 600:149–161, January 2004.

[556] R. L. Warnock and R. D. Ruth. Long-term bounds on nonlinear Hamiltonian motion. *Physica D Nonlinear Phenomena*, 56:188–215, May 1992.

[557] E. H. Wehner and W. E. Harris. From supermassive black holes to dwarf elliptical nuclei: A mass continuum. *Astrophys. J. Lett.*, 644:L17–L20, June 2006.

[558] A. E. Wehrle, S. C. Unwin, D. L. Jones, D. L. Meier, and B. G. Piner. Tracking the moving optical photocenters of active galaxies: Binary black holes, accretion disks and relativistic jet. In Michael Shao, editor, *Interferometry in Space*, volume 4852 of *Proceedings of the SPIE*, pages 152–160, February 2003.

[559] M. D. Weinberg. Orbital decay of satellite galaxies in spherical systems. *Astrophys. J.*, 300:93–111, January 1986.

[560] M. D. Weinberg. Self-gravitating response of a spherical galaxy to sinking satellites. *Mon. Not. R. Astron. Soc.*, 239:549–569, August 1989.

[561] N. N. Weinberg, M. Milosavljević, and A. M. Ghez. Stellar dynamics at the Galactic center with an extremely large telescope. *Astrophys. J.*, 622:878–891, April 2005.

[562] S. Weinberg. *Gravitation and Cosmology: Principles and Applications of the General Theory of Relativity*. Wiley, July 1972.

[563] M. L. White. Dynamical friction. *Astrophys. J.*, 109:159–163, January 1949.

[564] S. D. M. White. Simulations of sinking satellites. *Astrophys. J.*, 274:53–61, November 1983.

[565] A. Wilkinson and R. A. James. A stationary and a slowly rotating model of a triaxial elliptical galaxy. *Mon. Not. R. Astron. Soc.*, 199:171–196, April 1982.

[566] C. M. Will. *Theory and Experiment in Gravitational Physics*. Cambridge University Press, March 1993.

[567] C. M. Will. Testing the general relativistic "no-hair" theorems using the Galactic center black hole Sagittarius A*. *Astrophys. J. Lett.*, 674:L25–L28, February 2008.

[568] B. E. Woodgate, R. A. Kimble, C. W. Bowers, S. B. Kraemer, M. E. Kaiser, A. C. Danks, J. F. Grady, J. J. Loiacono, D. F. Hood, W. W. Meyer, C. N. van Houten, V. S. Argabright, R. L. Bybee, J. G. Timothy, M. M. Blouke, D. A. Dorn, M. Bottema, R. A. Woodruff, D. Michika, J. F. Sullivan, J. Hetlinger, R. Stocker, M. D. Brumfield, L. D. Feinberg, A. W. Delamere, D. Rose, H. W. Garner, D. J. Lindler, T. R. Gull, S. R. Heap, C. L. Joseph, R. F. Green, E. B. Jenkins, J. B. Linsky, J. B. Hutchings, H. W. Moos, A. Boggess, S. P. Maran, F. L. Roesler, D. E. Weistrop, G. Sonneborn, G. Bower, and J. Gardner. First results from the Space Telescope Imaging Spectrograph. In M. R. Descour and S. S. Shen, editors, *Imaging Spectrometry III*, volume 3118 of *Society of Photo-Optical Instrumentation Engineers (SPIE) Conference Series*, pages 2–12, January 1997.

[569] S. E. Woosley, A. Heger, and T. A. Weaver. The evolution and explosion of massive stars. *Rev. Mod. Phys.*, 74:1015–1071, November 2002.

[570] C. K.-S. Young, M. J. Currie, R. J. Dickens, A.-L. Luo, and T.-J. Zhang. A critical review of the evidence for M32 being a compact dwarf satellite of M31 rather than a more distant normal galaxy. *Chinese J. Astron. Astrophys.*, 8:369–384, August 2008.

[571] P. Young. Numerical models of star clusters with a central black hole. I - Adiabatic models. *Astrophys. J.*, 242:1232–1237, December 1980.

[572] P. J. Young. Capture of particles from plunge orbits by a black hole. *Phys. Rev. D*, 14:3281–3289, December 1976.

[573] P. J. Young. The black tide model of QSOs. II - Destruction in an isothermal sphere. *Astrophys. J.*, 215:36–52, July 1977.

[574] P. J. Young, G. A. Shields, and J. C. Wheeler. The black tide model of QSOs. *Astrophys. J.*, 212:367–382, March 1977.

[575] P. J. Young, J. A. Westphal, J. Kristian, C. P. Wilson, and F. P. Landauer. Evidence for a supermassive object in the nucleus of the galaxy M87 from SIT and CCD area photometry. *Astrophys. J.*, 221:721–730, May 1978.

[576] Q. Yu. Evolution of massive binary black holes. *Mon. Not. R. Astron. Soc.*, 331:935–958, April 2002.

[577] Q. Yu and S. Tremaine. Observational constraints on growth of massive black holes. *Mon. Not. R. Astron. Soc.*, 335:965–976, October 2002.

[578] Q. Yu and S. Tremaine. Ejection of hypervelocity stars by the (binary) black hole in the Galactic center. *Astrophys. J.*, 599:1129–1138, December 2003.

[579] H. S. Zapolsky and E. E. Salpeter. The mass-radius relation for cold spheres of low mass. *Astrophys. J.*, 158:809–813, November 1969.

[580] D. Zaritsky and S. D. M. White. Simulations of sinking satellites revisited. *Mon. Not. R. Astron. Soc.*, 235:289–296, November 1988.

[581] H. Zhao, M. G. Haehnelt, and M. J. Rees. Feeding black holes at galactic centres by capture from isothermal cusps. *New Astron.*, 7:385–394, October 2002.

[582] J.-H. Zhao, G. C. Bower, and W. M. Goss. Radio variability of Sagittarius A*—a 106 day cycle. *Astrophys. J. Lett.*, 547:L29–L32, January 2001.

[583] C. Zier and P. L. Biermann. Binary black holes and tori in AGN. I. Ejection of stars and merging of the binary. *Astron. Astrophys.*, 377:23–43, October 2001.

[584] S. Zucker, T. Alexander, S. Gillessen, F. Eisenhauer, and R. Genzel. Probing post-Newtonian physics near the galactic black hole with stellar redshift measurements. *Astrophys. J. Lett.*, 639:L21–L24, March 2006.

Index

Page numbers in boldface denote definitions or main references. "f," "n," or "t" following page reference denotes material in a figure caption, footnote, or table.